Functions of Matrices

Functions of Matrices

Theory and Computation

Nicholas J. Higham

The University of Manchester
Manchester, United Kingdom

siam. Society for Industrial and Applied Mathematics • Philadelphia

10 9 8 7 6 5 4 3 2 1

Library of Congress Cataloging-in-Publication Data

Higham, Nicholas J., 1961-
Functions of matrices : theory and computation / Nicholas J. Higham.
p. cm.
Includes bibliographical references and index.
ISBN 978-0-89871-646-7 1. Matrices. 2. Functions. 3. Factorization (Mathematics)
I. Title.

QA188.H53 2008
512.9'434–dc22

2007061811

To Françoise

Contents

List of Figures

List of Tables

Preface

Functions of matrices have been studied for as long as matrix algebra itself. Indeed, in his seminal *A Memoir on the Theory of Matrices* (1858), Cayley investigated the square root of a matrix, and it was not long before definitions of $f(A)$ for general f were proposed by Sylvester and others. From their origin in pure mathematics, matrix functions have broadened into a subject of study in applied mathematics, with widespread applications in science and engineering. Research on matrix functions involves matrix theory, numerical analysis, approximation theory, and the development of algorithms and software, so it employs a wide range of theory and methods and promotes an appreciation of all these important topics.

My first foray into $f(A)$ was as a graduate student when I became interested in the matrix square root. I have worked on matrix functions on and off ever since. Although there is a large literature on the subject, including chapters in several books (notably Gantmacher [203, 1959], Horn and Johnson [296, 1991], Lancaster and Tismenetsky [371, 1985], and Golub and Van Loan [224, 1996]), there has not previously been a book devoted to matrix functions. I started to write this book in 2003. In the intervening period interest in matrix functions has grown significantly, with new applications appearing and the literature expanding at a fast rate, so the appearance of this book is timely.

This book is a research monograph that aims to give a reasonably complete treatment of the theory of matrix functions and numerical methods for computing them, as well as an overview of applications. The theory of matrix functions is beautiful and nontrivial. I have strived for an elegant presentation with illuminating examples, emphasizing results of practical interest. I focus on three equivalent definitions of $f(A)$, based on the Jordan canonical form, polynomial interpolation, and the Cauchy integral formula, and use all three to develop the theory. A thorough treatment is given of problem sensitivity, based on the Fréchet derivative. The applications described include both the well known and the more speculative or recent, and differential equations and algebraic Riccati equations underlie many of them.

The bulk of the book is concerned with numerical methods and the associated issues of accuracy, stability, and computational cost. Both general purpose methods and methods for specific functions are covered. Little mention is made of methods that are numerically unstable or have exorbitant operation counts of order n^4 or higher; many methods proposed in the literature are ruled out for at least one of these reasons.

The focus is on theory and methods for general matrices, but a brief introduction to functions of structured matrices is given in Section 14.1. The problem of computing a function of a matrix times a vector, $f(A)b$, is of growing importance, though as yet numerical methods are relatively undeveloped; Chapter 13 is devoted to this topic.

One of the pleasures of writing this book has been to explore the many connections between matrix functions and other subjects, particularly matrix analysis and numerical analysis in general. These connections range from the expected, such as divided differences, the Kronecker product, and unitarily invariant norms, to the unexpected, which include the Mandelbrot set, the geometric mean, partial isometries, and the role of the Fréchet derivative beyond measuring problem sensitivity.

I have endeavoured to make this book more than just a monograph about matrix functions, and so it includes many useful or interesting facts, results, tricks, and techniques that have a (sometimes indirect) $f(A)$ connection. In particular, the book contains a substantial amount of matrix theory, as well as many historical references, some of which appear not to have previously been known to researchers in the area. I hope that the book will be found useful as a source of statements and applications of results in matrix analysis and numerical linear algebra, as well as a reference on matrix functions.

Four main themes pervade the book.

Role of the sign function. The matrix sign function has fundamental theoretical and algorithmic connections with the matrix square root, the polar decomposition, and, to a lesser extent, matrix pth roots. For example, a large class of iterations for the matrix square root can be obtained from corresponding iterations for the matrix sign function, and Newton's method for the matrix square root is mathematically equivalent to Newton's method for the matrix sign function.

Stability. The stability of iterations for matrix functions can be effectively defined and analyzed in terms of power boundedness of the Fréchet derivative of the iteration function at the solution. Unlike some earlier, more ad hoc analyses, no assumptions are required on the underlying matrix. General results (Theorems 4.18 and 4.19) simplify the analysis for idempotent functions such as the matrix sign function and the unitary polar factor.

Schur decomposition and Parlett recurrence. The use of a Schur decomposition followed by reordering and application of the block form of the Parlett recurrence yields a powerful general algorithm, with f-dependence restricted to the evaluation of f on the diagonal blocks of the Schur form.

Padé approximation. For transcendental functions the use of Padé approximants, in conjunction with an appropriate scaling technique that brings the matrix argument close to the origin, yields an effective class of algorithms whose computational building blocks are typically just matrix multiplication and the solution of multiple right-hand side linear systems. Part of the success of this approach rests on the several ways in which rational functions can be evaluated at a matrix argument, which gives the scope to find a good compromise between speed and stability.

In addition to surveying, unifying, and sometimes improving existing results and algorithms, this book contains new results. Some of particular note are as follows.

- Theorem 1.35, which relates $f(\alpha I_m + AB)$ to $f(\alpha I_n + BA)$ for $A \in \mathbb{C}^{m\times n}$ and $B \in \mathbb{C}^{n\times m}$ and is an analogue for general matrix functions of the Sherman–Morrison–Woodbury formula for the matrix inverse.

- Theorem 4.15, which shows that convergence of a scalar iteration implies convergence of the corresponding matrix iteration when applied to a Jordan block, under suitable assumptions. This result is useful when the matrix iteration can be block diagonalized using the Jordan canonical form of the underlying matrix, A. Nevertheless, we show in the context of Newton's method for the matrix square root that analysis via the Jordan canonical form of A does not always give the strongest possible convergence result. In this case a stronger result, Theorem 6.9, is obtained essentially by reducing the convergence analysis to the consideration of the behaviour of the powers of a certain matrix.

- Theorems 5.13 and 8.19 on the stability of essentially all iterations for the matrix sign function and the unitary polar factor, and the general results in Theorems 4.18 and 4.19 on which these are based.
- Theorems 6.14–6.16 on the convergence of the binomial, Pulay, and Visser iterations for the matrix square root.
- An improved Schur–Parlett algorithm for the matrix logarithm, given in Section 11.6, which makes use of improved implementations of the inverse scaling and squaring method in Section 11.5.

The Audience

The book's main audience is specialists in numerical analysis and applied linear algebra, but it will be of use to anyone who wishes to know something of the theory of matrix functions and state of the art methods for computing them. Much of the book can be understood with only a basic grounding in numerical analysis and linear algebra.

Using the Book

The book can be used as the basis for a course on functions of matrices at the graduate level. It is also a suitable reference for an advanced course on applied or numerical linear algebra, which might cover particular topics such as definitions and properties of $f(A)$, or the matrix exponential and logarithm. It can be used by instructors at all levels as a supplementary text from which to draw examples, historical perspective, statements of results, and exercises. The book, and the subject itself, are particularly well suited to self-study.

To a large extent the chapters can be read independently. However, it is advisable first to become familiar with Sections 1.1–1.3, the first section of Chapter 3 (*Conditioning*), and most of Chapter 4 (*Techniques for General Functions*).

The *Notes and References* are an integral part of each chapter. In addition to containing references, historical information, and further details, they include material not covered elsewhere in the chapter and should always be consulted, in conjunction with the index, to obtain the complete picture.

This book has been designed to be as easy to use as possible and is relatively self-contained. Notation is summarized in Appendix A, while Appendix B (*Background: Definitions and Useful Facts*) reviews basic terminology and needed results from matrix analysis and numerical analysis. When in doubt about the meaning of a term the reader should consult the comprehensive index. Appendix C provides a handy summary of operation counts for the most important matrix computation kernels. Each bibliography entry shows on which pages the item is cited, which makes browsing through the bibliography another route into the book's content.

The exercises, labelled "problems", are an important part of the book, and many of them are new. Solutions, or occasionally a reference to where a solution can be found, are given for almost every problem in Appendix E. Research problems given at the end of some sets of problems highlight outstanding open questions.

A Web page for the book can be found at

`http://www.siam.org/books/ot104`

It includes

- The Matrix Function Toolbox for MATLAB, described in Appendix D. This toolbox contains basic implementations of the main algorithms in the book.
- Updates relating to material in the book.
- A BIBTEX database `functions-of-matrices.bib` containing all the references in the bibliography.

Acknowledgments

A number of people have influenced my thinking about matrix functions. Discussions with Ralph Byers in 1984, when he was working on the matrix sign function and I was investigating the polar decomposition, first made me aware of connections between these two important tools. The work on the matrix exponential of Cleve Moler and Charlie Van Loan has been a frequent source of inspiration. Beresford Parlett's ideas on the exploitation of the Schur form and the adroit use of divided differences have been a guiding light. Charles Kenney and Alan Laub's many contributions to the matrix function arena have been important in my own research and are reported on many pages of this book. Finally, Nick Trefethen has shown me the importance of the Cauchy integral formula and has offered valuable comments on drafts at all stages.

I am grateful to several other people for providing valuable help, suggestions, or advice during the writing of the book:

> Rafik Alam, Awad Al-Mohy, Zhaojun Bai, Timo Betcke, Rajendra Bhatia, Tony Crilly, Philip Davies, Oliver Ernst, Andreas Frommer, Chun-Hua Guo, Gareth Hargreaves, Des Higham, Roger Horn, Bruno Iannazzo, Ilse Ipsen, Peter Lancaster, Jörg Liesen, Lijing Lin, Steve Mackey, Roy Mathias, Volker Mehrmann, Thomas Schmelzer, Gil Strang, Françoise Tisseur, and Andre Weideman.

Working with the SIAM staff on the publication of this book has been a pleasure. I thank, in particular, Elizabeth Greenspan (acquisitions), Sara Murphy (acquisitions), Lois Sellers (design), and Kelly Thomas (copy editing).

Research leading to this book has been supported by the Engineering and Physical Sciences Research Council, The Royal Society, and the Wolfson Foundation.

Manchester
December 2007

Nicholas J. Higham

Chapter 1
Theory of Matrix Functions

In this first chapter we give a concise treatment of the theory of matrix functions, concentrating on those aspects that are most useful in the development of algorithms.

Most of the results in this chapter are for general functions. Results specific to particular functions can be found in later chapters devoted to those functions.

1.1. Introduction

The term "function of a matrix" can have several different meanings. In this book we are interested in a definition that takes a scalar function f and a matrix $A \in \mathbb{C}^{n\times n}$ and specifies $f(A)$ to be a matrix of the same dimensions as A; it does so in a way that provides a useful generalization of the function of a scalar variable $f(z)$, $z \in \mathbb{C}$. Other interpretations of $f(A)$ that are not our focus here are as follows:

- Elementwise operations on matrices, for example $\sin A = (\sin a_{ij})$. These operations are available in some programming languages. For example, Fortran 95 supports "elemental operations" [423, 1999], and most of MATLAB's elementary and special functions are applied in an elementwise fashion when given matrix arguments. However, elementwise operations do not integrate well with matrix algebra, as is clear from the fact that the elementwise square of A is not equal to the matrix product of A with itself. (Nevertheless, the elementwise product of two matrices, known as the Hadamard product or Schur product, is a useful concept [294, 1990], [296, 1991, Chap. 5].)

- Functions producing a scalar result, such as the trace, the determinant, the spectral radius, the condition number $\kappa(A) = \|A\|\,\|A^{-1}\|$, and one particular generalization to matrix arguments of the hypergeometric function [359, 2006].

- Functions mapping $\mathbb{C}^{n\times n}$ to $\mathbb{C}^{m\times m}$ that do not stem from a scalar function. Examples include matrix polynomials with matrix coefficients, the matrix transpose, the adjugate (or adjoint) matrix, compound matrices comprising minors of a given matrix, and factors from matrix factorizations. However, as a special case, the polar factors of a matrix are treated in Chapter 8.

- Functions mapping $\mathbb{C}$ to $\mathbb{C}^{n\times n}$, such as the transfer function $f(t) = B(tI - A)^{-1}C$, for $B \in \mathbb{C}^{n\times m}$, $A \in \mathbb{C}^{m\times m}$, and $C \in \mathbb{C}^{m\times n}$.

Before giving formal definitions, we offer some motivating remarks. When $f(t)$ is a polynomial or rational function with scalar coefficients and a scalar argument, t, it is natural to define $f(A)$ by substituting A for t, replacing division by matrix inversion (provided that the matrices to be inverted are nonsingular), and replacing 1 by the identity matrix. Then, for example,

$$f(t) = \frac{1+t^2}{1-t} \quad \Rightarrow \quad f(A) = (I-A)^{-1}(I+A^2) \quad \text{if } 1 \notin \Lambda(A).$$

Here, $\Lambda(A)$ denotes the set of eigenvalues of A (the spectrum of A). Note that rational functions of a matrix commute, so it does not matter whether we write $(I-A)^{-1}(I+A^2)$ or $(I+A^2)(I-A)^{-1}$. If f has a convergent power series representation, such as

$$\log(1+t) = t - \frac{t^2}{2} + \frac{t^3}{3} - \frac{t^4}{4} + \cdots, \quad |t| < 1,$$

we can again simply substitute A for t to define

$$\log(I+A) = A - \frac{A^2}{2} + \frac{A^3}{3} - \frac{A^4}{4} + \cdots, \quad \rho(A) < 1. \tag{1.1}$$

Here, ρ denotes the spectral radius and the condition $\rho(A) < 1$ ensures convergence of the matrix series (see Theorem 4.7). In this ad hoc fashion, a wide variety of matrix functions can be defined. However, this approach has several drawbacks:

- In order to build up a general mathematical theory, we need a way of defining $f(A)$ that is applicable to arbitrary functions f.
- A particular formula may apply only for a restricted set of A, as in (1.1). If we *define* $f(A)$ from such a formula (rather than obtain the formula by applying suitable principles to a more general definition) we need to check that it is consistent with other definitions of the same function.
- For a multivalued function (multifunction), such as the logarithm or square root, it is desirable to classify *all possible* $f(A)$ that can be obtained by using different branches of the function and to identify any distinguished values.

For these reasons we now consider general definitions of functions of a matrix.

1.2. Definitions of $f(A)$

There are many equivalent ways of defining $f(A)$. We focus on three that are of particular interest. These definitions yield *primary* matrix functions; nonprimary matrix functions are discussed in Section 1.4.

1.2.1. Jordan Canonical Form

It is a standard result that any matrix $A \in \mathbb{C}^{n\times n}$ can be expressed in the Jordan canonical form

$$Z^{-1}AZ = J = \operatorname{diag}(J_1, J_2, \ldots, J_p), \tag{1.2a}$$

$$J_k = J_k(\lambda_k) = \begin{bmatrix} \lambda_k & 1 & & \\ & \lambda_k & \ddots & \\ & & \ddots & 1 \\ & & & \lambda_k \end{bmatrix} \in \mathbb{C}^{m_k\times m_k}, \tag{1.2b}$$

where Z is nonsingular and $m_1 + m_2 + \cdots + m_p = n$. The Jordan matrix J is unique up to the ordering of the blocks J_i, but the transforming matrix Z is not unique.

Denote by $\lambda_1, \ldots, \lambda_s$ the distinct eigenvalues of A and let n_i be the order of the largest Jordan block in which λ_i appears, which is called the *index* of λ_i.

We need the following terminology.

Definition 1.1.[1] *The function f is said to be defined on the spectrum of A if the values*

$$f^{(j)}(\lambda_i), \quad j = 0\colon n_i - 1, \quad i = 1\colon s$$

exist. These are called the values of the function f on the spectrum of A.

In most cases of practical interest f is given by a formula, such as $f(t) = e^t$. However, the following definition of $f(A)$ requires only the values of f on the spectrum of A; it does not require any other information about f. Indeed any $\sum_{i=1}^{s} n_i$ arbitrary numbers can be chosen and assigned as the values of f on the spectrum of A. It is only when we need to make statements about global properties such as continuity that we will need to assume more about f.

Definition 1.2 (matrix function via Jordan canonical form). *Let f be defined on the spectrum of $A \in \mathbb{C}^{n\times n}$ and let A have the Jordan canonical form* (1.2). *Then*

$$f(A) := Zf(J)Z^{-1} = Z\,\mathrm{diag}(f(J_k))Z^{-1}, \tag{1.3}$$

where

$$f(J_k) := \begin{bmatrix} f(\lambda_k) & f'(\lambda_k) & \cdots & \dfrac{f^{(m_k-1)}(\lambda_k)}{(m_k-1)!} \\ & f(\lambda_k) & \ddots & \vdots \\ & & \ddots & f'(\lambda_k) \\ & & & f(\lambda_k) \end{bmatrix}. \tag{1.4}$$

A simple example illustrates the definition. For the Jordan block $J = \left[\begin{smallmatrix} 1/2 & 1 \\ 0 & 1/2 \end{smallmatrix}\right]$ and $f(x) = x^3$, (1.4) gives

$$f(J) = \begin{bmatrix} f(1/2) & f'(1/2) \\ 0 & f(1/2) \end{bmatrix} = \begin{bmatrix} 1/8 & 3/4 \\ 0 & 1/8 \end{bmatrix},$$

which is easily verified to be J^3.

To provide some insight into this definition we make several comments. First, the definition yields an $f(A)$ that can be shown to be independent of the particular Jordan canonical form that is used; see Problem 1.1.

Second, note that if A is diagonalizable then the Jordan canonical form reduces to an eigendecomposition $A = ZDZ^{-1}$, with $D = \mathrm{diag}(\lambda_i)$ and the columns of Z eigenvectors of A, and Definition 1.2 yields $f(A) = Zf(D)Z^{-1} = Z\,\mathrm{diag}(f(\lambda_i))Z^{-1}$. Therefore for diagonalizable matrices $f(A)$ has the same eigenvectors as A and its eigenvalues are obtained by applying f to those of A.

[1] This is the terminology used by Gantmacher [203, 1959, Chap. 5] and Lancaster and Tismenetsky [371, 1985, Chap. 9]. Note that the values depend not just on the eigenvalues but also on the maximal Jordan block sizes n_i.

Finally, we explain how (1.4) can be obtained from Taylor series considerations. In (1.2b) write $J_k = \lambda_k I + N_k \in \mathbb{C}^{m_k \times m_k}$, where N_k is zero except for a superdiagonal of 1s. Note that for $m_k = 3$ we have

$$N_k = \begin{bmatrix} 0 & 1 & 0 \\ 0 & 0 & 1 \\ 0 & 0 & 0 \end{bmatrix}, \quad N_k^2 = \begin{bmatrix} 0 & 0 & 1 \\ 0 & 0 & 0 \\ 0 & 0 & 0 \end{bmatrix}, \quad N_k^3 = 0.$$

In general, powering N_k causes the superdiagonal of 1s to move a diagonal at a time towards the top right-hand corner, until at the m_kth power it disappears: $E_k^{m_k} = 0$; so N_k is nilpotent. Assume that f has a convergent Taylor series expansion

$$f(t) = f(\lambda_k) + f'(\lambda_k)(t - \lambda_k) + \cdots + \frac{f^{(j)}(\lambda_k)(t - \lambda_k)^j}{j!} + \cdots.$$

On substituting $J_k \in \mathbb{C}^{m_k \times m_k}$ for t we obtain the finite series

$$f(J_k) = f(\lambda_k)I + f'(\lambda_k)N_k + \cdots + \frac{f^{(m_k-1)}(\lambda_k)N_k^{m_k-1}}{(m_k - 1)!}, \tag{1.5}$$

since all powers of N_k from the m_kth onwards are zero. This expression is easily seen to agree with (1.4). An alternative derivation of (1.5) that does not rest on a Taylor series is given in the next section.

Definition 1.2 requires the function f to take well-defined values on the spectrum of A—including values associated with derivatives, where appropriate. Thus in the case of functions such as $\sqrt{t}$ and $\log t$ it is implicit that a single branch has been chosen in (1.4). Moreover, if an eigenvalue occurs in more than one Jordan block then the same choice of branch must be made in each block. If the latter requirement is violated then a *nonprimary* matrix function is obtained, as discussed in Section 1.4.

1.2.2. Polynomial Interpolation

The second definition is less obvious than the first, yet it has an elegant derivation and readily yields some useful properties. We first introduce some background on polynomials at matrix arguments.

The *minimal polynomial* of $A \in \mathbb{C}^{n \times n}$ is defined to be the unique monic polynomial ψ of lowest degree such that $\psi(A) = 0$. The existence of ψ is easily proved; see Problem 1.5. A key property is that the minimal polynomial divides any other polynomial p for which $p(A) = 0$. Indeed, by polynomial long division any such p can be written $p = \psi q + r$, where the degree of the remainder r is less than that of ψ. But $0 = p(A) = \psi(A)q(A) + r(A) = r(A)$, and this contradicts the minimality of the degree of ψ unless $r = 0$. Hence $r = 0$ and ψ divides p.

By considering the Jordan canonical form it is not hard to see that

$$\psi(t) = \prod_{i=1}^{s} (t - \lambda_i)^{n_i}, \tag{1.6}$$

where, as in the previous section, $\lambda_1, \ldots, \lambda_s$ are the distinct eigenvalues of A and n_i is the dimension of the largest Jordan block in which λ_i appears. It follows immediately that ψ is zero on the spectrum of A (in the sense of Definition 1.1).

For any $A \in \mathbb{C}^{n \times n}$ and any polynomial $p(t)$, it is obvious that $p(A)$ is defined (by substituting A for t) and that p is defined on the spectrum of A. Our interest in polynomials stems from the fact that the values of p on the spectrum of A determine $p(A)$.

Theorem 1.3. *For polynomials p and q and $A \in \mathbb{C}^{n\times n}$, $p(A) = q(A)$ if and only if p and q take the same values on the spectrum of A.*

Proof. Suppose that two polynomials p and q satisfy $p(A) = q(A)$. Then $d = p - q$ is zero at A so is divisible by the minimal polynomial ψ. In other words, d takes only the value zero on the spectrum of A, that is, p and q take the same values on the spectrum of A.

Conversely, suppose p and q take the same values on the spectrum of A. Then $d = p - q$ is zero on the spectrum of A and so must be divisible by the minimum polynomial ψ, in view of (1.6). Hence $d = \psi r$ for some polynomial r, and since $d(A) = \psi(A)r(A) = 0$, it follows that $p(A) = q(A)$. □

Thus it is a property of polynomials that the matrix $p(A)$ is completely determined by the values of p on the spectrum of A. It is natural to generalize this property to arbitrary functions and define $f(A)$ in such a way that $f(A)$ is completely determined by the values of f on the spectrum of A.

Definition 1.4 (matrix function via Hermite interpolation). *Let f be defined on the spectrum of $A \in \mathbb{C}^{n\times n}$ and let ψ be the minimal polynomial of A. Then $f(A) := p(A)$, where p is the polynomial of degree less than*

$$\sum_{i=1}^{s} n_i = \deg \psi$$

that satisfies the interpolation conditions

$$p^{(j)}(\lambda_i) = f^{(j)}(\lambda_i), \qquad j = 0\colon n_i - 1, \quad i = 1\colon s. \tag{1.7}$$

There is a unique such p and it is known as the Hermite interpolating polynomial.

An example is useful for clarification. Consider $f(t) = \sqrt{t}$ and

$$A = \begin{bmatrix} 2 & 2 \\ 1 & 3 \end{bmatrix}.$$

The eigenvalues are 1 and 4, so $s = 2$ and $n_1 = n_2 = 1$. We take $f(t)$ as the principal branch $t^{1/2}$ of the square root function and find that the required interpolant satisfying $p(1) = f(1) = 1$ and $p(4) = f(4) = 2$ is

$$p(t) = f(1)\frac{t-4}{1-4} + f(4)\frac{t-1}{4-1} = \frac{1}{3}(t+2).$$

Hence

$$f(A) = p(A) = \frac{1}{3}(A + 2I) = \frac{1}{3}\begin{bmatrix} 4 & 2 \\ 1 & 5 \end{bmatrix}.$$

It is easily checked that $f(A)^2 = A$. Note that the formula $A^{1/2} = (A + 2I)/3$ holds more generally for any diagonalizable $n \times n$ matrix A having eigenvalues 1 and/or 4 (and hence having a minimal polynomial that divides $\psi(t) = (t-1)(t-4)$)—including the identity matrix. We are not restricted to using the same branch of the square root function at each eigenvalue. For example, with $f(1) = 1$ and $f(4) = -2$ we obtain $p(t) = 2 - t$ and

$$f(A) = \begin{bmatrix} 0 & -2 \\ -1 & -1 \end{bmatrix}.$$

We make several remarks on this definition.

Remark 1.5. If q is a polynomial that satisfies the interpolation conditions (1.7) and some additional interpolation conditions (at the same or different λ_i) then q and the polynomial p of Definition 1.4 take the same values on the spectrum of A. Hence by Theorem 1.3, $q(A) = p(A) = f(A)$. Sometimes, in constructing a polynomial q for which $q(A) = f(A)$, it is convenient to impose more interpolation conditions than necessary—typically if the eigenvalues of A are known but the Jordan form is not (see the next remark, and Theorem 3.7, for example). Doing so yields a polynomial of higher degree than necessary but does not affect the ability of the polynomial to produce $f(A)$.

Remark 1.6. The Hermite interpolating polynomial p is given explicitly by the Lagrange–Hermite formula

$$p(t) = \sum_{i=1}^{s} \left[\left(\sum_{j=0}^{n_i-1} \frac{1}{j!} \phi_i^{(j)}(\lambda_i)(t-\lambda_i)^j \right) \prod_{j \neq i} (t-\lambda_j)^{n_j} \right], \tag{1.8}$$

where $\phi_i(t) = f(t)/\prod_{j\neq i}(t-\lambda_j)^{n_j}$. For a matrix with distinct eigenvalues ($n_i \equiv 1$, $s = n$) this formula reduces to the familiar Lagrange form

$$p(t) = \sum_{i=1}^{n} f(\lambda_i)\ell_i(t), \qquad \ell_i(t) = \prod_{\substack{j=1 \\ j\neq i}}^{n} \left(\frac{t-\lambda_j}{\lambda_i - \lambda_j} \right). \tag{1.9}$$

An elegant alternative to (1.8) is the Newton divided difference form

$$\begin{aligned} p(t) = f[x_1] + f[x_1,x_2](t-x_1) + f[x_1,x_2,x_3](t-x_1)(t-x_2) + \cdots \\ + f[x_1,x_2,\ldots,x_m](t-x_1)(t-x_2)\ldots(t-x_{m-1}), \end{aligned} \tag{1.10}$$

where $m = \deg \psi$ and the set $\{x_i\}_{i=1}^m$ comprises the distinct eigenvalues $\lambda_1,\ldots,\lambda_s$ with λ_i having multiplicity n_i. Here the $f[\ldots]$ denote divided differences, which are defined in Section B.16. Another polynomial q for which $f(A) = q(A)$ is given by (1.10) with $m = n$ and $\{x_i\}_{i=1}^n$ the set of all n eigenvalues of A:

$$\begin{aligned} q(t) = f[\lambda_1] + f[\lambda_1,\lambda_2](t-\lambda_1) + f[\lambda_1,\lambda_2,\lambda_3](t-\lambda_1)(t-\lambda_2) + \cdots \\ + f[\lambda_1,\lambda_2,\ldots,\lambda_n](t-\lambda_1)(t-\lambda_2)\ldots(t-\lambda_{n-1}). \end{aligned} \tag{1.11}$$

This polynomial is independent of the Jordan structure of A and is in general of higher degree than p. However, the properties of divided differences ensure that q and p take the same values on the spectrum of A, so $q(A) = p(A) = f(A)$.

Remark 1.7. This definition explicitly makes $f(A)$ a polynomial in A. It is important to note, however, that the polynomial p *depends on* A, through the values of f on the spectrum of A, so it is not the case that $f(A) \equiv q(A)$ for some fixed polynomial q independent of A.

Remark 1.8. If f is given by a power series, Definition 1.4 says that $f(A)$ is nevertheless expressible as a polynomial in A of degree at most $n-1$. Another way to arrive at this conclusion is as follows. The Cayley–Hamilton theorem says that any matrix satisfies its own characteristic equation: $q(A) = 0$,[2] where $q(t) = \det(tI - A)$ is the characteristic polynomial. This theorem follows immediately from the fact that the minimal polynomial ψ divides q (see Problem 1.18 for another proof). Hence the nth power of A, and inductively all higher powers, are expressible as a linear combination of I, A, ..., A^{n-1}. Thus any power series in A can be reduced to a polynomial in A of degree at most $n-1$. This polynomial is rarely of an elegant form or of practical interest; exceptions are given in (1.16) and Problem 10.13.

Remark 1.9. It is natural to ask whether $f(A)$ is real whenever A is real. By considering real, diagonal A, it is clear that for this condition to hold it is necessary that the scalar function f is real on the subset of the real line on which it is defined. Since the nonreal eigenvalues of a real matrix occur in complex conjugate pairs λ, $\overline{\lambda}$ it is reasonable also to assume that $f(\lambda)$, $f(\overline{\lambda})$ form a complex conjugate pair, and likewise for higher derivatives. The interpolation conditions (1.7) can be written in the form of a dual (confluent) Vandermonde system of equations whose solution is a vector comprising the coefficients of r. Considering, for a moment, a 2×2 real matrix with eigenvalues λ, $\overline{\lambda}$ ($\lambda \neq \overline{\lambda}$) this system is, under the assumption on f above,

$$\begin{bmatrix} 1 & \lambda \\ 1 & \overline{\lambda} \end{bmatrix} \begin{bmatrix} r_0 \\ r_1 \end{bmatrix} = \begin{bmatrix} f(\lambda) \\ f(\overline{\lambda}) \end{bmatrix} = \begin{bmatrix} f(\lambda) \\ \overline{f(\lambda)} \end{bmatrix}.$$

Premultiplying by the matrix $\left[\begin{smallmatrix} 1 & 1 \\ -i & i \end{smallmatrix}\right]/2$ yields the system

$$\begin{bmatrix} 1 & \operatorname{Re}\lambda \\ 0 & \operatorname{Im}\lambda \end{bmatrix} \begin{bmatrix} r_0 \\ r_1 \end{bmatrix} = \begin{bmatrix} \operatorname{Re} f(\lambda) \\ \operatorname{Im} f(\lambda) \end{bmatrix}$$

with a *real* coefficient matrix and right-hand side. We conclude that r has real coefficients and hence $f(A) = p(A)$ is real when A is real. This argument extends to real $n \times n$ matrices under the stated condition on f. As a particular example, we can conclude that if A is real and nonsingular with no eigenvalues on the negative real axis then A has a real square root and a real logarithm. For a full characterization of the existence of real square roots and logarithms see Theorem 1.23. Equivalent conditions to $f(A)$ being real for real A when f is analytic are given in Theorem 1.18.

Remark 1.10. We can derive directly from Definition 1.4 the formula (1.4) for a function of the Jordan block J_k in (1.2). It suffices to note that the interpolation conditions are $p^{(j)}(\lambda_k) = f^{(j)}(\lambda_k)$, $j = 0\colon m_k - 1$, so that the required Hermite interpolating polynomial is

$$p(t) = f(\lambda_k) + f'(\lambda_k)(t - \lambda_k) + \frac{f''(\lambda_k)(t-\lambda_k)^2}{2!} + \cdots + \frac{f^{(m_k-1)}(\lambda_k)(t-\lambda_k)^{m_k-1}}{(m_k-1)!},$$

and then to evaluate $p(J_k)$, making use of the properties of the powers of N_k noted in the previous section (cf. (1.5)).

1.2.3. Cauchy Integral Theorem

Perhaps the most concise and elegant definition of a function of a matrix is a generalization of the Cauchy integral theorem.

---

[2] It is incorrect to try to prove the Cayley–Hamilton theorem by "$q(A) = \det(AI - A) = 0$".

Definition 1.11 (matrix function via Cauchy integral). *For* $A \in \mathbb{C}^{n\times n}$,

$$f(A) := \frac{1}{2\pi i} \int_{\Gamma} f(z)(zI - A)^{-1}\, dz, \tag{1.12}$$

where f *is analytic on and inside a closed contour* Γ *that encloses* $\Lambda(A)$.

The integrand contains the resolvent, $(zI - A)^{-1}$, which is defined on Γ since Γ is disjoint from the spectrum of A.

This definition leads to short proofs of certain theoretical results and has the advantage that it can be generalized to operators.

1.2.4. Equivalence of Definitions

Our three definitions are equivalent, modulo the requirement in the Cauchy integral definition that f be analytic in a region of the complex plane containing the spectrum.

Theorem 1.12. *Definition* 1.2 (*Jordan canonical form*) *and Definition* 1.4 (*Hermite interpolation*) *are equivalent. If* f *is analytic then Definition* 1.11 (*Cauchy integral*) *is equivalent to Definitions* 1.2 *and* 1.4.

Proof. Definition 1.4 says that $f(A) = p(A)$ for a Hermite interpolating polynomial p satisfying (1.7). If A has the Jordan form (1.2) then $f(A) = p(A) = p(ZJZ^{-1}) = Zp(J)Z^{-1} = Z\,\mathrm{diag}(p(J_k))Z^{-1}$, just from elementary properties of matrix polynomials. But since $p(J_k)$ is completely determined by the values of p on the spectrum of J_k, and these values are a subset of the values of p on the spectrum of A, it follows from Remark 1.5 and Remark 1.10 that $p(J_k)$ is precisely (1.4). Hence the matrix $f(A)$ obtained from Definition 1.4 agrees with that given by Definition 1.2.

For the equivalence of Definition 1.11 with the other two definitions, see Horn and Johnson [296, 1991, Thm. 6.2.28]. □

We will mainly use (for theoretical purposes) Definitions 1.2 and 1.4. The polynomial interpolation definition, Definition 1.4, is well suited to proving basic properties of matrix functions, such as those in Section 1.3, while the Jordan canonical form definition, Definition 1.2, excels for solving matrix equations such as $X^2 = A$ and $e^X = A$. For many purposes, such as the derivation of the formulae in the next section, either of the definitions can be used.

In the rest of the book we will refer simply to "the definition of a matrix function".

1.2.5. Example: Function of Identity Plus Rank-1 Matrix

To illustrate the theory, and the consistency of the different ways of defining $f(A)$, it is instructive to consider the cases where A is a rank-1 matrix and a rank-1 perturbation of the identity matrix.

Consider, first, a rank-1 matrix $A = uv^*$. The interpolation definition provides the easiest way to obtain $f(A)$. We first need to determine the Jordan structure of A. If $v^*u \neq 0$ then A has an eigenpair (v^*u, u) and 0 is a semisimple eigenvalue of multiplicity $n-1$. The interpolation conditions (1.7) are therefore simply

$$p(v^*u) = f(v^*u), \quad p(0) = f(0),$$

and so

$$p(t) = \frac{t - v^*u}{0 - v^*u} f(0) + \frac{t - 0}{v^*u - 0} f(v^*u).$$

Hence

$$\begin{aligned} f(A) = p(A) &= -\frac{f(0)}{v^*u} uv^* + f(0)I + f(v^*u)\frac{uv^*}{v^*u} \\ &= f(0)I + \left(\frac{f(v^*u) - f(0)}{v^*u - 0}\right) uv^* \\ &= f(0)I + f[v^*u, 0]\, uv^*. \end{aligned} \tag{1.13}$$

We have manipulated the expression into this form involving a divided difference because it is suggestive of what happens when $v^*u = 0$. Indeed $f[0,0] = f'(0)$ and so when $v^*u = 0$ we may expect that $f(A) = f(0)I + f'(0)uv^*$. To confirm this formula, note that $v^*u = 0$ implies that the spectrum of A consists entirely of 0 and that $A^2 = (v^*u)uv^* = 0$. Hence, assuming $A \neq 0$, A must have one 2×2 Jordan block corresponding to the eigenvalue 0, with the other $n-2$ zero eigenvalues occurring in 1×1 Jordan blocks. The interpolation conditions (1.7) are therefore

$$p(0) = f(0), \quad p'(0) = f'(0),$$

and so $p(t) = f(0) + tf'(0)$. Therefore $p(A) = f(0)I + f'(0)uv^*$, as anticipated. To summarize, the formula

$$f(uv^*) = f(0)I + f[v^*u, 0]\, uv^* \tag{1.14}$$

is valid for all u and v. We could have obtained this formula directly by using the divided difference form (1.10) of the Hermite interpolating polynomial r, but the derivation above gives more insight.

We now show how the formula is obtained from Definition 1.2 when $v^*u \neq 0$ (for the case $v^*u = 0$ see Problem 1.15). The Jordan canonical form can be written as

$$A = [\, u \quad X \,]\, \mathrm{diag}(v^*u, 0, \dots, 0) \begin{bmatrix} v^*/(v^*u) \\ Y \end{bmatrix},$$

where X and Y are chosen so that $AX = 0$, $[\, u \quad X \,]$ is nonsingular, and

$$[\, u \quad X \,] \begin{bmatrix} v^*/(v^*u) \\ Y \end{bmatrix} = I. \tag{1.15}$$

Hence

$$f(A) = [\, u \quad X \,]\, \mathrm{diag}(f(v^*u), f(0), \dots, f(0)) \begin{bmatrix} v^*/(v^*u) \\ Y \end{bmatrix} = f(v^*u)\frac{uv^*}{v^*u} + f(0)XY.$$

But $XY = I - uv^*/(v^*u)$, from (1.15), and hence (1.13) is recovered.

If f has a power series expansion then (1.14) can also be derived by direct substitution into the power series, using $A^k = (v^*u)^{k-1}uv^*$.

The Cauchy integral definition (1.12) can also be used to derive (1.14) when f is analytic, by using the Sherman–Morrison formula (B.11).

Even in the rank-1 case issues of nonexistence are present. For f the square root, (1.14) provides the two square roots $uv^*/\sqrt{v^*u}$ for $v^*u \neq 0$. But if $v^*u = 0$ the formula breaks down because $f'(0)$ is undefined. In this case A has no square roots—essentially because the Jordan form of A has a block $\left[\begin{smallmatrix}0&1\\0&0\end{smallmatrix}\right]$, which has no square roots. Also note that if u and v are real, $f(uv^*)$ will be real only if $f[v^*u, 0]$ is real.

Analysis very similar to that above provides a formula for a function of the identity plus a rank-1 matrix that generalizes (1.14) (see Problem 1.16):

$$f(\alpha I + uv^*) = f(\alpha)I + f[\alpha + v^*u, \alpha]uv^*. \tag{1.16}$$

For a more general result involving a perturbation of arbitrary rank see Theorem 1.35.

1.2.6. Example: Function of Discrete Fourier Transform Matrix

Another interesting example is provided by the discrete Fourier transform (DFT) matrix

$$F_n = \frac{1}{n^{1/2}}\Big(\exp(-2\pi i(r-1)(s-1)/n)\Big)_{r,s=1}^{n} \in \mathbb{C}^{n\times n}. \tag{1.17}$$

F_n is a very special matrix: it is complex symmetric and unitary (and is a Vandermonde matrix based on the roots of unity). Let us see how to evaluate $f(F_n)$.

The DFT has the special property that $F_n^4 = I$, from which it follows that the minimal polynomial of F_n is $\psi(t) = t^4 - 1$ for $n \ge 4$. The interpolating polynomial in (1.7) therefore has degree 3 for all $n \ge 4$ and can be expressed in Lagrange form (1.9) as

$$\begin{aligned} p(t) = \frac{1}{4}\big[&f(1)(t+1)(t-i)(t+i) - f(-1)(t-1)(t-i)(t+i) \\ &+ if(i)(t-1)(t+1)(t+i) - if(-i)(t-1)(t+1)(t-i)\big]. \end{aligned} \tag{1.18}$$

Thus $f(A) = p(A)$, and in fact this formula holds even for $n = 1{:}3$, since incorporating extra interpolation conditions does not affect the ability of the interpolating polynomial to yield $f(A)$ (see Remark 1.5). This expression can be quickly evaluated in $O(n^2 \log n)$ operations because multiplication of a vector by F_n can be carried out in $O(n \log n)$ operations using the fast Fourier transform (FFT).

Because F_n is unitary and hence normal, F_n is unitarily diagonalizable: $F_n = QDQ^*$ for some unitary Q and diagonal D. (Indeed, any matrix with minimal polynomial $\psi(t)$ has distinct eigenvalues and so is diagonalizable.) Thus $f(F_n) = Qf(D)Q^*$. However, this formula requires knowledge of Q and D and so is much more complicated to use than (1.18).

1.3. Properties

The sign of a good definition is that it leads to the properties one expects or hopes for, as well as some useful properties that are less obvious. We collect some general properties that follow from the definition of $f(A)$.

Theorem 1.13. *Let $A \in \mathbb{C}^{n\times n}$ and let f be defined on the spectrum of A. Then*

(a) *$f(A)$ commutes with A;*

(b) *$f(A^T) = f(A)^T$;*

(c) *$f(XAX^{-1}) = Xf(A)X^{-1}$;*

(d) *the eigenvalues of $f(A)$ are $f(\lambda_i)$, where the λ_i are the eigenvalues of A;*

(e) *if X commutes with A then X commutes with $f(A)$;*

(f) *if $A = (A_{ij})$ is block triangular then $F = f(A)$ is block triangular with the same block structure as A, and $F_{ii} = f(A_{ii})$;*

(g) *if $A = \operatorname{diag}(A_{11}, A_{22}, \dots, A_{mm})$ is block diagonal then*

$$f(A) = \operatorname{diag}\bigl(f(A_{11}), f(A_{22}), \dots, f(A_{mm})\bigr);$$

(h) *$f(I_m \otimes A) = I_m \otimes f(A)$, where $\otimes$ is the Kronecker product;*

(i) *$f(A \otimes I_m) = f(A) \otimes I_m$.*

Proof. Definition 1.4 implies that $f(A)$ is a polynomial in A, $p(A)$ say. Then $f(A)A = p(A)A = Ap(A) = Af(A)$, which proves the first property. For (b) we have $f(A)^T = p(A)^T = p(A^T) = f(A^T)$, where the last equality follows from the fact that the values of f on the spectrum of A are the same as the values of f on the spectrum of A^T. (c) and (d) follow immediately from Definition 1.2. (e) follows from (c) when X is nonsingular; more generally it is obtained from $Xf(A) = Xp(A) = p(A)X = f(A)X$. For (f), $f(A) = p(A)$ is clearly block triangular and its ith diagonal block is $p(A_{ii})$. Since p interpolates f on the spectrum of A it interpolates f on the spectrum of each A_{ii}, and hence $p(A_{ii}) = f(A_{ii})$. (g) is a special case of (f). (h) is a special case of (g), since $I_m \otimes A = \operatorname{diag}(A, A, \dots, A)$. Finally, we have $A \otimes B = \Pi(B \otimes A)\Pi^T$ for a permutation matrix Π, and so

$$f(A\otimes I_m) = f(\Pi(I_m\otimes A)\Pi^T) = \Pi f(I_m\otimes A)\Pi^T = \Pi(I_m\otimes f(A))\Pi^T = f(A)\otimes I_m. \quad \square$$

Theorem 1.14 (equality of two matrix functions). *With the notation of Section 1.2, $f(A) = g(A)$ if and only if*

$$f^{(j)}(\lambda_i) = g^{(j)}(\lambda_i), \quad j = 0\colon n_i - 1, \quad i = 1\colon s.$$

Equivalently, $f(A) = 0$ if and only if

$$f^{(j)}(\lambda_i) = 0, \quad j = 0\colon n_i - 1, \quad i = 1\colon s.$$

Proof. This result is immediate from Definition 1.2 or Definition 1.4. $\square$

The next three results show how different functions interact in combination. It is worth emphasizing why these results are nontrivial. It is not immediate from any of the definitions of $f(A)$ how to evaluate at A a composite function, such as $f(t) = e^{-t}\sin(t)$ or $g(t) = t - (t^{1/2})^2$. Replacing "t" by "A" in these expressions needs to be justified, as does the deduction $g(A) = 0$ from $g(t) = 0$. However, in any polynomial (which may be an expression made up from other polynomials) the "$t \to A$" substitution is valid, and the proofs for general functions therefore work by reducing to the polynomial case. The first result concerns a sum or product of functions.

Theorem 1.15 (sum and product of functions). *Let f and g be functions defined on the spectrum of $A \in \mathbb{C}^{n\times n}$.*

(a) *If $h(t) = f(t) + g(t)$ then $h(A) = f(A) + g(A)$.*

(b) *If $h(t) = f(t)g(t)$ then $h(A) = f(A)g(A)$.*

Proof. Part (a) is immediate from any of the definitions of $h(A)$. For part (b), let p and q interpolate f and g on the spectrum of A, so that $p(A) = f(A)$ and $q(A) = g(A)$. By differentiating and using the product rule we find that the functions $h(t)$ and $r(t) = p(t)q(t)$ have the same values on the spectrum of A. Hence $h(A) = r(A) = p(A)q(A) = f(A)g(A)$. □

The next result generalizes the previous one and says that scalar functional relationships of a polynomial nature are preserved by matrix functions. For example $\sin^2(A) + \cos^2(A) = I$, $(A^{1/p})^p = A$, and $e^{iA} = \cos(A) + i\sin(A)$. Of course, generalizations of scalar identities that involve two or more noncommuting matrices may fail; for example, e^{A+B}, $e^A e^B$, and $e^B e^A$ are in general all different (see Section 10.1).

Theorem 1.16 (polynomial functional identities). *Let $Q(u_1,\dots,u_t)$ be a polynomial in $u_1,\dots,u_t$ and let $f_1,\dots,f_t$ be functions defined on the spectrum of $A \in \mathbb{C}^{n\times n}$. If $f(\lambda) = Q(f_1(\lambda),\dots,f_t(\lambda))$ takes zero values on the spectrum of A then $f(A) = Q(f_1(A),\dots,f_t(A)) = 0$.*

Proof. Let the polynomials $p_1,\dots,p_t$ interpolate $f_1,\dots,f_t$ on the spectrum of A. Then $p_i(A) = f_i(A)$, $i = 1\colon t$. Let $p(\lambda) = Q(p_1(\lambda),\dots,p_t(\lambda))$, and note that $p(\lambda)$ is a polynomial in λ. Since p_i and f_i take the same values on the spectrum of A, so do f and p. But f takes zero values on the spectrum of A, by assumption, and hence so does p. Therefore, by Theorem 1.14, $f(A) = p(A) = 0$. □

The next result concerns a composite function in which neither of the constituents need be a polynomial.

Theorem 1.17 (composite function). *Let $A \in \mathbb{C}^{n\times n}$ and let the distinct eigenvalues of A be $\lambda_1,\dots,\lambda_s$ with indices $n_1,\dots,n_s$. Let h be defined on the spectrum of A (so that the values $h^{(j)}(\lambda_i)$, $j = 0\colon n_i - 1$, $i = 1\colon s$ exist) and let the values $g^{(j)}(h(\lambda_i))$, $j = 0\colon n_i - 1$, $i = 1\colon s$ exist. Then $f(t) = g(h(t))$ is defined on the spectrum of A and $f(A) = g(h(A))$.*

Proof. Let $\mu_k = h(\lambda_k)$, $k = 1\colon s$. Since

$$f(\lambda_k) = g(\mu_k), \tag{1.19a}$$
$$f'(\lambda_k) = g'(\mu_k)h'(\lambda_k), \tag{1.19b}$$
$$\vdots$$
$$f^{(n_k-1)}(\lambda_k) = g^{(n_k-1)}(\mu_k)h'(\lambda_k)^{n_k-1} + \cdots + g'(\mu_k)h^{(n_k-1)}(\lambda_k), \tag{1.19c}$$

and all the derivatives on the right-hand side exist, f is defined on the spectrum of A.

Let $p(t)$ be any polynomial satisfying the interpolation conditions

$$p^{(j)}(\mu_i) = g^{(j)}(\mu_i), \qquad j = 0\colon n_i - 1, \quad i = 1\colon s. \tag{1.20}$$

From Definition 1.2 it is clear that the indices of the eigenvalues $\mu_1, \dots, \mu_s$ of $h(A)$ are at most $n_1, \dots, n_s$, so the values on the right-hand side of (1.20) contain the values of g on the spectrum of $B = h(A)$; thus $g(B)$ is defined and $p(B) = g(B)$. It now follows by (1.19) and (1.20) that the values of $f(t)$ and $p(h(t))$ coincide on the spectrum of A. Hence by applying Theorem 1.16 to $Q(f(t), h(t)) = f(t) - p(h(t))$ we conclude that

$$f(A) = p(h(A)) = p(B) = g(B) = g(h(A)),$$

as required. $\square$

The assumptions in Theorem 1.17 on g for $f(A)$ to exist are stronger than necessary in certain cases where a Jordan block of A splits under evaluation of h. Consider, for example, $g(t) = t^{1/3}$, $h(t) = t^2$, and $A = \left[\begin{smallmatrix} 0 & 1 \\ 0 & 0 \end{smallmatrix}\right]$. The required derivative $g'(0)$ in Theorem 1.17 does not exist, but $f(A) = (A^2)^{1/3} = 0$ nevertheless does exist. (A full description of the Jordan canonical form of $f(A)$ in terms of that of A is given in Theorem 1.36.)

Theorem 1.17 implies that $\exp(\log A) = A$, provided that log is defined on the spectrum of A. However, $\log(\exp(A)) = A$ does not hold unless the spectrum of A satisfies suitable restrictions, since the scalar relation $\log(e^t) = t$ is likewise not generally true in view of $e^t = e^{t+2k\pi i}$ for any integer k; see Problem 1.39.

Although $f(A^T) = f(A)^T$ always holds (Theorem 1.13 (b)), the property $f(A^*) = f(A)^*$ does not. The next result says essentially that for an analytic function f defined on a suitable domain that includes a subset S of the real line, $f(A^*) = f(A)^*$ holds precisely when f maps S back into the real line. This latter condition also characterizes when A real implies $f(A)$ real (cf. the sufficient conditions given in Remark 1.9).

Theorem 1.18 (Higham, Mackey, Mackey, and Tisseur). *Let f be analytic on an open subset $\Omega \subseteq \mathbb{C}$ such that each connected component of Ω is closed under conjugation. Consider the corresponding matrix function f on its natural domain in $\mathbb{C}^{n\times n}$, the set $\mathcal{D} = \{\, A \in \mathbb{C}^{n\times n} : \Lambda(A) \subseteq \Omega \,\}$. Then the following are equivalent:*

(a) $f(A^*) = f(A)^*$ *for all* $A \in \mathcal{D}$.

(b) $f(\overline{A}) = \overline{f(A)}$ *for all* $A \in \mathcal{D}$.

(c) $f(\mathbb{R}^{n\times n} \cap \mathcal{D}) \subseteq \mathbb{R}^{n\times n}$.

(d) $f(\mathbb{R} \cap \Omega) \subseteq \mathbb{R}$.

Proof. The first two properties are obviously equivalent, in view of Theorem 1.13 (b). Our strategy is therefore to show that (b) $\Rightarrow$ (c) $\Rightarrow$ (d) $\Rightarrow$ (b).

(b) $\Rightarrow$ (c): If $A \in \mathbb{R}^{n\times n} \cap \mathcal{D}$ then

$$\begin{aligned} f(A) &= f(\overline{A}) \quad (\text{since } A \in \mathbb{R}^{n\times n}) \\ &= \overline{f(A)} \quad (\text{given}), \end{aligned}$$

so $f(A) \in \mathbb{R}^{n\times n}$, as required.

(c) $\Rightarrow$ (d): If $\lambda \in \mathbb{R} \cap \Omega$ then $\lambda I \in \mathbb{R}^{n\times n} \cap \mathcal{D}$. So $f(\lambda I) \in \mathbb{R}^{n\times n}$ by (c), and hence, since $f(\lambda I) = f(\lambda)I$, $f(\lambda) \in \mathbb{R}$.

The argument that (d) $\Rightarrow$ (b) is more technical and involves complex analysis; see Higham, Mackey, Mackey, and Tisseur [283, 2005, Thm. 3.2]. $\square$

Our next result shows that although the definition of $f(A)$ utilizes only the values of f on the spectrum of A (the values assumed by f elsewhere in $\mathbb{C}$ being arbitrary), $f(A)$ is a continuous function of A under suitable assumptions on f and the domain.

Theorem 1.19 (continuity). *Let $\mathcal{D}$ be an open subset of $\mathbb{R}$ or $\mathbb{C}$ and let f be $n-1$ times continuously differentiable on $\mathcal{D}$. Then $f(A)$ is a continuous matrix function on the set of matrices $A \in \mathbb{C}^{n\times n}$ with spectrum in $\mathcal{D}$.*

Proof. See Horn and Johnson [296, 1991, Thm. 6.2.27 (1)], and Mathias [412, 1996, Lem. 1.1] for the conditions as stated here. $\square$

For continuity of $f(A)$ on the set of normal matrices just the continuity of f is sufficient [296, 1991, Thm. 6.2.37].

Our final result shows that under mild conditions to check the veracity of a matrix identity it suffices to check it for diagonalizable matrices.

Theorem 1.20. *Let f satisfy the conditions of Theorem 1.19. Then $f(A) = 0$ for all $A \in \mathbb{C}^{n\times n}$ with spectrum in $\mathcal{D}$ if and only if $f(A) = 0$ for all diagonalizable $A \in \mathbb{C}^{n\times n}$ with spectrum in $\mathcal{D}$.*

Proof. See Horn and Johnson [296, 1991, Thm. 6.2.27 (2)]. □

For an example of the use of Theorem 1.20 see the proof of Theorem 11.1. Theorem 1.13 (f) says that block triangular structure is preserved by matrix functions. An explicit formula can be given for an important instance of the block 2×2 case.

Theorem 1.21. *Let f satisfy the conditions of Theorem 1.19 with $\mathcal{D}$ containing the spectrum of*

$$A = \begin{array}{c} \\ {\scriptstyle n-1} \\ {\scriptstyle 1} \end{array} \!\!\begin{array}{c} \begin{array}{cc} {\scriptstyle n-1} & {\scriptstyle 1} \end{array} \\ \begin{bmatrix} B & c \\ 0 & \lambda \end{bmatrix} \end{array} \in \mathbb{C}^{n\times n}.$$

Then

$$f(A) = \begin{bmatrix} f(B) & g(B)c \\ 0 & f(\lambda) \end{bmatrix}, \tag{1.21}$$

where $g(z) = f[z, \lambda]$. In particular, if $\lambda \notin \Lambda(B)$ then $g(B) = (B - \lambda I)^{-1}(f(B) - f(\lambda)I)$.

Proof. We need only to demonstrate the formula for the (1,2) block F_{12} of $f(A)$. Equating (1,2) blocks in $f(A)A = Af(A)$ (Theorem 1.13 (a)) yields $BF_{12} + cf(\lambda) = f(B)c + F_{12}\lambda$, or $(B - \lambda I)F_{12} = (f(B) - f(\lambda)I)c$. If $\lambda \notin \Lambda(B)$ the result is proved. Otherwise, the result follows by a continuity argument: replace λ by $\lambda(\epsilon) = \lambda + \epsilon$, so that $\lambda(\epsilon) \notin \Lambda(B)$ for sufficiently small ϵ, let $\epsilon \to 0$, and use the continuity of divided differences and of $f(A)$. □

For an expression for a function of a general block 2×2 block triangular matrix see Theorem 4.12.

1.4. Nonprimary Matrix Functions

One of the main uses of matrix functions is for solving nonlinear matrix equations, $g(X) = A$. Two particular cases are especially important. We will call any solution of $X^2 = A$ a square root of A and any solution of $e^X = A$ a logarithm of A. We naturally turn to the square root and logarithm functions to solve the latter two equations. But for certain matrices A some of the solutions of $g(X) = A$ are not obtainable as a primary matrix function of A, that is, they cannot be produced by our (three equivalent) definitions of $f(A)$ (with $f = g^{-1}$ or otherwise). These X are examples of *nonprimary matrix functions.* Informally, a nonprimary matrix function is a "matrix equation solving function" that cannot be expressed as a primary matrix function; we will not try to make this notion precise.

Suppose we wish to find square roots of

$$A = \begin{bmatrix} 1 & 0 \\ 0 & 1 \end{bmatrix},$$

that is, solve $X^2 = A$. Taking $f(t) = \sqrt{t}$, the interpolation conditions in Definitions 1.4 are (with $s = 1$, $n_1 = 1$) simply $p(1) = \sqrt{1}$. The interpolating polynomial is therefore either $p(t) = 1$ or $p(t) = -1$, corresponding to the two square roots of 1, giving I and $-I$ as square roots of A. Both of these square roots are, trivially, polynomials in A. Turning to Definition 1.2, the matrix A is already in Jordan form with two 1×1 Jordan blocks, and the definition provides the same two square roots. However, if we ignore the prescription at the end of Section 1.2.1 about the choice of branches then we can obtain two more square roots,

$$\begin{bmatrix} -1 & 0 \\ 0 & 1 \end{bmatrix}, \quad \begin{bmatrix} 1 & 0 \\ 0 & -1 \end{bmatrix},$$

in which the two eigenvalues 1 have been sent to *different* square roots. Moreover, since $A = ZIZ^{-1}$ is a Jordan canonical form for any nonsingular Z, Definition 1.2 yields the square roots

$$Z\begin{bmatrix} -1 & 0 \\ 0 & 1 \end{bmatrix}Z^{-1}, \quad Z\begin{bmatrix} 1 & 0 \\ 0 & -1 \end{bmatrix}Z^{-1}, \tag{1.22}$$

and these formulae provide an infinity of square roots, because only for diagonal Z are the matrices in (1.22) independent of Z. Indeed, one infinite family of square roots of A comprises the Householder reflections

$$H(\theta) = \begin{bmatrix} \cos\theta & \sin\theta \\ \sin\theta & -\cos\theta \end{bmatrix}, \qquad \theta \in [0, 2\pi].$$

Definitions 1.2, 1.4, and 1.11 yield primary matrix functions. In most applications it is primary matrix functions that are of interest, and virtually all the existing theory and available methods are for such functions. Nonprimary matrix functions are obtained from Definition 1.2 when two equal eigenvalues in different Jordan blocks are mapped to different values of f; in other words, different branches of f are taken for different Jordan blocks with the same eigenvalue. The function obtained thereby depends on the matrix Z in (1.3). This possibility arises precisely when the function is multivalued and the matrix is derogatory, that is, the matrix has multiple eigenvalues and an eigenvalue appears in more than one Jordan block.

Unlike primary matrix functions, nonprimary ones are not expressible as polynomials in the matrix. However, a nonprimary function obtained from Definition 1.2, using the prescription in the previous paragraph, nevertheless commutes with the matrix. Such a function has the form $X = Z\,\mathrm{diag}(f_k(J_k))Z^{-1}$, where $A = Z\,\mathrm{diag}(J_k)Z^{-1}$ is a Jordan canonical form and where the notation f_k denotes that the branch of f taken depends on k. Then $XA = AX$, because $f_k(J_k)$ is a primary matrix function and so commutes with J_k.

But note that not all nonprimary matrix functions are obtainable from the Jordan canonical form prescription above. For example, $A = \left[\begin{smallmatrix} 0 & 0 \\ 0 & 0 \end{smallmatrix}\right]$ has the square root $X = \left[\begin{smallmatrix} 0 & 1 \\ 0 & 0 \end{smallmatrix}\right]$, and X is a Jordan block larger than the 1×1 Jordan blocks of A. This example also illustrates that a nonprimary function can have the same spectrum as a primary function, and so in general a nonprimary function cannot be identified from its spectrum alone.

Nonprimary functions can be needed when, for a matrix A depending on a parameter t, a smooth curve of functions $f(A(t))$ needs to be computed and eigenvalues of $A(t)$ coalesce. Suppose we wish to compute square roots of

$$G(\theta) = \begin{bmatrix} \cos\theta & \sin\theta \\ -\sin\theta & \cos\theta \end{bmatrix}$$

as θ varies from 0 to 2π. Since multiplication of a vector by $G(\theta)$ represents a rotation through θ radians clockwise, $G(\theta/2)$ is the natural square root. However, for $\theta = \pi$,

$$G(\pi) = \begin{bmatrix} -1 & 0 \\ 0 & -1 \end{bmatrix}, \qquad G(\pi/2) = \begin{bmatrix} 0 & 1 \\ -1 & 0 \end{bmatrix}.$$

The only primary square roots of $G(\pi)$ are $\pm iI$, which are nonreal. While it is nonprimary, $G(\pi/2)$ is the square root we need in order to produce a smooth curve of square roots.

An example of an application where nonprimary logarithms arise is the embeddability problems for Markov chains (see Section 2.3).

A primary matrix function with a nonprimary flavour is the matrix sign function (see Chapter 5), which for a matrix $A \in \mathbb{C}^{n\times n}$ is a (generally) nonprimary square root of I that depends on A.

Unless otherwise stated, $f(A)$ denotes a primary matrix function throughout this book.

1.5. Existence of (Real) Matrix Square Roots and Logarithms

If A is nonsingular, or singular with a semisimple zero eigenvalue, then the square root function is defined on the spectrum of A and so primary square roots exist. If A is singular with a defective zero eigenvalue then while it has no primary square roots it may have nonprimary ones. The existence of a square root of either type can be neatly characterized in terms of null spaces of powers of A.

Theorem 1.22 (existence of matrix square root). *$A \in \mathbb{C}^{n\times n}$ has a square root if and only if in the "ascent sequence" of integers $d_1, d_2, \ldots$ defined by*

$$d_i = \dim(\mathrm{null}(A^i)) - \dim(\mathrm{null}(A^{i-1}))$$

no two terms are the same odd integer.

Proof. See Cross and Lancaster [122, 1974] or Horn and Johnson [296, 1991, Cor. 6.4.13]. □

To illustrate, consider a Jordan block $J \in \mathbb{C}^{m\times m}$ with eigenvalue zero. We have $\dim(\mathrm{null}(J^0)) = 0$, $\dim(\mathrm{null}(J)) = 1$, $\dim(\mathrm{null}(J^2)) = 2, \ldots, \dim(\mathrm{null}(J^m)) = m$, and so the ascent sequence comprises m 1s. Hence J_k does not have a square root unless $m = 1$. However, the matrix

$$\begin{bmatrix} 0 & 1 & 0 \\ 0 & 0 & 0 \\ 0 & 0 & 0 \end{bmatrix} \tag{1.23}$$

has ascent sequence $2, 1, 0, \ldots$ and so does have a square root—for example, the matrix

$$\begin{bmatrix} 0 & 0 & 1 \\ 0 & 0 & 0 \\ 0 & 1 & 0 \end{bmatrix} \tag{1.24}$$

(which is the 3×3 Jordan block with eigenvalue 0 with rows and columns 2 and 3 interchanged).

Another important existence question is "If A is real does there exist a real $f(A)$, either primary or nonprimary?" For most common functions the answer is clearly yes, by considering a power series representation. For the square root and logarithm the answer is not obvious; the next result completes the partial answer to this question given in Remark 1.9 and Theorem 1.18.

Theorem 1.23 (existence of real square root and real logarithm).

(a) *$A \in \mathbb{R}^{n\times n}$ has a real square root if and only if it satisfies the condition of Theorem 1.22 and A has an even number of Jordan blocks of each size for every negative eigenvalue.*

(b) *The nonsingular matrix $A \in \mathbb{R}^{n\times n}$ has a real logarithm if and only if A has an even number of Jordan blocks of each size for every negative eigenvalue.*

(c) *If $A \in \mathbb{R}^{n\times n}$ has any negative eigenvalues then no primary square root or logarithm is real.*

Proof. For the last part consider the real Schur decomposition, $Q^TAQ = R$ (see Section B.5), where $Q \in \mathbb{R}^{n\times n}$ is orthogonal and $R \in \mathbb{R}^{n\times n}$ is upper quasi-triangular. Clearly, $f(A)$ is real if and only if $Q^Tf(A)Q = f(R)$ is real, and a primary matrix function $f(R)$ is block upper triangular with diagonal blocks $f(R_{ii})$. If A has a negative real eigenvalue then some R_{ii} is 1×1 and negative, making $f(R_{ii})$ nonreal for f the square root and logarithm.

The result of (b) is due to Culver [126, 1966], and the proof for (a) is similar; see also Horn and Johnson [296, 1991, Thms. 6.4.14, 6.4.15] and Nunemacher [451, 1989]. □

Theorem 1.23 implies that $-I_n$ has a real, nonprimary square root and logarithm for every even n. For some insight into part (a), note that if A has two Jordan blocks J of the same size then its Jordan matrix has a principal submatrix of the form $\left[\begin{smallmatrix} J & 0 \\ 0 & J \end{smallmatrix}\right] = \left[\begin{smallmatrix} 0 & I \\ J & 0 \end{smallmatrix}\right]^2$.

1.6. Classification of Matrix Square Roots and Logarithms

The theory presented above provides a means for identifying some of the solutions to nonlinear matrix equations such as $X^2 = A$, $e^X = A$, and $\cos(X) = A$, since in each case X can be expressed as a function of A. However, more work is needed to classify all the solutions. In particular, the possibility remains that there are solutions X that have a spectrum of the form required for a primary matrix function but that are not primary matrix functions according to our definition. This possibility can be ruled out when the inverse of the function of interest has a nonzero derivative on the spectrum of X.

We will concentrate on the matrix square root. Entirely analogous arguments apply to the logarithm, which we briefly discuss, and the matrix pth root, which is treated in Section 7.1. For f the square root function and $\lambda_k \neq 0$ we write

$$L_k^{(j_k)} \equiv L_k^{(j_k)}(\lambda_k) = f(J_k(\lambda_k)),$$

where $f(J_k(\lambda_k))$ is given in (1.4) and where $j_k = 1$ or 2 denotes the branch of f; thus $L_k^{(1)} = -L_k^{(2)}$. Our first result characterizes all square roots.

Theorem 1.24 (Gantmacher). *Let $A \in \mathbb{C}^{n\times n}$ be nonsingular with the Jordan canonical form* (1.2). *Then all solutions to $X^2 = A$ are given by*

$$X = ZU \operatorname{diag}(L_1^{(j_1)}, L_2^{(j_2)}, \dots, L_p^{(j_p)})U^{-1}Z^{-1}, \tag{1.25}$$

where U is an arbitrary nonsingular matrix that commutes with J.

Proof. Let X be any square root of A. Since A is nonsingular so is X, and hence the derivative of the function x^2 is nonzero at the eigenvalues of X. By Theorem 1.36, given that A has the Jordan canonical form $J = \operatorname{diag}(J_1(\lambda_1), J_2(\lambda_2), \dots, J_p(\lambda_p))$, X must have the Jordan canonical form

$$J_X = \operatorname{diag}(J_1(\mu_1), J_2(\mu_2), \dots, J_p(\mu_p)), \tag{1.26}$$

where $\mu_k^2 = \lambda_k$, $k = 1\colon p$.

Now consider the matrix

$$L = \operatorname{diag}(L_1^{(j_1)}, L_2^{(j_2)}, \dots, L_p^{(j_p)}), \tag{1.27}$$

where we choose the j_k so that $L_k^{(j_k)}$ has eigenvalue μ_k for each k. Since L is a square root of J, by the same argument as above L must have the Jordan canonical form J_X. Hence $X = WLW^{-1}$ for some nonsingular W. From $X^2 = A$ we have $WJW^{-1} = WL^2W^{-1} = ZJZ^{-1}$, which can be rewritten as $(Z^{-1}W)J = J(Z^{-1}W)$. Hence $U = Z^{-1}W$ is an arbitrary matrix that commutes with J, which completes the proof. □

The structure of the matrix U in Theorem 1.24 is described in the next result.

Theorem 1.25 (commuting matrices). *Let $A \in \mathbb{C}^{n\times n}$ have the Jordan canonical form* (1.2). *All solutions of $AX = XA$ are given by $X = ZWZ^{-1}$, where $W = (W_{ij})$ with $W_{ij} \in \mathbb{C}^{m_i\times m_j}$ (partitioned conformably with J in* (1.2)) *satisfies*

$$W_{ij} = \begin{cases} 0, & \lambda_i \neq \lambda_j, \\ T_{ij}, & \lambda_i = \lambda_j, \end{cases}$$

where T_{ij} is an arbitrary upper trapezoidal Toeplitz matrix which, for $m_i < m_j$, has the form $T_{ij} = [0, U_{ij}]$, where U_{ij} is square.

Proof. See Lancaster and Tismenetsky [371, 1985, Thm. 12.4.1]. □

Next we refine Theorem 1.24 to classify the square roots into primary and nonprimary square roots.

Theorem 1.26 (classification of square roots). *Let the nonsingular matrix $A \in \mathbb{C}^{n\times n}$ have the Jordan canonical form* (1.2) *with p Jordan blocks, and let $s \le p$ be the number of distinct eigenvalues of A. Then A has precisely 2^s square roots that are primary functions of A, given by*

$$X_j = Z \operatorname{diag}(L_1^{(j_1)}, L_2^{(j_2)}, \dots, L_p^{(j_p)})Z^{-1}, \quad j = 1\colon 2^s,$$

corresponding to all possible choices of $j_1, \dots, j_p$, $j_k = 1$ or 2, subject to the constraint that $j_i = j_k$ whenever $\lambda_i = \lambda_k$.

If $s < p$, A has nonprimary square roots. They form parametrized families

$$X_j(U) = ZU \operatorname{diag}(L_1^{(j_1)}, L_2^{(j_2)}, \dots, L_p^{(j_p)})U^{-1}Z^{-1}, \quad j = 2^s + 1{:}\, 2^p,$$

where $j_k = 1$ or 2, U is an arbitrary nonsingular matrix that commutes with J, and for each j there exist i and k, depending on j, such that $\lambda_i = \lambda_k$ while $j_i \neq j_k$.

Proof. The proof consists of showing that for the square roots (1.25) for which $j_i = j_k$ whenever $\lambda_i = \lambda_k$,

$$U \operatorname{diag}(L_1^{(j_1)}, L_2^{(j_2)}, \dots, L_p^{(j_p)})U^{-1} = \operatorname{diag}(L_1^{(j_1)}, L_2^{(j_2)}, \dots, L_p^{(j_p)}),$$

that is, U commutes with the block diagonal matrix in the middle. This commutativity follows from the explicit form for U provided by Theorem 1.25 and the fact that upper triangular Toeplitz matrices commute. □

Theorem 1.26 shows that the square roots of a nonsingular matrix fall into two classes. The first class comprises finitely many primary square roots, which are "isolated", being characterized by the fact that the sum of any two of their eigenvalues is nonzero. The second class, which may be empty, comprises a finite number of parametrized families of matrices, each family containing infinitely many square roots sharing the same spectrum.

Theorem 1.26 has two specific implications of note. First, if $\lambda_k \neq 0$ then the two upper triangular square roots of $J_k(\lambda_k)$ given by (1.4) with f the square root function are the only square roots of $J_k(\lambda_k)$. Second, if A is nonsingular and nonderogatory, that is, none of the s distinct eigenvalues appears in more than one Jordan block, then A has precisely 2^s square roots, each of which is a primary function of A.

There is no analogue of Theorems 1.24 and 1.26 for singular A. Indeed the Jordan block structure of a square root (when one exists) can be very different from that of A. The search for square roots X of a singular matrix is aided by Theorem 1.36 below, which helps identify the possible Jordan forms of X; see Problem 1.29.

Analogous results, with analogous proofs, hold for the matrix logarithm.

Theorem 1.27 (Gantmacher). *Let $A \in \mathbb{C}^{n\times n}$ be nonsingular with the Jordan canonical form (1.2). Then all solutions to $e^X = A$ are given by*

$$X = ZU \operatorname{diag}(L_1^{(j_1)}, L_2^{(j_2)}, \dots, L_p^{(j_p)})U^{-1}Z^{-1},$$

where

$$L_k^{(j_k)} = \log(J_k(\lambda_k)) + 2j_k\pi i I_{m_k}; \tag{1.28}$$

$\log(J_k(\lambda_k))$ denotes (1.4) with the f the principal branch of the logarithm, defined by $\operatorname{Im}(\log(z)) \in (-\pi, \pi]$; j_k is an arbitrary integer; and U is an arbitrary nonsingular matrix that commutes with J. □

Theorem 1.28 (classification of logarithms). *Let the nonsingular matrix $A \in \mathbb{C}^{n\times n}$ have the Jordan canonical form (1.2) with p Jordan blocks, and let $s \leq p$ be the number of distinct eigenvalues of A. Then $e^X = A$ has a countable infinity of solutions that are primary functions of A, given by*

$$X_j = Z \operatorname{diag}(L_1^{(j_1)}, L_2^{(j_2)}, \dots, L_p^{(j_p)})Z^{-1},$$

where $L_1^{(j_1)}$ *is defined in* (1.28), *corresponding to all possible choices of the integers* $j_1, \ldots, j_p$, *subject to the constraint that* $j_i = j_k$ *whenever* $\lambda_i = \lambda_k$.

If $s < p$ *then* $e^X = A$ *has nonprimary solutions. They form parametrized families*

$$X_j(U) = ZU \operatorname{diag}(L_1^{(j_1)}, L_2^{(j_2)}, \ldots, L_p^{(j_p)})U^{-1}Z^{-1},$$

where j_k *is an arbitrary integer,* U *is an arbitrary nonsingular matrix that commutes with* J, *and for each* j *there exist* i *and* k, *depending on* j, *such that* $\lambda_i = \lambda_k$ *while* $j_i \neq j_k$. □

1.7. Principal Square Root and Logarithm

Among the square roots and logarithms of a matrix, the principal square root and principal logarithm are distinguished by their usefulness in theory and in applications. We denote by $\mathbb{R}^-$ the closed negative real axis.

Theorem 1.29 (principal square root). *Let* $A \in \mathbb{C}^{n\times n}$ *have no eigenvalues on* $\mathbb{R}^-$. *There is a unique square root* X *of* A *all of whose eigenvalues lie in the open right half-plane, and it is a primary matrix function of* A. *We refer to* X *as the* principal square root *of* A *and write* $X = A^{1/2}$. *If* A *is real then* $A^{1/2}$ *is real.*

Proof. Note first that a nonprimary square root of A, if one exists, must have eigenvalues μ_i and μ_j with $\mu_i = -\mu_j$, and hence the eigenvalues cannot all lie in the open right half-plane. Therefore only a primary square root can have spectrum in the open right half-plane. Since A has no eigenvalues on $\mathbb{R}^-$, it is clear from Theorem 1.26 that there is precisely one primary square root of A whose eigenvalues all lie in the open right half-plane. Hence the existence and uniqueness of $A^{1/2}$ is established. That $A^{1/2}$ is real when A is real follows from Theorem 1.18 or Remark 1.9. □

See Problem 1.27 for an extension of Theorem 1.29 that allows A to be singular.

Corollary 1.30. *A Hermitian positive definite matrix* $A \in \mathbb{C}^{n\times n}$ *has a unique Hermitian positive definite square root.*

Proof. By Theorem 1.29 the only possible Hermitian positive definite square root is $A^{1/2}$. That $A^{1/2}$ is Hermitian positive definite follows from the expression $A^{1/2} = QD^{1/2}Q^*$, where $A = QDQ^*$ is a spectral decomposition (Q unitary, D diagonal), with D having positive diagonal entries. □

For a proof of the corollary from first principles see Problem 1.41.

Theorem 1.31 (principal logarithm). *Let* $A \in \mathbb{C}^{n\times n}$ *have no eigenvalues on* $\mathbb{R}^-$. *There is a unique logarithm* X *of* A *all of whose eigenvalues lie in the strip* $\{ z : -\pi < \operatorname{Im}(z) < \pi \}$. *We refer to* X *as the* principal logarithm *of* A *and write* $X = \log(A)$. *If* A *is real then its principal logarithm is real.*

Proof. The proof is entirely analogous to that of Theorem 1.29. □

1.8. $f(AB)$ and $f(BA)$

Although the matrices AB and BA are generally different, their Jordan structures are closely related. We show in this section that for arbitrary functions f, $f(AB)$ and $f(BA)$ also enjoy a close relationship—one that can be exploited both in theory and computationally. Underlying all these relations is the fact that for any polynomial p, and any A and B for which the products AB and BA are defined,

$$Ap(BA) = p(AB)A. \tag{1.29}$$

This equality is trivial for monomials and follows immediately for general polynomials.

First we recap a result connecting the Jordan structures of AB and BA. We denote by $z_i(X)$ the nonincreasing sequence of the sizes $z_1, z_2, \ldots,$ of the Jordan blocks corresponding to the zero eigenvalues of the square matrix X.

Theorem 1.32 (Flanders). *Let $A \in \mathbb{C}^{m\times n}$ and $B \in \mathbb{C}^{n\times m}$. The nonzero eigenvalues of AB are the same as those of BA and have the same Jordan structure. For the zero eigenvalues (if any), $|z_i(AB) - z_i(BA)| \le 1$ for all i, where the shorter sequence is appended with zeros as necessary, and any such set of inequalities is attained for some A and B. If $m \ne n$ then the larger (in dimension) of AB and BA has a zero eigenvalue of geometric multiplicity at least $|m - n|$.*

Proof. See Problem 1.43. □

Theorem 1.33. *Let $A \in \mathbb{C}^{n\times n}$ and $B \in \mathbb{C}^{m\times m}$ and let f be defined on the spectrum of both A and B. Then there is a single polynomial p such that $f(A) = p(A)$ and $f(B) = p(B)$.*

Proof. Let p be the Hermite interpolating polynomial satisfying the union of the interpolation conditions (1.7) for A with those for B. Let r be the Hermite interpolating polynomial to f on the spectrum of A. Then p and r take the same values on the spectrum of A, so $f(A) := r(A) = p(A)$. By the same argument with A and B interchanged, $f(B) = p(B)$, as required. □

Corollary 1.34. *Let $A \in \mathbb{C}^{m\times n}$ and $B \in \mathbb{C}^{n\times m}$ and let f be defined on the spectra of both AB and BA. Then*

$$Af(BA) = f(AB)A. \tag{1.30}$$

Proof. By Theorem 1.33 there is a single polynomial p such that $f(AB) = p(AB)$ and $f(BA) = p(BA)$. Hence, using (1.29),

$$Af(BA) = Ap(BA) = p(AB)A = f(AB)A. \quad \square$$

When A and B are square and A, say, is nonsingular, another proof of Corollary 1.34 is as follows: $AB = A(BA)A^{-1}$ so $f(AB) = Af(BA)A^{-1}$, or $f(AB)A = Af(BA)$.

As a special case of the corollary, when AB (and hence also BA) has no eigenvalues on $\mathbb{R}^-$ (which implies that A and B are square, in view of Theorem 1.32),

$$A(BA)^{1/2} = (AB)^{1/2}A.$$

In fact, this equality holds also when AB has a semisimple zero eigenvalue and the definition of $A^{1/2}$ is extended as in Problem 1.27.

Corollary 1.34 is useful for converting $f(AB)$ into $f(BA)$ within an expression, and vice versa; see, for example, (2.26), the proof of Theorem 6.11, and (8.5). However, when $m > n$, (1.30) cannot be directly solved to give an expression for $f(AB)$ in terms of $f(BA)$, because (1.30) is an underdetermined system for $f(AB)$. The next result gives such an expression, and in more generality.

Theorem 1.35. *Let $A \in \mathbb{C}^{m\times n}$ and $B \in \mathbb{C}^{n\times m}$, with $m \geq n$, and assume that BA is nonsingular. Let f be defined on the spectrum of $\alpha I_m + AB$, and if $m = n$ let f be defined at α. Then*

$$f(\alpha I_m + AB) = f(\alpha)I_m + A(BA)^{-1}\big(f(\alpha I_n + BA) - f(\alpha)I_n\big)B. \tag{1.31}$$

Proof. Note first that by Theorem 1.32, the given assumption on f implies that f is defined on the spectrum of $\alpha I_n + BA$ and at α.

Let $g(t) = f[\alpha + t, \alpha] = t^{-1}(f(\alpha + t) - f(\alpha))$, so that $f(\alpha + t) = f(\alpha) + tg(t)$. Then, using Corollary 1.34,

$$\begin{aligned} f(\alpha I_m + AB) &= f(\alpha)I_m + ABg(AB) \\ &= f(\alpha)I_m + Ag(BA)B \\ &= f(\alpha)I_m + A(BA)^{-1}\big(f(\alpha I_n + BA) - f(\alpha)I_n\big)B, \end{aligned}$$

as required. □

This result is of particular interest when $m > n$, for it converts the $f(\alpha I_m + AB)$ problem—a function evaluation of an $m \times m$ matrix—into the problem of evaluating f and the inverse on $n \times n$ matrices. Some special cases of the result are as follows.

(a) With $n = 1$, we recover (1.16) (albeit with the restriction $v^*u \neq 0$).

(b) With f the inverse function and $\alpha = 1$, (1.31) yields, after a little manipulation, the formula $(I + AB)^{-1} = I - A(I + BA)^{-1}B$, which is often found in textbook exercises. This formula in turn yields the Sherman–Morrison–Woodbury formula (B.12) on writing $A + UV^* = A(I + A^{-1}U \cdot V^*)$. Conversely, when f is analytic we can obtain (1.31) by applying the Sherman–Morrison–Woodbury formula to the Cauchy integral formula (1.12). However, Theorem 1.35 does not require analyticity.

As an application of Theorem 1.35, we now derive a formula for $f(\alpha I_n + uv^* + xy^*)$, where $u, v, x, y \in \mathbb{C}^n$, thereby extending (1.16) to the rank-2 case. Write

$$uv^* + xy^* = \begin{bmatrix} u & x \end{bmatrix} \begin{bmatrix} v^* \\ y^* \end{bmatrix} \equiv AB.$$

Then

$$C := BA = \begin{bmatrix} v^*u & v^*x \\ y^*u & y^*x \end{bmatrix} \in \mathbb{C}^{2\times 2}.$$

Hence

$$f(\alpha I_n + uv^* + xy^*) = f(\alpha)I_n + \begin{bmatrix} u & x \end{bmatrix} C^{-1}\big(f(\alpha I_2 + C) - f(\alpha)I_2\big) \begin{bmatrix} v^* \\ y^* \end{bmatrix}. \tag{1.32}$$

The evaluation of both C^{-1} and $f(\alpha I_2 + C)$ can be done explicitly (see Problem 1.9 for the latter), so (1.32) gives a computable formula that can, for example, be used for testing algorithms for the computation of matrix functions.

1.9. Miscellany

In this section we give a selection of miscellaneous results that either are needed elsewhere in the book or are of independent interest.

The first result gives a complete description of the Jordan canonical form of $f(A)$ in terms of that of A. In particular, it shows that under the action of f a Jordan block $J(\lambda)$ splits into at least two smaller Jordan blocks if $f'(\lambda) = 0$.

Theorem 1.36 (Jordan structure of $f(A)$). *Let $A \in \mathbb{C}^{n\times n}$ with eigenvalues λ_k, and let f be defined on the spectrum of A.*

(a) *If $f'(\lambda_k) \neq 0$ then for every Jordan block $J(\lambda_k)$ in A there is a Jordan block of the same size in $f(A)$ associated with $f(\lambda_k)$.*

(b) *Let $f'(\lambda_k) = f''(\lambda_k) = \cdots = f^{(\ell-1)}(\lambda_k) = 0$ but $f^{(\ell)}(\lambda_k) \neq 0$, where $\ell \geq 2$, and consider a Jordan block $J(\lambda_k)$ of size r in A.*

(i) *If $\ell \geq r$, $J(\lambda_k)$ splits into r 1×1 Jordan blocks associated with $f(\lambda_k)$ in $f(A)$.*

(ii) *If $\ell \leq r-1$, $J(\lambda_k)$ splits into the following Jordan blocks associated with $f(\lambda_k)$ in $f(A)$:*

- *$\ell - q$ Jordan blocks of size p,*
- *q Jordan blocks of size $p+1$,*

where $r = \ell p + q$ with $0 \leq q \leq \ell - 1$, $p > 0$.

Proof. We prove just the first part. From Definition 1.2 it is clear that f either preserves the size of a Jordan block $J_k(\lambda_k) \in \mathbb{C}^{m_k\times m_k}$ of A—that is, $f(J_k(\lambda_k))$ has Jordan form $J_k(f(\lambda_k)) \in \mathbb{C}^{m_k\times m_k}$—or splits $J_k(\lambda_k)$ into two or more smaller blocks, each with eigenvalue $f(\lambda_k)$. When $f'(\lambda_k) \neq 0$, (1.4) shows that $f(J_k(\lambda_k)) - f(\lambda_k)I$ has rank $m_k - 1$, which implies that f does not split the block $J_k(\lambda_k)$. When $f'(\lambda_k) = 0$, it is clear from (1.4) that $f(J_k(\lambda_k)) - f(\lambda_k)I$ has rank at most $m_k - 2$, which implies that $f(J_k(\lambda_k))$ has at least two Jordan blocks. For proofs of the precise splitting details, see Horn and Johnson [296, 1991, Thm. 6.2.25] or Lancaster and Tismenetsky [371, 1985, Thm. 9.4.7]. □

To illustrate the result, consider the matrix

$$A = \begin{bmatrix} 0 & 1 & 0 & 0 \\ 0 & 0 & 1 & 0 \\ 0 & 0 & 0 & 1 \\ 0 & 0 & 0 & 0 \end{bmatrix},$$

which is in Jordan form with one Jordan block of size 4. Let

$$f(A) = A^3 = \begin{bmatrix} 0 & 0 & 0 & 1 \\ 0 & 0 & 0 & 0 \\ 0 & 0 & 0 & 0 \\ 0 & 0 & 0 & 0 \end{bmatrix}.$$

Clearly $f(A)$ has Jordan form comprising two 1×1 blocks and one 2×2 block. We have $f'(0) = f''(0) = 0$ and $f'''(0) \neq 0$. Applying Theorem 1.36 (b) with $\ell = 3$, $r = 4$, $p = 1$, $q = 1$, the theorem correctly predicts $\ell - q = 2$ Jordan blocks of size

1 and $q = 1$ Jordan block of size 2. For an example of a Jordan block splitting with $f(X) = X^2$, see the matrices (1.23) and (1.24).

Theorem 1.36 is useful when trying to solve nonlinear matrix equations, because once the Jordan form of $f(A)$ is known it narrows down the possible Jordan forms of A; see, e.g., Problems 1.30 and 1.51.

We noted in Section 1.4 that a nonprimary function of a derogatory A may commute with A but is not a polynomial in A. The next result shows that all matrices that commute with A are polynomials in A precisely when A is nonderogatory—that is, when no eigenvalue appears in more than one Jordan block in the Jordan canonical form of A.

Theorem 1.37. *Every matrix that commutes with $A \in \mathbb{C}^{n\times n}$ is a polynomial in A if and only if A is nonderogatory.*

Proof. This result is a consequence of Theorem 1.25. See Lancaster and Tismenetsky [371, 1985, Prop. 12.4.1] for the details. □

While commuting with A is not sufficient to be a polynomial in A, commuting with *every* matrix that commutes with A is sufficient.

Theorem 1.38. *$B \in \mathbb{C}^{n\times n}$ commutes with every matrix that commutes with $A \in \mathbb{C}^{n\times n}$ if and only if B is a polynomial in A.*

Proof. See Horn and Johnson [296, 1991, Thm. 4.4.19]. □

The following result is useful for finding solutions of a nonlinear matrix equation of the form $f(X) = A$.

Theorem 1.39. *Consider the equation $f(X) = A \in \mathbb{C}^{n\times n}$.*

(a) *If A is upper triangular and nonderogatory then any solution X is upper triangular.*

(b) *If A is a single Jordan block $J(\lambda)$ then any solution X is upper triangular with constant diagonal elements $x_{ii} \equiv \xi$, where $f(\xi) = \lambda$.*

(c) *If the equation with $A = \theta I$ has a solution X that is not a multiple of I then there are infinitely many solutions to the equation.*

Proof.

(a) The nonderogatory matrix $A = f(X)$ commutes with X so, by Theorem 1.37, X is a polynomial in A, which means that X is upper triangular.

(b) This follows from the proof of (a) on noting that a polynomial in $J(\lambda)$ has constant diagonal.

(c) Since $f(X) = \theta I$, for any nonsingular Z we have $\theta I = Z^{-1}f(X)Z = f(Z^{-1}XZ)$, so $Z^{-1}XZ$ is a solution. The result now follows from the fact that any matrix other than a scalar multiple of the identity shares its Jordan canonical form with infinitely many other matrices. □

The next result shows that a family of pairwise commuting matrices can be simultaneously unitarily triangularized.

Theorem 1.40. *If $A_1, A_2, \ldots, A_k \in \mathbb{C}^{n\times n}$ satisfy $A_iA_j = A_jA_i$ for all i and j then there exists a unitary $U \in \mathbb{C}^{n\times n}$ such that U^*A_iU is upper triangular for all i.*

Proof. See Horn and Johnson [295, 1985, Thm. 2.3.3]. □

We denote by $\lambda_i(A)$ the ith eigenvalue of A in some given ordering.

Corollary 1.41. *Suppose $A, B \in \mathbb{C}^{n\times n}$ and $AB = BA$. Then for some ordering of the eigenvalues of A, B, and AB we have $\lambda_i(A \mathrm{op} B) = \lambda_i(A) \mathrm{op} \lambda_i(B)$, where* op $= +$, $-$, *or* $*$.

Proof. By Theorem 1.40 there exists a unitary U such that $U^*AU = T_A$ and $U^*BU = T_B$ are both upper triangular. Thus $U^*(A \text{ op } B)U = T_A \text{ op } T_B$ is upper triangular with diagonal elements $(T_A)_{ii} \text{ op } (T_B)_{ii}$, as required. □

This corollary will be used in Section 11.1. Note that for any A and B we have $\mathrm{trace}(A+B) = \mathrm{trace}(A) + \mathrm{trace}(B)$ and $\det(AB) = \det(A)\det(B)$, but the conclusion of the corollary for commuting A and B is much stronger.

Related to Theorem 1.40 and Corollary 1.41 are the following characterizations of A and B for which "$\lambda_i(p(A,B)) = p(\lambda_i(A), \lambda_i(B))$".

Theorem 1.42 (McCoy). *For $A, B \in \mathbb{C}^{n\times n}$ the following conditions are equivalent.*

(a) *There is an ordering of the eigenvalues such that $\lambda_i(p(A,B)) = p(\lambda_i(A), \lambda_i(B))$ for all polynomials of two variables $p(x,y)$.*

(b) *There exists a unitary $U \in \mathbb{C}^{n\times n}$ such that U^*AU and U^*BU are upper triangular.*

(c) *$p(A,B)(AB - BA)$ is nilpotent for all polynomials $p(x,y)$ of two variables.*

Theorem 1.43. *$A \in \mathbb{C}^{n\times n}$ is unitary if and only if $A = e^{iH}$ for some Hermitian H. In this representation H can be taken to be Hermitian positive definite.*

Proof. The Schur decomposition of A has the form $A = QDQ^*$ with Q unitary and $D = \mathrm{diag}(\exp(i\theta_j)) = \exp(i\Theta)$, where $\Theta = \mathrm{diag}(\theta_j) \in \mathbb{R}^{n\times n}$. Hence $A = Q\exp(i\Theta)Q^* = \exp(iQ\Theta Q^*) = \exp(iH)$, where $H = H^*$. Without loss of generality we can take $\theta_j > 0$, whence H is positive definite. □

Theorem 1.44. *$A \in \mathbb{C}^{n\times n}$ has the form $A = e^S$ with S real and skew-symmetric if and only if A is real orthogonal with* $\det(A) = 1$.

Proof. "$\Rightarrow$": If S is real and skew-symmetric then A is real, $A^TA = e^{-S}e^S = I$, and $\det(e^S) = \exp(\sum \lambda_i(S)) = \exp(0) = 1$, since the eigenvalues of S are either zero or occur in pure imaginary complex conjugate pairs.

"$\Leftarrow$": If A is real orthogonal then it has the real Schur decomposition $A = QDQ^T$ with Q orthogonal and $D = \mathrm{diag}(D_{ii})$, where each D_{ii} is 1, -1, or of the form $\left[\begin{smallmatrix} a_j & b_j \\ -b_j & a_j \end{smallmatrix}\right]$ with $a_j^2 + b_j^2 = 1$. Since $\det(A) = 1$, there is an even number of -1s, and so we can include the -1 blocks among the $\left[\begin{smallmatrix} a_j & b_j \\ -b_j & a_j \end{smallmatrix}\right]$ blocks. It is easy to show that

$$\begin{bmatrix} a_j & b_j \\ -b_j & a_j \end{bmatrix} \equiv \begin{bmatrix} \cos\theta_j & \sin\theta_j \\ -\sin\theta_j & \cos\theta_j \end{bmatrix} = \exp\left(\begin{bmatrix} 0 & \theta_j \\ -\theta_j & 0 \end{bmatrix}\right) =: \exp(\Theta_j). \tag{1.33}$$

We now construct a skew-symmetric K such that $D = e^K$: K has the same block structure as D, $k_{ii} = 0$ if $d_{ii} = 1$, and the other blocks have the form Θ_j in (1.33). Hence $A = Qe^KQ^T = e^{QKQ^T} = e^S$, where S is real and skew-symmetric. □

Theorem 1.45. *For* $A \in \mathbb{C}^{n\times n}$, $\det(e^A) = \exp(\operatorname{trace}(A))$.

Proof. We have

$$\det(e^A) = \prod_{i=1}^{n} \lambda_i(e^A) = \prod_{i=1}^{n} e^{\lambda_i(A)} = e^{\lambda_1(A)+\cdots+\lambda_n(A)} = \exp\big(\operatorname{trace}(A)\big). \qquad \square$$

Note that another way of expressing Theorem 1.45 is that for *any* logarithm of a nonsingular X, $\det(X) = \exp(\operatorname{trace}(\log(X)))$.

1.10. A Brief History of Matrix Functions

Sylvester (1814–1897) [465, 2006] coined the term "matrix" in 1850 [553, 1850]. Cayley (1821–1895) [121, 2006], in his *A Memoir on the Theory of Matrices* [99, 1858], was the first to investigate the algebraic properties of matrices regarded as objects of study in their own right (in contrast with earlier work on bilinear and quadratic forms). Matrix theory was subsequently developed by Cayley, Sylvester, Frobenius, Kronecker, Weierstrass, and others; for details, see [253, 1974], [254, 1977], [255, 1977], [463, 1985].

The study of functions of matrices began in Cayley's 1858 memoir, which treated the square roots of 2×2 and 3×3 matrices, and he later revisited these cases in [100, 1872]. Laguerre [367, 1867], and later Peano [467, 1888], defined the exponential of a matrix via its power series. The interpolating polynomial definition of $f(A)$ was stated by Sylvester [557, 1883] for $n\times n$ A with distinct eigenvalues λ_i, in the form

$$f(A) = \sum_{i=1}^{n} f(\lambda_i) \prod_{j\neq i} \frac{A - \lambda_j I}{\lambda_i - \lambda_j}.$$

Buchheim gave a derivation of the formula [84, 1884] and then generalized it to multiple eigenvalues using Hermite interpolation [85, 1886].

Weyr [614, 1887] defined $f(A)$ using a power series for f and showed that the series converges if the eigenvalues of A lie within the radius of convergence of the series. Hensel [258, 1926] obtained necessary and sufficient conditions for convergence when one or more eigenvalues lies on the circle of convergence (see Theorem 4.7).

Metzler [424, 1892] defined the transcendental functions e^A, $\log(A)$, $\sin(A)$, and $\arcsin(A)$, all via power series.

The Cauchy integral representation was anticipated by Frobenius [195, 1896], who states that if f is analytic then $f(A)$ is the sum of the residues of $(zI-A)^{-1}f(z)$ at the eigenvalues of A. Poincaré [473, 1899] uses the Cauchy integral representation, and this way of defining $f(A)$ was proposed in a letter from Cartan to Giorgi, circa 1928 [216, 1928].

The Jordan canonical form definition is due to Giorgi [216, 1928]; Cipolla [109, 1932] extended it to produce nonprimary matrix functions.

Probably the first book (actually a booklet) to be written on matrix functions is that of Schwerdtfeger [513, 1938]. With the same notation as in Definitions 1.2 and 1.4 he defines

$$f(A) = \sum_{i=1}^{s} A_i \sum_{j=0}^{n_i-1} \frac{f^{(j)}(\lambda_i)}{j!}(A-\lambda_i I)^j,$$

where the A_i are the Frobenius covariants: $A_i = Z\operatorname{diag}(g_i(J_k))Z^{-1}$, where $g_i(J_k) = I$ if λ_i is an eigenvalue of J_k and $g_i(J_k) = 0$ otherwise, where $A = Z\operatorname{diag}(J_k)Z^{-1}$ is the

Jordan canonical form. This is just a rewritten form of the expression for $f(A)$ given by Definition 1.2 or by the Lagrange–Hermite formula (1.8). It can be restated as

$$f(A) = \sum_{i=1}^{s} \sum_{j=0}^{n_i-1} f^{(j)}(\lambda_i) Z_{ij},$$

where the Z_{ij} depend on A but not on f. For more details on these formulae see Horn and Johnson [296, 1991, pp. 401–404, 438] and Lancaster and Tismenetsky [371, 1985, Sec. 9.5].

The equivalence of all the above definitions of $f(A)$ (modulo their different levels of generality) was first shown by Rinehart [493, 1955] (see the quote at the end of the chapter).

One of the earliest uses of matrices in practical applications was by Frazer, Duncan, and Collar of the Aerodynamics Department of the National Physical Laboratory (NPL), England, who were developing matrix methods for analyzing flutter (unwanted vibrations) in aircraft. Their book *Elementary Matrices and Some Applications to Dynamics and Differential Equations* [193, 1938] emphasizes the important role of the matrix exponential in solving differential equations and was "the first to employ matrices as an engineering tool" [71, 1987], and indeed "the first book to treat matrices as a branch of applied mathematics" [112, 1978].

Early books with substantial material on matrix functions are Turnbull and Aitken [579, 1932, Sec. 6.6–6.8]; Wedderburn [611, 1934, Chap. 8], which has a useful bibliography arranged by year, covering 1853–1933; MacDuffee [399, 1946, Chap. IX], which gives a concise summary of early work with meticulous attribution of results; Ferrar [184, 1951, Chap. 5]; and Hamburger and Grimshaw [245, 1951]. Papers with useful historical summaries include Afriat [5, 1959] and Heuvers and Moak [259, 1987].

Interest in computing matrix functions grew rather slowly following the advent of the digital computer. As the histogram on page 379 indicates, the literature expanded rapidly starting in the 1970s, and interest in the theory and computation of matrix functions shows no signs of abating, spurred by the growing number of applications. A landmark paper is Moler and Van Loan's "Nineteen Dubious Ways to Compute the Exponential of a Matrix" [437, 1978], [438, 2003], which masterfully organizes and assesses the many different ways of approaching the e^A problem. In particular, it explains why many of the methods that have been (and continue to be) published are unsuitable for finite precision computation.

The "problem solving environments" MATLAB, Maple, and Mathematica have been invaluable for practitioners using matrix functions and numerical analysts developing algorithms for computing them. The original 1978 version of MATLAB included the capability to evaluate the exponential, the logarithm, and several other matrix functions. The availability of matrix functions in MATLAB and it competitors has undoubtedly encouraged the use of succinct, matrix function-based solutions to problems in science and engineering.

1.11. Notes and References

The theory of functions of a matrix is treated in a number of books, of which several are of particular note. The most encyclopedic treatment is given by Horn and Johnson [296, 1991, Chap. 6], who devote a chapter of 179 pages to the subject. A more concise but very elegant exposition emphasizing the interpolation definition is given by Lancaster and Tismenetsky [371, 1985, Chap. 9]. A classic reference is Gantmacher [203, 1959, Chap. 5]. Golub and Van Loan [224, 1996, Chap. 11] briefly treat the theory before turning to computational matters. Linear algebra and matrix analysis textbooks with a significant component on $f(A)$ include Cullen [125, 1972], Pullman [481, 1976], and Meyer [426, 2000].

For more details on the Jordan canonical form see Horn and Johnson [295, 1985, Chap. 3] and Lancaster and Tismenetsky [371, 1985, Chap. 6].

Almost every textbook on numerical analysis contains a treatment of polynomial interpolation for distinct nodes, including the Lagrange form (1.9) and the Newton divided difference form (1.10). Textbook treatments of Hermite interpolation are usually restricted to once-repeated nodes; for the general case see, for example, Horn and Johnson [296, 1991, Sec. 6.1.14] and Stoer and Bulirsch [542, 2002, Sec. 2.1.5].

For the theory of functions of operators (sometimes called the holomorphic functional calculus), see Davies [133, 2007], Dunford and Schwartz [172, 1971], [171, 1988], and Kato [337, 1976].

Functions of the DFT matrix, and in particular fractional powers, are considered by Dickinson and Steiglitz [151, 1982], who obtain a formula equivalent to (1.18). Much has been written about fractional transforms, mainly in the engineering literature; for the fractional discrete cosine transform, for example, see Cariolaro, Erseghe, and Kraniauskas [96, 2002].

Theorems 1.15–1.17 can be found in Lancaster and Tismenetsky [371, 1985, Sec. 9.7].

Theorem 1.18 is from Higham, Mackey, Mackey, and Tisseur [283, 2005, Thm. 3.2]. The sufficient condition of Remark 1.9 and the equivalence (c) $\equiv$ (d) in Theorem 1.18 can be found in Richter [491, 1950].

Different characterizations of the reality of $f(A)$ for real A can be found in Evard and Uhlig [179, 1992, Sec. 4] and Horn and Piepmeyer [298, 2003].

The terminology "primary matrix function" has been popularized through its use by Horn and Johnson [296, 1991, Chap. 6], but the term was used much earlier by Rinehart [495, 1960] and Cullen [125, 1972].

A number of early papers investigate square roots and pth roots of (singular) matrices, including Taber [561, 1890], Metzler [424, 1892], Frobenius [195, 1896], Kreis [363, 1908], Baker [40, 1925], and Richter [491, 1950], and Wedderburn's book also treats the topic [611, 1934, Secs. 8.04–8.06].

Theorem 1.24 is a special case of a result of Gantmacher for pth roots [203, 1959, Sec. 8.6]. Theorem 1.26 is from Higham [268, 1987]. Theorem 1.27 is from [203, 1959, Sec. 8.8].

Theorem 1.32 is proved by Flanders [188, 1951]. Alternative proofs are given by Thompson [566, 1968] and Horn and Merino [297, 1995, Sec. 6]; see also Johnson and Schreiner [321, 1996].

We derived Theorem 1.35 as a generalization of (1.16) while writing this book; our original proof is given in Problem 1.45. Harris [249, 1993, Lem. 2] gives the result for $\alpha = 0$ and f a holomorphic function, with the same method of proof that we have given. The special case of Theorem 1.35 with f the exponential and $\alpha = 0$ is given by Celledoni and Iserles [102, 2000].

Formulae for a rational function of a general matrix plus a rank-1 perturbation, $r(C + uv^*)$, are derived by Bernstein and Van Loan [61, 2000]. These are more complicated and less explicit than (1.31), though not directly comparable with it since C need not be a multiple of the identity. The formulae involve the coefficients of r and so cannot be conveniently applied to an arbitrary function f by using "$f(A) = p(A)$

for some polynomial p."

Theorem 1.42 is due to McCoy [415, 1936]. See also Drazin, Dungey, and Gruenberg [164, 1951] for a more elementary proof and the discussions of Taussky [564, 1957], [565, 1988]. A complete treatment of simultaneous triangularization is given in the book by Radjavi and Rosenthal [483, 2000].

Problems

The only way to learn mathematics is to do mathematics.
— PAUL R. HALMOS, *A Hilbert Space Problem Book* (1982)

1.1. Show that the value of $f(A)$ given by Definition 1.2 is independent of the particular Jordan canonical form that is used.

1.2. Let J_k be the Jordan block (1.2b). Show that

$$f(-J_k) = \begin{bmatrix} f(-\lambda_k) & -f'(-\lambda_k) & \cdots & (-1)^{m_k-1}\dfrac{f^{(m_k-1)}(-\lambda_k)}{(m_k-1)!} \\ & f(-\lambda_k) & \ddots & \vdots \\ & & \ddots & -f'(-\lambda_k) \\ & & & f(-\lambda_k) \end{bmatrix}. \tag{1.34}$$

1.3. (Cullen [125, 1972, Thm. 8.9]) Define $f(A)$ by the Jordan canonical form definition. Use Theorem 1.38 and the property $f(XAX^{-1}) = Xf(A)X^{-1}$ to show that $f(A)$ is a polynomial in A.

1.4. (a) Let $A \in \mathbb{C}^{n\times n}$ have an eigenvalue λ and corresponding eigenvector x. Show that $(f(\lambda), x)$ is a corresponding eigenpair for $f(A)$.

(b) Suppose A has constant row sums α, that is, $Ae = \alpha e$, where $e = [1, 1, \ldots, 1]^T$. Show that $f(A)$ has row sums $f(\alpha)$. Deduce the corresponding result for column sums.

1.5. Show that the minimal polynomial ψ of $A \in \mathbb{C}^{n\times n}$ exists, is unique, and has degree at most n.

1.6. (Turnbull and Aitken [579, 1932, p. 75]) Show that if $A \in \mathbb{C}^{n\times n}$ has minimal polynomial $\psi(A) = A^2 - A - I$ then $(I - \frac{1}{3}A)^{-1} = \frac{3}{5}(A + 2I)$.

1.7. (Pullman [481, 1976, p. 56]) The matrix

$$A = \begin{bmatrix} -2 & 2 & -2 & 4 \\ -1 & 2 & -1 & 1 \\ 0 & 0 & 1 & 0 \\ -2 & 1 & -1 & 4 \end{bmatrix}$$

has minimal polynomial $\psi(t) = (t-1)^2(t-2)$. Find $\cos(\pi A)$.

1.8. Find the characteristic polynomial and the minimal polynomial of the nonzero rank-1 matrix $uv^* \in \mathbb{C}^{n\times n}$.

1.9. Use (1.11) to give an explicit formula for $f(A)$ for $A \in \mathbb{C}^{2\times 2}$ requiring knowledge only of the eigenvalues of A.

1.10. Let $J = ee^T \in \mathbb{R}^{n\times n}$ denote the matrix of 1s. Show using Definition 1.4 that

$$f(\alpha I + \beta J) = f(\alpha)I + n^{-1}(f(\alpha + n\beta) - f(\alpha))J.$$

1.11. What are the interpolation conditions (1.7) for the polynomial p such that $p(A) = A$?

1.12. Let $A \in \mathbb{C}^{n\times n}$ have only two distinct eigenvalues, λ_1 and λ_2, both semisimple. Obtain an explicit formula for $f(A)$.

1.13. Show using each of the three definitions (1.2), (1.4), and (1.11) of $f(A)$ that $AB = BA$ implies $f(A)B = Bf(A)$.

1.14. For a given $A \in \mathbb{C}^{n\times n}$ and a given function f explain how to reliably compute in floating point arithmetic a polynomial p such that $f(A) = p(A)$.

1.15. Show how to obtain the formula (1.14) from Definition 1.2 when $v^*u = 0$ with $uv^* \neq 0$.

1.16. Prove the formula (1.16) for $f(\alpha I + uv^*)$. Use this formula to derive the Sherman–Morrison formula (B.11).

1.17. Use (1.16) to obtain an explicit formula for $f(A)$ for $A = \left[\begin{smallmatrix} \lambda I_{n-1} & c \\ 0 & \lambda \end{smallmatrix}\right] \in \mathbb{C}^{n\times n}$. Check your result against Theorem 1.21.

1.18. (Schwerdtfeger [513, 1938]) Let p be a polynomial and $A \in \mathbb{C}^{n\times n}$. Show that $p(A) = 0$ if and only if $p(t)(tI - A)^{-1}$ is a polynomial in t. Deduce the Cayley–Hamilton theorem.

1.19. Cayley actually discovered a more general version of the Cayley–Hamilton theorem, which appears in a letter to Sylvester but not in any of his published work [120, 1978], [121, 2006, p. 470], [464, 1998, Letter 44]. Prove his general version: if $A, B \in \mathbb{C}^{n\times n}$, $AB = BA$, and $f(x, y) = \det(xA - yB)$ then $f(B, A) = 0$. Is the commutativity condition necessary?

1.20. Let f satisfy $f(-z) = \pm f(z)$. Show that $f(-A) = \pm f(A)$ whenever the primary matrix functions $f(A)$ and $f(-A)$ are defined. (Hint: Problem 1.2 can be used.)

1.21. Let $P \in \mathbb{C}^{n\times n}$ be idempotent ($P^2 = P$). Show that $f(aI + bP) = f(a)I + (f(a + b) - f(a))P$.

1.22. Is $f(A) = A^*$ possible for a suitable choice of f? Consider, for example, $f(\lambda) = \overline{\lambda}$.

1.23. Verify the Cauchy integral formula (1.12) in the case $f(\lambda) = \lambda^j$ and $A = J_n(0)$, the Jordan block with zero eigenvalue.

1.24. Show from first principles that for $\lambda_k \neq 0$ a Jordan block $J_k(\lambda_k)$ has exactly two upper triangular square roots. (There are in fact only two square roots of any form, as shown by Theorem 1.26.)

1.25. (Davies [131, 2007]) Let $A \in \mathbb{C}^{n\times n}$ ($n > 1$) be nilpotent of index n (that is, $A^n = 0$ but $A^{n-1} \neq 0$). Show that A has no square root but that $A + cA^{n-1}$ is a square root of A^2 for any $c \in \mathbb{C}$. Describe all such A.

1.26. Suppose that $X \in \mathbb{C}^{n\times n}$ commutes with $A \in \mathbb{C}^{n\times n}$, and let A have the Jordan canonical form $Z^{-1}AZ = \mathrm{diag}(J_1, J_2, \dots, J_p) = J$. Is $Z^{-1}XZ$ block diagonal with blocking conformable with that of J?

1.27. (Extension of Theorem 1.29.) Let $A \in \mathbb{C}^{n\times n}$ have no eigenvalues on $\mathbb{R}^-$ except possibly for a semisimple zero eigenvalue. Show that there is a unique square root X of A that is a primary matrix function of A and whose nonzero eigenvalues lie in the open right half-plane. Show that if A is real then X is real.

1.28. Investigate the square roots of the upper triangular matrix

$$A = \begin{bmatrix} 0 & 1 & 1 \\ & 1 & 1 \\ & & 0 \end{bmatrix}.$$

1.29. Find all the square roots of the matrix

$$A = \begin{bmatrix} 0 & 1 & 0 \\ 0 & 0 & 0 \\ 0 & 0 & 0 \end{bmatrix}$$

(which is the matrix in (1.23)). Hint: use Theorem 1.36.

1.30. Show that if $A \in \mathbb{C}^{n\times n}$ has a defective zero eigenvalue (i.e., a zero eigenvalue appearing in a Jordan block of size greater than 1) then A does not have a square root that is a polynomial in A.

1.31. The symmetric positive definite matrix A with $a_{ij} = \min(i,j)$ has a square root X with

$$x_{ij} = \begin{cases} 0, & i+j \le n, \\ 1, & i+j > n. \end{cases}$$

For example,

$$\begin{bmatrix} 0 & 0 & 0 & 1 \\ 0 & 0 & 1 & 1 \\ 0 & 1 & 1 & 1 \\ 1 & 1 & 1 & 1 \end{bmatrix}^2 = \begin{bmatrix} 1 & 1 & 1 & 1 \\ 1 & 2 & 2 & 2 \\ 1 & 2 & 3 & 3 \\ 1 & 2 & 3 & 4 \end{bmatrix}.$$

Is X a primary square root of A? Explain how X fits in with the theory of matrix square roots.

1.32. Show that any square root or logarithm X of $A \in \mathbb{C}^{n\times n}$ (primary or nonprimary) commutes with A. Show further that if A is nonderogatory then X is a polynomial in A.

1.33. Find a logarithm of the upper triangular Toeplitz matrix

$$A = \begin{bmatrix} 1 & \frac{1}{2!} & \frac{1}{3!} & \cdots & \frac{1}{(n-1)!} \\ & 1 & \frac{1}{2!} & \ddots & \vdots \\ & & 1 & \ddots & \vdots \\ & & & \ddots & \frac{1}{2!} \\ & & & & 1 \end{bmatrix}.$$

Hence find all the logarithms of A.

1.34. Let $A \in \mathbb{C}^{n\times n}$ have no eigenvalues on $\mathbb{R}^-$. Show that $A^{1/2} = e^{\frac{1}{2}\log A}$, where the logarithm is the principal logarithm.

1.35. Let $A, B \in \mathbb{C}^{n\times n}$ and $AB = BA$. Is it true that $(AB)^{1/2} = A^{1/2}B^{1/2}$ when the square roots are defined?

1.36. (Hille [290, 1958]) Show that if $e^A = e^B$ and no two elements of $\Lambda(A)$ differ by a nonzero integer multiple of $2\pi i$ then $AB = BA$. Given an example to show that this conclusion need not be true without the assumption on $\Lambda(A)$.

1.37. Show that if $e^A = e^B$ and no eigenvalue of A differs from an eigenvalue of B by a nonzero integer multiple of $2\pi i$ then $A = B$.

1.38. Let $A \in \mathbb{C}^{n\times n}$ be nonsingular. Show that if f is an even function ($f(z) = f(-z)$ for all $z \in \mathbb{C}$) then $f(\sqrt{A})$ is the same for all choices of square root (primary or nonprimary). Show that if f is an odd function ($f(-z) = -f(z)$ for all $z \in \mathbb{C}$) then $\sqrt{A}^{\pm 1} f(\sqrt{A})$ is the same for all choices of square root.

1.39. Show that for $A \in \mathbb{C}^{n\times n}$, $\log(e^A) = A$ if and only if $|\operatorname{Im}(\lambda_i)| < \pi$ for every eigenvalue λ_i of A, where log denotes the principal logarithm. (Since $\rho(A) \le \|A\|$ for any consistent norm, $\|A\| < \pi$ is sufficient for the equality to hold.)

1.40. Let $A, B \in \mathbb{C}^{n\times n}$ and let f and g be functions such that $g(f(A)) = A$ and $g(f(B)) = B$. Assume also that B and $f(A)$ are nonsingular. Show that $f(A)f(B) = f(B)f(A)$ implies $AB = BA$. For example, if the spectra of A and B lie in the open right half-plane we can take $f(x) = x^2$ and $g(x) = x^{1/2}$, or if $\rho(A) < \pi$ and $\rho(B) < \pi$ we can take $f(x) = e^x$ and $g(x) = \log x$ (see Problem 1.39).

1.41. Give a proof from first principles (without using the theory of matrix functions developed in this chapter) that a Hermitian positive definite matrix $A \in \mathbb{C}^{n\times n}$ has a unique Hermitian positive definite square root.

1.42. Let $A \in \mathbb{C}^{n\times n}$ have no eigenvalues on $\mathbb{R}^-$. Given that A has a square root X with eigenvalues in the open right half-plane and that X is a polynomial in A, show from first principles, and without using any matrix decompositions, that X is the *unique* square root with eigenvalues in the open right half-plane.

1.43. Prove the first and last parts of Theorem 1.32. (For the rest, see the sources cited in the Notes and References.)

1.44. Give another proof of Corollary 1.34 for $m \ne n$ by using the identity

$$\begin{bmatrix} AB & 0 \\ B & 0 \end{bmatrix} \begin{bmatrix} I_m & A \\ 0 & I_n \end{bmatrix} = \begin{bmatrix} I_m & A \\ 0 & I_n \end{bmatrix} \begin{bmatrix} 0 & 0 \\ B & BA \end{bmatrix} \tag{1.35}$$

(which is (1.36) below with $\alpha = 0$). What additional hypotheses are required for this proof?

1.45. Give another proof of Theorem 1.35 based on the identity

$$\begin{bmatrix} AB + \alpha I_m & 0 \\ B & \alpha I_n \end{bmatrix} \begin{bmatrix} I_m & A \\ 0 & I_n \end{bmatrix} = \begin{bmatrix} I_m & A \\ 0 & I_n \end{bmatrix} \begin{bmatrix} \alpha I_m & 0 \\ B & BA + \alpha I_n \end{bmatrix}. \tag{1.36}$$

What additional hypotheses are required for this proof?

1.46. Show that Corollary 1.34 can be obtained from Theorem 1.35.

1.47. Can (1.31) be generalized to $f(D+AB)$ with $D \in \mathbb{C}^{m\times m}$ diagonal by "replacing αI by D"?

1.48. (Klosinski, Alexanderson, and Larson [355, 1991]) If A and B are $n\times n$ matrices does $ABAB = 0$ imply $BABA = 0$?

1.49. Let $A \in \mathbb{C}^{m\times n}$ and $B \in \mathbb{C}^{n\times m}$. Show that $\det(I_m + AB) = \det(I_n + BA)$.

1.50. (Borwein, Bailey, and Girgensohn [77, 2004, p. 216]) Does the equation $\sin A = \left[\begin{smallmatrix} 1 & 1996 \\ 0 & 1 \end{smallmatrix}\right]$ have a solution? (This was Putnam Problem 1996-B4.)

1.51. Show that the equation

$$\cosh(A) = \begin{bmatrix} 1 & a & a & \dots & a \\ & 1 & a & \dots & a \\ & & 1 & \dots & \vdots \\ & & & \ddots & a \\ & & & & 1 \end{bmatrix} \in \mathbb{C}^{n\times n}$$

has no solutions for $a \neq 0$ and $n > 1$.

1.52. An interesting application of the theory of matrix functions is to the Drazin inverse of $A \in \mathbb{C}^{n\times n}$, which can be defined as the unique matrix A^D satisfying $A^D A A^D = A^D$, $AA^D = A^D A$, $A^{k+1}A^D = A^k$, where k is the index of A (see Section B.2). If $A \in \mathbb{C}^{n\times n}$ has index k then it can be written

$$A = P\begin{bmatrix} B & 0 \\ 0 & N \end{bmatrix} P^{-1}, \tag{1.37}$$

where B is nonsingular and N is nilpotent of index k (and hence has dimension at least k), and then

$$A^D = P\begin{bmatrix} B^{-1} & 0 \\ 0 & 0 \end{bmatrix} P^{-1}.$$

(a) For what function f is $A^D = f(A)$?

(b) Show that if p is a polynomial such that for B in (1.37), $B^{-1} = p(B)$, then $A^D = A^k p(A)^{k+1}$.

(c) Determine $(uv^*)^D$, for nonzero $u, v \in \mathbb{C}^n$.

1.53. How might the definition of $f(A)$ be extended to rectangular matrices?

After developing some properties of "linear transformations" in earlier papers,
Cayley finally wrote "A Memoir on the Theory of Matrices" in 1858
in which a matrix is considered as a single mathematical quantity.
This paper gives Cayley considerable claim to the honor of
introducing the modern concept of matrix,
although the name is due to Sylvester (1850).

— CYRUS COLTON MACDUFFEE, *Vectors and Matrices* (1943)

It will be convenient to introduce here a notion . . .
namely that of the latent roots *of a matrix. . .*
There results the important theorem that the latent roots of any function of a matrix
are respectively the same functions of the latent roots of the matrix itself.

— J. J. SYLVESTER, *On the Equation to the*
Secular Inequalities in the Planetary Theory (1883)

There have been proposed in the literature since 1880
eight distinct definitions of a matric function,
by Weyr, Sylvester and Buchheim,
Giorgi, Cartan, Fantappiè, Cipolla,
Schwerdtfeger and Richter . . .
All of the definitions except those of Weyr and Cipolla
are essentially equivalent.

— R. F. RINEHART, *The Equivalence of Definitions of a Matric Function* (1955)

I have not thought it necessary to undertake the
labour of a formal proof of the theorem in the
general case of a matrix of any degree.[3]

— ARTHUR CAYLEY, *A Memoir on the Theory of Matrices* (1858)

On reaching the chapter on functions of matrices I found that,
starting from a few 'well-known' facts,
the theory unfolded itself naturally and easily,
but that only patches of it here and there appeared to have been published before.

— W. L. FERRAR, *Finite Matrices* (1951)

If one had to identify the two most important topics
in a mathematics degree programme,
they would have to be calculus and matrix theory.
Noncommutative multiplication underlies the whole of quantum theory
and is at the core of some of the most exciting current research
in both mathematics and physics.

— E. BRIAN DAVIES, *Science in the Looking Glass* (2003)

[3] Re the Cayley–Hamilton theorem.

Chapter 2
Applications

Functions of matrices play an important role in many applications. We describe some examples in this chapter.

It is important to bear in mind that while the appearance of $f(A)$ in a formula may be natural and useful from the point of view of theory, it does not always mean that it is necessary or desirable to compute $f(A)$ in this context, as is well known for $f(A) = A^{-1}$. Rearranging or reformulating an expression may remove or modify the $f(A)$ term, and it is always worth exploring these possibilities. Here are two examples in which computation of a matrix square root can be avoided.

- For Hermitian positive definite A, the quantity $\|A^{1/2}x\|_2$ can be computed as $(x^*Ax)^{1/2}$.

- The generalized eigenvalue problem $Ax = \lambda Bx$, with A Hermitian and B Hermitian positive definite, can be rewritten as $B^{-1/2}AB^{-1/2}(B^{1/2}x) = \lambda(B^{1/2}x)$, which is a standard Hermitian eigenvalue problem $Cy = \lambda y$. This reduction is useful for theoretical purposes [461, 1998, Sec. 15.10], but for practical computation the reduction is usually accomplished with a Cholesky factorization $B = R^*R$, for which the reduced problem is $R^{-*}AR^{-1}(Rx) = \lambda(Rx)$ [137, 2001], [538, 2001, Sec. 3.4]. The Cholesky factorization is generally much less expensive to compute than the square root.

2.1. Differential Equations

Differential equations provide a rich source of $f(A)$ problems, because of the fundamental role that the exponential function plays in linear differential equations. The classic scalar problem

$$\frac{dy}{dt} = ay, \qquad y(0) = c$$

has solution $y(t) = e^{at}c$, while the analogous vector problem

$$\frac{dy}{dt} = Ay, \qquad y(0) = c, \qquad y \in \mathbb{C}^n, \ A \in \mathbb{C}^{n\times n}, \tag{2.1}$$

has solution $y(t) = e^{At}c$. More generally, with suitable assumptions on the smoothness of f, the solution to the inhomogeneous system

$$\frac{dy}{dt} = Ay + f(t,y), \qquad y(0) = c, \qquad y \in \mathbb{C}^n, \ A \in \mathbb{C}^{n\times n} \tag{2.2}$$

satisfies

$$y(t) = e^{At}c + \int_0^t e^{A(t-s)} f(s,y)\, ds, \tag{2.3}$$

which is an explicit formula for y in the case that f is independent of y. These formulae do not necessarily provide the best way to compute the solutions numerically; the large literature on the numerical solution of ordinary differential equations (ODEs) provides alternative techniques. However, the matrix exponential is explicitly used in certain methods—in particular the exponential integrators described in the next subsection.

In the special case $f(t, y) = b$, (2.3) is

$$y(t) = e^{At}c + e^{At}\left[-A^{-1}e^{-As}b\right]_0^t, \tag{2.4}$$

which can be reworked into the form

$$y(t) = t\psi_1(tA)(b + Ac) + c, \tag{2.5}$$

where

$$\psi_1(z) = \frac{e^z - 1}{z} = 1 + \frac{z}{2!} + \frac{z^2}{3!} + \cdots. \tag{2.6}$$

The expression (2.5) has the advantage over (2.4) that it is valid when A is singular.

Some matrix differential equations have solutions expressible in terms of the matrix exponential. For example, the solution of

$$\frac{dY}{dt} = AY + YB, \qquad Y(0) = C$$

is easily verified to be

$$Y(t) = e^{At}Ce^{Bt}.$$

Trigonometric matrix functions, as well as matrix roots, arise in the solution of second order differential equations. For example, the problem

$$\frac{d^2y}{dt^2} + Ay = 0, \qquad y(0) = y_0, \quad y'(0) = y_0' \tag{2.7}$$

has solution

$$y(t) = \cos(\sqrt{A}t)y_0 + \left(\sqrt{A}\right)^{-1}\sin(\sqrt{A}t)y_0', \tag{2.8}$$

where $\sqrt{A}$ denotes any square root of A (see Problems 2.2 and 4.1). The solution exists for all A. When A is singular (and $\sqrt{A}$ possibly does not exist) this formula is interpreted by expanding $\cos(\sqrt{A}t)$ and $\left(\sqrt{A}\right)^{-1}\sin(\sqrt{A}t)$ as power series in A.

2.1.1. Exponential Integrators

Many semidiscretized partial differential equations (PDEs) naturally take the form (2.2) with A representing a spatially discretized linear operator and $f(t, y)$ containing the nonlinear terms. Exponential integrators are a broad class of methods for (2.2) that treat the linear term exactly and integrate the remaining part of the solution (the integral in (2.3)) numerically, using an explicit scheme. They are based on the premise that most of the difficulty (e.g., stiffness) lies with the matrix A and not with the nonlinear term f.

Exponential integrators, which date back to the 1960s, have the property that they use the exponential (or related) function of the Jacobian of the differential equation or an approximation to it. They have attracted renewed interest since the late

1990s due principally to advances in numerical linear algebra that have made efficient implementation of the methods possible.

A simple example of an exponential integrator is the exponential time differencing (ETD) Euler method

$$y_{n+1} = e^{hA} y_n + h\psi_1(hA) f(t_n, y_n), \tag{2.9}$$

where $y_n \approx y(t_n)$, $t_n = nh$, and h is a stepsize. The function ψ_1, defined in (2.6), is one of a family of functions $\{\psi_k\}$ that plays an important role in these methods; for more on the ψ_k see Section 10.7.4. The method (2.9) requires the computation of the exponential and ψ_1 (or, at least, their actions on a vector) at each step of the integration.

For an overview of exponential integrators see Minchev and Wright [431, 2005], and see LeVeque [380, 2007] for a concise textbook treatment. A few key papers are Cox and Matthews [118, 2002], Hochbruck, Lubich, and Selhofer [292, 1998], Kassam and Trefethen [336, 2005], and Schmelzer and Trefethen [503, 2006], and a MATLAB toolbox is described by Berland, Skaflestad, and Wright [59, 2007].

2.2. Nuclear Magnetic Resonance

Two-dimensional nuclear magnetic resonance (NMR) spectroscopy is a tool for determining the structure and dynamics of molecules in solution. The basic theory for the nuclear Overhauser effect experiment specifies that a matrix of intensities $M(t)$ is related to a symmetric, diagonally dominant matrix R, known as the relaxation matrix, by the Solomon equations

$$\frac{dM}{dt} = -RM, \qquad M(0) = I.$$

Hence $M(t) = e^{-Rt}$. This relation is used in both directions: in simulations and testing to compute $M(t)$ given R, and in the inverse problem to determine R from observed intensities. The latter problem can be solved with a matrix logarithm evaluation, but in practice not all the m_{ij} are known and estimation methods, typically based on least squares approximations and requiring matrix exponential evaluations, are used.

2.3. Markov Models

The matrix exponential and logarithm play an important role in Markov models, which are used in a variety of different subjects. Consider a time-homogeneous continuous-time Markov process in which individuals move among n states. The *transition probability matrix* $P(t) \in \mathbb{R}^{n\times n}$ has (i,j) entry equal to the probability that an individual starting in state i at time 0 will be in state j at time t. The row sums of P are all 1, so P is a stochastic matrix. Associated with the process is the *transition intensity matrix* $Q \in \mathbb{R}^{n\times n}$, which is related to P by

$$P(t) = e^{Qt}.$$

The elements of Q satisfy

$$q_{ij} \geq 0, \quad i \neq j, \qquad \sum_{j=1}^{n} q_{ij} = 0, \quad i = 1\colon n. \tag{2.10}$$

For any such Q, e^{Qt} is nonnegative for all $t \geq 0$ (see Theorem 10.29) and has unit row sums, so is stochastic.

Now consider a discrete-time Markov process with transition probability matrix P in which the transition probabilities are independent of time. We can ask whether $P = e^Q$ for some intensity matrix Q, that is, whether the process can be regarded as a discrete manifestation of an underlying time-homogeneous Markov process. If such a Q exists it is called a *generator* and P is said to be *embeddable*. Necessary and sufficient conditions for the existence of a generator for general n are not known. Researchers in sociology [525, 1976], statistics [331, 1985], and finance [315, 2001] have all investigated this embeddability problem. A few interesting features are as follows:

- If P has distinct, real positive eigenvalues then the only real logarithm, and hence the only candidate generator, is the principal logarithm.
- P may have one or more real negative eigenvalues, so that the principal logarithm is undefined, yet a generator may still exist. For example, consider the matrix [525, 1976, Ex. 10]

$$P = \frac{1}{3}\begin{bmatrix} 1+2x & 1-x & 1-x \\ 1-x & 1+2x & 1-x \\ 1-x & 1-x & 1+2x \end{bmatrix}, \qquad x = -e^{-2\sqrt{3}\pi} \approx -1.9 \times 10^{-5}.$$

  P is diagonalizable, with eigenvalues 1, x, and x. Every primary logarithm is complex, since it cannot have complex conjugate eigenvalues. Yet the nonprimary logarithm

$$Q = 2\sqrt{3}\pi \begin{bmatrix} -2/3 & 1/2 & 1/6 \\ 1/6 & -2/3 & 1/2 \\ 1/2 & 1/6 & -2/3 \end{bmatrix}$$

  is a generator.
- More than one generator may exist.

Suppose a given transition matrix $P \equiv P(1)$ has a generator $Q = \log(P)$. Then Q can be used to construct $P(t)$ at other times, through $P(t) = \exp(Qt)$. For example, if P is the transition matrix for the time period of one year then the transition matrix for a month is $P(1/12) = e^{\log(P)/12}$. However, it is more direct and efficient to compute $P(1/12)$ as $P^{1/12}$, thus avoiding the computation of a generator. Indeed, the standard inverse scaling and squaring method for the principal logarithm of a matrix requires the computation of a matrix root, as explained in Section 11.5. Similarly, the transition matrix for a week can be computed directly as $P^{1/52}$.

This use of matrix roots is suggested by Waugh and Abel [610, 1967], mentioned by Israel, Rosenthal, and Wei [315, 2001], and investigated in detail by Kreinin and Sidelnikova [362, 2001]. The latter authors, who are motivated by credit risk models, address the problems that the principal root and principal logarithm of P may have the wrong sign patterns; for example, the root may have negative elements, in which case it is not a transition matrix. They show how to optimally adjust these matrices to achieve the required properties—a process they term regularization. Their preferred method for obtaining transition matrices for short times is to regularize the appropriate matrix root. Questions of existence and uniqueness of stochastic roots of stochastic matrices arise (see Problem 7.17).

2.4. Control Theory

In control theory a linear dynamical system can be expressed as a continuous-time system

$$\begin{aligned} \frac{dx}{dt} &= Fx(t) + Gu(t), \qquad & F \in \mathbb{C}^{n\times n}, \quad G \in \mathbb{C}^{n\times m}, \\ y &= Hx(t) + Ju(t), \qquad & H \in \mathbb{C}^{p\times n}, \quad J \in \mathbb{C}^{p\times m}, \end{aligned}$$

or as the corresponding discrete-time state-space system

$$\begin{aligned} x_{k+1} &= Ax_k + Bu_k, \qquad & A \in \mathbb{C}^{n\times n}, \quad B \in \mathbb{C}^{n\times m}, \\ y_k &= Hx_k + Ju_k. \end{aligned}$$

Here, x is the state vector and u and y are the input and output vectors, respectively. The connection between the two forms is given by

$$A = e^{F\tau}, \qquad B = \left(\int_0^\tau e^{Ft} dt\right) G,$$

where τ is the sampling period. Therefore the matrix exponential and logarithm are needed for these conversions. In the MATLAB Control System Toolbox [413] the functions `c2d` and `d2c` carry out the conversions, making use of MATLAB's `expm` and `logm` functions.

We turn now to algebraic equations arising in control theory and the role played by the matrix sign function. The matrix sign function was originally introduced by Roberts [496] in 1971 as a tool for solving the Lyapunov equation and the algebraic Riccati equation. It is most often defined in terms of the Jordan canonical form $A = ZJZ^{-1}$ of $A \in \mathbb{C}^{n\times n}$. If we arrange that

$$J = \begin{bmatrix} J_1 & 0 \\ 0 & J_2 \end{bmatrix},$$

where the eigenvalues of $J_1 \in \mathbb{C}^{p\times p}$ lie in the open left half-plane and those of $J_2 \in \mathbb{C}^{q\times q}$ lie in the open right half-plane, then

$$\operatorname{sign}(A) = Z \begin{bmatrix} -I_p & 0 \\ 0 & I_q \end{bmatrix} Z^{-1}. \tag{2.11}$$

The sign function is undefined if A has an eigenvalue on the imaginary axis. Note that $\operatorname{sign}(A)$ is a primary matrix function corresponding to the scalar sign function

$$\operatorname{sign}(z) = \begin{cases} 1, & \operatorname{Re} z > 0, \\ -1, & \operatorname{Re} z < 0, \end{cases} \qquad z \in \mathbb{C},$$

which maps z to the nearest square root of unity. For more on the sign function, including alternative definitions and iterations for computing it, see Chapter 5.

The utility of the sign function is easily seen from Roberts' observation that the Sylvester equation

$$AX + XB = C, \qquad A \in \mathbb{C}^{m\times m},\ B \in \mathbb{C}^{n\times n},\ C \in \mathbb{C}^{m\times n},$$

is equivalent to the equation

$$\begin{bmatrix} A & -C \\ 0 & -B \end{bmatrix} = \begin{bmatrix} I_m & X \\ 0 & I_n \end{bmatrix} \begin{bmatrix} A & 0 \\ 0 & -B \end{bmatrix} \begin{bmatrix} I_m & X \\ 0 & I_n \end{bmatrix}^{-1}.$$

If $\operatorname{sign}(A) = I$ and $\operatorname{sign}(B) = I$ then

$$\operatorname{sign}\left(\begin{bmatrix} A & -C \\ 0 & -B \end{bmatrix}\right) = \begin{bmatrix} I_m & X \\ 0 & I_n \end{bmatrix} \begin{bmatrix} I_m & 0 \\ 0 & -I_n \end{bmatrix} \begin{bmatrix} I_m & -X \\ 0 & I_n \end{bmatrix} = \begin{bmatrix} I_m & -2X \\ 0 & -I_n \end{bmatrix}, \quad (2.12)$$

so the solution X can be read from the $(1,2)$ block of the sign of the block upper triangular matrix $\left[\begin{smallmatrix} A & -C \\ 0 & -B \end{smallmatrix}\right]$. The conditions that $\operatorname{sign}(A)$ and $\operatorname{sign}(B)$ are identity matrices are certainly satisfied for the Lyapunov equation, in which $B = A^*$, in the common case where A is positive stable, that is, $\operatorname{Re}\lambda_i(A) > 0$ for all i.

Consider now the algebraic Riccati equation

$$XFX - A^*X - XA - G = 0, \quad (2.13)$$

where all matrices are $n \times n$ and F and G are Hermitian. The desired solution is Hermitian and stabilizing, in the sense that the spectrum of $A - FX$ lies in the open left half-plane. Such a solution exists and is unique under suitable conditions that we will not describe; see [349, 1989] and [370, 1995, Chap. 22] for details. The equation can be written in the equivalent form

$$W = \begin{bmatrix} A^* & G \\ F & -A \end{bmatrix} = \begin{bmatrix} X & -I_n \\ I_n & 0 \end{bmatrix} \begin{bmatrix} -(A-FX) & -F \\ 0 & (A-FX)^* \end{bmatrix} \begin{bmatrix} X & -I_n \\ I_n & 0 \end{bmatrix}^{-1}. \quad (2.14)$$

By assumption, $A - FX$ has eigenvalues with negative real part. Hence we can apply the sign function to (2.14) to obtain

$$\operatorname{sign}(W) = \begin{bmatrix} X & -I_n \\ I_n & 0 \end{bmatrix} \begin{bmatrix} I_n & Z \\ 0 & -I_n \end{bmatrix} \begin{bmatrix} X & -I_n \\ I_n & 0 \end{bmatrix}^{-1}$$

for some Z. Writing $\operatorname{sign}(W) - I_{2n} = [M_1 \ M_2]$, where $M_1, M_2 \in \mathbb{C}^{2n\times n}$, the latter equation becomes

$$[\,M_1 \quad M_2\,] \begin{bmatrix} X & -I_n \\ I_n & 0 \end{bmatrix} = \begin{bmatrix} X & -I_n \\ I_n & 0 \end{bmatrix} \begin{bmatrix} 0 & Z \\ 0 & -2I_n \end{bmatrix}, \quad (2.15)$$

which gives

$$M_1 X = -M_2,$$

which is a system of $2n^2$ equations in n^2 unknowns. The system is consistent, by construction, and by rewriting (2.15) as

$$\begin{bmatrix} 0 & I_n \\ -I_n & X \end{bmatrix} [\,M_1 \quad M_2\,] = \begin{bmatrix} 0 & Z \\ 0 & -2I_n \end{bmatrix} \begin{bmatrix} 0 & I_n \\ -I_n & X \end{bmatrix}$$

we see that $[-I_n \ X]M_1 = 2I_n$, which implies that M_1 has full rank. To summarize: the Riccati equation (2.13) can be solved by computing the sign of a $2n \times 2n$ matrix and then solving an overdetermined but consistent system; the latter can be done using a QR factorization.

2.5. The Nonsymmetric Eigenvalue Problem

The matrix sign function, defined in the previous section, can be used to count how many eigenvalues of a matrix lie in particular regions of the complex plane and to obtain the corresponding invariant subspaces. From (2.11) we have $\operatorname{trace}(\operatorname{sign}(A)) = q - p$, and since $p + q = n$ this gives the formulae

$$p = \frac{1}{2}\big(n - \operatorname{trace}(\operatorname{sign}(A))\big), \qquad q = \frac{1}{2}\big(n + \operatorname{trace}(\operatorname{sign}(A))\big)$$

for the number of eigenvalues lying in the open left half-plane and open right half-plane, respectively. Moreover, writing Z in (2.11) as $Z = [Z_1\ Z_2]$ with $Z_1 \in \mathbb{C}^{n\times p}$ and $Z_2 \in \mathbb{C}^{n\times q}$, then $I + \operatorname{sign}(A) = 2[Z_1\ Z_2]\operatorname{diag}(0, I_q)Z^{-1} = 2Z_2Z^{-1}(p+1\colon n, :)$, so the columns of $(I + \operatorname{sign}(A))/2$ span the invariant subspace corresponding to the eigenvalues of A in the open right half-plane (indeed, $(I+\operatorname{sign}(A))/2$ is a projector onto this subspace—see Theorem 5.1). In fact, as the next theorem shows, the eigenvalue problem for A can be deflated into two smaller problems for the eigenvalues in the two half-planes.

Theorem 2.1 (spectrum splitting via sign function). *Let $A \in \mathbb{R}^{n\times n}$ have no pure imaginary eigenvalues and define $W = (\operatorname{sign}(A) + I)/2$. Let*

$$Q^T W \Pi = \begin{array}{c} \\ q \\ n-q \end{array}\!\!\begin{array}{c} \begin{array}{cc} q & n-q \end{array} \\ \begin{bmatrix} R_{11} & R_{12} \\ 0 & 0 \end{bmatrix} \end{array}$$

be a rank-revealing QR factorization, where Π is a permutation matrix and $q = \operatorname{rank}(W)$. Then

$$Q^T A Q = \begin{array}{c} \\ q \\ n-q \end{array}\!\!\begin{array}{c} \begin{array}{cc} q & n-q \end{array} \\ \begin{bmatrix} A_{11} & A_{12} \\ 0 & A_{22} \end{bmatrix} \end{array},$$

where the eigenvalues of A_{11} lie in the open right half-plane and those of A_{22} lie in the open left half-plane.

Proof. See Problem 2.3. □

The number of eigenvalues in more complicated regions of the complex plane can be counted by suitable sequences of matrix sign evaluations. For example, assuming that A has no eigenvalues lying on the edges of the relevant regions:

- The number of eigenvalues of A lying in the vertical strip $\operatorname{Re} z \in (\xi_1, \xi_2)$ is $\frac{1}{2}\operatorname{trace}\big(\operatorname{sign}(A - \xi_1 I) - \operatorname{sign}(A - \xi_2 I)\big)$.
- Let ξ_w, ξ_e, ξ_n, ξ_s be complex numbers at the corners of a rectangle oriented at $\pi/4$ to the axes. With $N(\xi) = \operatorname{sign}\big((A - \xi I)^2\big)$, the number of eigenvalues of A lying inside the rectangle is

$$\frac{1}{4}\operatorname{trace}\big(N(\xi_w) + N(\xi_e) - N(\xi_n) - N(\xi_s)\big).$$

For details, and analogues of the deflation in Theorem 2.1, see Howland [302, 1983] and Bai and Demmel [27, 1993]. By combining these techniques it is possible to set up a divide and conquer algorithm for a partial or complete set of eigenvalues [27, 1993], [29, 1998].

The use of the matrix sign function as described in this section is particularly attractive in the context of high-performance computing.

2.6. Orthogonalization and the Orthogonal Procrustes Problem

In many applications a matrix that should be orthogonal turns out not to be because of errors of measurement, truncation, or rounding. The question then arises how best to orthogonalize the matrix. One approach is to apply Gram–Schmidt orthogonalization or, equivalently, to compute the orthogonal factor in a QR factorization. An alternative, "optimal orthogonalization", is to replace A by the nearest orthogonal matrix. More generally, let $A \in \mathbb{C}^{m\times n}$ $(m \geq n)$ and define the distance to the nearest matrix with orthonormal columns by

$$\min\{\, \|A - Q\| : Q^*Q = I \,\}.$$

For the 2- and Frobenius norms an optimal Q is U in the polar decomposition $A = UH$, where $U^*U = I$ and H is Hermitian positive semidefinite; see Theorem 8.4. The factor H can be written as $H = (A^*A)^{1/2}$ and so if A is of full rank then $U = A(A^*A)^{-1/2}$. An important advantage of the unitary polar factor U over the QR factor Q is that it is basis-independent: if we transform $A \to W_1AW_2$ with the W_i unitary then U transforms likewise to W_1UW_2 (since $W_1AW_2 = W_1UW_2 \cdot W_2^*HW_2$ is a polar decomposition), but Q does not change in such a predictable way. The polar factors are not functions of A according to our definition, because of the appearance of A^* in these relations. Nevertheless, there are strong connections with the matrix square root and the matrix sign function and so we devote Chapter 8 to the polar decomposition.

Optimal orthogonalization is used in a number of applications. In aerospace computations the direction cosine matrix (DCM) $D(t) \in \mathbb{R}^{3\times 3}$ describes the rotation of a coordinate system relative to a reference coordinate system. It satisfies the matrix ODE

$$\frac{dD}{dt} = SD, \quad S = -S^T, \quad D(0) \text{ orthogonal.}$$

This system is typically solved by ODE methods that do not preserve orthogonality, yet $D(t) = \exp(St)D(0)$ is orthogonal (see Theorem 1.44). Approximate DCMs therefore need to be reorthogonalized periodically; see Mao [405, 1986] and the references therein. Optimal orthogonalization is also used in the numerical solution of more general matrix ODEs with orthogonal solutions:

$$Y'(t) = F(t, Y(t)), \qquad Y(0)^TY(0) = I, \tag{2.16}$$

where $Y(t) \in \mathbb{R}^{m\times n}$, $m \geq n$, and $Y(t)^TY(t) = I$ for all $t > 0$; see Hairer, Lubich, and Wanner [239, 2002, Sec. 4.4], D. J. Higham [262, 1997], and Sofroniou and Spaletta [534, 2002].

In quantum chemistry orthogonalization using the unitary polar factor is called Löwdin orthogonalization; see Bhatia and Mukherjea [67, 1986], Goldstein and Levy [221, 1991], and Jansik et al. [318, 2007]. An application of the polar decomposition to determining the orientation of "parallel spherical wrist" robots is described by Vertechy and Parenti-Castelli [601, 2006]. Moakher [434, 2002] shows that a certain geometric mean of a set of rotation matrices can be expressed in terms of the orthogonal polar factor of their arithmetic mean.

The polar decomposition is also used in computer graphics as a convenient way of decomposing a 3×3 or 4×4 linear transformation into simpler component parts (see Shoemake and Duff [520, 1992]) and in continuum mechanics for representing the deformation gradient as the product of a rotation tensor and a stretch tensor (see Bouby, Fortuné, Pietraszkiewicz, and Vallée [78, 2005]).

The orthogonal Procrustes problem is to solve

$$\min\{ \|A - BQ\|_F : Q \in \mathbb{C}^{n\times n},\ Q^*Q = I \}, \tag{2.17}$$

where $A, B \in \mathbb{C}^{m\times n}$; thus a unitary matrix is required that most nearly transforms a rectangular matrix B into a matrix A of the same dimensions in a least squares sense. A solution is given by the unitary polar factor of B^*A; see Theorem 8.6. The orthogonal Procrustes problem is a well-known and important problem in factor analysis and in multidimensional scaling in statistics; see the books by Gower and Dijksterhuis [226, 2004] and Cox and Cox [119, 1994]. In these applications the matrices A and B represent sets of experimental data, or multivariate samples, and it is necessary to determine whether the sets are equivalent up to rotation. An important variation of (2.17) requires Q to be a "pure rotation", that is, $\det(Q) = 1$; one application area is shape analysis, as discussed by Dryden and Mardia [168, 1998]. Many other variations of the orthogonal Procrustes problem exist, including those involving two-sided transformations, permutation transformations, and symmetric transformations, but the solutions have weaker connections with the polar decomposition and with matrix functions.

The polar decomposition and the orthogonal Procrustes problem both arise in numerical methods for computing analytic singular value decompositions, as explained by Mehrmann and Rath [419, 1993].

2.7. Theoretical Particle Physics

Lattice quantum chromodynamics (QCD) is a research area of physics that has in recent years made extensive use of matrix functions. QCD is a physical theory that describes the strong interactions between quarks as the constituents of matter. Lattice QCD formulates the theory on a four dimensional space-time lattice, and its numerical simulations currently occupy large amounts of high-performance computer time.

An important recent development in lattice QCD is the overlap-Dirac operator of Neuberger [446, 1998], the study of which requires the solution of n-dimensional linear systems of the form

$$\bigl(G - \mathrm{sign}(H)\bigr)x = b. \tag{2.18}$$

Here, $G = \mathrm{diag}(\pm 1)$, H is sparse, complex, and Hermitian, and n is extremely large: of order perhaps 10^6. For Hermitian A, the matrix sign function can be written as $\mathrm{sign}(A) = Q\,\mathrm{diag}(\mathrm{sign}(\lambda_i))Q^*$, where $A = Q\,\mathrm{diag}(\lambda_i)Q^*$ is a spectral decomposition with Q unitary. The system (2.18) is currently solved by Krylov subspace techniques, which require matrix–vector products with the coefficient matrix. Hence a key step in the solution of (2.18) is the computation of $\mathrm{sign}(H)c$ for given vectors c. In view of the huge dimension of H, this product must be computed without forming the dense matrix $\mathrm{sign}(H)$. A variety of methods have been proposed for this computation, including Krylov techniques and methods based on rational matrix sign function approximations. For details, see van den Eshof, Frommer, Lippert, Schilling, and Van der Vorst [585, 2002], the articles in Frommer, Lippert, Medeke, and Schilling [196, 2000], and the summary in Frommer and Simoncini [198, 2008]. A collection of relevant MATLAB functions is given by Boriçi [75].

In Monte Carlo QCD simulations, $\det(A)$ must be estimated for a large, sparse, symmetric positive definite matrix A known as the fermion matrix. One approach makes use of the identity, for symmetric positive definite A,

$$\log(\det(A)) = \operatorname{trace}(\log(A)),$$

where log denotes the principal logarithm (see Theorem 1.31); for such A this is equivalent to Theorem 1.45. This identity converts the problem into one of estimating the diagonal elements of $\log(A)$. See, for example, Bai, Fahey, and Golub [32, 1996] and Thron, Dong, Liu, and Ying [567, 1997].

Another area where determinants must be evaluated is computational quantum field theory [378, 2003]. Motivated by this application, Ipsen and Lee [310, 2003] exploit the relation $\det(A) = \exp(\operatorname{trace}(\log(A)))$ (Theorem 1.45) to derive numerical approximations to the determinant.

Quantum Monte Carlo (QMC) simulations with the Hubbard model of particle interactions rely on a number of fundamental linear algebra operations. In particular, solving linear systems $Mx = b$ is the key computational kernel in QMC and forming M requires the computation of matrix exponentials. For details see Bai, Chen, Scalettar, and Yamazaki [26, 2007].

2.8. Other Matrix Functions

One matrix function generates the need for another. In testing the quality of a computed approximation $\widehat{f}(A) \approx f(A)$ we may want to form $f^{-1}(\widehat{f}(A)) - A$ or to test how closely an identity involving f is satisfied. Thus e^A may generate the need for $\log(A)$ and $\cos(A)$ for $\sin(A)$. Moreover, methods for different functions are often interrelated. The inverse scaling and squaring algorithm for the matrix logarithm requires matrix square roots (see Section 11.5), while one method for computing the square root of a symmetric positive definite matrix employs the polar decomposition of a related matrix (see Section 6.8.4).

2.9. Nonlinear Matrix Equations

One reason for developing a theory of matrix functions is to aid the solution of nonlinear matrix equations. Ideally, closed form solutions can be found in terms of an appropriate matrix function.

The algebraic Riccati equation (2.13) and its special cases form an important class of nonlinear matrix equations. We examine the special case

$$XAX = B, \qquad A, B \in \mathbb{C}^{n\times n}. \tag{2.19}$$

This equation has the solution

$$X = B(AB)^{-1/2}, \tag{2.20}$$

provided that AB has no eigenvalues on $\mathbb{R}^-$, as is easily verified by substitution. Here, any square root of AB can be taken. The formula (2.20) can be rewritten in alternative forms. Corollary 1.34, implies that $X = (BA)^{-1/2}B$. Other equivalent forms are $A^{-1}(AB)^{1/2}$ and $(BA)^{1/2}A^{-1}$. We can also derive a more symmetric expression for X as follows:

$$\begin{aligned} X &= B\big(B^{-1/2}\cdot B^{1/2}AB^{1/2}\cdot B^{1/2}\big)^{-1/2} \\ &= BB^{-1/2}\big(B^{1/2}AB^{1/2}\big)^{-1/2}B^{1/2} \\ &= B^{1/2}\big(B^{1/2}AB^{1/2}\big)^{-1/2}B^{1/2}, \end{aligned} \tag{2.21}$$

where again any square root can be taken, and again this expression is easily verified to satisfy (2.19). Another expression for X is as the (1,2) block of $\mathrm{sign}(\left[\begin{smallmatrix} 0 & B \\ A & 0 \end{smallmatrix}\right])$, which follows from the sign-based solution of (2.13) and also from Theorem 5.2. As this example illustrates, there may be several ways to express the solutions to a matrix equation in terms of matrix functions.

If A and B are Hermitian positive definite then there is a unique Hermitian positive definite solution to (2.19), given by any of the expressions above, where the Hermitian positive definite square root is always taken. The uniqueness follows from writing (2.19) as $Y^2 = C$, where $Y = A^{1/2}XA^{1/2}$ and $C = A^{1/2}BA^{1/2}$ and using the fact that a Hermitian positive definite matrix has a unique Hermitian positive definite square root (Corollary 1.30). (Note that this approach leads directly to (2.21).) In this case formulae that are more computationally efficient than those above are available; see Problem 2.7 and Algorithm 6.22.

An equation that generalizes the scalar quadratic in a different way to the algebraic Riccati equation is the quadratic matrix equation

$$AX^2 + BX + C = 0, \qquad A, B, C \in \mathbb{C}^{n\times n}. \tag{2.22}$$

Unfortunately, there is no closed-form expression for X in general, and the theory of such equations is nontrivial; see Higham and Kim [280, 2000], and the references therein. A special case in which the usual quadratic formula generalizes is when $A = I$, B commutes with C, and $B^2 - 4C$ has a square root. Then we can complete the square in the usual way to obtain

$$X = -\frac{1}{2}B + \frac{1}{2}(B^2 - 4C)^{1/2},$$

where the square root can be any primary square root. One way of solving (2.22) is by Newton's method [281, 2001]. If a sufficiently good starting matrix is not available then continuation can be used. Assume that a solution $X(t)$ to the quadratic matrix equation $AX(t)^2 + tBX(t) + C = 0$ exists for $t \in [0,1]$. Then $X(0) = \sqrt{-A^{-1}C}$ and we can solve for $X(t_{k+1})$ by using Newton's method with $X(t_k)$ as the starting matrix, for $0 = t_0 < t_1 < t_2 < \cdots < t_m = 1$. The aim is to take the number of steps m as small as possible such that each application of Newton's method converges quickly.

We mention two applications in which the quadratic (2.22) arises. Quasi-birth-death processes are two-dimensional Markov chains with a block tridiagonal transition probability matrix. They are widely used as stochastic models in telecommunications, computer performance, and inventory control. Analysis using the matrix-geometric method leads to three quadratic matrix equations whose elementwise minimal nonnegative solutions can be used to characterize most of the features of the Markov chain. Excellent references are the books by Bini, Latouche, and Meini [70, 2005] and Latouche and Ramaswami [373, 1999].

A second application is the solution of the quadratic eigenvalue problem

$$Q(\lambda)x = (\lambda^2 A + \lambda B + C)x = 0, \qquad A, B, C \in \mathbb{C}^{n\times n}, \tag{2.23}$$

which arises in the analysis of damped structural systems and vibration problems [369, 1966], [570, 2001]. The standard approach is to reduce (2.23) to a generalized eigenproblem (GEP) $Gx = \lambda Hx$ of twice the dimension, $2n$. This "linearized" problem can be further converted to a standard eigenvalue problem of dimension $2n$ under suitable nonsingularity conditions on the coefficients A, B, and C. However, if we can find a solution X of the associated quadratic matrix equation (2.22) then we can write

$$\lambda^2 A + \lambda B + C = -(B + AX + \lambda A)(X - \lambda I), \tag{2.24}$$

and so the eigenvalues of (2.23) are those of X together with those of the GEP $(B + AX)x = -\lambda Ax$, both of which are $n \times n$ problems. Bridges and Morris [82, 1984] employ this approach in the solution of differential eigenproblems.

For a less obvious example of where a matrix function arises in a nonlinear matrix equation, consider the problem of finding an orthogonal $Q \in \mathbb{R}^{n\times n}$ such that

$$Q - Q^T = S, \qquad S = -S^T \in \mathbb{R}^{n\times n} \text{ given}, \tag{2.25}$$

which arises in the analysis of the dynamics of a rigid body [94, 2003] and in the solution of algebraic Riccati equations [309, 1984]. The equation can be rewritten $Q - Q^{-1} = S$, which implies both $Q^2 - I = QS$ and $Q^2 - I = SQ$. Hence $Q^2 - \frac{1}{2}(QS + SQ) - I = 0$, or $(Q - \frac{1}{2}S)^2 = I + S^2/4$. Thus Q is of the form $Q = \frac{1}{2}S + \sqrt{I + S^2/4}$ for some square root. Using the theory of matrix square roots the equation (2.25) can then be fully analyzed.

There is little in the way of numerical methods for solving general nonlinear matrix equations $f(X) = A$, other than Newton's method. By using the Jordan canonical form it is usually possible to determine and classify all solutions (as we did in Section 1.6 for the square root and logarithm), but this approach is usually not feasible computationally; see Horn and Johnson [296, 1991, Cor. 6.2.12, Sec. 6.4] for details.

Finally, we note that nonlinear matrix equations provide useful test problems for optimization and nonlinear least squares solvers, especially when a reference solution can be computed by matrix function techniques. Some matrix square root problems are included in the test collection maintained by Fraley on Netlib [191] and in the CUTEr collection [225, 2003].

2.10. Geometric Mean

The geometric mean of positive scalars can be generalized to Hermitian positive definite matrices in various ways, which to a greater or lesser extent possess the properties one would like of a mean. Let $A, B \in \mathbb{C}^{n\times n}$ be Hermitian positive definite. The *geometric mean* $A \# B$ is defined as the unique Hermitian positive definite solution to $XA^{-1}X = B$, or (cf. (2.19)–(2.21))

$$X = B^{1/2}(B^{-1/2}AB^{-1/2})^{1/2}B^{1/2} = B(B^{-1}A)^{1/2} = (AB^{-1})^{1/2}B, \tag{2.26}$$

where the last equality can be seen using Corollary 1.34. The geometric mean has the properties (see Problem 2.5)

$$A \# A = A, \tag{2.27a}$$

$$(A \# B)^{-1} = A^{-1} \# B^{-1}, \tag{2.27b}$$

$$A \# B = B \# A, \tag{2.27c}$$

$$A \# B \le \frac{1}{2}(A + B), \tag{2.27d}$$

all of which generalize properties of the scalar geometric mean $a \# b = \sqrt{ab}$. Here, $X \ge 0$ denotes that the Hermitian matrix X is positive semidefinite; see Section B.12.

The geometric mean also satisfies the extremal property

$$A \# B = \max \left\{ X = X^* : \begin{bmatrix} A & X \\ X & B \end{bmatrix} \geq 0 \right\}. \tag{2.28}$$

The geometric mean yields the solution to more general equations than $XA^{-1}X = B$. For example, if A and B are Hermitian positive definite then the unique Hermitian positive definite solution to $XA^{-1}X \pm X - B = 0$ is $X = \frac{1}{2}(\mp A + A \# (A+4B))$ [391, 2007, Thm. 3.1].

Another definition of geometric mean of Hermitian positive definite matrices A and B is

$$E(A, B) = \exp(\tfrac{1}{2}(\log(A) + \log(B))), \tag{2.29}$$

where log is the principal logarithm. This is called the log-Euclidean mean by Arsigny, Fillard, Pennec, and Ayache [20, 2007], who investigate its properties.

2.11. Pseudospectra

Pseudospectra are not so much an application of matrix functions as objects with intimate connections to them. The *ϵ-pseudospectrum* of $A \in \mathbb{C}^{n\times n}$ is defined, for a given $\epsilon > 0$ and a subordinate matrix norm, to be the set

$$\Lambda_\epsilon(A) = \{\, z : z \text{ is an eigenvalue of } A + E \text{ for some } E \text{ with } \|E\| < \epsilon \,\}. \tag{2.30}$$

It can also be represented, in terms of the resolvent $(zI - A)^{-1}$, as

$$\Lambda_\epsilon(A) = \{\, z : \|(zI - A)^{-1}\| > \epsilon^{-1} \,\}.$$

The resolvent therefore provides a link between pseudospectra and matrix functions, through Definition 1.11.

Pseudospectra provide a means for judging the sensitivity of the eigenvalues of a matrix to perturbations in the matrix elements. For example, the 0.01-pseudospectrum indicates the uncertainty in the eigenvalues if the elements are known to only two decimal places. More generally, pseudospectra have much to say about the behaviour of a matrix. They provide a way of describing the effects of nonnormality on processes such as matrix powering and exponentiation.

Matrix functions and pseudospectra have several features in common: they are applicable in a wide variety of situations, they are "uninteresting" for normal matrices, and they are nontrivial to compute.

For more on pseudospectra we can do no better than refer the reader to the ultimate reference on the subject: Trefethen and Embree [573, 2005].

2.12. Algebras

While this book is concerned with matrices over the real or complex fields, some of the results and algorithms are applicable to more general algebras, and thereby provide tools for working with these algebras. For example, the GluCat library [217] is a generic library of C++ templates that implements universal Clifford algebras over the real and complex fields. It includes algorithms for the exponential, logarithm, square root, and trigonometric functions, all based on algorithms for matrices.

2.13. Sensitivity Analysis

Ideally a numerical algorithm returns not only an approximate solution but also an estimate or bound for the error in that solution. Producing an a priori error bound for an algorithm can be very difficult, as it involves analysis of truncation errors and rounding errors and their propagation. A separate question, usually easier to answer, is how sensitive is the solution of the problem to perturbations in the data. Knowledge of problem sensitivity can be crucial in applications since it gives insight into whether the problem has been well formulated, allows prediction of the effects of inaccuracy in the data, and indicates the best accuracy that any algorithm can be expected to provide in floating point arithmetic. Sensitivity is determined by the derivative of the function that maps the input data to the solution. For matrix functions the appropriate derivative is the Fréchet derivative, and its norm determines a condition number for the problem, as explained in the next chapter. Thus every $f(A)$ problem gives rise to the related problem of characterizing, computing, and estimating the Fréchet derivative of f and its norm.

2.14. Other Applications

Finally, we describe some more speculative or less well established applications.

2.14.1. Boundary Value Problems

Schmitt [504, 1990] proposes a symmetric difference scheme for linear, stiff, or singularly perturbed constant coefficient boundary value problems of first order based on the stability function $f(z) = z + (1 + z^2)^{1/2}$. The function f agrees with e^z up to terms in z of second order, and the idea is that f may bring more favourable stability properties than a polynomial or rational stability function when the eigenvalues of the Jacobian matrix vary greatly in magnitude and location. Implementing the method requires computing on each step of the integration a matrix square root $(\omega^2 I + h_k^2 A^2)^{1/2}$, where ω is a parameter, h_k a stepsize, and A is the Jacobian of the system. Schmitt uses the (unscaled) Denman–Beavers iteration (6.15) to compute the square roots.

2.14.2. Semidefinite Programming

Semidefinite programming problems are a class of constrained optimization problems in which the variable is a symmetric positive semidefinite matrix, $X \in \mathbb{R}^{n\times n}$. Various algorithms are available for the solution of such problems. An algorithm suggested by Kanzow and Nagel [333, 2002] requires the computation of square roots of symmetric positive semidefinite matrices. The authors observe that the condition $XS = 0$, with X and S symmetric positive semidefinite, which arises in the conditions for X to be an optimal solution, can be expressed as $\phi(X,S) = 0$, where $\phi(X,S) = X + S - (X^2 + S^2)^{1/2}$. Their algorithm involves line searches that involve repeated evaluation of $\phi(X,S)$ at arguments that may differ only slightly.

2.14.3. Matrix Sector Function

The matrix sector function, introduced by Shieh, Tsay, and Wang [519, 1984], is a generalization of the matrix sign function. For a given p, the matrix sector function

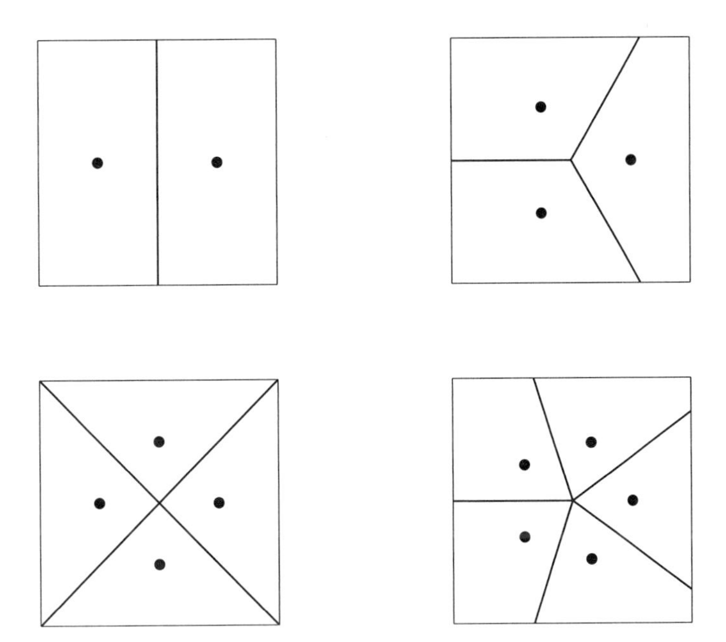

Figure 2.1. *The dots are the pth roots of unity and the lines the sector boundaries, illustrated for* $p = 2\colon 5$. *The scalar sector function* $\mathrm{sect}_p(z)$ *maps* $z \in \mathbb{C}$ *to the nearest pth root of unity.* $p = 2$ *gives the sign function.*

can be defined via the Jordan canonical form, in an analogous way as for the sign function in Section 2.4, but now mapping each eigenvalue to the nearest pth root of unity. For $p = 2$, the matrix sign function is obtained. More precisely, for $A \in \mathbb{C}^{n\times n}$ having no eigenvalues with argument $(2k+1)\pi/p$, $k = 0\colon p-1$, the matrix p-sector function can be defined by $\mathrm{sect}_p(A) = A(A^p)^{-1/p}$ (where the principal pth root is taken; see Theorem 7.2). Figure 2.1 illustrates the scalar sector function. The sector function has attracted interest in the control theory literature because it can be used to determine the number of eigenvalues in a specific sector and to obtain the corresponding invariant subspace [358, 1995]. However, a good numerical method for computing the matrix sector function is currently lacking.

2.14.4. Matrix Disk Function

Let $A \in \mathbb{C}^{n\times n}$ have Jordan canonical form $A = ZJZ^{-1}$ with $J = \mathrm{diag}(J_1, J_2)$, where the eigenvalues of J_1 are inside the unit disk and the eigenvalues of J_2 are outside the unit disk. The matrix disk function is defined by $\mathrm{disk}(A) = Z\,\mathrm{diag}(I, 0)Z^{-1}$; if A has an eigenvalue on the unit circle then $\mathrm{disk}(A)$ is undefined. An alternative representation is

$$\mathrm{disk}(A) = \frac{1}{2}\big(I - \mathrm{sign}\big((A-I)^{-1}(A+I)\big)\big).$$

The matrix disk function was introduced in the same paper by Roberts [496] that introduced the matrix sign function. It can be used to obtain invariant subspaces in an analogous way as for the matrix sign function. For more details, see Benner and Byers [53, 1997], Benner, Byers, Mehrmann, and Xu [54, 2000], Sun and Quintana-Ortí [551, 2004], and the references therein.

2.14.5. The Average Eye in Optics

In optics there is interest in defining an average of the optical characteristics of a set of eyes, or more generally a set of optical systems. The first-order character of an optical system is characterized by a transference matrix $T = \left[\begin{smallmatrix} S & \delta \\ 0 & 1 \end{smallmatrix}\right] \in \mathbb{R}^{5\times 5}$, where $S \in \mathbb{R}^{4\times 4}$ is symplectic, that is, $S^TJS = J$, where $J = \left[\begin{smallmatrix} 0 & I_2 \\ -I_2 & 0 \end{smallmatrix}\right]$. A straightforward average $m^{-1}\sum_{i=1}^m T_i$ of m transference matrices T_i is not in general a transference matrix. Harris [250, 2005] proposes as a suitable average the matrix $\exp(m^{-1}\sum_{i=1}^m \log(T_i))$, where the log is the principal logarithm (cf. the geometric mean (2.29)); that this average is a transference matrix follows from the facts that the principal logarithm of a symplectic matrix with no eigenvalues on $\mathbb{R}^-$ is Hamiltonian and the exponential of a Hamiltonian matrix is symplectic [152, 1996], [251, 2006]

2.14.6. Computer Graphics

In computer graphics a transformation can be represented as a matrix, usually of dimension 3 or 4, and multiplication by the matrix effects the transformation. Alexa [9, 2002] defines a "commutative addition" operation $A \oplus B = \exp(\log(A) + \log(B))$, using which he constructs linear combinations of transformations. If A and B commute this is just the usual matrix product $A \oplus B = AB = BA$. One use of the operation is to interpolate between two transformations A and B via $tA \oplus (1-t)B$, $t \in [0,1]$. Note that this is the same mathematical idea as in Section 2.14.5. The interpolation problem with A and B rotation matrices is discussed by Gallier and Xu [199, 2002].

2.14.7. Bregman Divergences

Matrix nearness problems ask for the distance from a given matrix to the nearest matrix with a certain property, and for that nearest matrix. The nearest unitary matrix problem mentioned in Section 2.6 is of this type. The use of a Bregman divergence in place of a matrix norm is proposed by Dhillon and Tropp [150, 2007]. The Bregman divergence of $X \in \mathbb{C}^{n\times n}$ from $Y \in \mathbb{C}^{n\times n}$ is defined by $D(X,Y) = \psi(X) - \psi(Y) - \langle \nabla\psi(Y), X - Y\rangle$, where $\psi : \mathbb{C}^{n\times n} \to \mathbb{R}^+$ is strictly convex and $\langle\cdot\rangle$ is an inner product. A particular instance applying to Hermitian positive definite matrices is the von Neumann divergence $D(X,Y) = \operatorname{trace}\big(X(\log(X) - \log(Y)) - X + Y\big)$, the use of which leads to the need to evaluate expressions such as $\exp(\log(Y) + W)$. Thus Bregman divergences provide another application of matrix functions.

2.14.8. Structured Matrix Interpolation

Numerical solution of a matrix differential equation $dY/dt = f(t,Y)$, $Y(0)$ given, leads to a sequence of pairs (t_i, Y_i), where $Y_i \approx Y(t_i)$. In order to approximate Y at non-mesh points t it is necessary to interpolate the data, which can be done by standard procedures. However, if $Y(t)$ has structure the interpolated matrices may not possess that structure. Dieci, Morini, Papini, and Pasquali [155, 1999] propose interpolating $(t_i, f(Y_i))$, to obtain the polynomial $P(t)$ and then taking $g(t) = f^{-1}(P(t))$ as the interpolation function. The idea is to choose f so that the required structure is enforced. In [155, 1999] the use of $f(Y) = \log(Y)$ is advocated for orthogonal, symplectic, and Hermitian positive definite matrices.

2.14.9. The Lambert W Function and Delay Differential Equations

The solutions of the equation $se^s = a$ are given by $s = W_k(a)$, where $W_k(z)$ is the kth branch of the Lambert W function. This function is a "log-like" function and arises in many different situations, ranging from the enumeration of trees in combinatorics to the explicit solution of various differential equations. The matrix Lambert W function can be defined via (1.2) or (1.4). One application of the matrix function is to delay differential equations. Consider the model problem

$$y'(t) = Ay(t-1), \quad t \geq 0, \qquad A \in \mathbb{C}^{n\times n},$$

where

$$y(t) = g(t), \quad -1 \leq t \leq 0,$$

for a given function g. If we look for solutions $y(t) = \exp(tS)c$ for some constant $c \in \mathbb{C}^n$ then we are led to the matrix equation $S\exp(S) = A$, and hence to $S = W_k(A)$. The general solution to the problem can then be expressed as $y(t) = \sum_{k=-\infty}^{\infty} e^{W_k(A)t}c_k$, where the vectors c_k are determined by g.

For more details of the scalar Lambert W function see Corless, Gonnet, Hare, Jeffrey, and Knuth [115, 1996]. The matrix Lambert W function is analyzed by Corless, Ding, Higham, and Jeffrey [114, 2007], who show that as a primary matrix function it does not yield all solutions of $S\exp(S) = A$, which is analogous to the fact that the primary matrix logarithm does not provide all solutions of $e^X = A$ (Theorem 1.28). For the application to delay differential equation see Jarlebring and Damm [320, 2007] and the references therein, and Heffernan and Corless [256, 2006]. A good reference on the numerical solution of delay differential equations is Bellen and Zennaro [50, 2003].

2.15. Notes and References

Good references for matrix differential equations are Bellman [51, 1970, Chap. 10], Gantmacher [203, 1959, Sec. 5.5], and Godunov [219, 1997]. The expression (2.5) involving ψ_1 can be found in Schwerdtfeger [513, 1938, p. 44].

Our description of NMR is based on Havel, Najfeld, and Yang [252, 1994], [445, 1995]; see also Levitt [383, 2001]. Our description of the need to compute roots of transition matrices is adapted from Guo and Higham [233, 2006]; see also Problem 7.15.

For more on continuous-time and discrete-time linear dynamical systems see, for example, Chen [106, 1999, Chap. 4] or Franklin, Powell, and Workman [192, 1998].

Further details on solution of Riccati equations via the matrix sign function can be found in Byers [88, 1987], Gardiner [204, 1997], Lancaster and Rodman [370, 1995, Chap. 22], and Laub [374, 1991]. See also Benner and Quintana-Ortí [55, 1999] for the generalized Lyapunov equation $A^*XE + E^*XA - G = 0$ and Benner, Quintana-Ortí, and Quintana-Ortí [56, 2006] for the Sylvester equation. With Newton's method for computing it, the matrix sign function is a versatile way of solving these equations; it can be adapted to compute low rank solutions efficiently and to exploit the hierarchical $\mathcal{H}$-matrix storage format. See Baur and Benner [47, 2006] and Grasedyck, Hackbusch, and Khoromskij [228, 2003].

Theorem 2.1 is based on a result of Lin and Zmijewski [392, 1991], incorporating refinements by Bai and Demmel [27, 1993].

The idea of using the matrix sign function to compute eigensystems in divide and conquer fashion was first investigated by Denman and Beavers in the early 1970s; see [146, 1976] and the references therein. However, in these early papers nonunitary transformations are used, in contrast to the approach in Section 2.5.

Early references for (2.19) and formulae of the form (2.20) are Frobenius [195, 1896], Baker [40, 1925], and Turnbull and Aitken [579, 1932, p. 152].

For more on geometric means of positive definite matrices see Bhatia [65, 2007, Chap. 4], Lawson and Lim [377, 2001], Ando, Li, and Mathias [15, 2004], and Moakher [435, 2005]. Proofs of (2.28) can be found in Bhatia [65, 2007, Thm. 4.1.3] and Ando [13, 1979, Thm. 2], [14, 1998, Thm. 2.8]. The geometric mean (2.26) appears to have been first introduced by Pusz and Woronowicz [482, 1975].

The factorization (2.24) and other properties of matrix polynomials are treated by Davis [140, 1981], Dennis, Traub, and Weber [147, 1976], Gohberg, Lancaster, and Rodman [220, 1982], Lancaster [369, 1966], and Lancaster and Tismenetsky [371, 1985].

Problems

Though mathematics is much easier to watch than do,
it is a most unrewarding spectator sport.
— CHARLES G. CULLEN, *Matrices and Linear Transformations* (1972)

2.1. Derive the formula (2.3) for the solution of problem (2.2).

2.2. Reconcile the fact that the initial value problem (2.7) has a unique solution with the observation that $\cos(\sqrt{A}t)y_0 + (\sqrt{A})^{-1}\sin(\sqrt{A}t)y_0'$ is a solution for *any* square root $\sqrt{A}$ of A.

2.3. Prove Theorem 2.1.

2.4. Show that if the Hermitian positive definite matrices A and B commute then the geometric means $\#$ in (2.26) and E in (2.29) are given by $A \# B = A^{1/2}B^{1/2} = E(A,B)$.

2.5. Prove the relations (2.27) satisfied by the geometric mean $\#$.

2.6. (Bhatia [65, 2007, p. 111]) Show that for Hermitian positive definite $A, B \in \mathbb{C}^{2\times 2}$,

$$A \# B = \frac{\sqrt{\alpha\beta}}{\sqrt{\det(\alpha^{-1}A + \beta^{-1}B)}}(\alpha^{-1}A + \beta^{-1}B),$$

with $\alpha^2 = \det(A)$, $\beta^2 = \det(B)$.

2.7. Consider the Riccati equation $XAX = B$, where A and B are Hermitian positive definite. Show that the Hermitian positive definite solution X can be computed as $R^{-1}(RBR^*)^{1/2}R^{-*}$, where $A = R^*R$ is a Cholesky factorization.

Whenever there is too much talk of applications,
one can rest assured that the theory has very few of them.

— GIAN-CARLO ROTA, *Indiscrete Thoughts* (1997)

The 1930's . . . saw the diversification of aeroelastic studies. . .
and the first practical use of the hitherto
recondite branch of pure mathematics known as "matrices".

— A. R. COLLAR, *The First Fifty Years of Aeroelasticity* (1978)

The main computational challenge in the
implementation of any exponential integrator is the need for
fast and computationally stable evaluations
of the exponential and the related ψ-functions.

— BORISLAV V. MINCHEV and WILL M. WRIGHT, *A Review of Exponential Integrators for First Order Semi-linear Problems* (2005)

Prior to about 1978, solving an algebraic Riccati equation (ARE)
was something that was most definitely to be avoided,
and with good reason.
Existing techniques were often quite unreliable.

— ALAN J. LAUB, *Invariant Subspace Methods for the Numerical Solution of Riccati Equations* (1991)

This work is intended for graphics
where standard matrix packages only offer elementary matrix operations.
For this reason, implementations are provided using only
matrix inversion, multiplication, and addition.

— MARC ALEXA, *Linear Combination of Transformations* (2002)

Since W is such a simple function,
we would expect by Pareto's principle
(eighty percent of your work is accomplished with twenty percent of your tools)
that W would have many applications.
In fact this is the case, although the presence of W often goes unrecognized.

— ROBERT M. CORLESS, GASTON H. GONNET, D. E. G. HARE, DAVID J. JEFFREY, and DONALD E. KNUTH, *On the Lambert W Function* (1996)

Chapter 3
Conditioning

In practice most data are inexact or uncertain. Computations with exact data are subject to rounding errors, and the rounding errors in an algorithm can often be interpreted as being equivalent to perturbations in the data, through the process of backward error analysis. Therefore whether the data are exact or inexact, it is important to understand the sensitivity of matrix functions to perturbations in the data. Sensitivity is measured by condition numbers. This chapter is concerned with defining appropriate condition numbers and showing how to evaluate or estimate them efficiently. The condition numbers can be expressed in terms of the norm of the Fréchet derivative, so we investigate in some detail the properties of the Fréchet derivative.

3.1. Condition Numbers

It is useful to begin by recalling how condition numbers are defined for scalar functions, $f(x)$. The standard definition of (relative) condition number is

$$\operatorname{cond}_{\mathrm{rel}}(f,x) := \lim_{\epsilon\to 0} \sup_{|\Delta x|\le\epsilon|x|} \left| \frac{f(x+\Delta x)-f(x)}{\epsilon f(x)} \right|,$$

which measures by how much, at most, small changes in the data can be magnified in the function value, when both changes are measured in a relative sense. Assuming for simplicity that f is twice continuously differentiable, $f(x+\Delta x)-f(x) = f'(x)\Delta x + o(\Delta x)$, which can be rewritten as

$$\frac{f(x+\Delta x)-f(x)}{f(x)} = \left(\frac{xf'(x)}{f(x)}\right)\frac{\Delta x}{x} + o(\Delta x).$$

(Recall that $h = o(\epsilon)$ means that $\|h\|/\epsilon \to 0$ as $\epsilon \to 0$.) It is then immediate that

$$\operatorname{cond}_{\mathrm{rel}}(f,x) = \left| \frac{xf'(x)}{f(x)} \right|. \tag{3.1}$$

The definition of condition number extends readily to matrix functions $f : \mathbb{C}^{n\times n} \to \mathbb{C}^{n\times n}$. We define the relative condition number by[4]

$$\operatorname{cond}_{\mathrm{rel}}(f,X) := \lim_{\epsilon\to 0} \sup_{\|E\|\le\epsilon\|X\|} \frac{\|f(X+E)-f(X)\|}{\epsilon\|f(X)\|}, \tag{3.2}$$

[4]The definition is applicable to arbitrary functions f. Most of the results of Section 3.2 onwards assume f is a primary matrix function as defined in Chapter 1. We will use the condition number (3.2) for a more general f in Section 8.2.

where the norm is any matrix norm. This definition implies that

$$\frac{\|f(X+E)-f(X)\|}{\|f(X)\|} \le \mathrm{cond}_{\mathrm{rel}}(f,X)\frac{\|E\|}{\|X\|} + o(\|E\|), \tag{3.3}$$

and so provides an approximate perturbation bound for small perturbations E.

Some care is needed in interpreting (3.3) for functions not defined throughout $\mathbb{C}^{n\times n}$. The definition (3.2) is clearly valid as long as f is defined in a neighbourhood of X. The bound (3.3) is therefore valid for $X+E$ in that neighbourhood. An example is given in Section 5.1 that shows how blindly invoking (3.3) can lead to a patently incorrect bound.

A corresponding absolute condition number, in which the change in the data and the function are measured in an absolute sense, is defined by

$$\mathrm{cond}_{\mathrm{abs}}(f,X) := \lim_{\epsilon\to 0} \sup_{\|E\|\le\epsilon} \frac{\|f(X+E)-f(X)\|}{\epsilon}. \tag{3.4}$$

Note that

$$\mathrm{cond}_{\mathrm{rel}}(f,X) = \mathrm{cond}_{\mathrm{abs}}(f,X)\frac{\|X\|}{\|f(X)\|}, \tag{3.5}$$

so the two condition numbers differ by just a constant factor. Usually, it is the relative condition number that is of interest, but it is more convenient to state results for the absolute condition number.

To obtain explicit expressions analogous to (3.1) we need an appropriate notion of derivative for matrix functions. The *Fréchet derivative* of a matrix function $f : \mathbb{C}^{n\times n} \to \mathbb{C}^{n\times n}$ at a point $X \in \mathbb{C}^{n\times n}$ is a linear mapping

$$\begin{array}{ccc} \mathbb{C}^{n\times n} & \stackrel{L}{\longrightarrow} & \mathbb{C}^{n\times n} \\ E & \longmapsto & L(X,E) \end{array}$$

such that for all $E \in \mathbb{C}^{n\times n}$

$$f(X+E) - f(X) - L(X,E) = o(\|E\|). \tag{3.6}$$

The Fréchet derivative may not exist, but if it does it is unique (see Problem 3.3). The notation $L(X,E)$ can be read as "the Fréchet derivative of f at X in the direction E", or "the Fréchet derivative of f at X applied to the matrix E". If we need to show the dependence on f we will write $L_f(X,E)$. When we want to refer to the mapping at X and not its value in a particular direction we will write $L(X)$. In the case $n=1$ we have, trivially, $L(x,e) = f'(x)e$, and more generally if X and E commute then $L(X,E) = f'(X)E = Ef'(X)$ (see Problem 3.8).

The absolute and relative condition numbers can be expressed in terms of the norm of $L(X)$, which is defined by

$$\|L(X)\| := \max_{Z\ne 0} \frac{\|L(X,Z)\|}{\|Z\|}. \tag{3.7}$$

Theorem 3.1 (Rice). *The absolute and relative condition numbers are given by*

$$\mathrm{cond}_{\mathrm{abs}}(f,X) = \|L(X)\|, \tag{3.8}$$

$$\mathrm{cond}_{\mathrm{rel}}(f,X) = \frac{\|L(X)\|\,\|X\|}{\|f(X)\|}. \tag{3.9}$$

Proof. In view of (3.5), it suffices to prove (3.8). From (3.4) and (3.6), and using the linearity of L, we have

$$\begin{aligned}
\operatorname{cond}_{\mathrm{abs}}(f, X) &= \lim_{\epsilon\to 0} \sup_{\|E\|\le\epsilon} \frac{\|f(X+E) - f(X)\|}{\epsilon} \\
&= \lim_{\epsilon\to 0} \sup_{\|E\|\le\epsilon} \left\| \frac{L(X,E) + o(\|E\|)}{\epsilon} \right\| \\
&= \lim_{\epsilon\to 0} \sup_{\|E\|\le\epsilon} \big\| L(X, E/\epsilon) + o(\|E\|)/\epsilon \big\| \\
&= \sup_{\|Z\|\le 1} \|L(X,Z)\|.
\end{aligned}$$

Finally, the sup can be replaced by a max, since we are working on a finite dimensional vector space, and the maximum is attained with $\|Z\| = 1$. □

To illustrate, for $f(X) = X^2$ we have $f(X+E) - f(X) = XE + EX + E^2$, so

$$L_{x^2}(X, E) = XE + EX \tag{3.10}$$

and $\|L_{x^2}(X)\| \le 2\|X\|$.

It is usually not straightforward to obtain an explicit formula or representation for the Fréchet derivative, not least because matrix multiplication is not commutative, but in later chapters we will see how this can be done for certain functions.

3.2. Properties of the Fréchet Derivative

The condition number of f is essentially the norm of the Fréchet derivative (see Theorem 3.1). In this section we describe some properties of the Fréchet derivative that will be useful in bounding or estimating the condition number.

Related to the Fréchet derivative is the *directional* or *Gâteaux* derivative

$$G(X,E) = \lim_{t\to 0} \frac{f(X+tE) - f(X)}{t} = \frac{d}{dt}\Big|_{t=0} f(X+tE). \tag{3.11}$$

If the Fréchet derivative exists at X then it is equal to the Gâteaux derivative (see Problem 3.4). But the converse is not true: the existence of directional derivatives in all directions is a weaker notion of differentiability than the existence of the Fréchet derivative. However, if the Gâteaux derivative exists, is a linear function of E, and is continuous in X, then it is also the Fréchet derivative. (See the works cited in the Notes and References.) Depending on the context, (3.6) or (3.11) may be the more useful expression to work with.

We begin with four rules that show how to obtain the Fréchet derivatives of a sum or product and of composite and inverse functions. In the following four results $f, g, h : \mathbb{C}^{n\times n} \to \mathbb{C}^{n\times n}$.

Theorem 3.2 (sum rule). *If g and h are Fréchet differentiable at A then so is $f = \alpha g + \beta h$ and $L_f(A,E) = \alpha L_g(A,E) + \beta L_h(A,E)$.*

Proof. The proof is immediate from the definition of the Fréchet derivative. □

Theorem 3.3 (product rule). *If g and h are Fréchet differentiable at A then so is $f = gh$ and $L_f(A,E) = L_g(A,E)h(A) + g(A)L_h(A,E)$.*

Proof. We have

$$\begin{aligned} f(A+E) &= g(A+E)h(A+E) \\ &= \big(g(A) + L_g(A,E) + o(\|E\|)\big)\big(h(A) + L_h(A,E) + o(\|E\|)\big) \\ &= f(A) + L_g(A,E)h(A) + g(A)L_h(A,E) + o(\|E\|). \qquad \square \end{aligned}$$

Theorem 3.4 (chain rule). *Let h and g be Fréchet differentiable at A and $h(A)$, respectively, and let $f = g \circ h$ (i.e., $f(A) = g(h(A))$). Then f is Fréchet differentiable at A and $L_f = L_g \circ L_h$, that is, $L_f(A,E) = L_g\big(h(A), L_h(A,E)\big)$.*

Proof.

$$\begin{aligned} f(A+E) - f(A) &= g(h(A+E)) - g(h(A)) \\ &= g\big(h(A) + L_h(A,E) + o(\|E\|)\big) - g\big(h(A)\big) \\ &= g\big(h(A)\big) + L_g\big(h(A), L_h(A,E) + o(\|E\|)\big) + o(\|E\|) - g\big(h(A)\big) \\ &= L_g\big(h(A), L_h(A,E)\big) + o(\|E\|). \qquad \square \end{aligned}$$

The following theorem says that the Fréchet derivative of the inverse function is the inverse of the Fréchet derivative of the function.

Theorem 3.5 (derivative of inverse function). *Let f and f^{-1} both exist and be continuous in an open neighbourhood of X and $f(X)$, respectively, and assume L_f exists and is nonsingular at X. Then $L_{f^{-1}}$ exists at $Y = f(X)$ and $L_{f^{-1}}(Y,E) = L_f^{-1}(X,E)$, or equivalently, $L_f\left(X, L_{f^{-1}}(Y,E)\right) = E$. Hence $\|L_{f^{-1}}(Y)\| = \|L_f^{-1}(X)\|$.*

Proof. For the existence see Dieudonné [159, 1960, Thm. 8.2.3]. The formulae can be obtained by applying the chain rule to the relation $f(f^{-1}(Y)) = Y$, which gives the equality $L_f\big(X, L_{f^{-1}}(Y,E)\big) = E$. $\square$

To illustrate Theorem 3.5 we take $f(x) = x^2$ and $f^{-1}(x) = x^{1/2}$. The theorem says that $L_{x^2}\left(X, L_{x^{1/2}}(X^2,E)\right) = E$, i.e., using (3.10), $XL_{x^{1/2}}(X^2,E) + L_{x^{1/2}}(X^2,E)X = E$. In other words, $L = L_{x^{1/2}}(A,E)$ is the solution of the Sylvester equation $A^{1/2}L + LA^{1/2} = E$.

The following theorem will also be very useful. For the rest of this section $\mathcal{D}$ denotes an open subset of $\mathbb{R}$ or $\mathbb{C}$.

Theorem 3.6 (Mathias). *Let f be $2n-1$ times continuously differentiable on $\mathcal{D}$. Let $A(t) \in \mathbb{C}^{n\times n}$ be differentiable at $t=0$ and assume that the spectrum of $A(t)$ is contained in $\mathcal{D}$ for all t in some neighbourhood of 0. Then, with $A = A(0)$,*

$$f\left(\begin{bmatrix} A & A'(0) \\ 0 & A \end{bmatrix}\right) = \begin{bmatrix} f(A) & \frac{d}{dt}\Big|_{t=0} f(A(t)) \\ 0 & f(A) \end{bmatrix}. \tag{3.12}$$

Proof. Let $\epsilon \neq 0$ and define

$$U = \begin{bmatrix} I & \epsilon^{-1}I \\ 0 & I \end{bmatrix}.$$

Then

$$f\left(\begin{bmatrix} A(0) & \frac{A(\epsilon)-A(0)}{\epsilon} \\ 0 & A(\epsilon) \end{bmatrix}\right) = Uf\left(U^{-1}\begin{bmatrix} A(0) & \frac{A(\epsilon)-A(0)}{\epsilon} \\ 0 & A(\epsilon) \end{bmatrix}U\right)U^{-1}$$

$$= Uf\left(\begin{bmatrix} A(0) & 0 \\ 0 & A(\epsilon) \end{bmatrix}\right)U^{-1}$$

$$= U\begin{bmatrix} f(A(0)) & 0 \\ 0 & f(A(\epsilon)) \end{bmatrix}U^{-1}$$

$$= \begin{bmatrix} f(A) & \frac{f(A(\epsilon))-f(A(0))}{\epsilon} \\ 0 & f(A(\epsilon)) \end{bmatrix}.$$

Since f is $2n-1$ times continuously differentiable the $2n \times 2n$ matrix on the left-hand side is defined for small enough ϵ, and since f is a continuous matrix function on the $2n \times 2n$ matrices with spectrum in $\mathcal{D}$ by Theorem 1.19, the limit as $\epsilon \to 0$ exists and is

$$f\left(\begin{bmatrix} A(0) & A'(0) \\ 0 & A(0) \end{bmatrix}\right).$$

The result now follows by taking the limit of the matrix on the right. □

The assumption that f is $2n-1$ times continuously differentiable is needed in Theorem 3.6 to cater for the "worst case" Jordan structure of the argument of f in (3.12); see Problem 3.5.

If A is Hermitian, or more generally normal, then the differentiability condition on f in all the results of this section can be relaxed to f being just once continuously differentiable, as a result of the fact that all Jordan blocks of A are 1×1.

We actually need only a special case of (3.12). Letting $A(t) = A + tE$, we have from (3.12),

$$f\left(\begin{bmatrix} A & E \\ 0 & A \end{bmatrix}\right) = \begin{bmatrix} f(A) & \frac{d}{dt}\Big|_{t=0} f(A+tE) \\ 0 & f(A) \end{bmatrix}. \tag{3.13}$$

The next result is useful because it implies that the Gâteaux derivative of a general function f at a particular point agrees with the Gâteaux derivative of a certain polynomial at that point, and polynomials are easier to work with.

Theorem 3.7 (Horn and Johnson). *Let f be $2n-1$ times continuously differentiable on $\mathcal{D}$. Let $A(t) \in \mathbb{C}^{n\times n}$ be differentiable at $t = 0$ and assume that the spectrum of $A(t)$ is contained in $\mathcal{D}$ for all t in some neighbourhood of 0. Then, with $A = A(0)$,*

$$\frac{d}{dt}\Big|_{t=0} f(A(t)) = \frac{d}{dt}\Big|_{t=0} p_{A\oplus A}(A(t)), \tag{3.14}$$

where $p_{A\oplus A}$ interpolates f and its derivatives at the zeros of the characteristic polynomial of $A \oplus A \equiv \operatorname{diag}(A, A)$, that is,

$$p_{A\oplus A}^{(j)}(\lambda_i) = f^{(j)}(\lambda_i), \qquad j = 0\colon 2r_i - 1, \quad i = 1\colon s, \tag{3.15}$$

where $\lambda_1, \ldots, \lambda_s$ are the distinct eigenvalues of A, with algebraic multiplicities $r_1, \ldots, r_s$, respectively.

Proof. Define

$$B = \begin{bmatrix} A & A'(0) \\ 0 & A \end{bmatrix}.$$

Theorem 3.6 shows that

$$\frac{d}{dt}\bigg|_{t=0} f(A(t)) = [f(B)]_{12} = [p(B)]_{12} = \frac{d}{dt}\bigg|_{t=0} p(A(t)),$$

where the second equality holds for any polynomial p that takes the same values as f on the spectrum of B. By its definition (3.15), $p_{A\oplus A}$ is such as polynomial. (Note that $p_{A\oplus A}$ may satisfy more interpolation conditions than are required in order to take the same values as f on the spectrum of B. The polynomial $p_{A\oplus A}$ is essentially an "overestimate" that has the required properties and can be defined without knowledge of the Jordan structure of B; see Remark 1.5.) □

With the aid of the previous two results we can now identify sufficient conditions on f for the Fréchet derivative to exist and be continuous.

Theorem 3.8 (existence and continuity of Fréchet derivative). *Let f be $2n-1$ times continuously differentiable on $\mathcal{D}$. For $X \in \mathbb{C}^{n\times n}$ with spectrum in $\mathcal{D}$ the Fréchet derivative $L(X,E)$ exists and is continuous in the variables X and E.*

Proof. Since f has $2n-1$ continuous derivatives, we know from Theorem 1.19 that f is a continuous matrix function on the set of $2n\times 2n$ matrices with spectrum in $\mathcal{D}$. Also, the map from a $2n\times 2n$ matrix to its $(1\colon n, n+1\colon 2n)$ submatrix is continuous. Therefore since (3.13) shows the Gâteaux derivative $G(X,E)$ to be the composition of the two maps just described, it is continuous in X and E. Moreover, $G(X,E)$ is a linear function of E. This follows from (3.13) when f is a polynomial, and then in general from Theorem 3.7 with $A(t) = X+tE$, which shows that f and the polynomial $p_{X\oplus X}$ have the same Gâteaux derivative at X. But the linearity in E and continuity in X of G imply that G is the Fréchet derivative, L. □

Under the conditions of Theorem 3.8 we can rewrite (3.13) as

$$f\left(\begin{bmatrix} X & E \\ 0 & X \end{bmatrix}\right) = \begin{bmatrix} f(X) & L(X,E) \\ 0 & f(X) \end{bmatrix}. \tag{3.16}$$

The significance of this formula is that it converts the problem of evaluating the Fréchet derivative in a particular direction to that of computing a single matrix function—albeit for a matrix of twice the dimension. This is useful both in theory and in practice.

We now find the eigenvalues of the Fréchet derivative. An eigenpair (λ, V) of $L(X)$ comprises a scalar λ, the eigenvalue, and a nonzero matrix $V \in \mathbb{C}^{n\times n}$, the eigenvector, such that $L(X,V) = \lambda V$.

Since L is a linear operator

$$\mathrm{vec}(L(X,E)) = K(X)\,\mathrm{vec}(E) \tag{3.17}$$

for some $K(X) \in \mathbb{C}^{n^2\times n^2}$ that is independent of E. We refer to $K(X)$ as the Kronecker form of the Fréchet derivative. If (λ, V) is an eigenpair of $L(X)$ then $K(X)v = \lambda v$, where $v = \mathrm{vec}(V)$, so (λ, v) is an eigenpair of $K(X)$ in the usual matrix sense. The following lemma identifies eigenpairs of $L(X)$.

Recall that the divided difference $f[\lambda,\mu]$ is defined by

$$f[\lambda,\mu] = \begin{cases} \dfrac{f(\lambda)-f(\mu)}{\lambda-\mu}, & \lambda \neq \mu, \\ f'(\lambda), & \lambda = \mu. \end{cases}$$

Theorem 3.9 (eigenvalues of Fréchet derivative). *Let f be $2n-1$ times continuously differentiable on $\mathcal{D}$ and let $X \in \mathbb{C}^{n\times n}$ have spectrum in $\mathcal{D}$. The eigenvalues of the Fréchet derivative L of f at X are*

$$f[\lambda_i,\lambda_j], \qquad i,j = 1\colon n,$$

where the λ_i are the eigenvalues of X. If u_i and v_j are nonzero vectors such that $Xu_i = \lambda_i u_i$ and $v_j^T X = \lambda_j v_j^T$, then $u_i v_j^T$ is an eigenvector of $L(X)$ corresponding to $f[\lambda_i,\lambda_j]$.

Proof. Suppose, first, that f is a polynomial: $f(t) = \sum_{k=0}^m a_k t^k$. Then (see the more general Problem 3.6)

$$L(X,E) = \sum_{k=1}^m a_k \sum_{j=1}^k X^{j-1} E X^{k-j}.$$

Hence, using (B.16),

$$K(X) = \sum_{k=1}^m a_k \sum_{j=1}^k (X^{k-j})^T \otimes X^{j-1}.$$

From (B.17) the eigenvalues of $K(X)$ are, for $p,q = 1\colon n$,

$$\begin{aligned} \sum_{k=1}^m a_k \sum_{j=1}^k \lambda_p^{k-j}\lambda_q^{j-1} &= \begin{cases} \displaystyle\sum_{k=1}^m a_k \frac{\lambda_p^k - \lambda_q^k}{\lambda_p - \lambda_q}, & \lambda_p \neq \lambda_q, \\ \displaystyle\sum_{k=1}^m a_k k \lambda_p^{k-1}, & \lambda_p = \lambda_q, \end{cases} \\ &= \begin{cases} \dfrac{f(\lambda_p)-f(\lambda_q)}{\lambda_p-\lambda_q}, & \lambda_p \neq \lambda_q, \\ f'(\lambda_p), & \lambda_p = \lambda_q, \end{cases} \\ &= f[\lambda_p,\lambda_q]. \end{aligned}$$

It is easy to show that $v_j \otimes u_i$ is an eigenvector of $K(X)$ corresponding to the eigenvalue $f[\lambda_i,\lambda_j]$ and that this corresponds to an eigenvector $u_i v_j^T$ of $L(X)$.

Now consider a general function f. Theorem 3.7 implies that the Fréchet derivative L of f at X is the same as that of the polynomial $p_{X\oplus X}$. We can therefore use the first part to deduce that the eigenvalues of $L(X)$ are the numbers $p_{X\oplus X}[\lambda_i,\lambda_j]$, $i,j = 1\colon n$. But by definition, $p_{X\oplus X}$ has the same values as f on the spectrum of X, so $p_{X\oplus X}[\lambda_i,\lambda_j] = f[\lambda_i,\lambda_j]$ for all i and j. The form of the eigenvectors follows from the first part. □

Problem 3.10 shows how to identify eigenpairs of $L(X)$ without employing the Kronecker form. However, with that approach it is difficult to show that all the eigenvalues of $L(X)$ have been accounted for. Note that Theorem 3.9 does not necessarily identify all the eigenvectors of $L(X)$; see Problem 3.11.

Theorem 3.9 enables us to deduce when the Fréchet derivative is nonsingular.

Corollary 3.10. *Let f be $2n-1$ times continuously differentiable on $\mathcal{D}$. The Fréchet derivative L of f at a matrix $X \in \mathbb{C}^{n\times n}$ with eigenvalues $\lambda_i \in \mathcal{D}$ is nonsingular when $f'(\lambda_i) \neq 0$ for all i and $f(\lambda_i) = f(\lambda_j) \Rightarrow \lambda_i = \lambda_j$.*

The next result shows that the Fréchet derivative in any direction at a diagonal matrix is formed simply by Hadamard multiplication by the matrix of divided differences of the eigenvalues. Here, $\circ$ denotes the Hadamard (or Schur) product of $A, B \in \mathbb{C}^{n\times n}$: $A \circ B = (a_{ij}b_{ij})$.

Theorem 3.11 (Daleckiĭ and Kreĭn). *Let f be $2n-1$ times continuously differentiable on $\mathcal{D}$. Let $D = \operatorname{diag}(\lambda_i) \in \mathbb{C}^{n\times n}$ be diagonal and $\lambda_i \in \mathcal{D}$ for all i. Then*

$$L(D,E) = (f[\lambda_i,\lambda_j]e_{ij}) = (f[\lambda_i,\lambda_j]) \circ E. \tag{3.18}$$

Proof. From (3.16) we have

$$f\left(\begin{bmatrix} D & E \\ 0 & D \end{bmatrix}\right) = \begin{bmatrix} f(D) & L(D,E) \\ 0 & f(D) \end{bmatrix},$$

and using the fact that $f(A)$ commutes with A (Theorem 1.13 (a)) we obtain

$$\begin{bmatrix} f(D) & L(D,E) \\ 0 & f(D) \end{bmatrix}\begin{bmatrix} D & E \\ 0 & D \end{bmatrix} = \begin{bmatrix} D & E \\ 0 & D \end{bmatrix}\begin{bmatrix} f(D) & L(D,E) \\ 0 & f(D) \end{bmatrix}.$$

Equating the (1,2) blocks of this equation gives

$$f(D)E - Ef(D) = DL(D,E) - L(D,E)D,$$

or, since $D = \operatorname{diag}(\lambda_i)$,

$$(f(\lambda_i) - f(\lambda_j))e_{ij} = (\lambda_i - \lambda_j)(L(D,E))_{ij}, \qquad i,j = 1\colon n.$$

If the λ_i are distinct then the latter equation immediately gives (3.18). If the λ_i are not distinct then consider $D + \operatorname{diag}(1,2,\dots,n)\epsilon$, which has distinct eigenvalues for ϵ sufficiently small and positive, and for which (3.18) therefore holds. Letting $\epsilon \to 0$, since $L(D,E)$ is continuous in D and E by Theorem 3.8, (3.18) holds for D by continuity. $\square$

Corollary 3.12. *Let f be $2n-1$ times continuously differentiable on $\mathcal{D}$ and let $X \in \mathbb{C}^{n\times n}$ have spectrum in $\mathcal{D}$ and be diagonalizable: $X = ZDZ^{-1}$, $D = \operatorname{diag}(\lambda_i)$. Then*

$$L(X,E) = ZL(D,Z^{-1}EZ)Z^{-1} = Z\big[(f[\lambda_i,\lambda_j]) \circ Z^{-1}EZ\big]Z^{-1}. \qquad \square$$

The final result is an analogue for the Fréchet derivative of the fact that the 2-norm of a diagonal matrix equals its spectral radius.

Corollary 3.13. *Under the conditions of Theorem* 3.11, $\|L(D)\|_F = \max_{i,j} |f[\lambda_i,\lambda_j]|$.

Proof. Using (3.7) and Theorem 3.11 we have

$$\begin{aligned}\|L(D)\|_F &= \max_{Z\neq 0}\frac{\|L(D,Z)\|_F}{\|Z\|_F}\\ &= \max_{Z\neq 0}\frac{\|(f[\lambda_i,\lambda_j])\circ Z\|_F}{\|Z\|_F}\\ &= \max_{\mathrm{vec}(Z)\neq 0}\frac{\|\operatorname{diag}(f[\lambda_i,\lambda_j])\operatorname{vec}(Z)\|_2}{\|\operatorname{vec}(Z)\|_2}\\ &= \max_{i,j}|f[\lambda_i,\lambda_j]|. \qquad \square\end{aligned}$$

3.3. Bounding the Condition Number

Our aim in this section is to develop bounds for the condition number. We assume throughout that f is $2n-1$ times continuously differentiable on an open subset of $\mathbb{R}$ or $\mathbb{C}$, which by Theorem 3.8 implies that the Fréchet derivative exists.

Let λ and E be an eigenvalue and corresponding eigenvector of $L(X)$. Then $L(X,E)=\lambda E$ and hence, from (3.7),

$$\|L(X)\| \geq |\lambda|. \tag{3.19}$$

Note that this is essentially the standard result (B.8) that no eigenvalue of a matrix can exceed any norm of the matrix.

Theorem 3.14. *For any norm,*

$$\mathrm{cond_{abs}}(f,X) \geq \max_{\lambda,\mu\in\Lambda(X)}|f[\lambda,\mu]|.$$

Proof. From (3.8), $\mathrm{cond_{abs}}(f,X)=\|L(X)\|$. The bound of the theorem is obtained by maximizing over all the eigenvalues, using (3.19) and Theorem 3.9. $\square$

By specializing to the Frobenius norm we can obtain an upper bound for the condition number. Here we need the matrix condition number with respect to inversion, $\kappa(Z)=\|Z\|\|Z^{-1}\|$ for $Z\in\mathbb{C}^{n\times n}$.

Theorem 3.15. *Let $X\in\mathbb{C}^{n\times n}$ be diagonalizable: $X=ZDZ^{-1}$, $D=\operatorname{diag}(\lambda_i)$. Then, for the Frobenius norm,*

$$\mathrm{cond_{abs}}(f,X) \leq \kappa_2(Z)^2\max_{\lambda,\mu\in\Lambda(X)}|f[\lambda,\mu]|.$$

Proof. By Corollary 3.12 we have $L(X,E)=ZL(D,\widetilde{E})Z^{-1}$, where $\widetilde{E}=Z^{-1}EZ$. Hence, using (B.7),

$$\|L(X,E)\|_F \leq \kappa_2(Z)\|L(D,\widetilde{E})\|_F \leq \kappa_2(Z)\|L(D)\|_F\|\widetilde{E}\|_F \leq \kappa_2(Z)^2\|L(D)\|_F\|E\|_F.$$

Now D is diagonal, so by Corollary 3.13 $\|L(D)\|_F=\max_{\lambda,\mu\in\Lambda(D)}|f[\lambda,\mu]|$. $\square$

Corollary 3.16. *Let $X\in\mathbb{C}^{n\times n}$ be normal. Then, for the Frobenius norm,*

$$\mathrm{cond_{abs}}(f,X) = \max_{\lambda,\mu\in\Lambda(X)}|f[\lambda,\mu]|.$$

Proof. A normal matrix is diagonalizable by a unitary similarity. Hence we can take $\kappa_2(Z) = 1$ in Theorem 3.15, and Theorem 3.14 then shows that the upper and lower bounds are equalities. □

The theorems show that for diagonalizable X, $\mathrm{cond}(f, X)$ is governed principally by two factors: the maximum first order divided difference on the eigenvalues and (possibly, since it appears only in an upper bound) the nonnormality of X, as measured by the minimum of $\kappa_2(Z)$ over all diagonalizing transformations Z. The divided difference terms may be innocuous, such as for $f(x) = x^2$, for which $f[\lambda, \mu] = \lambda + \mu$, or potentially large, such as for $f(x) = x^{1/2}$, for which $f[\lambda, \mu] = (\sqrt{\lambda} + \sqrt{\mu})^{-1}$. The nonnormality term can of course be arbitrarily large.

3.4. Computing or Estimating the Condition Number

The essential problem in computing or estimating the absolute or relative condition number of f at X is to compute or estimate $\|L(X)\|$. For the Frobenius norm the link (3.17) between the operator $L(X)$ and the $n^2 \times n^2$ Kronecker matrix $K(X)$ yields

$$\begin{aligned}\|L(X)\|_F &= \max_{E\neq 0} \frac{\|L(X,E)\|_F}{\|E\|_F} = \max_{E\neq 0} \frac{\|\operatorname{vec}(L(X,E))\|_2}{\|\operatorname{vec}(E)\|_2} \\ &= \max_{E\neq 0} \frac{\|K(X)\operatorname{vec}(E)\|_2}{\|\operatorname{vec}(E)\|_2} \\ &= \|K(X)\|_2 = \|K(X)^*K(X)\|_2^{1/2} = \lambda_{\max}\big(K(X)^*K(X)\big)^{1/2}. \qquad (3.20)\end{aligned}$$

Given the ability to compute $L(X, E)$ we can therefore compute the condition number exactly in the Frobenius norm by explicitly forming $K(X)$.

Algorithm 3.17 (exact condition number). Given $X \in \mathbb{C}^{n\times n}$ and a function f and its Fréchet derivative this algorithm computes $\mathrm{cond}_{\mathrm{rel}}(f, X)$ in the Frobenius norm.

```
1 for j = 1:n
2     for i = 1:n
3         Compute Y = L(X, e_i e_j^T).
4         K(:, (j − 1)n + i) = vec(Y)
5     end
6 end
7 cond_rel(f, X) = ‖K‖_2 ‖X‖_F / ‖f(X)‖_F
```

Cost: $O(n^5)$ flops, assuming $f(X)$ and $L(X, E)$ cost $O(n^3)$ flops.

For large n, Algorithm 3.17 is prohibitively expensive and so the condition number must be estimated rather than computed exactly. In practice, what is needed is an estimate that is of the correct order of magnitude—more than one correct significant digit is not needed.

No analogous equalities to (3.20) hold for the 1-norm, but we can bound the ratio of the 1-norms of $L(X)$ and $K(X)$.

Lemma 3.18. *For $X \in \mathbb{C}^{n\times n}$ and any function f,*

$$\frac{\|L(X)\|_1}{n} \le \|K(X)\|_1 \le n\|L(X)\|_1.$$

Proof. For $E \in \mathbb{C}^{n\times n}$ we have $\|E\|_1 \le \|\operatorname{vec}(E)\|_1 \le n\|E\|_1$ (with equality on the left for $E = ee_1^T$ and on the right for $E = ee^T$). Hence, using (3.17),

$$\frac{1}{n}\frac{\|L(X,E)\|_1}{\|E\|_1} \le \frac{\|K(X)\operatorname{vec}(E)\|_1}{\|\operatorname{vec}(E)\|_1} \le n\frac{\|L(X,E)\|_1}{\|E\|_1}.$$

Maximizing over all E gives the result. □

One of the uses of the condition number is to estimate the error in a computed result produced by a backward stable method. Since rounding error bounds invariably contain pessimistic constants that are quadratic or cubic in n, the agreement between $\|L(X)\|_1$ and $\|K(X)\|_1$ to within a factor n is sufficient for it to be reasonable to use the latter quantity to estimate the former.

In considering how to estimate $\|K(X)\|$ we treat first the Frobenius norm. In view of (3.20) the power method can be applied. We first state the power method for estimating the 2-norm of a general matrix.

Algorithm 3.19 (power method). Given $A \in \mathbb{C}^{n\times n}$ this algorithm uses the power method applied to A^*A to produce an estimate $\gamma \le \|A\|_2$.

```
1  Choose a nonzero starting vector z_0 ∈ C^n
2  for k = 0: ∞
3      w_{k+1} = A z_k
4      z_{k+1} = A^* w_{k+1}
5      γ_{k+1} = ||z_{k+1}||_2 / ||w_{k+1}||_2
6      if converged, γ = γ_{k+1}, quit, end
7  end
```

In practice we would normalize w_k and z_k to unit 2-norm after computing them, to avoid subsequent overflow and underflow. We have omitted the normalizations to avoid cluttering the algorithm.

To analyze convergence we can exploit the fact that we are applying the power method to the Hermitian positive semidefinite matrix A^*A. Let A have singular values $\sigma_1 = \cdots = \sigma_p > \sigma_{p+1} \ge \cdots \ge \sigma_n$ $(1 \le p \le n)$. It is straightforward to show that $\gamma_k \to \|A\|_2$ linearly as $k \to \infty$ provided that z_0 has a nonzero component in the space spanned by the right singular vectors corresponding to $\sigma_1, \dots, \sigma_p$.

Note that given $w_{k+1} = Az_k$ and $z_{k+1} = A^*w_{k+1} = A^*Az_k$ we have three lower bounds for $\|A\|_2$, given in

$$\|A\|_2 \ge \max\left[\frac{\|w_{k+1}\|_2}{\|z_k\|_2}, \frac{\|z_{k+1}\|_2}{\|w_{k+1}\|_2}, \left(\frac{\|z_{k+1}\|_2}{\|z_k\|_2}\right)^{1/2}\right]. \tag{3.21}$$

Now

$$\|w_{k+1}\|_2^2 = w_{k+1}^* w_{k+1} = z_k^* A^* w_{k+1} = z_k^* z_{k+1},$$

so $\|w_{k+1}\|_2^2 \le \|z_k\|_2 \|z_{k+1}\|_2$, which implies

$$\frac{\|w_{k+1}\|_2}{\|z_k\|_2} \le \frac{\|z_{k+1}\|_2}{\|w_{k+1}\|_2}, \qquad \frac{\|z_{k+1}\|_2}{\|z_k\|_2} \le \frac{\|z_{k+1}\|_2^2}{\|w_{k+1}\|_2^2}.$$

Hence γ in Algorithm 3.19 is the best estimate obtainable from the lower bounds in (3.21).

To estimate $\|L(X)\|_F$ we simply apply Algorithm 3.19 to $A = K(X)$. The resulting algorithm can be written entirely in terms of $L(X)$ and $L^\star(X)$, the adjoint of $L(X)$ defined with respect to the inner product $\langle X, Y\rangle = \operatorname{trace}(Y^*X)$. When $X \in \mathbb{R}^{n\times n}$ and $f: \mathbb{R}^{n\times n} \to \mathbb{R}^{n\times n}$, the adjoint is given by $L^\star(X) = L(X^T)$. In the complex case, $L_f^\star(X) = L_{\overline{f}}(X^*)$, where $\overline{f}(z) := \overline{f(\overline{z})}$, so that if f has a power series representation then $\overline{f}$ is obtained by conjugating the coefficients.

Algorithm 3.20 (power method on Fréchet derivative). Given $X \in \mathbb{C}^{n\times n}$ and the Fréchet derivative L of a function f, this algorithm uses the power method to produce an estimate $\gamma \le \|L(X)\|_F$.

1 Choose a nonzero starting matrix $Z_0 \in \mathbb{C}^{n\times n}$
2 for $k = 0\colon\infty$
3 $\quad W_{k+1} = L(X, Z_k)$
4 $\quad Z_{k+1} = L^\star(X, W_{k+1})$
5 $\quad \gamma_{k+1} = \|Z_{k+1}\|_F/\|W_{k+1}\|_F$
6 $\quad$ if converged, $\gamma = \gamma_{k+1}$, quit, end
7 end

A random Z_0 is a reasonable choice in Algorithm 3.20. A possible expression for "converged" in the convergence tests of Algorithms 3.19 and 3.20 is

$$k > \text{it\_max} \text{ or } |\gamma_{k+1} - \gamma_k| \le \text{tol}\,\gamma_{k+1},$$

where it_max is the maximum allowed number of iterations and tol is a relative convergence tolerance. Since an estimate of just the correct order of magnitude is required, tol $= 10^{-1}$ or 10^{-2} may be suitable. However, since linear convergence can be arbitrarily slow it is difficult to construct a truly reliable convergence test.

An alternative to the power method for estimating the largest eigenvalue of a Hermitian matrix is the Lanczos algorithm. Mathias [408, 1992] gives a Lanczos-based analogue of Algorithm 3.20. He shows that the Lanczos approach generates estimates at least as good as those from Algorithm 3.20 at similar cost, but notes that for obtaining order of magnitude estimates the power method is about as good as Lanczos.

Turning to the 1-norm, we need the following algorithm.

Algorithm 3.21 (LAPACK matrix norm estimator). Given $A \in \mathbb{C}^{n\times n}$ this algorithm computes γ and $v = Aw$ such that $\gamma \le \|A\|_1$ with $\|v\|_1/\|w\|_1 = \gamma$ (w is not returned). For $z \in \mathbb{C}$, $\operatorname{sign}(z) = z/|z|$ if $z \ne 0$ and $\operatorname{sign}(0) = 1$.[5]

1 $v = A(n^{-1}e)$
2 if $n = 1$, quit with $\gamma = |v_1|$, end
3 $\gamma = \|v\|_1$
4 $\xi = \operatorname{sign}(v)$
5 $x = A^*\xi$
6 $k = 2$
7 repeat
8 $\qquad j = \min\{\, i\colon |x_i| = \|x\|_\infty \,\}$
9 $\qquad v = Ae_j$

[5]This definition of sign is different from that used in Chapter 5.

10 $\quad\quad \overline{\gamma} = \gamma$
11 $\quad\quad \gamma = \|v\|_1$
12 $\quad\quad$ if (A is real and $\mathrm{sign}(v) = \pm\xi$) or $\gamma \le \overline{\gamma}$, goto line 17, end
13 $\quad\quad \xi = \mathrm{sign}(v)$
14 $\quad\quad x = A^*\xi$
15 $\quad\quad k = k + 1$
16 until ($\|x\|_\infty = x_j$ or $k > 5$)
17 $x_i = (-1)^{i+1}\left(1 + \frac{i-1}{n-1}\right), \quad i = 1\colon n$
18 $x = Ax$
19 if $2\|x\|_1/(3n) > \gamma$ then
20 $\quad v = x$
21 $\quad \gamma = 2\|x\|_1/(3n)$
22 end

Algorithm 3.21 is the basis of all the condition number estimation in LAPACK and is used by MATLAB's `rcond` function. MATLAB's `normest1` implements a block generalization of Algorithm 3.21 due to Higham and Tisseur [288, 2000] that iterates with an $n \times t$ matrix where $t \ge 1$; for $t = 1$, Algorithm 3.21 (without lines 17–22) is recovered.

Key properties of Algorithm 3.21 are that it typically requires 4 or 5 matrix–vector products, it frequently produces an exact estimate ($\gamma = \|A\|_1$), it can produce an arbitrarily poor estimate on specially constructed "counterexamples", but it almost invariably produces an estimate correct to within a factor 3. Thus the algorithm is a very reliable means of estimating $\|A\|_1$.

We can apply Algorithm 3.21 with $A = K(X)$ and thereby estimate $\|L(X)\|_1$.

Algorithm 3.22 (LAPACK matrix norm estimator on Fréchet derivative). Given a matrix $X \in \mathbb{C}^{n\times n}$ this algorithm uses the LAPACK norm estimator to produce an estimate γ of $\|L(X)\|_1$, given the ability to compute $L(X,E)$ and $L^\star(X,E)$ for any E. More precisely, $\gamma \le \|K(X)\|_1$, where $\|K(X)\|_1 \in \left[n^{-1}\|L(X)\|_1, n\|L(X)\|_1\right]$.

1 Apply Algorithm 3.21 to the matrix $A := K(X)$,
noting that $Ay \equiv \mathrm{vec}(L(X,E))$ and $A^*y \equiv \mathrm{vec}(L^\star(X,E))$, where $\mathrm{vec}(E) = y$.

Advantages of Algorithm 3.22 over Algorithm 3.20 are a "built-in" starting matrix and convergence test and a more predictable number of iterations.

Algorithms 3.20 and 3.22 both require two Fréchet derivative evaluations per iteration. One possibility is to approximate these derivatives by finite differences, using, from (3.11),

$$L(X,E) \approx \frac{f(X+tE) - f(X)}{t} =: \Delta_f(X,t,E) \tag{3.22}$$

for a small value of t. The choice of t is a delicate matter—more so than for scalar finite difference approximations because the effect of rounding errors on the evaluation of $f(X+tE)$ is more difficult to predict. A rough guide to the choice of t can be developed by balancing the truncation and rounding errors. For a sufficiently smooth f, (3.6) implies $f(X+tE) - f(X) - L(X,tE) = O(t^2\|E\|^2)$. Hence the truncation error $\Delta_f(X,t,E) - L(X,E)$ can be estimated by $t\|E\|^2$. For the evaluation of $f(X)$ we have *at best* $fl(f(X)) = f(X) + E$, where $\|E\| \le u\|f(X)\|$. Hence

$$\|fl(\Delta_f(X,t,E)) - \Delta_f(X,t,E)\| \le u(\|f(X+tE)\| + \|f(X)\|)/t \approx 2u\|f(X)\|/t.$$

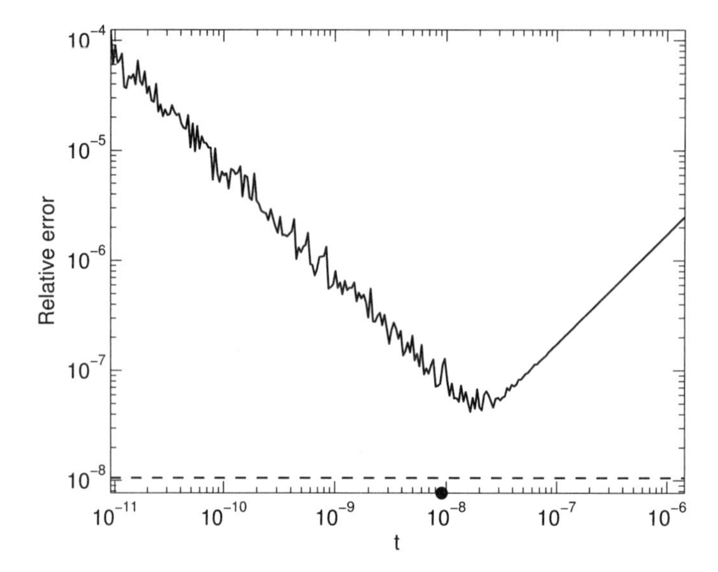

Figure 3.1. *Relative errors in the Frobenius norm for the finite difference approximation* (3.22) *with* $f(A) = e^X$, X *and* E *random matrices from the normal* $(0, 1)$ *distribution, and* 250 *different* t. *The dotted line is* $u^{1/2}$ *and the circle on the x-axis denotes* t_{opt} *in* (3.23).

The error due to rounding is therefore estimated by $u\|f(X)\|/t$. The natural choice of t is that which minimizes the maximum of the error estimates, that is, t for which $t\|E\|^2 = u\|f(X)\|/t$, or

$$t_{\text{opt}} = \left(\frac{u\|f(X)\|}{\|E\|^2} \right)^{1/2}. \tag{3.23}$$

The minimum is $u^{1/2}\|f(X)\|^{1/2}\|E\|$. In practice, t_{opt} is usually fairly close to minimizing the overall error. Figure 3.1 shows a typical example.

For methods for computing $f(X)$ that employ a Schur decomposition, $X = QTQ^*$, an efficient way to obtain the function evaluation $f(X + G)$ in (3.22), where $G \equiv tE$, has been suggested by Mathias [411, 1995]. The idea is to write $f(X + G) = Qf(T + \widetilde{G})Q^*$, where $\widetilde{G} = Q^*GQ$, and then reduce $\widetilde{G}$ to upper triangular form by nearly unitary transformations, exploiting the fact that $\widetilde{G}$ is small. Then the underlying method is used to evaluate f at the triangular matrix. See [411, 1995] for details.

Finally, we briefly mention a probabilistic approach to condition estimation. To first order in t, (3.22) gives $\|\Delta_f(X, t, E)\|_F \le \|L(X)\|_F\|E\|_F$. Kenney and Laub [346, 1994] propose choosing E with elements from the normal (0,1) distribution, scaling so that $\|E\|_F = 1$, and then approximating $\|L(X)\|_F$ by $\phi_n\|\Delta_f(X, t, E)\|_F$, where ϕ_n is a constant such that the expected value of $\|\Delta_f(X, t, E)\|_F$ is $\|L(X)\|_F$. They give an explicit formula for ϕ_n and show how to evaluate the probability that the estimate is within a certain factor of $\|L(X)\|_F$, assuming that the $O(t)$ error in (3.22) can be ignored. Several independent E can be used in order to get a better estimate. For example, with two E the estimate is within a factor 5 of $\|L(X)\|_F$ with probability at least 0.9691 (for all n). The main weaknesses of this approach are that the theory applies only to real matrices, it is unclear how small t must be for the theory to be valid, and the method is expensive if many significant digits are required with high probability. Probabilistic estimates for the power method and Lanczos method are also available; see Dixon [160, 1983], Kuczyński and Woźniakowski [365, 1992], and Van Dorsselaer, Hochstenbach, and Van der Vorst [590, 2000].

3.5. Notes and References

For more details on the Fréchet derivative, and on connections between Fréchet and Gâteaux derivatives, see, for example, Aubin and Ekeland [23, 1984, Sec. 1.4], Atkinson and Han [22, 2005, Sec. 5.3], Bhatia [64, 1997, Sec. X.4], or Ortega and Rheinboldt [453, 2000, Sec. 3.1].

There is a literature on matrix differential calculus aimed at application areas and not focusing on matrix functions as defined in this book. See, for example, the book by Magnus and Neudecker [402, 1999] concerned particularly with statistics and psychometrics.

Our definitions (3.2) and (3.4) of condition number, and Theorem 3.1, are special cases of definitions and results of Rice [487, 1966].

Theorem 3.6 is from Mathias [412, 1996], where it is stated in a form that requires f to be only $2m-1$ times continuously differentiable, where m is the size of the largest Jordan block of $A(t)$, for all t in some neighbourhood of 0. The identity (3.13) is proved by Najfeld and Havel [445, 1995, Thm. 4.11] under the assumption that f is analytic. Theorem 3.7 is from Horn and Johnson [296, 1991, Thm. 6.6.14], with conditions modified as in [412, 1996].

Theorem 3.9 appears to be new in the form stated. A weaker version that assumes f has a power series expansion and does not show that all eigenvalues of $L(X)$ are accounted for is given by Kenney and Laub [340, 1989].

Theorem 3.11 is due to Daleckiĭ and Kreĭn [129, 1965], [130, 1965]. Presentations of this and more general results can be found in Bhatia [64, 1997, Sec. V.3] and Horn and Johnson [296, 1991, Sec. 6.6].

Theorem 3.14 and Corollary 3.16 are obtained by Kenney and Laub [340, 1989] in the case where f has a convergent power series representation. Their proofs work with $K(X)$, which in this case has an explicit representation in terms of Kronecker products; see Problem 3.6. That Corollary 3.16 holds without this restriction on f is noted by Mathias [411, 1995].

Convergence analysis for the power method for computing an eigenpair of a general matrix B can be found in Golub and Van Loan [224, 1996, Sec. 7.3], Stewart [538, 2001, Sec. 2.1], Watkins [607, 2002, Sec. 5.3], and Wilkinson [616, 1965, Sec. 9.3]. In these analyses the assumption of a dominant eigenvalue is needed to guarantee convergence; for Algorithm 3.19, with $B \equiv A^*A$, no such assumption is needed because eigenvalues of B of maximal modulus are necessarily equal.

The power method in Algorithm 3.20 is suggested by Kenney and Laub [340, 1989].

Algorithm 3.21 was developed by Higham [270, 1988]. The algorithm is based on a p-norm power method of Boyd [79, 1974], also investigated by Tao [563, 1984] and derived independently for the 1-norm by Hager [237, 1984]. For more details see Higham [270, 1988], [271, 1990], and [276, 2002, Chap. 15].

Problems

3.1. Evaluate the Fréchet derivatives $L(X,E)$ of $F(X)=I$, $F(X)=X$, and $F(X)=\cos(X)$, assuming in the last case that $XE=EX$.

3.2. Show that if $X=QTQ^*$ is a Schur decomposition then $L(X,E)=QL(T,Q^*EQ)Q^*$.

3.3. Show that the Fréchet derivative is unique.

3.4. Prove (3.11), namely that the Fréchet derivative is a directional derivative.

3.5. Let $A=\left[\begin{smallmatrix} A_{11} & A_{12} \\ 0 & A_{22}\end{smallmatrix}\right]$, where $A_{11}\in\mathbb{C}^{n_1\times n_1}$ and $A_{22}\in\mathbb{C}^{n_2\times n_2}$ with $n=n_1+n_2$. What is the maximum size of a Jordan block of A?

3.6. Let the power series $f(x)=\sum_{i=0}^{\infty}a_i x^i$ have radius of convergence r. Show that for $X,E\in\mathbb{C}^{n\times n}$ with $\|X\|<r$, the Fréchet derivative

$$L(X,E)=\sum_{i=1}^{\infty}a_i\sum_{j=1}^{i}X^{j-1}EX^{i-j}, \tag{3.24}$$

and hence that $K(X)$ in (3.17) is given by

$$K(X)=\sum_{i=1}^{\infty}a_i\sum_{j=1}^{i}(X^T)^{i-j}\otimes X^{j-1}.$$

3.7. Show that if f has a power series expansion with real coefficients then $L(X^*,E)=L(X,E^*)^*$.

3.8. Show that if X and E commute then $L(X,E)=f'(X)E=Ef'(X)$, where f' denotes the derivative of the scalar function f.

3.9. (Stickel [541, 1987], Rinehart [494, 1956]) Suppose that f is analytic on and inside a closed contour Γ that encloses $\Lambda(X)$. Show that the Fréchet derivative of f is given by

$$L(X,E)=\frac{1}{2\pi i}\int_{\Gamma}f(z)(zI-X)^{-1}E(zI-X)^{-1}\,dz.$$

Deduce that if $XE=EX$ then $L(X,E)=f'(X)E=Ef'(X)$, where f' denotes the derivative of the scalar function f.

3.10. Consider any two eigenvalues λ and μ of $X\in\mathbb{C}^{n\times n}$, with corresponding right and left eigenvectors u and v, so that $Xu=\lambda u$ and $v^TX=\mu v^T$. Show directly (without using Kronecker products or reducing to the case that f is a polynomial) that uv^T is an eigenvector of $L(X)$ with corresponding eigenvalue $f[\lambda,\mu]$.

3.11. (Research problem) Determine the Jordan form of the Fréchet derivative $L(X)$ of f in terms of that of X. To see that this question is nontrivial, note that for $f(X)=X^2$ and $X=\left[\begin{smallmatrix} 0 & 1 \\ 0 & 0\end{smallmatrix}\right]$, $L(X)$ has one Jordan block of size 3 and one of size 1, both for the eigenvalue 0.

Chapter 4
Techniques for General Functions

Many different techniques are available for computing or approximating matrix functions, some of them very general and others specialized to particular functions. In this chapter we survey a variety of techniques applicable to general functions f. We begin with the basic tasks of evaluating polynomial and rational functions, and address the validity of matrix Taylor series and the truncation error when a finite number of terms are summed. Then we turn to methods based on similarity transformations, concentrating principally on the use of the Schur decomposition and evaluation of a function of a triangular matrix. Matrix iterations are an important tool for certain functions, such as matrix roots. We discuss termination criteria, show how to define stability in terms of a Fréchet derivative, and explain how a convergence result for a scalar iteration can be translated into a convergence result for the corresponding matrix iteration. Finally, we discuss preprocessing, which may be beneficial before applying a particular algorithm, and present several bounds on $\|f(A)\|$. Many of the methods for specific f described later in the book make use of one or more of the techniques treated in this chapter.

We describe the cost of an algorithm in one of two ways. If, as is often the case, the algorithm is expressed at the matrix level, then we count the number of matrix operations. Thus

- M denotes a matrix multiplication, AB,
- I denotes a matrix inversion, A^{-1},
- D denotes a "matrix division", that is, the solution of a multiple right-hand side linear system $AX = B$,

where the matrices A and B are $n \times n$. Other operation counts are given in terms of flops, where a flop denotes any of the four elementary operations on scalars $+$, $-$, $*$, $/$. The costs in flops of various matrix operations are summarized in Appendix C, where some comments on the relevance and interpretation of these different measures are given.

It is worth clarifying the terms "method" and "algorithm". For us, a method is usually a general technique that, when the details are worked out, can lead to more than one algorithm. An algorithm is a specific automatic procedure that we name "Algorithm" and specify using pseudocode.

4.1. Matrix Powers

A basic requirement in some methods for matrix functions is to compute a power of a matrix. A sequence of successive powers A^2, A^3, $\ldots$, should be computed in the obvious way: by repeated multiplication by A. But if a single power, A^m, is needed, it is not necessary to compute all the lower powers first. Instead, repeated squaring can be used, as described in the following algorithm. The initial while loop simply ensures that multiplication of a power of A with the identity matrix is avoided.

Algorithm 4.1 (binary powering). This algorithm evaluates $X = A^m$ for $A \in \mathbb{C}^{n\times n}$.

```
1  Let m = Σ_{i=0}^t β_i 2^i be the binary representation of m, with β_t ≠ 0.
2  P = A
3  i = 0
4  while β_i = 0
5          P = P^2
6          i = i + 1
7  end
8  X = P
9  for j = i + 1: t
10     P = P^2
11     if β_j = 1
12        X = XP
13     end
14 end
```

Cost: $(\lfloor \log_2 m\rfloor + \mu - 1)M \le 2\lfloor \log_2 m\rfloor M$, where $\mu \le \lceil \log_2 m\rceil$ is the number of 1s in the binary representation of m.

A special case of binary powering is repeated squaring to form A^{2^k}, which is used in the scaling and squaring method for the matrix exponential; see Section 10.3.

4.2. Polynomial Evaluation

Many methods for computing matrix functions require the evaluation of a matrix polynomial

$$p_m(X) = \sum_{k=0}^{m} b_k X^k, \qquad X \in \mathbb{C}^{n\times n}. \tag{4.1}$$

The economics of the evaluation are rather different than for scalar polynomials. Whereas Horner's method (nested multiplication) is almost always used in the scalar case, for matrix polynomials there are four competing methods.

First, p_m can be evaluated by Horner's method.

Algorithm 4.2 (Horner's method). This algorithm evaluates the polynomial (4.1) by Horner's method.

```
1  S_{m-1} = b_m X + b_{m-1} I
2  for k = m - 2: -1: 0
3     S_k = X S_{k+1} + b_k I
4  end
5  p_m = S_0
```

Cost: $(m-1)M$.

Horner's method is not suitable when m is not known at the start of the evaluation, as is often the case when a truncated power series is to be summed. In this case p_m can be evaluated by explicitly forming each power of X.

Algorithm 4.3 (evaluate polynomial via explicit powers). This algorithm evaluates the polynomial (4.1) by explicitly forming matrix powers.

```
1  P = X
2  S = b_0 I + b_1 X
3  for k = 2: m
4      P = PX
5      S = S + b_k P
6  end
7  P_m = S
```

Cost: $(m-1)M$.

Note that while Algorithms 4.2 and 4.3 have the same cost for matrices, when X is a scalar Algorithm 4.3 is twice as expensive as Algorithm 4.2.

Another method factorizes the polynomial $p_m(x) = b_m(x-\xi_1)\dots(x-\xi_m)$ and then evaluates this factorized form at the matrix X.

Algorithm 4.4 (evaluate polynomial in factored form). This algorithm evaluates the polynomial (4.1) given the roots $\xi_1,\dots,\xi_m$ of p_m.

```
1  S = X - ξ_m I
2  for k = m - 1: -1: 1
3      S = S(X - ξ_k I)
4  end
5  p_m = S
```

Cost: $(m-1)M$.

One drawback to Algorithm 4.4 is that some of the roots ξ_j may be complex and so complex arithmetic can be required even when the polynomial and X are real. In such situations the algorithm can be adapted in an obvious way to employ a factorization of p_m into real linear and quadratic factors.

The fourth and least obvious method is that of Paterson and Stockmeyer [466, 1973], [224, 1996, Sec. 11.2.4], in which p_m is written as

$$p_m(X) = \sum_{k=0}^{r} B_k \cdot (X^s)^k, \qquad r = \lfloor m/s \rfloor, \tag{4.2}$$

where s is an integer parameter and

$$B_k = \begin{cases} b_{sk+s-1}X^{s-1} + \cdots + b_{sk+1}X + b_{sk}I, & k = 0: r-1, \\ b_m X^{m-sr} + \cdots + b_{sr+1}X + b_{sr}I, & k = r. \end{cases}$$

The powers $X^2,\dots,X^s$ are computed; then (4.2) is evaluated by Horner's method, with each B_k formed when needed. The two extreme cases are $s = 1$, which reduces to Algorithm 4.2, and $s = m$, which reduces to Algorithm 4.3. As an example, for $m = 6$ and $s = 3$ we have

$$p_6(X) = \underbrace{b_6 I}_{B_2}(X^3)^2 + \underbrace{(b_5X^2 + b_4X + b_3I)}_{B_1}X^3 + \underbrace{(b_2X^2 + b_1X + b_0I)}_{B_0},$$

Table 4.1. *Number of matrix multiplications required by the Paterson–Stockmeyer method and Algorithms* 4.2 *and* 4.3 *to evaluate a degree m matrix polynomial.*

m	2	3	4	5	6	7	8	9	10	11	12	13	14	15	16
PS method	1	2	2	3	3	4	4	4	5	5	5	6	6	6	6
Algs 4.2/4.3	1	2	3	4	5	6	7	8	9	10	11	12	13	14	15

which can be evaluated in $3M$, compared with the $5M$ required for Horner's method. Note that (4.2) is a polynomial with matrix coefficients, and the cost of evaluation given the B_k and X^s is rM. The total cost of evaluating p_m is

$$\big(s + r - 1 - f(s,m)\big)M, \qquad f(s,m) = \begin{cases} 1 & \text{if } s \text{ divides } m, \\ 0 & \text{otherwise.} \end{cases} \tag{4.3}$$

This quantity is approximately minimized by $s = \sqrt{m}$, so we take for s either $\lfloor \sqrt{m} \rfloor$ or $\lceil \sqrt{m}\, \rceil$; it can be shown that both choices yield the same operation count. As an extreme case, this method evaluates A^{q^2} as $(A^q)^q$, which clearly requires much less work than the previous two methods for large q, though for a single high power of A binary powering (Algorithm 4.1) is preferred.

Table 4.1 shows the cost of the Paterson–Stockmeyer method for $m = 2\colon 16$; for each $m \ge 4$ it requires strictly fewer multiplications than Algorithms 4.2 and 4.3. For each m in the table it can be shown that both choices of s minimize (4.3) [247, 2005].

Unfortunately, the Paterson–Stockmeyer method requires $(s + 2)n^2$ elements of storage. This can be reduced to $4n^2$ by computing p_m a column at a time, as shown by Van Loan [596, 1979], though the cost of evaluating p_m then increases to $(2s + r - 3 - f(s,m))M$. The value $s = \sqrt{m/2}$ approximately minimizes the cost of Van Loan's variant, and it then costs about 40% more than the original method.

It is important to understand the effect of rounding errors on these four polynomial evaluation methods. The next theorem provides error bounds for three of the methods. For matrices, absolute values and inequalities are defined componentwise. We write $\widetilde{\gamma}_n = cnu/(1-cnu)$, where u is the unit roundoff and c is a small integer constant whose precise value is unimportant. For details of our model of floating point arithmetic see Section B.15.

Theorem 4.5. *The computed polynomial $\widehat{p}_m$ obtained by applying Algorithm* 4.2, *Algorithm* 4.3, *or the Paterson–Stockmeyer method to p_m in* (4.1) *satisfies*

$$|p_m - \widehat{p}_m| \le \widetilde{\gamma}_{mn}\, \widetilde{p}_m(|X|),$$

where $\widetilde{p}_m(X) = \sum_{k=0}^m |b_k| X^k$. Hence $\|p_m - \widehat{p}_m\|_{1,\infty} \le \widetilde{\gamma}_{mn}\, \widetilde{p}_m(\|X\|_{1,\infty})$. □

The bound of the theorem is pessimistic in the sense that inequalities such as $|X^j| \le |X|^j$ are used in the derivation. But the message of the theorem is clear: if there is significant cancellation in forming p_m then the error in the computed result can potentially be large. This is true even in the case of scalar x. Indeed, Stegun and Abramowitz [535, 1956] presented the now classic example of evaluating $e^{-5.5}$ from a truncated Taylor series and showed how cancellation causes a severe loss of accuracy in floating point arithmetic; they also noted that computing $e^{5.5}$ and reciprocating avoids the numerical problems.

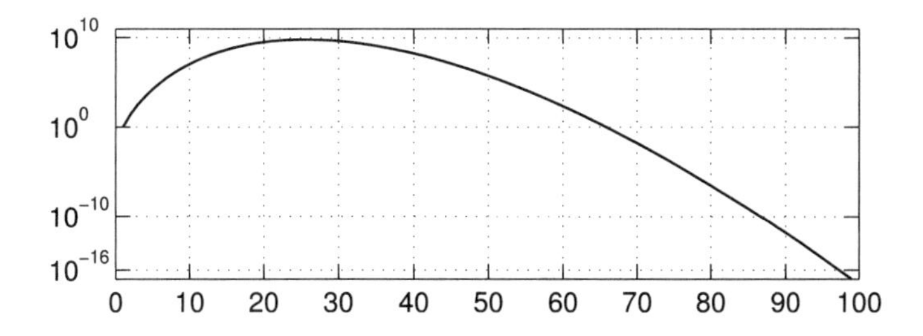

Figure 4.1. 2-*norms of first* 99 *terms in Taylor series of* e^A, *for* A *in* (4.4) *with* $\alpha = 25$.

For a matrix example we take

$$A = \begin{bmatrix} 0 & \alpha \\ -\alpha & 0 \end{bmatrix}, \tag{4.4}$$

for which

$$e^A = \begin{bmatrix} \cos\alpha & \sin\alpha \\ -\sin\alpha & \cos\alpha \end{bmatrix}.$$

We took $\alpha = 25$ and summed the first 99 terms of the Taylor series for e^A using Algorithm 4.3; at this point adding further terms makes no difference to the computed sum. The computed sum $\widehat{X}$ has error $\|e^A - \widehat{X}\|_2 = 1.5 \times 10^{-7}$, which represents a loss of 9 significant digits in all components of X. This loss of significance can be understood with the aid of Figure 4.1, which shows that the terms in the series grow rapidly in norm, reaching a maximum of order 10^{10}. Since all the elements of e^A are of order 1, there is clearly massive cancellation in the summation, and based on the size of the maximum term a loss of about 10 significant digits would be expected. Turning to Theorem 4.5, the 2-norm of $u\widetilde{p}_m(|X|)$ is 8×10^{-6}, so the upper bound of the theorem is reasonably sharp in this example if we set the constant $\widetilde{\gamma}_{mn}$ to u.

Unlike in the scalar example of Stegun and Abramowitz, computing e^{-A} and then inverting does not help in this example, since $e^{-A} = e^{A^T}$ in this case and so the Taylor series is merely transposed.

Error analysis for Algorithm 4.4 is essentially the same as error analysis of a matrix product (see [276, 2002, Secs. 3.7, 18.2]).

Theorem 4.6. *The computed polynomial* $\widehat{p}_m$ *obtained by applying Algorithm* 4.4 *to* p_m *in* (4.1) *satisfies*

$$|p_m - \widehat{p}_m| \le \widetilde{\gamma}_{mn}\, |b_m||X - \xi_1 I| \dots |X - \xi_m I|. \qquad \square \tag{4.5}$$

Note that this theorem assumes the ξ_j are known exactly. The ξ_k can be ill conditioned functions of the coefficients b_k, so in practice the errors in computing the ξ_j could have a significant effect. The main thing to note about the bound (4.5) is that it depends on the ordering of the ξ_k, since the matrices $|X - \xi_k I|$ do not commute with each other in general. An ordering that has been suggested is the Leja ordering [485, 1991]; see the discussion in [444, 1992]. We will not consider Algorithm 4.4 further because the special nature of the polynomials in matrix function applications tends to make the other algorithms preferable.

4.3. Taylor Series

A basic tool for approximating matrix functions is the Taylor series. We begin with a theorem that guarantees the validity of a matrix Taylor series if the eigenvalues of the "increment" lie within the radius of convergence of the associated scalar Taylor series.

Theorem 4.7 (convergence of matrix Taylor series). *Suppose f has a Taylor series expansion*

$$f(z) = \sum_{k=0}^{\infty} a_k (z-\alpha)^k \qquad \left(a_k = \frac{f^{(k)}(\alpha)}{k!} \right) \tag{4.6}$$

with radius of convergence r. If $A \in \mathbb{C}^{n\times n}$ then $f(A)$ is defined and is given by

$$f(A) = \sum_{k=0}^{\infty} a_k (A-\alpha I)^k \tag{4.7}$$

if and only if each of the distinct eigenvalues $\lambda_1, \dots, \lambda_s$ of A satisfies one of the conditions

(a) $|\lambda_i - \alpha| < r$,

(b) $|\lambda_i - \alpha| = r$ *and the series for $f^{(n_i-1)}(\lambda)$ (where n_i is the index of λ_i) is convergent at the point $\lambda = \lambda_i$, $i = 1\colon s$.*

Proof. It is easy to see from Definition 1.2 that it suffices to prove the theorem for a Jordan block, $A = J(\lambda) = \lambda I + N \in \mathbb{C}^{n\times n}$, where N is strictly upper triangular. Let $f_m(z) = \sum_{k=0}^{m} a_k (z-\alpha)^k$. We have

$$\begin{aligned}
f_m(J(\lambda)) &= \sum_{k=0}^{m} a_k \big((\lambda-\alpha)I + N\big)^k \\
&= \sum_{k=0}^{m} a_k \sum_{i=0}^{k} \binom{k}{i} (\lambda-\alpha)^{k-i} N^i \\
&= \sum_{i=0}^{m} N^i \sum_{k=i}^{m} a_k \binom{k}{i} (\lambda-\alpha)^{k-i} \\
&= \sum_{i=0}^{m} \frac{N^i}{i!} \sum_{k=i}^{m} a_k\, k(k-1)\dots(k-i+1)(\lambda-\alpha)^{k-i} \\
&= \sum_{i=0}^{m} \frac{N^i}{i!} f_m^{(i)}(\lambda) = \sum_{i=0}^{\min(m,n-1)} \frac{N^i}{i!} f_m^{(i)}(\lambda).
\end{aligned}$$

Evidently, $\lim_{m\to\infty} f_m(J(\lambda))$ exists if and only if $\lim_{m\to\infty} f_m^{(i)}(\lambda)$ exists for $i = 1\colon n-1$, which is essentially the statement of case (b), because if the series for f differentiated term by term $n_i - 1$ times converges at λ then so does the series differentiated j times for $j = 0\colon n_i - 1$. Case (a) follows from the standard result in complex analysis that a power series differentiated term by term converges within the radius of convergence of the original series. $\square$

The four most important matrix Taylor series are

$$\begin{aligned}\exp(A) &= I + A + \frac{A^2}{2!} + \frac{A^3}{3!} + \cdots,\\ \cos(A) &= I - \frac{A^2}{2!} + \frac{A^4}{4!} - \frac{A^6}{6!} + \cdots,\\ \sin(A) &= A - \frac{A^3}{3!} + \frac{A^5}{5!} - \frac{A^7}{7!} + \cdots,\\ \log(I+A) &= A - \frac{A^2}{2} + \frac{A^3}{3} - \frac{A^4}{4} + \cdots, \quad \rho(A) < 1,\end{aligned}$$

the first three series having infinite radii of convergence. These series can be used to approximate the respective functions, by summing a suitable finite number of terms. Two types of error need to be addressed: truncation errors, and rounding errors in the floating point evaluation. Truncation errors are bounded in the following result.

Theorem 4.8 (Taylor series truncation error bound). *Suppose f has the Taylor series expansion* (4.6) *with radius of convergence r. If $A \in \mathbb{C}^{n\times n}$ with $\rho(A - \alpha I) < r$ then for any matrix norm*

$$\Big\| f(A) - \sum_{k=0}^{s-1} a_k (A - \alpha I)^k \Big\| \le \frac{1}{s!} \max_{0\le t\le 1} \big\| (A-\alpha I)^s f^{(s)}(\alpha I + t(A - \alpha I)) \big\|. \tag{4.8}$$

Proof. See Mathias [409, 1993, Cor. 2]. Note that this bound does not contain a factor depending on n, unlike the 2-norm version of the bound in [224, 1996, Thm. 11.2.4]. Moreover, the norm need not be consistent. □

In order to apply this theorem we need to bound the term $\max_{0\le t\le 1} \|Z^s f^{(s)}(\alpha I + tZ)\|_\infty$. For certain f this is straightforward. We illustrate using the cosine function. With $\alpha = 0$, $s = 2k+2$, and

$$T_{2k}(A) = \sum_{i=0}^{2k} \frac{(-1)^i}{(2i)!} A^{2i},$$

the bound of Theorem 4.8 is, for the ∞-norm,

$$\begin{aligned}\|\cos(A) - T_{2k}(A)\|_\infty &\le \frac{1}{(2k+2)!} \max_{0\le t\le 1} \|A^{2k+2} \cos^{(2k+2)}(tA)\|_\infty\\ &\le \frac{1}{(2k+2)!} \|A^{2k+2}\|_\infty \max_{0\le t\le 1} \|\cos^{(2k+2)}(tA)\|_\infty.\end{aligned}$$

Now

$$\begin{aligned}\max_{0\le t\le 1} \|\cos^{(2k+2)}(tA)\|_\infty &= \max_{0\le t\le 1} \|\cos(tA)\|_\infty\\ &\le 1 + \frac{\|A\|_\infty^2}{2!} + \frac{\|A\|_\infty^4}{4!} + \cdots = \cosh(\|A\|_\infty),\end{aligned}$$

and so the error in the truncated Taylor series approximation to the matrix cosine satisfies the bound

$$\|\cos(A) - T_{2k}(A)\|_\infty \le \frac{\|A^{2k+2}\|_\infty}{(2k+2)!} \cosh(\|A\|_\infty). \tag{4.9}$$

We also need to bound the error in evaluating $T_{2k}(A)$ in floating point arithmetic. From Theorem 4.5 we have that if $T_{2k}(A)$ is evaluated by any of the methods of Section 4.2 then the computed $\widehat{T}_{2k}$ satisfies

$$\|T_{2k} - \widehat{T}_{2k}\|_\infty \le \widetilde{\gamma}_{kn} \cosh(\|A\|_\infty).$$

Hence

$$\frac{\|\cos(A) - \widehat{T}_{2k}\|_\infty}{\|\cos(A)\|_\infty} \le \left(\frac{\|A^{2k+2}\|_\infty}{(2k+2)!} + \widetilde{\gamma}_{kn}\right)\frac{\cosh(\|A\|_\infty)}{\|\cos(A)\|_\infty}. \tag{4.10}$$

We can draw two conclusions. First, for maximum accuracy we should choose k so that

$$\frac{\|A^{2k+2}\|_\infty}{(2k+2)!} \approx \widetilde{\gamma}_{kn}.$$

Second, no matter how small the truncation error, the total relative error can potentially be as large as $\widetilde{\gamma}_{kn} \cosh(\|A\|_\infty)/\|\cos(A)\|_\infty$, and this quantity can be guaranteed to be of order $\widetilde{\gamma}_{kn}$ only if $\|A\|_\infty \lesssim 1$. The essential problem is that if $\|A\|_\infty \gg 1$ then there can be severe cancellation in summing the series.

If $\|A\|_\infty \le 1$ then, using $\|\cos(A)\|_\infty \ge 1 - (\cosh(\|A\|_\infty) - 1) = 2 - \cosh(\|A\|_\infty)$, we have

$$0.45 \le 2 - \cosh(1) \le \|\cos(A)\|_\infty \le \cosh(1) \le 1.55, \tag{4.11}$$

which gives

$$\frac{\cosh(\|A\|_\infty)}{\|\cos(A)\|_\infty} \le 3.4.$$

We conclude that a relative error $\|\cos(A) - \widehat{T}_{2k}\|_\infty/\|\cos(A)\|_\infty$ of order $\widetilde{\gamma}_{kn}$ is guaranteed if $\|A\|_\infty \le 1$ and k is sufficiently large. In fact, since $18! \approx 6 \times 10^{15}$, $k = 8$ suffices in IEEE standard double precision arithmetic, for which the unit roundoff $u \approx 1.1 \times 10^{-16}$.

4.4. Rational Approximation

Rational functions

$$r_{km}(x) = \frac{p_{km}(x)}{q_{km}(x)},$$

where p_{km} and q_{km} are polynomials in x of degree at most k and m, respectively, are powerful tools for approximation. An advantage over polynomials is that they are better able than polynomials to mimic the behaviour of general nonlinear functions.

When a scalar approximation $f(x) \approx r_{km}(x)$ is translated into a matrix approximation $f(A) \approx r_{km}(A)$, the scalar approximation is essentially being applied on the spectrum of A. However, a good approximation on the spectrum does not imply a good approximation overall. Suppose, for simplicity, that A is diagonalizable with $A = XDX^{-1}$, $D = \mathrm{diag}(\lambda_i)$. Defining $e(x) = f(x) - r_{km}(x)$, we have $e(A) = Xe(D)X^{-1}$ and so

$$\|e(A)\| \le \kappa(X)\|e(D)\|.$$

If A is normal, in which case we can take X unitary and hence $\kappa_2(X) = 1$, this bound is very satisfactory. But in general $\kappa(X)$ can be large and $\|e(A)\|$ can be large even though $\|e(D)\|$ is small. Therefore, to produce useful error bounds either rational approximations need to be applied to a restricted class of A or approximation errors must be analyzed at the matrix level.

The main classes of rational approximation used in computing matrix functions are those described in the following two subsections.

Let $\mathcal{R}_{k,m}$ denote the space of rational functions with numerator and denominator of degrees at most k and m, respectively.

4.4.1. Best L_∞ Approximation

The rational function r is a *best* L_∞ (or *minimax*, or *Chebyshev*) approximation to f on $[a,b]$ from $\mathcal{R}_{k,m}$ if

$$\|r(x) - f(x)\|_\infty = \min_{s\in\mathcal{R}_{k,m}} \|s(x) - f(x)\|_\infty,$$

where $\|g\|_\infty = \max_{x\in[a,b]} |g(x)|$. Best L_∞ rational approximations can be constructed using the Remez algorithm, a standard algorithm in approximation theory [477, 1981], [609, 1980]. These approximations are usually employed for Hermitian matrices only, so that error bounds for the scalar problem translate directly into error bounds at the matrix level.

In this book we will discuss best L_∞ approximations for the matrix sign function (Section 5.9) and the matrix exponential (Section 10.7.1).

4.4.2. Padé Approximation

For a given scalar function $f(x)$ the rational function $r_{km}(x) = p_{km}(x)/q_{km}(x)$ is a $[k/m]$ *Padé approximant* of f if $r_{k,m} \in \mathcal{R}_{k,m}$, $q_{km}(0) = 1$, and

$$f(x) - r_{km}(x) = O(x^{k+m+1}). \tag{4.12}$$

If a $[k/m]$ Padé approximant exists then it is unique; see Problem 4.2. It is usually required that p_{km} and q_{km} have no common zeros, so that p_{km} and q_{km} are unique. For a given f, k, and m, a $[k/m]$ Padé approximant might not exist, though for certain f existence has been proved for all k and m.

The condition (4.12) shows that r_{km} reproduces the first $k+m+1$ terms of the Taylor series of f about the origin, and of course if $m=0$ then r_{km} is precisely a truncated Taylor series.

Continued fraction representations

$$f(x) = b_0 + \cfrac{a_1 x}{b_1 + \cfrac{a_2 x}{b_2 + \cfrac{a_3 x}{b_3 + \cdots}}}$$

are intimately connected with Padé approximation and provide a convenient way of obtaining them. Specifically, if $b_1 = b_2 = \cdots = 1$ and the a_i are all nonzero then the convergents

$$r_m(x) \equiv r_{mm}(x) = b_0 + \cfrac{a_1 x}{b_1 + \cfrac{a_2 x}{b_2 + \cfrac{a_3 x}{b_3 + \cdots + \cfrac{a_{2m-1} x}{b_{2m-1} + \cfrac{a_{2m} x}{b_{2m}}}}}}. \tag{4.13}$$

Table 4.2. *Number of matrix multiplications required by the Paterson–Stockmeyer method to evaluate both $p_{mm}(A)$ and $q_{mm}(A)$. The second row indicates whether this cost is achieved for $s = \lfloor\sqrt{2m}\rfloor$ (F) or $s = \lceil\sqrt{2m}\rceil$ (C), or both.*

m	2	3	4	5	6	7	8	9	10	11	12	13	14	15	16
	1	2	3	4	4	5	5	6	6	7	7	8	8	8	9
	FC	C	F	F	F	C	FC	C	C	F	F	F	F	F	C

are the [0/0], [1/0], [1/1], [2/1], [2/2], ... Padé approximants of f [38, 1975, Sec 4.D], [39, 1996, Thm. 4.2.1].

Padé approximants are of particular interest in matrix function approximation for three main reasons:

- they can potentially produce approximations of a given accuracy with lower computational cost than a polynomial approximation such as a truncated Taylor series;
- while for good accuracy they require x to be near the origin, which can for some important functions be arranged by a suitable scaling, they do not require x to be real;
- the theory of Padé approximation is very well developed, and Padé approximants of some important functions are known explicitly, sometimes in several representations.

Details of Padé approximants to specific functions will be given in later chapters.

Padé approximants can be computed symbolically with Maple, Mathematica, or MATLAB with the Extended Symbolic Math Toolbox.

4.4.3. Evaluating Rational Functions

An important question is how to evaluate a given rational function at a matrix argument. Several possibilities exist, corresponding to different representations of the rational.

Suppose, first, that r_{km} is specified by its numerator and denominator polynomials p_{km} and q_{km}. The obvious approach is to evaluate p_{km} and q_{km} by any of the methods discussed in Section 4.2 and then to compute $r_{km}(A) = q_{km}(A)^{-1}p_{km}(A)$ by solving a multiple right-hand side linear system. In some cases it is possible to exploit similarity between the coefficients of p_{km} and q_{km} in order to reduce the work.

When the Paterson–Stockmeyer method is used to evaluate $p_{mm}(A)$ and $q_{mm}(A)$, some savings can be made. Referring to the description in Section 4.2, the powers $X^2, \dots, X^s$ can be computed once and used in both the p_{mm} and the q_{mm} evaluations. The cost of evaluating r_{mm} is then $\big(s+2r-1-2f(s,m)\big)M+D$, where f is defined in (4.3), and this quantity is approximately minimized by $s = \sqrt{2m}$. We therefore take for s whichever of $\lfloor\sqrt{2m}\rfloor$ and $\lceil\sqrt{2m}\rceil$ yields the smaller operation count. Table 4.2 shows the cost of the evaluation for $m = 2\colon 16$.

Consider now the continued fraction representation (4.13). This expansion can be evaluated at the matrix X in two ways: top down or bottom up. Top-down evaluation (which converts the continued fraction to rational form) is effected by the following recurrence, which dates back to Wallis (1655).

Algorithm 4.9 (continued fraction, top-down). This algorithm evaluates the continued fraction (4.13) in top-down fashion at the matrix $X \in \mathbb{C}^{n\times n}$.

```
1 P_{-1} = I, Q_{-1} = 0, P_0 = b_0 I, Q_0 = I
2 for j = 1: 2m
3     P_j = b_j P_{j-1} + a_j X P_{j-2}
4     Q_j = b_j Q_{j-1} + a_j X Q_{j-2}
5 end
6 r_m = P_{2m} Q_{2m}^{-1}
```

Cost: $2(2m-2)M+D$. (Note that for $j = 1, 2$ the multiplications XP_{j-2} and XQ_{j-2} are trivial, since P_{j-2} and Q_{j-2} are multiples of I.)

Using bottom-up evaluation, $r_m(X)$ is evaluated as follows.

Algorithm 4.10 (continued fraction, bottom-up). This algorithm evaluates the continued fraction (4.13) in bottom-up fashion at the matrix $X \in \mathbb{C}^{n\times n}$.

```
1 Y_{2m} = (a_{2m}/b_{2m}) X
2 for j = 2m - 1: -1: 1
3     Solve (b_j I + Y_{j+1}) Y_j = a_j X for Y_j.
4 end
5 r_m = b_0 I + Y_1
```

Cost: $(2m-1)D$.

The top-down evaluation is computationally expensive, but it is well suited to situations in which the a_j and b_j are independent of m and the whole sequence $r_1(X)$, $r_2(X)$, ..., needs to be evaluated; in this case the bottom-up evaluation has to start afresh each time.

Another representation of r_m in (4.13) is in partial fraction form:

$$r_m(x) = \sum_{j=1}^{m} \frac{\alpha_j^{(m)} x}{1 + \beta_j^{(m)} x}. \tag{4.14}$$

The coefficients $\beta_j^{(m)}$ are minus the reciprocals of the roots of the denominator polynomial $q_{mm}(x)$ and so may be complex; in this case an alternative partial fraction with quadratic denominators could be considered. The cost of evaluating (4.14) at the matrix X is just mD, but on a parallel computer the m terms in (4.14) can be evaluated in parallel.

Of course, the numerical stability of these different methods of evaluation needs to be considered along with the computational cost. Since the stability depends very much on the function f, we delay further consideration until later sections on particular f.

4.5. Diagonalization

A wide class of methods for evaluating matrix functions is based on exploiting the relation $f(ZBZ^{-1}) = Zf(B)Z^{-1}$ (Theorem 1.13 (c)). The idea is to factor $A = ZBZ^{-1}$, with B of a form that allows easy computation of $f(B)$. Then $f(A) = Zf(B)Z^{-1}$ is readily obtained.

The most obvious choice of similarity is a diagonalizing one, assuming A is indeed diagonalizable: $A = Z\,\mathrm{diag}(\lambda_i)Z^{-1}$, where the λ_i are the eigenvalues of A and the columns of Z are eigenvectors. Then $f(A) = Z\,\mathrm{diag}(f(\lambda_i))Z^{-1}$. Listing 4.1 lists a MATLAB function `funm_ev` implementing this formula. Here is an example of how the function is used to compute the matrix square root:

```
>> A = [3 -1; 1 1]; X = funm_ev(A,@sqrt)
X =
   1.7678e+000 -3.5355e-001
   3.5355e-001  1.0607e+000

>> norm(A-X^2)
ans =
   9.9519e-009
```

Given that A has norm of order 1 and the unit roundoff $u \approx 10^{-16}$, the residual $\|A - X^2\|_2 \approx 10^{-8}$ of the computed X is disappointing—especially considering that MATLAB's `sqrtm` function achieves a residual of order u:

```
>> Y = sqrtm(A); norm(A-Y^2)
ans =
   6.4855e-016
```

The explanation lies with the ill conditioning of the matrix Z:

```
>> [Z,D] = eig(A); cond(Z)
ans =
   9.4906e+007
```

That $\kappa_2(Z)u$ is roughly the size of the residual is no coincidence. Suppose the only error in the process is an error E in evaluating $f(B)$. Then we obtain

$$\widetilde{f}(A) = Z(f(B) + E)Z^{-1} = f(A) + ZEZ^{-1}$$

and

$$\|\widetilde{f}(A) - f(A)\| \le \|Z\|\,\|E\|\,\|Z^{-1}\| = \kappa(Z)\,\|E\|.$$

When B is diagonal and Gaussian elimination with partial pivoting is used in the evaluation, we should interpret $\kappa(Z)$ as

$$\min\{\,\kappa(ZD) : D \text{ diagonal, nonsingular}\,\}, \tag{4.15}$$

which for any p-norm is approximately achieved (and exactly achieved when $p = 1$) when ZD has columns of unit p-norm; see [276, 2002, Thm. 7.5, Sec. 9.8]. The conclusion is that we must expect errors proportional to $\kappa(Z)$ in our computed function. Since the conditioning of $f(A)$ is not necessarily related to $\kappa(Z)$, this diagonalization method may be numerically unstable.

Diagonalization was used to compute certain matrix functions in the original Fortran version of MATLAB ("Classic MATLAB", 1978–1984), which was designed for teaching purposes:

```
      < M A T L A B >
   Version of 01/10/84
```

Listing 4.1: MATLAB function funm_ev.

```
function F = funm_ev(A,fun)
%FUNM_EV   Evaluate general matrix function via eigensystem.
%   F = FUNM_EV(A,FUN) evaluates the function FUN at the
%   square matrix A using the eigensystem of A.
%   This function is intended for diagonalizable matrices only
%   and can be numerically unstable.

[V,D] = eig(A);
F = V * diag(feval(fun,diag(D))) / V;
```

```
 HELP is available

<>
help fun

 FUN    For matrix arguments  X , the  functions  SIN,  COS,  ATAN,
        SQRT,  LOG,  EXP and X**p are computed using eigenvalues  D
        and eigenvectors  V .  If  <V,D> =  EIG(X)   then   f(X)  =
        V*f(D)/V  .   This method may give inaccurate results if  V
        is badly conditioned.  Some idea of  the  accuracy  can  be
        obtained by comparing  X**1  with  X .
        For vector arguments,  the  function  is  applied  to  each
        component.
```

If A is not diagonalizable we can in theory evaluate f from Definition 1.2 based on the Jordan canonical form. However, the Jordan canonical form cannot be reliably computed in floating point arithmetic; even if it could, the similarity transformation that produces it can again be very ill conditioned.

The diagonalization approach can be recommended only when the diagonalizing transformation is guaranteed to be well conditioned. An important class of matrices for which this guarantee holds is the class of normal matrices: $A \in \mathbb{C}^{n\times n}$ for which $A^*A = AA^*$. This class includes orthogonal matrices, symmetric matrices, skew-symmetric matrices, and their complex analogues the unitary, Hermitian, and skew-Hermitian matrices. In fact, the normal matrices are precisely those that are diagonalizable by a unitary matrix, and unitary matrices are perfectly conditioned in the 2-norm.

One minor problem with the diagonalization approach is that if A is real with some complex eigenvalues and $f(A)$ is real, then rounding errors may cause the computed $f(A)$ to have a tiny nonzero imaginary part. This problem can be overcome by discarding the computed imaginary part, but of course numerical instability may potentially produce a *large* spurious imaginary part. (For symmetric matrices the eigensystem is real and these considerations do not apply.) One way to overcome this problem of nonreal computed $f(A)$ for real A is to employ a block diagonalization $A = ZDZ^{-1}$ in which $D, Z \in \mathbb{R}^{n\times n}$ and D has 2×2 diagonal blocks corresponding to complex conjugate pairs of eigenvalues. This approach falls into the class of block diagonalization methods considered in Section 4.7. Another approach is described in the next section.

In summary, for a normal matrix, computing $f(A)$ via a (unitary) diagonalization is the method of choice if the diagonalization can be computed. In particular, this method is recommended if A is symmetric or Hermitian.

4.6. Schur Decomposition and Triangular Matrices

We saw in the last section the importance of using well conditioned similarity transformations in order to maintain numerical stability. Taking this view to the extreme, we can restrict to unitary transformations. In general, the closest one can go to diagonal form via unitary similarities is the Schur triangular form. The Schur decomposition factors $A \in \mathbb{C}^{n\times n}$ as

$$Q^*AQ = T,$$

where $Q \in \mathbb{C}^{n\times n}$ is unitary and $T \in \mathbb{C}^{n\times n}$ is upper triangular. The eigenvalues of A appear on the diagonal of T.

The Schur decomposition can be computed with perfect backward stability by the QR algorithm, and hence it is a standard tool in numerical linear algebra. For us, the significance of the decomposition is that, since $f(A) = Qf(T)Q^*$, it reduces the $f(A)$ problem to that of computing $f(T)$. We therefore now turn our attention to functions of triangular matrices.

It is easy to see from any of the definitions that $f(T)$ is upper triangular with diagonal elements $f(t_{ii})$. In fact, explicit formulae are available for all the elements of $f(T)$.

Theorem 4.11 (function of triangular matrix). *Let $T \in \mathbb{C}^{n\times n}$ be upper triangular and suppose that f is defined on the spectrum of T. Then $F = f(T)$ is upper triangular with $f_{ii} = f(t_{ii})$ and*

$$f_{ij} = \sum_{(s_0,\ldots,s_k)\in S_{ij}} t_{s_0,s_1} t_{s_1,s_2} \ldots t_{s_{k-1},s_k} f[\lambda_{s_0},\ldots,\lambda_{s_k}],$$

where $\lambda_i = t_{ii}$, S_{ij} is the set of all strictly increasing sequences of integers that start at i and end at j, and $f[\lambda_{s_0},\ldots,\lambda_{s_k}]$ is the kth order divided difference of f at $\lambda_{s_0},\ldots,\lambda_{s_k}$.

Proof. See Davis [139, 1973], Descloux [148, 1963], or Van Loan [592, 1975]. □

It is worth examining the case $n = 2$ in Theorem 4.11. For $\lambda_1 \neq \lambda_2$ we have

$$f\left(\begin{bmatrix} \lambda_1 & t_{12} \\ 0 & \lambda_2 \end{bmatrix}\right) = \begin{bmatrix} f(\lambda_1) & t_{12}\dfrac{f(\lambda_2)-f(\lambda_1)}{\lambda_2-\lambda_1} \\ 0 & f(\lambda_2) \end{bmatrix}. \tag{4.16}$$

In the case $\lambda_1 = \lambda_2 = \lambda$ we have, using (B.27),

$$f\left(\begin{bmatrix} \lambda & t_{12} \\ 0 & \lambda \end{bmatrix}\right) = \begin{bmatrix} f(\lambda) & t_{12}f'(\lambda) \\ 0 & f(\lambda) \end{bmatrix}. \tag{4.17}$$

(This formula is a special case of (3.16).)

We can also use Theorem 4.11 to check formula (1.4) for a function of a Jordan block, $T = J_k(\lambda_k)$. Since the only nonzero off-diagonal elements of T are 1s on the superdiagonal, we have, using (B.27) again,

$$f_{ij} = t_{i,i+1} \dots t_{j-1,j}\, f[\underbrace{\lambda_k, \lambda_k, \dots, \lambda_k}_{j-i+1 \text{ times}}] = \frac{f^{(j-i)}(\lambda_k)}{(j-i)!}, \qquad i < j.$$

An expression for a function of an arbitrary block 2×2 block upper triangular matrix is given in the following result, which generalizes (4.16) and (4.17).

Theorem 4.12 (Kenney and Laub). *Let f be $2n-1$ times continuously differentiable and let*

$$A = \begin{bmatrix} A_{11} & A_{12} \\ 0 & A_{22} \end{bmatrix}, \quad D = \begin{bmatrix} A_{11} & 0 \\ 0 & A_{22} \end{bmatrix}, \quad N = \begin{bmatrix} 0 & A_{12} \\ 0 & 0 \end{bmatrix}.$$

Then $f(A) = f(D) + L(D,N)$ (i.e., the $o(\cdot)$ term in (3.6) is zero).

Proof. If f is a polynomial of degree m then the result is trivial if $m \le 1$ and otherwise follows from the fact that $f(D+N) - f(D) - L(D,N)$ comprises sums of products of the form $X_1 X_2 \dots X_p$, $p \ge 2$, where each X_i is either D or N and at least two of the X_i are N. Each of these products is zero. For general f, Theorem 3.7 implies that the Fréchet derivative $L(D,N)$ of f is the same as that of the polynomial $p_{D \oplus D}$. The result therefore follows from the first part. □

An interesting property revealed by the theorem is that the (1,2) block of $f(A)$ depends only linearly on the (1,2) block of A.

To compute $f(T)$ via Theorem 4.11 would cost $O(2^n)$ flops, which is prohibitively expensive. A method that avoids the combinatorial explosion in the formulae of the theorem was derived by Parlett [460, 1976]. He notes that $f(T)$ commutes with T (see Theorem 1.13 (a)) and that, since the diagonal of $F = f(T)$ is known, the equation $TF = FT$ can be solved for the off-diagonal elements of F. Indeed, equating (i,j) elements in this equation for $i < j$ yields

$$\sum_{k=i}^{j} t_{ik} f_{kj} = \sum_{k=i}^{j} f_{ik} t_{kj}$$

or

$$f_{ij}(t_{ii} - t_{jj}) = t_{ij}(f_{ii} - f_{jj}) + \sum_{k=i+1}^{j-1} (f_{ik} t_{kj} - t_{ik} f_{kj}), \tag{4.18}$$

which gives, if $t_{ii} \ne t_{jj}$,

$$f_{ij} = t_{ij} \frac{f_{ii} - f_{jj}}{t_{ii} - t_{jj}} + \sum_{k=i+1}^{j-1} \frac{f_{ik} t_{kj} - t_{ik} f_{kj}}{t_{ii} - t_{jj}}, \qquad i < j.$$

The right-hand side depends only on the elements to the left of f_{ij} and below it. Hence this recurrence enables F to be computed either a superdiagonal at a time, starting with the diagonal, or a column at a time, from the second column to the last, moving up each column.

Algorithm 4.13 (Parlett recurrence). Given an upper triangular $T \in \mathbb{C}^{n\times n}$ with distinct diagonal elements and a function f defined on the spectrum of T, this algorithm computes $F = f(T)$ using Parlett's recurrence.

1 $f_{ii} = f(t_{ii}),\ i = 1\colon n$
2 for $j = 2\colon n$
3 for $i = j-1\colon -1\colon 1$
4 $$f_{ij} = t_{ij}\frac{f_{ii} - f_{jj}}{t_{ii} - t_{jj}} + \Big(\sum_{k=i+1}^{j-1} f_{ik}t_{kj} - t_{ik}f_{kj}\Big) \Big/ (t_{ii} - t_{jj})$$
5 end
6 end

Cost: $2n^3/3$ flops.

Parlett's recurrence has a major drawback: it breaks down when $t_{ii} = t_{jj}$ for some $i \neq j$, that is, when T has repeated eigenvalues. In this situation (4.18) provides no information about f_{ij}.

A way around this difficulty is to employ a block form of the recurrence, also described by Parlett [459, 1974]. Let $T = (T_{ij})$ be block upper triangular with square diagonal blocks, possibly of different sizes. We will assume that T is also triangular, though this is not necessary to derive the recurrence. Then $F = (F_{ij})$ has the same block structure and equating (i, j) blocks in $TF = FT$ leads to

$$T_{ii}F_{ij} - F_{ij}T_{jj} = F_{ii}T_{ij} - T_{ij}F_{jj} + \sum_{k=i+1}^{j-1} (F_{ik}T_{kj} - T_{ik}F_{kj}), \qquad i < j. \tag{4.19}$$

This recurrence can be used to compute F a block superdiagonal at a time or a block column at a time, provided we can evaluate the diagonal blocks $F_{ii} = f(T_{ii})$ and solve the Sylvester equations (4.19) for the F_{ij}. The Sylvester equation (4.19) is nonsingular if and only if T_{ii} and T_{jj} have no eigenvalue in common (see Section B.14). Therefore in order to use this block recurrence we need first to reorder the matrix T so that no two diagonal blocks have an eigenvalue in common; here, reordering means applying a unitary similarity transformation to permute the diagonal elements whilst preserving triangularity. Reordering can be achieved by standard techniques and we return to this topic in Section 9.3.

Algorithm 4.14 (block Parlett recurrence). Given a triangular matrix $T = (T_{ij}) \in \mathbb{C}^{n\times n}$ partitioned in block form with no two diagonal blocks having an eigenvalue in common, and a function f defined on the spectrum of T, this algorithm computes $F = f(T)$ using the block form of Parlett's recurrence.

1 $F_{ii} = f(T_{ii}),\ i = 1\colon n$
2 for $j = 2\colon n$
3 for $i = j-1\colon -1\colon 1$
4 Solve for F_{ij} the Sylvester equation
$T_{ii}F_{ij} - F_{ij}T_{jj} = F_{ii}T_{ij} - T_{ij}F_{jj} + \sum_{k=i+1}^{j-1}(F_{ik}T_{kj} - T_{ik}F_{kj})$
5 end
6 end

Cost: Dependent on the block sizes and f, in general.

In Algorithm 4.14, computing $F_{ii} = f(T_{ii})$ for a block of dimension greater than 1 is a nontrivial problem that we pursue in Section 9.1.

The block recurrence can be used in conjunction with the real Schur decomposition of $A \in \mathbb{R}^{n\times n}$,

$$Q^T AQ = T,$$

where $Q \in \mathbb{R}^{n\times n}$ is orthogonal and $T \in \mathbb{R}^{n\times n}$ is quasi upper triangular, that is, block upper triangular with 1×1 or 2×2 diagonal blocks, with any 2×2 diagonal blocks having complex conjugate eigenvalues. When $f(A)$ is real, this enables it to be computed entirely in real arithmetic.

We turn now to numerical considerations. In Listing 4.2 we give a function `funm_simple` that employs Algorithm 4.13 in conjunction with an initial Schur reduction to triangular form. The function is in principle applicable to any matrix with distinct eigenvalues. It is very similar to the function `funm` in versions 6.5 (R14) and earlier of MATLAB; version 7 of MATLAB introduced a new `funm` that implements Algorithm 9.6 described in Section 9.4, which itself employs Algorithm 4.14.

Function `funm_simple` often works well. For example the script M-file

```
format rat, A = gallery('parter',4), format short
evals = eig(A)'
X = real(funm_simple(A,@sqrt))
res = norm(A-X^2)
```

produces the output

```
A =
         2            -2           -2/3          -2/5
       2/3             2             -2          -2/3
       2/5           2/3              2            -2
       2/7           2/5            2/3             2
evals =
   1.5859 - 2.0978i   1.5859 + 2.0978i   2.4141 - 0.7681i
   2.4141 + 0.7681i
X =
    1.4891   -0.6217   -0.3210   -0.2683
    0.2531    1.5355   -0.5984   -0.3210
    0.1252    0.2678    1.5355   -0.6217
    0.0747    0.1252    0.2531    1.4891
res =
  1.6214e-014
```

A major weakness of `funm_simple` is demonstrated by the following experiment. Let A be the 8×8 triangular matrix with $a_{ii} \equiv 1$ and $a_{ij} \equiv -1$ for $j > i$, which is MATLAB's `gallery('triw',8)`. With f the exponential, Table 4.3 shows the normwise relative errors for A and two small perturbations of A, one full and one triangular. The condition number of $f(A)$ (see Chapter 3) is about 2 in each case, so we would expect to be able to compute $f(A)$ accurately. For A itself, `funm_simple` yields an error of order 1, which is expected since it is unable to compute any of the superdiagonal elements of $f(A)$. For A plus the random full perturbation (which, being full, undergoes the Schur reduction) the eigenvalues are distinct and at distance at least 10^{-2} apart. But nevertheless, `funm_simple` loses 6 significant digits of accuracy. For the third matrix, in which the perturbation is triangular and the eigenvalues are

Listing 4.2: MATLAB function funm_simple.

```
function F = funm_simple(A,fun)
%FUNM_SIMPLE Simplified Schur-Parlett method for function of a matrix.
%   F = FUNM_SIMPLE(A,FUN) evaluates the function FUN at the
%   square matrix A by the Schur-Parlett method using the scalar
%   Parlett recurrence (and hence without blocking or reordering).
%   This function is intended for matrices with distinct eigenvalues
%   only and can be numerically unstable.
%   FUNM should in general be used in preference.

n = length(A);

[Q,T] = schur(A,'complex');   % Complex Schur form.
F = diag(feval(fun,diag(T))); % Diagonal of F.

% Compute off-diagonal of F by scalar Parlett recurrence.
for j=2:n
   for i = j-1:-1:1
      s = T(i,j)*(F(i,i)-F(j,j));
      if j-i >= 2
         k = i+1:j-1;
         s = s + F(i,k)*T(k,j) - T(i,k)*F(k,j);
      end
      d = T(i,i) - T(j,j);
      if d ~= 0
         F(i,j) = s/d;
      end
   end
end

F = Q*F*Q';
```

about 10^{-9} apart, funm_simple is spectacularly inaccurate: it produces a computed answer with elements increasing away from the diagonal to a maximum of 10^{44} in the top right corner. On the other hand, MATLAB's funm, for which the errors are shown in the second column, produces answers accurate to almost full machine precision.

The conclusion of this experiment is that it is not just repeated eigenvalues that are bad for the scalar Parlett recurrence. Close eigenvalues can lead to a severe loss of accuracy, and even when the eigenvalues are far apart the recurrence can produce more inaccurate answers than expected. Indeed it is a well-known phenomenon in numerical analysis that near confluence is much more dangerous and difficult to deal with than exact confluence [326, 1972]. The block recurrence in Algorithm 4.14 is therefore not a panacea: while the recurrence will always run to completion under the distinct spectra condition, when two diagonal blocks have close spectra we can expect difficulties.

We return to the Schur–Parlett approach in Chapter 9, where we develop a practical algorithm that performs well in finite precision arithmetic with only minimal assumptions on the matrix and function.

Table 4.3. *Errors* $\|e^A - \widehat{F}\|/\|e^A\|$ *for* $\widehat{F}$ *from* `funm_simple` *for* $A =$ `gallery('triw',8)`.

	`funm_simple`	`funm`
A	1.0e0	5.0e-16
$A +$ `rand(8)*1e-8`	4.6e-10	3.4e-15
$A +$ `triu(rand(8))*1e-8`	4.5e44	4.1e-16

4.7. Block Diagonalization

An alternative to diagonalization is block diagonalization, in which we compute $A = XDX^{-1}$, where D is block diagonal. As we saw in Section 4.5 it is important for numerical stability reasons that X be well conditioned. We assume that a block diagonalization is computed in the usual way by first computing the Schur form and then eliminating off-diagonal blocks by solving Sylvester equations [48, 1979], [224, 1996, Sec. 7.6.3], [375, 1997]. In order to guarantee a well conditioned X a bound must be imposed on the condition of the individual transformations; this bound will be a parameter in the algorithm.

Since $f(A) = Xf(D)X^{-1}$, the problem reduces to computing $f(D)$ and hence to computing $f(D_{ii})$ for each diagonal block D_{ii}. The D_{ii} are triangular but no particular eigenvalue distribution is guaranteed, because of the limitations on the condition of the transformations; therefore $f(D_{ii})$ is still a nontrivial calculation. When A is upper triangular, it is even possible that the block diagonalization procedure leaves A unchanged. We therefore prefer the Schur–Parlett method to be described in Chapter 9.

The Schur–Parlett method and the block diagonalization method are closely related. Both employ a Schur decomposition, both solve Sylvester equations, and both must compute $f(T_{ii})$ for triangular blocks T_{ii}. Parlett and Ng [462, 1985, Sec. 5] show that the two methods are mathematically equivalent, differing only in the order in which two commuting Sylvester operators are applied. See Problem 4.3.

4.8. Interpolating Polynomial and Characteristic Polynomial

The definition $f(A) = p(A)$, where A is the Hermite interpolating polynomial p in (1.7) or a polynomial such as that in (1.11) that satisfies additional interpolation conditions, suggests a numerical method. However, the method is impractical in general—for two reasons. First, it requires $O(n)$ matrix multiplications and hence $O(n^4)$ flops to produce $f(A)$, even given the polynomial p in monomial or divided difference form, whereas most methods require only $O(n^3)$ flops. Second, the numerical stability of this approach is highly dubious, and no error analysis is available to justify its usage. For one specialized context in which a viable numerical method can be built from this approach, see Section 10.4.1.

Frequently in the literature methods have been proposed for computing $f(A)$ that first compute the characteristic polynomial of A and then make use of this polynomial in one way or another. We have eschewed all such methods in this book because the characteristic polynomial cannot be reliably computed in floating point arithmetic. Hence we do not believe that any method based on the characteristic polynomial is a viable practical method, except possibly for very special A. We give an example to illustrate the numerical dangers. Consider computation of the inverse of a nonsingular

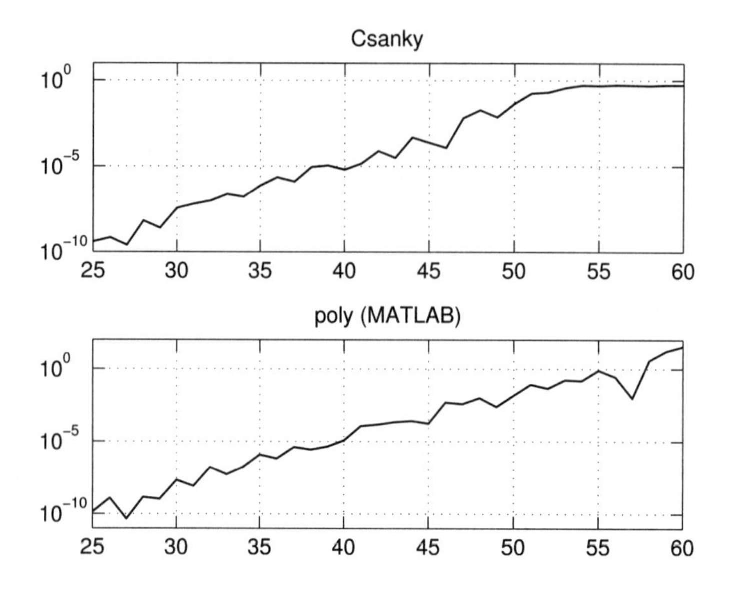

Figure 4.2. *Relative errors for inversion of* $A = 3I_n$, $n = 25\colon 60$, *via the characteristic polynomial.*

$A \in \mathbb{C}^{n\times n}$ with characteristic polynomial $\det(A - xI) = x^n + c_1x^{n-1} + \cdots + c_n$. By the Cayley–Hamilton theorem,

$$A^{-1} = -\frac{1}{c_n}\left(A^{n-1} + \sum_{i=1}^{n-1} c_i A^{n-i-1}\right).$$

The coefficients c_i can be obtained by computing the eigenvalues λ_i of A and then using recurrences obtained from the factorization $\det(xI - A) = \prod_{i=1}^{n}(x - \lambda_i)$; this method is implemented in the MATLAB function `poly`. Alternatively, the c_i can be obtained by computing the traces $s_k = \mathrm{trace}(A^k)$, $k = 1\colon n$ and then solving the Newton identities [301, 1970, p. 37], [311, 1966, p. 324]

$$\begin{bmatrix} 1 & & & & & \\ s_1 & 2 & & & & \\ s_2 & s_1 & 3 & & & \\ s_3 & s_2 & s_1 & 4 & & \\ \vdots & \ddots & \ddots & \ddots & \ddots & \\ s_{n-1} & \dots & s_3 & s_2 & s_1 & n \end{bmatrix} \begin{bmatrix} c_1 \\ c_2 \\ \vdots \\ \vdots \\ \vdots \\ c_n \end{bmatrix} = \begin{bmatrix} -s_1 \\ -s_2 \\ \vdots \\ \vdots \\ \vdots \\ -s_n \end{bmatrix}. \tag{4.20}$$

The method based on the latter system was proposed for parallel computation by Csanky [124, 1976] (it takes $O(\log^2 n)$ time on $O(n^4)$ processors) but in fact goes back at least to Bingham [68, 1941]. Figure 4.2 plots the ∞-norm relative errors when $A = 3I_n$ is inverted by these two approaches. All accuracy is lost by the time $n = 55$, despite A being perfectly conditioned.

4.9. Matrix Iterations

A number of matrix functions $f(A)$ are amenable to computation by iteration:

$$X_{k+1} = g(X_k). \tag{4.21}$$

For the iterations used in practice, X_0 is not arbitrary but is a fixed function of A—usually $X_0 = I$ or $X_0 = A$. The iteration function g may or may not depend on A.

Considerations of computational cost usually dictate that g is a polynomial or rational function. Rational g require the solution of linear systems with multiple right-hand sides, or even explicit matrix inversion. On modern computers with hierarchical memories, matrix multiplication is usually much faster than solving a matrix equation or inverting a matrix, so iterations that are multiplication-rich, which means having a polynomial g, are preferred. The drawback is that such iterations usually have weaker convergence properties than rational iterations.

A standard means of deriving matrix iterations is to apply Newton's method to an algebraic equation satisfied by $f(A)$ and then to choose X_0 so that the iteration formula is simplified. In subsequent chapters we will study iterations that can be derived in this way for computing the matrix sign function, the matrix square root, matrix pth roots, and the polar decomposition.

It is important to keep in mind that results and intuition from scalar nonlinear iterations do not necessarily generalize to the matrix case. For example, standard convergence conditions expressed in terms of derivatives of g at a fixed point in the scalar case do not directly translate into analogous conditions on the Frechét and higher order derivatives in the matrix case.

4.9.1. Order of Convergence

One of the key properties of a matrix iteration (4.21) is its order (or rate) of convergence. If X_k is a sequence of matrices converging to X_* we say the convergence is of *order* p if p is the largest number such that

$$\|X_* - X_{k+1}\| \le c\|X_* - X_k\|^p \tag{4.22}$$

for all sufficiently large k, for some positive constant c. The iteration (4.21) is said to have order p if the convergent sequences it generates have order p. *Linear convergence* corresponds to $p = 1$ and *quadratic convergence* to $p = 2$. The convergence is called *superlinear* if $\lim_{k\to\infty} \|X_* - X_{k+1}\|/\|X_* - X_k\| = 0$.

The convergence of a sequence breaks into two parts: the initial phase in which the error is reduced safely below 1, and the asymptotic phase in which (4.22) guarantees convergence to zero. The order of convergence tells us about the rate of convergence in the asymptotic phase, but it has nothing to say about how many iterations are taken up by the first phase. This is one reason why a higher order of convergence is not necessarily better. Other factors to consider when comparing iterations with different orders of convergence are as follows.

- The overall efficiency of an iteration depends on the computational cost of evaluating X_{k+1} as well as the convergence behaviour. Two successive steps of a quadratically convergent iteration (such as Newton's method) define a quartically convergent iteration, so any other quartic iteration needs to have cost per iteration no larger than twice that of the quadratic one if it is to be worth considering.

- In practice we often need to scale an iteration, that is, introduce scaling parameters to reduce the length of the initial convergence phase. The higher the order the less opportunity there is to scale (relative to the amount of computation).
- Numerical stability considerations may rule out certain iterations from consideration for practical use.

4.9.2. Termination Criteria

An important issue is how to terminate a matrix iteration. The first question that must be addressed is the aim of the test: to stop when X_k has relative error below a tolerance or to stop when some suitable residual (such as $X_k^2 - A$ for the square root function) is below the tolerance. The relevant aim will in general be function and problem dependent.

Often, a stopping test is based on the relative difference

$$\delta_{k+1} = \frac{\|X_{k+1} - X_k\|}{\|X_{k+1}\|}$$

between two successive iterates, which is regarded as an approximation to the relative error $\|X_{k+1} - X_*\|/\|X_*\|$ in X_{k+1}, where $X_* = \lim_{k\to\infty} X_k$. However, the relative difference is actually approximating the relative error in X_k, $\|X_k - X_*\|/\|X_*\|$. Indeed

$$X_{k+1} - X_* = (X_{k+1} - X_k) + (X_k - X_*),$$

and for sufficiently fast convergence $\|X_{k+1} - X_*\| \ll \|X_k - X_*\|$ and hence the two terms on the right-hand side are of roughly equal norm (and their largest elements are of opposite signs). To obtain more insight, consider a quadratically convergent method, for which

$$\|X_{k+1} - X_*\| \le c\|X_k - X_*\|^2 \tag{4.23}$$

close to convergence, where c is a constant. Now

$$\begin{aligned}\|X_k - X_*\| &\le \|X_k - X_{k+1}\| + \|X_{k+1} - X_*\| \\ &\le \|X_k - X_{k+1}\| + c\|X_k - X_*\|^2.\end{aligned}$$

"Solving" for $\|X_k - X_*\|$ and substituting in (4.23) gives

$$\begin{aligned}\|X_{k+1} - X_*\| &\le c\left(\frac{\|X_{k+1} - X_k\|}{1 - c\|X_k - X_*\|}\right)^2 \\ &\le 2c\|X_{k+1} - X_k\|^2,\end{aligned} \tag{4.24}$$

for small enough $\|X_k - X_*\|$, so the error in X_{k+1} is bounded in terms of the *square* of $\|X_{k+1} - X_k\|$. The conclusion is that a stopping test that accepts X_{k+1} when δ_{k+1} is of the order of the desired relative error may terminate one iteration too late.

The analysis above suggests an alternative stopping test for a quadratically convergent iteration. From (4.24) we expect $\|X_{k+1} - X_*\|/\|X_{k+1}\| \le \eta$ if

$$\|X_{k+1} - X_k\| \le \left(\frac{\eta\|X_{k+1}\|}{2c}\right)^{1/2}. \tag{4.25}$$

The value of c in (4.23) is usually known. The test (4.25) can be expected to terminate one iteration earlier than the test $\delta_{k+1} \le \eta$, but it can potentially terminate too soon if rounding errors vitiate (4.23) or if the test is satisfied before the iteration enters the regime where (4.23) is valid.

A stopping test based on the distance between successive iterates is dangerous because in floating point arithmetic there is typically no guarantee that δ_k (or indeed the relative error) will reach a given tolerance. One solution is to stop when a preassigned number of iterations have passed. An alternative, suitable for any quadratically or higher order convergent iteration, is to terminate once the relative change has not decreased by a factor at least 2 since the previous iteration. A reasonable convergence test is therefore to terminate the iteration at X_{k+1} when

$$\delta_{k+1} \le \eta \quad \text{or} \quad \delta_{k+1} \ge \delta_k/2, \tag{4.26}$$

where η is the desired relative error. This test can of course be combined with others, such as (4.25).

4.9.3. Convergence

Convergence analysis for the matrix iterations in this book will be done in two general ways. The first is to work entirely at the matrix level. This is our preference, when it is possible, because it tends to give the most insight into the behaviour of the iteration. The second approach breaks into three parts:

- show, if possible, that the iteration converges for an arbitrary matrix if and only if it converges when applied to the Jordan blocks of the matrix (why this is not always possible is explained in the comments after Theorem 6.9);
- prove convergence of the scalar iteration when applied to the eigenvalues (which implies convergence of the diagonal of the Jordan blocks);
- prove convergence of the off-diagonals of the Jordan blocks.

The last step can be done in some generality, as our next theorem shows.

Consider, for example, the iteration

$$X_{k+1} = \frac{1}{2}(X_k + X_k^{-1}A), \qquad X_0 = A \in \mathbb{C}^{n\times n}, \tag{4.27}$$

which is an analogue for matrices of Heron's method for the square root of a scalar. As is well known, for $n = 1$ the iteration converges to the principal square root of a if a does not lie on $\mathbb{R}^-$. If A has the Jordan canonical form $A = ZJZ^{-1}$ then it is easy to see that the iterates from (4.27) are given by $X_k = ZJ_kZ^{-1}$, where $J_{k+1} = \frac{1}{2}(J_k + J_k^{-1}J)$, $J_0 = J$. Hence J_k has the same block diagonal structure as J, and convergence reduces to the case where J is a single Jordan block. The next result allows us to deduce convergence of the X_k when A has no eigenvalues on $\mathbb{R}^-$. This result is quite general and will be of use for other iterations in later chapters.

Theorem 4.15. *Let $g(x,t)$ be a rational function of both of its arguments. Let the scalar sequence generated by $x_{k+1} = g(x_k, \lambda)$, $x_0 = \phi_0(\lambda)$ converge to $x_* = f(\lambda)$, where ϕ_0 is a rational function and $\lambda \in \mathbb{C}$, and assume that $|\frac{\partial g}{\partial x}(x_*, \lambda)| < 1$ (i.e., x_* is an attracting fixed point of the iteration). Then the matrix sequence generated by $X_{k+1} = g(X_k, J(\lambda))$, $X_0 = \phi_0(J(\lambda))$, where $J(\lambda) \in \mathbb{C}^{m\times m}$ is a Jordan block, converges to a matrix X_* with $(X_*)_{ii} \equiv f(\lambda)$.*

Proof. Let $\phi_k(t) = g(\phi_{k-1}(t), t)$, $k \geq 1$, so that ϕ_k is a rational function of t and $x_k = \phi_k(\lambda)$. Upper triangular Toeplitz structure is preserved under inversion and multiplication, and hence under any rational function. Hence

$$X_k = g(X_{k-1}, J(\lambda)) = \phi_k(J(\lambda)) = \begin{bmatrix} a_1(k) & a_2(k) & \dots & a_m(k) \\ & a_1(k) & \ddots & \vdots \\ & & \ddots & a_2(k) \\ & & & a_1(k) \end{bmatrix}.$$

Clearly, the diagonal elements of X_k satisfy $a_1(k) = \phi_k(\lambda) = x_k$ and hence tend to x_*. It remains to show that the elements in the strictly upper triangular part of X_k converge.

From (1.4) we know that

$$a_j(k) = \frac{1}{(j-1)!} \frac{d^{j-1}}{dt^{j-1}} \phi_k(t)\Big|_\lambda .$$

For $j = 2$, using the chain rule,

$$\begin{aligned} a_2(k) &= \frac{d\phi_k}{dt}(t)\Big|_\lambda = \frac{d}{dt} g(\phi_{k-1}(t), t)\Big|_\lambda \\ &= \left[\frac{\partial g}{\partial x}(\phi_{k-1}(t), t)\,\phi'_{k-1}(t) + \frac{\partial g}{\partial t}(\phi_{k-1}(t), t)\right]\Bigg|_\lambda \\ &= \frac{\partial g}{\partial x}(x_{k-1}, \lambda)\phi'_{k-1}(\lambda) + \frac{\partial g}{\partial t}(x_{k-1}, \lambda) \\ &= \frac{\partial g}{\partial x}(x_{k-1}, \lambda)a_2(k-1) + \frac{\partial g}{\partial t}(x_{k-1}, \lambda). \end{aligned}$$

Since $|\frac{\partial g}{\partial x}(x_*, \lambda)| < 1$, by Problem 4.7 it follows that

$$a_2(k) \to \frac{\partial g}{\partial t}(x_*, \lambda)/(1 - \frac{\partial g}{\partial x}(x_*, \lambda)) \quad \text{as } k \to \infty. \tag{4.28}$$

As an induction hypothesis suppose that $a_i(k)$ has a limit as $k \to \infty$ for $i = 2{:}\,j-1$. Then by the chain rule

$$\begin{aligned} a_j(k) &= \frac{1}{(j-1)!} \frac{d^{j-1}}{dt^{j-1}} \phi_k(t)\Big|_\lambda = \frac{1}{(j-1)!} \frac{d^{j-1}}{dt^{j-1}} g(\phi_{k-1}(t), t)\Big|_\lambda \\ &= \frac{1}{(j-1)!}\left[\frac{d^{j-2}}{dt^{j-2}}\left(\frac{\partial g}{\partial x}(\phi_{k-1}(t), t)\,\phi'_{k-1}(t) + \frac{\partial g}{\partial t}(\phi_{k-1}(t), t)\right)\right]\Big|_\lambda \\ &= \frac{1}{(j-1)!}\left[\frac{\partial g}{\partial x}(\phi_{k-1}(t), t)\frac{d^{j-1}}{dt^{j-1}}\phi_{k-1}(t) + \tau_j^{(k)}(\phi_{k-1}(t), t)\right]\Big|_\lambda \\ &= \frac{\partial g}{\partial x}(x_{k-1}, \lambda)a_j(k-1) + \frac{1}{(j-1)!}\tau_j^{(k)}(x_{k-1}, \lambda), \end{aligned} \tag{4.29}$$

where $\tau_j^{(k)}(x_{k-1}, \lambda)$ is a sum of terms comprising products of one or more elements $a_i(k-1)$, $i = 1{:}\,j-1$, and derivatives of g evaluated at (x_{k-1}, λ), the number of terms in the sum depending on j but not k. By the inductive hypothesis, and since

$x_{k-1} \to x_*$ as $k \to \infty$, $\tau_j^{(k)}(x_{k-1}, \lambda)$ has a limit as $k \to \infty$. Hence by Problem 4.7, $a_j(k)$ has a limit as $k \to \infty$, as required. □

The notation "$g(x_k, \lambda)$" in Theorem 4.15 may seem unnecessarily complicated. It is needed in order for the theorem to allow either or both of the possibilities that

(a) the iteration function g depends on A,

(b) the starting matrix X_0 depends on A.

For the Newton iteration (4.27), both g and X_0 depend on A, while for the sign iteration in (4.30) below only X_0 depends on A.

Notice that Theorem 4.15 does not specify the off-diagonal elements of the limit matrix X_*. These can usually be deduced from the equation $X_* = g(X_*, A)$ together with knowledge of the eigenvalues of X_*. For example, for the Newton square root iteration we know that $X_* = \frac{1}{2}(X_* + X_*^{-1}A)$, or $X_*^2 = A$, and, from knowledge of the scalar iteration, that X_* has spectrum in the open right half-plane provided that A has no eigenvalues on $\mathbb{R}^-$. It follows that X_* is the principal square root (see Theorem 1.29). A variant of Theorem 4.15 exists that assumes that f is analytic and guarantees convergence to $X_* = f(J(\lambda))$; see Iannazzo [307, 2007].

In the special case where the iteration function does not depend on the parameter λ, the limit matrix must be diagonal, even though the starting matrix is a Jordan block.

Corollary 4.16 (Iannazzo). *Under the conditions of Theorem 4.15, if the iteration function g does not depend on λ then the limit X_* of the matrix iteration is diagonal.*

Proof. In the notation of the proof of Theorem 4.15 we need to show that $a_j(k) \to 0$ as $k \to \infty$ for $j = 2\colon m$. By assumption, $\partial g/\partial t \equiv 0$ and so $a_2(k) \to 0$ as $k \to \infty$ by (4.28). An inductive proof using (4.29) then shows that $a_j(k) \to 0$ as $k \to \infty$ for $j = 3\colon m$. □

4.9.4. Numerical Stability

If X_0 is a function of A then the iterates from (4.21) are all functions of A and hence commute with A, and indeed commutativity properties are frequently used when deriving an iteration and in proving convergence. In finite precision arithmetic, rounding errors cause a loss of commutativity that can manifest itself as numerical instability. Looked at another way, iterations for matrix functions are often not self-correcting, so arbitrary errors can potentially propagate unstably. We will therefore pay careful attention to stability when deriving and analyzing matrix iterations.

The stability properties of superficially similar matrix iterations can be very different, as we now illustrate. Consider the Newton square root iteration (4.27). As noted in the previous section, for any A with no eigenvalues on $\mathbb{R}^-$, X_k converges quadratically to $A^{1/2}$ (see also Theorem 6.9). Table 4.4 shows the relative errors and $(1,1)$ elements of the iterates when the iteration is carried out in IEEE single precision arithmetic with A the Wilson matrix

$$A = \begin{bmatrix} 10 & 7 & 8 & 7 \\ 7 & 5 & 6 & 5 \\ 8 & 6 & 10 & 9 \\ 7 & 5 & 9 & 10 \end{bmatrix},$$

Table 4.4. *Square root iteration* (4.27) *and sign iteration* (4.30) *applied to Wilson matrix in single precision arithmetic.* $(A^{1/2})_{11} = 2.389$ *to four significant figures, while* $\mathrm{sign}(A) = I$.

	Square root		Sign	
	$\frac{\|A^{1/2}-X_k\|_2}{\|A^{1/2}\|_2}$	$(X_k)_{11}$	$\|I - Y_k\|_2$	$(Y_k)_{11}$
1	1.84e0	5.50e0	4.83e1	1.75e1
2	5.97e-1	3.36e0	2.36e1	8.90e0
3	1.12e-1	2.57e0	1.13e1	4.67e0
4	5.61e-3	2.40e0	5.21e0	2.61e0
5	4.57e-3	2.40e0	2.19e0	1.63e0
6	1.22e-1	2.21e0	7.50e-1	1.20e0
7	3.26e0	7.20e0	1.61e-1	1.04e0
8	8.74e1	−1.26e2	1.11e-2	1.00e0
9	2.33e3	3.41e3	6.12e-5	1.00e0
10	1.91e4	2.79e4	9.78e-10	1.00e0
11	1.97e4	−2.87e4	0	1.00e0
12	8.45e3	−1.23e4	0	1.00e0
13	1.08e3	−1.58e3	0	1.00e0
14	8.37e3	1.22e4	0	1.00e0

which is symmetric positive definite and moderately ill conditioned, with $\kappa_2(A) \approx 2984$. For such a benign matrix, we might expect no difficulty in computing $A^{1/2}$. The computed X_k behave as they would in exact arithmetic up to around iteration 4, but thereafter the iterates rapidly diverge. It turns out that iteration (4.27) is unstable unless A has very closely clustered eigenvalues. The instability is related to the fact that (4.27) fails to converge for some matrices X_0 in a neighbourhood of $A^{1/2}$; see Section 6.4 for the relevant analysis.

Let us now modify the square root iteration by replacing A in the iteration formula by I:

$$Y_{k+1} = \frac{1}{2}(Y_k + Y_k^{-1}), \qquad Y_0 = A. \tag{4.30}$$

For any A for which $\mathrm{sign}(A)$ is defined, Y_k converges quadratically to $\mathrm{sign}(A)$. (Note, for consistency with (4.27), that $\mathrm{sign}(A)$ is one of the square roots of I.) Iteration (4.30) is stable for all A (see Theorem 5.13), and Table 4.4 confirms the stability for the Wilson matrix.

In order to understand fully the behaviour of a matrix iteration in finite precision arithmetic we would like to

- bound or estimate the minimum relative error $\|X_* - \widehat{X}_k\|/\|X_*\|$ over the computed iterates $\widehat{X}_k$, and determine whether this error is consistent with the conditioning of the problem,

- determine how the iteration propagates errors introduced by roundoff and, in particular, whether the error growth is bounded or unbounded.

The first task is very difficult for most iterations, though the conditioning can be determined, as shown in the previous chapter. To make progress on this task we examine what happens when X_0 is very close to X_*, which leads to the notion of limiting accuracy. The second task, which is related to the first, is also difficult when considered over the whole iteration, but near to convergence the propagation of errors is more amenable to analysis.

We will use the following definition of stability of an iteration. We write $L^i(X)$ to denote the ith power of the Fréchet derivative L at X, defined as i-fold composition; thus $L^3(X,E) \equiv L\big(X, L(X, L(X,E))\big)$.

Definition 4.17 (stability). *Consider an iteration $X_{k+1} = g(X_k)$ with a fixed point X. Assume that g is Fréchet differentiable at X. The iteration is* stable in a neighborhood of X *if the Fréchet derivative $L_g(X)$ has bounded powers, that is, there exists a constant c such that $\|L_g^i(X)\| \le c$ for all $i > 0$.*

Note that stability concerns behaviour close to convergence and so is an asymptotic property. We will reserve the term "numerical stability" for global properties of the iteration.

To see the relevance of the definition, let $X_0 = X + E_0$, with arbitrary E_0 of sufficiently small norm, and let $E_k := X_k - X$. Then, by the definition of Fréchet derivative,

$$X_k = g(X_{k-1}) = g(X + E_{k-1}) = g(X) + L_g(X, E_{k-1}) + o(\|E_{k-1}\|).$$

So, since $g(X) = X$,

$$E_k = L_g(X, E_{k-1}) + o(\|E_{k-1}\|). \tag{4.31}$$

This relation can be recurred to give

$$E_k = L_g^k(X, E_0) + \sum_{i=0}^{k-1} L_g^i\big(X, o(\|E_{k-1-i}\|)\big),$$

so for a stable iteration,

$$\|E_k\| \le c\|E_0\| + c\sum_{i=0}^{k-1} o(\|E_{k-1-i}\|) \le c\|E_0\| + kc \cdot o(\|E_0\|).$$

The definition therefore ensures that in a stable iteration sufficiently small errors introduced near a fixed point have a *bounded effect*, to first order, on succeeding iterates.

To test for stability we need to find the Fréchet derivative of the iteration function g and then determine the behaviour of its powers, possibly by working with the Kronecker matrix form of the Fréchet derivative. Useful here is the standard result that a linear operator on $\mathbb{C}^{n\times n}$ (or its equivalent $n^2 \times n^2$ matrix) is power bounded if its spectral radius is less than 1 (see Problem 4.6) and not power bounded if its spectral radius exceeds 1. The latter property can be seen from the fact that $\|L_g^k(X)\| \ge |\lambda|^k$ for any eigenvalue λ of $L_g(X)$ (cf. (3.19)). We will sometimes find the pleasing situation that $L_g(X)$ is idempotent, that is, $L_g^2(X,E) = L_g(X, L_g(X,E)) = L_g(X,E)$, in which case power boundedness is immediate.

Note that any superlinearly convergent scalar iteration $x_{k+1} = g(x_k)$ has zero derivative g' at a fixed point, so for such scalar iterations convergence implies stability. For matrix iterations, however, the Fréchet derivative is not generally zero at a fixed point. For example, for the sign iteration (4.30) it is easy to see that at a fixed point Y we have $L_g(Y,E) = \frac{1}{2}(E - YEY)$ (see Theorem 5.13). Although $Y^2 = I$, $YE \ne EY$

in general and so $L_g(Y, E) \neq 0$. This emphasizes that stability is a more subtle and interesting issue for matrices than in the scalar case.

An important special case is when the underlying function f has the property that $f(f(A)) = f(A)$ for all A for which f is defined, that is, f is idempotent. For such an f we can show that $L_f(X)$ is idempotent at $X = f(X)$.

Theorem 4.18. *Let f be an idempotent function that is Fréchet differentiable at $X = f(X)$. Then $L_f(X)$ is idempotent.*

Proof. Let $h(t) = f(f(t))$. By the chain rule (Theorem 3.4),

$$L_h(X, E) = L_f(f(X), L_f(X, E)) = L_f(X, L_f(X, E)).$$

But because f is idempotent, $h(A) \equiv f(A)$ and so $L_h(X, E) = L_f(X, E)$. Hence $L_f(X, L_f(X, E)) = L_f(X, E)$, which shows that $L_f(X)$ is idempotent. $\square$

Stability is determined by the Fréchet derivative of the iteration function, not that of f. However, these two derivatives are one and the same if we add to the condition of Theorem 4.18 the conditions that in the iteration $X_{k+1} = g(X_k)$ the function g is independent of the starting matrix X_0 and that the iteration is superlinearly convergent when started sufficiently close to a fixed point. These conditions are satisfied for the matrix sign function and the unitary polar factor, and the corresponding iterations of interest, but not for matrix roots (since $(A^{1/p})^{1/p} \neq A^{1/p}$ and the iteration function g invariably depends on X_0).

Theorem 4.19. *Let f be an idempotent function that is Fréchet differentiable at $X \equiv f(X)$ with Fréchet derivative $L_f(X)$. Let $X_{k+1} = g(X_k)$ be superlinearly convergent to $f(X_0)$ for all X_0 sufficiently close to X and assume that g is independent of X_0. Then the Fréchet derivative of g at X is $L_g(X) = L_f(X)$.*

Proof. For sufficiently small E, $f(X + E)$ is defined and

$$f(X + E) = f(X) + L_f(X, E) + o(\|E\|) = X + L_f(X, E) + o(\|E\|).$$

Hence $f(X + E) - (X + E) = L_f(X, E) - E + o(\|E\|) = O(\|E\|)$, since $L_f(X, E)$ is linear in E. Since the iteration is (locally) superlinearly convergent,

$$\|f(X + E) - g(X + E)\| = o(\|f(X + E) - (X + E)\|) = o(O(\|E)\|) = o(\|E\|).$$

Hence

$$\begin{aligned} g(X + E) - g(X) &= g(X + E) - X \\ &= f(X + E) - X + o(\|E\|)) \\ &= L_f(X, E) + o(\|E\|), \end{aligned}$$

which shows that $L_g(X, E) = L_f(X, E)$. $\square$

The significance of Theorems 4.18 and 4.19 is that when they are applicable they tell us that *all* superlinearly convergent iterations are stable, without the need to compute Fréchet derivatives and test their power boundedness.

For the iterations described earlier in this section:

- $f(A) = A^{1/2}$ is not idempotent and the iteration function in (4.27) depends on $X_0 = A$, so the theorems are not applicable.

- $f(A) = \operatorname{sign}(A)$ is idempotent and Fréchet differentiable and the iteration (4.30) is quadratically convergent with iteration function independent of $Y_0 = A$, so the theorems apply and the iteration is therefore stable. See Section 5.7 for details.

The Fréchet derivative also allows us to estimate the limiting accuracy of an iteration.

Definition 4.20 (limiting accuracy). *For an iteration $X_{k+1} = g(X_k)$ with a fixed point X the* (relative) limiting accuracy *is $u\|L_g(X)\|$.*

To interpret the definition, consider one iteration applied to the rounded exact solution, $X_0 = X + E_0$, where $\|E_0\| \lesssim u\|X\|$. From (4.31) we have

$$\|E_1\| \lesssim \|L_g(X, E_0)\| \le \|L_g(X)\| \, \|E_0\| \lesssim u\|L_g(X)\| \, \|X\|,$$

and so the limiting accuracy is a bound for the relative error $\|X - X_1\|/\|X\|$. We can therefore think of the limiting accuracy as the smallest error we can reasonably expect to achieve in floating point arithmetic once—and indeed *if*—the iteration enters an $O(u)$ neighbourhood of a fixed point. Limiting accuracy is once again an asymptotic property.

While stability corresponds to the boundedness of the powers of $L_g(X)$, which depends only on the eigenvalues of $L_g(X)$, the limiting accuracy depends on the norm of $L_g(X)$, and so two stable iterations can have quite different limiting accuracy. However, neither of these two notions necessarily gives us a reliable estimate of the accuracy of a computed solution or the size of its residual. Clearly, an unstable iteration may never achieve its limiting accuracy, because instability may prevent it reaching the region of uncertainty around the solution whose size limiting accuracy measures.

Finally, it is important to note that the Fréchet derivative analysis treats the propagation of errors by the *exact iteration*. In practice, rounding errors are incurred during the evaluation of the iteration formula, and these represent another source of error that is dependent on how the formula is evaluated. For example, as noted in Section 4.4.3, rational functions can be evaluated in several ways and these potentially have quite different numerical stability properties.

Analysis based on the Fréchet derivative at the solution will prove to be informative for many of the iterations considered in this book, but because of the limitations explained above this analysis cannot always provide a complete picture of the numerical stability of a matrix iteration.

4.10. Preprocessing

In an attempt to improve the accuracy of an $f(A)$ algorithm we can preprocess the data. Two available techniques are argument reduction (or translation) and balancing. Both aim to reduce the norm of the matrix, which is important when Taylor or Padé approximants—most accurate near the origin—are to be applied.

Argument reduction varies slightly depending on the function. For the matrix exponential, $e^{A-\mu I} = e^{-\mu}e^{A}$, so any multiple of I can be subtracted from A. For trigonometric functions such as the cosine,

$$\cos(A - \pi j I) = (-1)^j \cos(A), \qquad j \in \mathbb{Z}, \tag{4.32}$$

so we can subtract integer multiples of π. Norm-minimizing real shifts are known for three particular norms. (The optimal 1-norm shift is obtained from the optimal ∞-norm shift using $\|A\|_1 = \|A^*\|_\infty$.)

Theorem 4.21 (norm-minimizing shifts). *Let* $A \in \mathbb{C}^{n\times n}$.

(a) $\min_{\mu\in\mathbb{R}} \|A - \mu I\|_F$ *is attained for* $\mu = n^{-1}\operatorname{trace}(A)$.

(b) *If* $a_{ii} \in \mathbb{R}$ *for all* i *then*

$$\min_{\mu\in\mathbb{R}} \|A - \mu I\|_\infty = \frac{1}{2}\bigl(\max_i(a_{ii} + r_i) + \max_i(-a_{ii} + r_i)\bigr), \tag{4.33}$$

where $r_i = \sum_{j\neq i} |a_{ij}|$, *and the optimal* μ *is*

$$\mu = \frac{1}{2}\bigl(\max_i(a_{ii} + r_i) - \max_i(-a_{ii} + r_i)\bigr).$$

Proof. (a) We have $\|A-\mu I\|_F = \|\operatorname{vec}(A)-\operatorname{vec}(I)\mu\|_2$, so the problem is essentially a 1-variable linear least squares problem. The normal equations are $\operatorname{vec}(I)^*\operatorname{vec}(I)\mu = \operatorname{vec}(I)^*\operatorname{vec}(A)$, or $n\mu = \operatorname{trace}(A)$, as required.

(b) We have

$$\begin{aligned}\|A\|_\infty &= \max_i \sum_j |a_{ij}| = \max_i \max(a_{ii} + r_i, -a_{ii} + r_i)\\ &= \max\bigl(\max_i(a_{ii} + r_i), \max_i(-a_{ii} + r_i)\bigr).\end{aligned}$$

The inner max terms in the last expression are the rightmost point and the negative of the leftmost point in the union of the Gershgorin discs for A. It is clear that the optimal shift μ must make these extremal points equidistant from the origin; hence it must satisfy $\max_i(a_{ii} - \mu + r_i) = \max_i(-a_{ii} + \mu + r_i)$, that is, $\mu = \frac{1}{2}(\max_i(a_{ii} + r_i) - \max_i(-a_{ii} + r_i))$. The formula (4.33) follows on substitution of the optimal μ. □

In the computation of scalar elementary functions it is well known that special techniques must be used to avoid severe loss of accuracy in argument reduction for large arguments [441, 1997, Chap. 8]. The standard techniques are not directly applicable to the matrix case, so we must recognize that argument reduction is potentially a significant source of error.

Balancing is a heuristic that attempts to equalize the norms of the ith row and ith column, for each i, by a diagonal similarity transformation. It is known that balancing in the 2-norm is equivalent to minimizing $\|D^{-1}AD\|_F$ over all nonsingular diagonal D [454, 1960]. Nowadays, "balancing" is synonymous with the balancing algorithms in LAPACK [12] and MATLAB [414], which compute $B = D^{-1}AD$, where D is a permuted diagonal matrix with diagonal elements powers of the machine base chosen so that the 1-norms of the ith row and ith column of B are of similar magnitude for all i. Balancing is an $O(n^2)$ calculation that can be performed without roundoff. It is not guaranteed to reduce the norm, so it is prudent to replace A by the balanced B only if $\|B\| < \|A\|$.

Balancing can be combined with argument reduction. Since $\operatorname{trace}(A) = \operatorname{trace}(D^{-1}AD)$ and the balancing transformation is independent of the diagonal elements of the matrix, argument reduction in the Frobenius norm yields the same shift before balancing as after. Therefore it makes no difference in which order these two operations are done.

In some special cases we can say more about argument reduction and balancing. Recall from Section 2.3 that an intensity matrix is a matrix $Q \in \mathbb{R}^{n\times n}$ such that $q_{ij} \ge 0$ for $i \ne j$ and $\sum_{j=1}^n q_{ij} = 0$, $i = 1\colon n$.

Corollary 4.22 (Melloy and Bennett). *Let $Q \in \mathbb{R}^{n\times n}$ be an intensity matrix. Then*

$$\min_{\mu\in\mathbb{R}} \|Q + \mu I\|_\infty = \max_i |q_{ii}| = \|Q\|_\infty/2.$$

The minimum is attained for $\mu = \max_i |q_{ii}|$, for which $Q + \mu I$ is nonnegative with row sums all equal to μ.

Proof. In the notation of Theorem 4.21 (b), we have $q_{ii} + r_i = 0$, so $-q_{ii} + r_i = 2|q_{ii}|$, and the result then follows from the theorem. □

Corollary 4.22 shows that for an intensity matrix the optimal shift reduces the ∞-norm by a factor 2. We now show that the shifted matrix is already of minimal ∞-norm under diagonal similarities. We need the following more general result.

Theorem 4.23 (Ström). *For $A \in \mathbb{C}^{n\times n}$,*

$$\inf\{\, \|D^{-1}AD\|_\infty : D = \mathrm{diag}(d_i) \text{ is nonsingular}\} = \rho(|A|). \tag{4.34}$$

If $|A|x = \rho(|A|)x$ with $x > 0$ then $D = \mathrm{diag}(x)$ achieves the lower bound.

Proof. For any $A \in \mathbb{C}^{n\times n}$ and nonsingular diagonal D we have

$$\|D^{-1}AD\|_\infty = \|\,|D^{-1}AD|\,\|_\infty = \|D^{-1}|A|D\|_\infty \ge \rho(D^{-1}|A|D) = \rho(|A|). \tag{4.35}$$

If x satisfies the stated conditions then with $D = \mathrm{diag}(x)$ we have

$$|D^{-1}AD|e = D^{-1}|A|De = D^{-1}|A|x = \rho(|A|)D^{-1}x = \rho(|A|)e.$$

Taking the ∞-norm gives $\|D^{-1}AD\|_\infty = \rho(|A|)$, as required.

If A is irreducible then from the Perron–Frobenius theory (Theorem B.6) we know that a positive eigenvector x of $|A|$ exists corresponding to an eigenvalue $\rho(|A|)$, and so the lower bound (4.35) is attained. If A is reducible then for any $\epsilon > 0$ we can choose B so that $\|B\|_\infty < \epsilon$ and $|A| + |B|$ is irreducible. Then

$$\begin{aligned}
\inf_D \|D^{-1}AD\|_\infty = \inf_D \|D^{-1}|A|D\|_\infty &\le \inf_D \|D^{-1}(|A| + |B|)D\|_\infty \\
&= \rho(|A| + |B|) \\
&\le \rho(|A|) + \rho(|B|) \le \rho(|A|) + \|B\|_\infty \\
&< \rho(|A|) + \epsilon,
\end{aligned}$$

where we have used (B.8) and (B.13), and (4.34) follows. □

Corollary 4.24. *Let $A \in \mathbb{R}^{n\times n}$ be a scalar multiple of a stochastic matrix. Then diagonal similarities cannot reduce the ∞-norm of A.*

Proof. The result follows from Theorem 4.23 because $Ae = \mu e$ for some μ with $|\mu| = \|A\|_\infty = \rho(A) = \rho(|A|)$, so $\|A\|_\infty$ is already minimal. □

Corollary 4.24 shows that any nonnegative matrix with equal row sums has minimal ∞-norm under diagonal similarities. In particular, this is true of the optimally shifted intensity matrix. To summarize: to preprocess an intensity matrix it suffices to shift it as in Corollary 4.22.

4.11. Bounds for $\|f(A)\|$

For both theoretical and practical purposes it is useful to be able to bound the norm of $f(A)$. We give a variety of different bounds in this section, beginning with one based on the Jordan canonical form.

Theorem 4.25. *Let $A \in \mathbb{C}^{n\times n}$ have the Jordan canonical form (1.2), with distinct eigenvalues $\lambda_1, \dots, \lambda_s$. Then*

$$\|f(A)\|_p \le n_{\max}\, \kappa_p(Z) \max_{\substack{i=1:s\\ j=0:n_i-1}} \left| \frac{f^{(j)}(\lambda_i)}{j!} \right|, \qquad p = 1\colon\infty, \tag{4.36}$$

where $n_{\max} = \max_{i=1:s} n_i$.

Proof. The proof is straightforward from Definition 1.2 and on using $\|B\|_p \le \|B\|_1^{1/p}\|B\|_\infty^{1-1/p}$ [276, 2002, Sec. 6.3]. □

If $A = Z\,\mathrm{diag}(\lambda_i)Z^{-1}$ is diagonalizable then (4.36) simplifies to

$$\|f(A)\|_p \le \kappa_p(Z) f(\rho(A)). \tag{4.37}$$

In both (4.36) and (4.37), $\kappa_p(Z)$ should be interpreted as the minimum over all Jordan canonical forms. If A is normal then Z can be taken to be unitary and these bounds are then an equality for $p = 2$: $\|f(A)\|_2 = f(\rho(A))$. Thus only the nonnormal case is interesting.

A pseudospectral analogue of (4.37) is as follows. Recall that the ϵ-pseudospectrum is defined by (2.30).

Theorem 4.26 (Trefethen and Embree). *Let $A \in \mathbb{C}^{n\times n}$ and $\epsilon > 0$. Let f be analytic on and inside a closed contour Γ_ϵ that encloses $\Lambda_\epsilon(A)$. Then*

$$\|f(A)\|_2 \le \frac{L_\epsilon}{2\pi\epsilon} \max_{z\in\Gamma_\epsilon} |f(z)|, \tag{4.38}$$

where L_ϵ is the arc length of Γ_ϵ. In particular,

$$\|f(A)\|_2 \le \frac{\rho_\epsilon}{\epsilon} \max_{\theta\in[0,2\pi]} |f(\rho_\epsilon e^{i\theta})|, \tag{4.39}$$

where $\rho_\epsilon = \rho_\epsilon(A) = \max\{\, |z| : z \in \Lambda_\epsilon(A)\,\}$ is the pseudospectral radius.

Proof. To prove (4.38) let $\Gamma := \Gamma_\epsilon$ in the Cauchy integral (1.12), take norms, and use the fact that $\|(zI - A)^{-1}\| \le \epsilon^{-1}$ on and outside $\Lambda_\epsilon(A)$. The bound (4.39) is obtained from (4.38) by taking Γ_ϵ to be a circle with centre 0 and radius ρ_ϵ. □

The next result does not involve a similarity taking A to Jordan form but it needs knowledge of the derivatives of f on a larger region than just the spectrum. Denote by $\mathrm{conv}(S)$ the convex hull of the set S.

Theorem 4.27 (Young). *Let $A \in \mathbb{C}^{n\times n}$ and let f be analytic in a convex open set containing $\Lambda(A)$. Then*

$$\|f(A)\|_F \le \left[\sum_{i=0}^{n-1} \binom{n}{i+1} \left(\frac{\|A\|_F^i \sup_{z\in\mathrm{conv}(\Lambda(A))} |f^{(i)}(z)|}{i!} \right)^2 \right]^{1/2}. \qquad \square \tag{4.40}$$

A bound can also be given in terms of the Schur decomposition.

Theorem 4.28 (Golub and Van Loan). *Let $Q^*AQ = T = \mathrm{diag}(\lambda_i) + N$ be a Schur decomposition of $A \in \mathbb{C}^{n\times n}$, where N is strictly upper triangular. If $f(z)$ is analytic on a closed convex set Ω containing $\Lambda(A)$ then*

$$\|f(A)\|_F \le \left\| \sum_{i=0}^{n-1} \omega_i \frac{|N|^i}{i!} \right\|_F \le \max_{0\le i\le n-1} \frac{\omega_i}{i!} \, \|(I-|N|)^{-1}\|_F, \tag{4.41}$$

where $\omega_i = \sup_{z\in\Omega} |f^{(i)}(z)|$.

Proof. Let $S_{ij}^{(r)}$ denote the set of all strictly increasing sequences of $r+1$ integers that start at i and end at j, and note that with S_{ij} defined as in Theorem 4.11,

$$S_{ij} = \bigcup_{r=1}^{j-i} S_{ij}^{(r)}.$$

From Theorem 4.11, $F = f(T)$ is given by, for $j \ge i$,

$$f_{ij} = \sum_{r=1}^{j-i} \sum_{(s_0,\dots,s_r)\in S_{ij}^{(r)}} t_{s_0,s_1} t_{s_1,s_2} \dots t_{s_{r-1},s_r} f[\lambda_{s_0},\dots,\lambda_{s_r}].$$

Using (B.28) we have

$$|f_{ij}| \le \sum_{r=1}^{j-i} \frac{\omega_r}{r!} \sum_{(s_0,\dots,s_r)\in S_{ij}^{(r)}} |t_{s_0,s_1}||t_{s_1,s_2}| \dots |t_{s_{r-1},s_r}|.$$

Now it can be shown that $|N|^r = (n_{ij}^{(r)})$ satisfies

$$n_{ij}^{(r)} = \begin{cases} 0, & j < i+r, \\ \sum_{(s_0,\dots,s_r)\in S_{ij}^{(r)}} |t_{s_0,s_1}||t_{s_1,s_2}| \dots |t_{s_{r-1},s_r}|, & j \ge i+r. \end{cases}$$

Hence for $j > i$,

$$|f_{ij}| \le \sum_{r=1}^{j-i} \frac{\omega_r}{r!} (|N|^r)_{ij} = \sum_{r=1}^{n-1} \frac{\omega_r}{r!} (|N|^r)_{ij}.$$

Extending the summation to $r = 0$ correctly bounds $|f_{ii}| \le \omega_0$. Hence $|f(T)| \le \sum_{r=0}^{n-1} \omega_r |N|^r/r!$. This yields the first bound in (4.41) on using the unitary invariance of the Frobenius norm. The second inequality follows from $I + |N| + \dots + |N|^{n-1} = (I - |N|)^{-1}$. □

A weaker variant of Theorem 4.28 is given by Descloux [148, 1963]: under the same conditions,

$$\|f(A)\|_2 \le \sum_{i=0}^{n-1} \frac{\omega_i}{i!} \|N\|_F^i. \tag{4.42}$$

The quantity $\Delta(A) := \|N\|_F$ is the departure from normality, introduced by Henrici [257, 1962]. It can be expressed as $\Delta(A)^2 = \|A\|_F^2 - \sum_{i=1}^n |\lambda_i(A)|^2$ and is zero when A is normal. Another bound involving the departure from normality is one of Gil that is in a similar vein to the bound of Theorem 4.27.

Theorem 4.29 (Gil). *Let $A \in \mathbb{C}^{n\times n}$ and let f be analytic in a convex open set containing $\Lambda(A)$. Then*

$$\|f(A)\|_2 \le \sum_{i=0}^{n-1} \sup_{z\in\operatorname{conv}(\Lambda(A))} |f^{(i)}(z)|\, \Delta(A)^i\,(i!)^{-3/2}. \qquad \square \tag{4.43}$$

4.12. Notes and References

For more on the theory of Padé approximation see Baker [38, 1975], Baker and Graves-Morris [39, 1996], Gragg [227, 1972], or Sidi [521, 2003, Chap. 17]. Some authors use an alternative definition of the $[k/m]$ Padé approximant of f in which $f(x)-r_{km}(x)$ is required not necessarily to be of order x^{k+m+1} but rather to be of the highest possible order; with this definition the approximant always exists but the order of the error is not immediately apparent.

Theorem 4.5 (a less general version of which is stated in [278, 2005]) can be proved using techniques from [276, 2002].

Theorem 4.7 is due to Hensel [258, 1926]. Weyr [614, 1887] had previously proved part (a) of the theorem.

Calvetti, Gallopoulos, and Reichel [90, 1995] give a detailed discussion and extensive bibliography on matrix partial fraction expansions and they investigate "incomplete" expansions in an aim to obtain improved numerical stability properties.

Theorem 4.12 is given by Kenney and Laub [348, 1998].

Parlett's recurrence is derived independently by Filipponi [186, 1981].

In his thesis, Kågström [323, 1977] handles confluency in Parlett's recurrence by a different means than by using the block form. He uses the scalar recurrence, and if t_{ii} is equal or very close to t_{jj} he invokes the formula for f_{ij} in Theorem 4.11, replacing the divided difference by a derivative.

Theorem 4.15 generalizes, and is modeled on, a result of Kenney and Laub [343, 1991, Lem. 5.1] that applies to rational iterations for the matrix sign function. The theorem can also be proved using complex analysis, as shown by Iannazzo [307, 2007], who pointed out Corollary 4.16. An early use of the Jordan form in convergence analysis is in Laasonen's analysis of the Newton square root iteration [366, 1958].

The framework given here for stability analysis, based on Fréchet derivatives, was initiated by Cheng, Higham, Kenney, and Laub [108, 2001] and developed by Higham, Mackey, Mackey, and Tisseur [283, 2005]. Our treatment here expands significantly on that in [283, 2005]. Theorem 4.19 is new. In a different style of analysis used by Higham [267, 1986] and employed in subsequent papers by various authors, perturbation expansions are developed for the iterations, from which scalar error amplification factors are identified under the assumption that the matrix A is diagonalizable. The approach used here leads to shorter and more insightful analysis, especially when the Fréchet derivative is idempotent at the solution.

A more general treatment of matrix argument reduction for periodic functions is given by Ng [448, 1984], in which different eigenvalues can undergo different shifts. His algorithm for shifting makes use of a Schur decomposition, and he applies it to the matrix exponential, for which the period is $2\pi i$.

Ward's implementation of the scaling and squaring algorithm for computing the matrix exponential [606, 1977] uses both argument reduction and balancing.

Part (b) of Theorem 4.21 is stated without proof by Arioli, Codenotti, and Fassino [17, 1996] for an unspecified class of A.

Corollary 4.22 is due to Melloy and Bennett [422, 1993]. Further analysis of balancing can be found in Ström [545, 1972] and Fenner and Loizou [183, 1977]. Theorem 4.23 is from Ström [545, 1972]. Corollary 4.24 generalizes a result of Melloy and Bennett [422, 1993] stated for a shifted intensity matrix.

Theorem 4.27 is from Young [620, 1981]. Theorem 4.26 is from Trefethen and Embree [573, 2005]. Theorem 4.28 is from Golub and Van Loan [224, 1996, Thm. 11.2.2] and it appeared first in Van Loan [592, 1975, Thm. 4]. Theorem 4.29 is from Gil [214, 1993]. For some other approaches to bounding $\|f(A)\|$ see Crouzeix [123, 2004] and Greenbaum [230, 2004].

Problems

Since the purpose of mathematics is to solve problems, it is impossible to judge one's progress without breaking a lance on a few problems from stage to stage.

— RICHARD BELLMAN, *Introduction to Matrix Analysis* (1970)

4.1. Let $A \in \mathbb{C}^{m\times n}$, $B \in \mathbb{C}^{n\times m}$, and $C = \left[\begin{smallmatrix} 0 & A \\ B & 0 \end{smallmatrix}\right] \in \mathbb{C}^{(m+n)\times(m+n)}$. Show that if $f(z) = \sum_{i=0}^{\infty} a_i z^i$ then (within the radius of convergence)

$$f(C) = \begin{bmatrix} \sum_{i=0}^{\infty} a_{2i}(AB)^i & A\sum_{i=0}^{\infty} a_{2i+1}(BA)^i \\ B\sum_{i=0}^{\infty} a_{2i+1}(AB)^i & \sum_{i=0}^{\infty} a_{2i}(BA)^i \end{bmatrix}.$$

Deduce the form of e^C. Hence obtain the solution (2.8) to (2.7).

4.2. Show that if a $[k/m]$ Padé approximant to $f(x)$ exists then it is unique.

4.3. Let

$$T = \begin{bmatrix} T_{11} & T_{12} \\ 0 & T_{22} \end{bmatrix} \in \mathbb{C}^{n\times n}$$

be block upper triangular with $\Lambda(T_{11}) \cap \Lambda(T_{22}) = \emptyset$. Let f be defined on the spectrum of T and let

$$F = f(T) = \begin{bmatrix} f(T_{11}) & F_{12} \\ 0 & f(T_{22}) \end{bmatrix}.$$

The block Parlett recurrence yields F_{12} as the solution of the Sylvester equation

$$T_{11}F_{12} - F_{12}T_{22} = f(T_{11})T_{12} - T_{12}f(T_{22}). \tag{4.44}$$

But F_{12} can also be obtained by block diagonalization. Show that if

$$T_{11}X - XT_{22} = T_{12} \tag{4.45}$$

then

$$\begin{bmatrix} T_{11} & 0 \\ 0 & T_{22} \end{bmatrix} = \begin{bmatrix} I & -X \\ 0 & I \end{bmatrix}^{-1} \begin{bmatrix} T_{11} & T_{12} \\ 0 & T_{22} \end{bmatrix} \begin{bmatrix} I & -X \\ 0 & I \end{bmatrix},$$

and deduce that

$$F_{12} = f(T_{11})X - Xf(T_{22}). \tag{4.46}$$

Show that the formulae (4.44) and (4.46) are equivalent.

4.4. (Parlett [459, 1974]) Let $S, T \in \mathbb{C}^{n\times n}$ be upper triangular and let $X = f(T^{-1}S)$, $Y = f(ST^{-1})$. Show that

$$SX - YS = 0, \qquad TX - YT = 0, \tag{4.47}$$

and hence show how to compute X and Y together, by a finite recurrence, without explicitly forming $T^{-1}S$. When does your recurrence break down?

4.5. Consider the conditions of Theorem 4.15 under the weaker assumption that $|\frac{\partial g}{\partial x}(x_*, \lambda)| = 1$. Construct examples with $n = 2$ to show that the corresponding matrix iteration may or may not converge.

4.6. Show that $A \in \mathbb{C}^{n\times n}$ is power bounded (that is, for any norm there exists a constant c such that $\|A^k\| \le c$ for all $k \ge 0$) if $\rho(A) < 1$. Give a necessary and sufficient condition for A to be power bounded.

4.7. (Elsner [176, 1970]) Consider the recurrence $y_{k+1} = c_k y_k + d_k$, where $c_k \to c$ and $d_k \to d$ as $k \to \infty$, with $|c| < 1$. Show that $\lim_{k\to\infty} y_k = d/(1-c)$.

4.8. (Kahan [326, 1972]) Show that $A \in \mathbb{C}^{n\times n}$ is nilpotent of index k if and only if $\operatorname{trace}(A^i) = 0$, $i = 1\colon k$.

4.9. (RESEARCH PROBLEM) Develop bounds for $\|f(A) - r(A)\|$ for nonnormal A and r a best L_∞ approximation or Padé approximant, for any suitable norm. Some bounds for particular f and Padé approximants can be found in later chapters.

4.10. (RESEARCH PROBLEM) Develop new bounds on $\|f(A)\|$ to add to those in Section 4.11.

Chapter 5
Matrix Sign Function

The scalar sign function is defined for $z \in \mathbb{C}$ lying off the imaginary axis by

$$\operatorname{sign}(z) = \begin{cases} 1, & \operatorname{Re} z > 0, \\ -1, & \operatorname{Re} z < 0. \end{cases}$$

The matrix sign function can be obtained from any of the definitions in Chapter 1. Note that in the case of the Jordan canonical form and interpolating polynomial definitions, the derivatives $\operatorname{sign}^{(k)}(z)$ are zero for $k \geq 1$. Throughout this chapter, $A \in \mathbb{C}^{n\times n}$ is assumed to have no eigenvalues on the imaginary axis, so that $\operatorname{sign}(A)$ is defined. Note that this assumption implies that A is nonsingular.

As we noted in Section 2.4, if $A = ZJZ^{-1}$ is a Jordan canonical form arranged so that $J = \operatorname{diag}(J_1, J_2)$, where the eigenvalues of $J_1 \in \mathbb{C}^{p\times p}$ lie in the open left half-plane and those of $J_2 \in \mathbb{C}^{q\times q}$ lie in the open right half-plane, then

$$\operatorname{sign}(A) = Z \begin{bmatrix} -I_p & 0 \\ 0 & I_q \end{bmatrix} Z^{-1}. \tag{5.1}$$

Two other representations have some advantages. First is the particularly concise formula (see (5.5))

$$\operatorname{sign}(A) = A(A^2)^{-1/2}, \tag{5.2}$$

which generalizes the scalar formula $\operatorname{sign}(z) = z/(z^2)^{1/2}$. Recall that $B^{1/2}$ denotes the principal square root of B (see Section 1.7). Note that A having no pure imaginary eigenvalues is equivalent to A^2 having no eigenvalues on $\mathbb{R}^-$. Next, $\operatorname{sign}(A)$ has the integral representation (see Problem 5.3)

$$\operatorname{sign}(A) = \frac{2}{\pi} A \int_0^\infty (t^2 I + A^2)^{-1}\, dt. \tag{5.3}$$

Some properties of $\operatorname{sign}(A)$ are collected in the following theorem.

Theorem 5.1 (properties of the sign function). *Let $A \in \mathbb{C}^{n\times n}$ have no pure imaginary eigenvalues and let $S = \operatorname{sign}(A)$. Then*

(a) $S^2 = I$ *(S is involutory);*

(b) *S is diagonalizable with eigenvalues ± 1;*

(c) $SA = AS$*;*

(d) *if A is real then S is real;*

(e) *$(I+S)/2$ and $(I-S)/2$ are projectors onto the invariant subspaces associated with the eigenvalues in the right half-plane and left half-plane, respectively.*

Proof. The properties follow from (5.1)–(5.3). Of course, properties (c) and (d) hold more generally for matrix functions, as we know from Chapter 1 (see Theorem 1.13 (a) and Theorem 1.18). □

Although $\operatorname{sign}(A)$ is a square root of the identity matrix, it is not equal to I or $-I$ unless the spectrum of A lies entirely in the open right half-plane or open left half-plane, respectively. Hence, in general, $\operatorname{sign}(A)$ is a nonprimary square root of I. Moreover, although $\operatorname{sign}(A)$ has eigenvalues ± 1, its norm can be arbitrarily large.

The early appearance of this chapter in the book is due to the fact that the sign function plays a fundamental role in iterative methods for matrix roots and the polar decomposition. The definition (5.2) might suggest that the sign function is a "special case" of the square root. The following theorem, which provides an explicit formula for the sign of a block 2×2 matrix with zero diagonal blocks, shows that, if anything, the converse is true: the square root can be obtained from the sign function (see (5.4)). The theorem will prove useful in the next three chapters.

Theorem 5.2 (Higham, Mackey, Mackey, and Tisseur). *Let $A, B \in \mathbb{C}^{n\times n}$ and suppose that AB (and hence also BA) has no eigenvalues on $\mathbb{R}^-$. Then*

$$\operatorname{sign}\left(\begin{bmatrix} 0 & A \\ B & 0 \end{bmatrix}\right) = \begin{bmatrix} 0 & C \\ C^{-1} & 0 \end{bmatrix},$$

where $C = A(BA)^{-1/2}$.

Proof. The matrix $P = \left[\begin{smallmatrix} 0 & A \\ B & 0 \end{smallmatrix}\right]$ cannot have any eigenvalues on the imaginary axis, because if it did then $P^2 = \left[\begin{smallmatrix} AB & 0 \\ 0 & BA \end{smallmatrix}\right]$ would have an eigenvalue on $\mathbb{R}^-$. Hence $\operatorname{sign}(P)$ is defined and

$$\begin{aligned} \operatorname{sign}(P) = P(P^2)^{-1/2} &= \begin{bmatrix} 0 & A \\ B & 0 \end{bmatrix} \begin{bmatrix} AB & 0 \\ 0 & BA \end{bmatrix}^{-1/2} \\ &= \begin{bmatrix} 0 & A \\ B & 0 \end{bmatrix} \begin{bmatrix} (AB)^{-1/2} & 0 \\ 0 & (BA)^{-1/2} \end{bmatrix} \\ &= \begin{bmatrix} 0 & A(BA)^{-1/2} \\ B(AB)^{-1/2} & 0 \end{bmatrix} =: \begin{bmatrix} 0 & C \\ D & 0 \end{bmatrix}. \end{aligned}$$

Since the square of the matrix sign of any matrix is the identity,

$$I = (\operatorname{sign}(P))^2 = \begin{bmatrix} 0 & C \\ D & 0 \end{bmatrix}^2 = \begin{bmatrix} CD & 0 \\ 0 & DC \end{bmatrix},$$

so $D = C^{-1}$. Alternatively, Corollary 1.34 may be used to see more directly that $CD = A(BA)^{-1/2}B(AB)^{-1/2}$ is equal to I. □

A special case of the theorem, first noted by Higham [274, 1997], is

$$\operatorname{sign}\left(\begin{bmatrix} 0 & A \\ I & 0 \end{bmatrix}\right) = \begin{bmatrix} 0 & A^{1/2} \\ A^{-1/2} & 0 \end{bmatrix}. \tag{5.4}$$

In addition to the association with matrix roots and the polar decomposition (Chapter 8), the importance of the sign function stems from its applications to Riccati equations (Section 2.4), the eigenvalue problem (Section 2.5), and lattice QCD (Section 2.7).

In this chapter we first give perturbation theory for the matrix sign function and identify appropriate condition numbers. An expensive, but stable, Schur method for $\mathrm{sign}(A)$ is described. Then Newton's method and a rich Padé family of iterations, having many interesting properties, are described and analyzed. How to scale and how to terminate the iterations are discussed. Then numerical stability is considered, with the very satisfactory conclusion that all sign iterations of practical interest are stable. Numerical experiments illustrating these various features are presented. Finally, best L_∞ rational approximation via Zolotarev's formulae, of interest for Hermitian matrices, is described.

As we will see in Chapter 8, the matrix sign function has many connections with the polar decomposition, particularly regarding iterations for computing it. Some of the results and ideas in Chapter 8 are applicable, with suitable modification, to the sign function, but are not discussed here to avoid repetition. See, for example, Problem 8.26.

5.1. Sensitivity and Conditioning

Associated with the matrix sign function is the *matrix sign decomposition*

$$A = SN, \qquad S = \mathrm{sign}(A), \quad N = (A^2)^{1/2}. \tag{5.5}$$

To establish the decomposition note that $N = S^{-1}A = SA$. Since S commutes with A, $N^2 = A^2$, and since the spectrum of SA lies in the open right half-plane, $N = (A^2)^{1/2}$.

The matrix sign factor N is useful in characterizing the Fréchet derivative of the matrix sign function.

Let $S + \Delta S = \mathrm{sign}(A + \Delta A)$, where the sign function is assumed to be defined in a ball of radius $\|\Delta A\|$ about A. The definition (3.6) of Fréchet derivative says that

$$\Delta S - L(A, \Delta A) = o(\|\Delta A\|), \tag{5.6}$$

where $L(A, \Delta A)$ is the Fréchet derivative of the matrix sign function at A in the direction ΔA. Now from $(A + \Delta A)(S + \Delta S) = (S + \Delta S)(A + \Delta A)$ we have

$$A\Delta S - \Delta S A = S\Delta A - \Delta A S + \Delta S \Delta A - \Delta A \Delta S = S\Delta A - \Delta A S + o(\|\Delta A\|), \tag{5.7}$$

since $\Delta S = O(\|\Delta A\|)$. Moreover, $(S + \Delta S)^2 = I$ gives

$$S\Delta S + \Delta S S = -\Delta S^2 = o(\|\Delta A\|).$$

Premultiplying (5.7) by S and using the latter equation gives

$$N\Delta S + \Delta S N = \Delta A - S\Delta A S + o(\|\Delta A\|). \tag{5.8}$$

Theorem 5.3 (Kenney and Laub). *The Fréchet derivative $L = L_{\mathrm{sign}}(A, \Delta A)$ of the matrix sign function satisfies*

$$NL + LN = \Delta A - S\Delta A S, \tag{5.9}$$

where $A = SN$ is the matrix sign decomposition.

Proof. Since the eigenvalues of N lie in the open right half-plane, the Sylvester equation (5.9) has a unique solution L which is a linear function of ΔA and, in view of (5.8), differs from $\Delta S = \mathrm{sign}(A + \Delta A) - S$ by $o(\|\Delta A\|)$. Hence (5.6) implies that $L = L(A, \Delta A)$. □

By applying the vec operator and using the relation (B.16) we can rewrite (5.9) as

$$P\,\mathrm{vec}(L) = (I_{n^2} - S^T \otimes S)\,\mathrm{vec}(\Delta A),$$

where

$$P = I \otimes N + N^T \otimes I.$$

Hence

$$\begin{aligned}\max_{\|\Delta A\|_F=1} \|L(A, \Delta A)\|_F &= \max_{\|\Delta A\|_F=1} \|P^{-1}(I_{n^2} - S^T \otimes S)\,\mathrm{vec}(\Delta A)\|_2 \\ &= \|P^{-1}(I_{n^2} - S^T \otimes S)\|_2.\end{aligned}$$

The (relative) condition number of $\mathrm{sign}(A)$ in the Frobenius norm is therefore

$$\kappa_{\mathrm{sign}}(A) := \mathrm{cond}_{\mathrm{rel}}(\mathrm{sign}, A) = \|P^{-1}(I_{n^2} - S^T \otimes S)\|_2 \frac{\|A\|_F}{\|S\|_F}. \tag{5.10}$$

If $S = I$, which means that all the eigenvalues of A are in the open right half-plane, then $\mathrm{cond}(S) = 0$, which corresponds to the fact that the eigenvalues remain in this half-plane under sufficiently small perturbations of A.

To gain some insight into the condition number, suppose that A is diagonalizable: $A = ZDZ^{-1}$, where $D = \mathrm{diag}(\lambda_i)$. Then $S = ZD_SZ^{-1}$ and $N = ZD_NZ^{-1}$, where $D_S = \mathrm{diag}(\sigma_i)$ and $D_N = \mathrm{diag}(\sigma_i\lambda_i)$, with $\sigma_i = \mathrm{sign}(\lambda_i)$. Hence

$$\kappa_{\mathrm{sign}}(A) = \|(Z^{-T} \otimes Z) \cdot (I \otimes D_N + D_N \otimes I)^{-1}(I_{n^2} - D_S^T \otimes D_S) \cdot (Z^T \otimes Z^{-1})\|_2 \frac{\|A\|_F}{\|S\|_F}.$$

The diagonal matrix in the middle has elements $(1 - \sigma_i\sigma_j)/(\sigma_i\lambda_i + \sigma_j\lambda_j)$, which are either zero or of the form $2/|\lambda_i - \lambda_j|$. Hence

$$\kappa_{\mathrm{sign}}(A) \le 2\kappa_2(Z)^2 \max\left\{ \frac{1}{|\lambda_i - \lambda_j|} : \mathrm{Re}\,\lambda_i\,\mathrm{Re}\,\lambda_j < 0 \right\} \frac{\|A\|_F}{\|S\|_F}. \tag{5.11}$$

Equality holds in this bound for normal A, for which Z can be taken to unitary. The gist of (5.11) is that the condition of S is bounded in terms of the minimum distance between eigenvalues across the imaginary axis and the square of the condition of the eigenvectors. Note that (5.11) is precisely the bound obtained by applying Theorem 3.15 to the matrix sign function.

One of the main uses of κ_{sign} is to indicate the sensitivity of $\mathrm{sign}(A)$ to perturbations in A, through the perturbation bound (3.3), which we rewrite here for the sign function as

$$\frac{\|\mathrm{sign}(A + E) - \mathrm{sign}(A)\|_F}{\|\mathrm{sign}(A)\|_F} \le \kappa_{\mathrm{sign}}(A) \frac{\|E\|_F}{\|A\|_F} + o(\|E\|_F). \tag{5.12}$$

This bound is valid as long as $\mathrm{sign}(A+tE)$ is defined for all $t \in [0, 1]$. It is instructive to see what can go wrong when this condition is not satisfied. Consider the example, from [347, 1995],

$$A = \mathrm{diag}(1, -\epsilon^2), \qquad E = \mathrm{diag}(0, 2\epsilon^2), \qquad 0 < \epsilon \ll 1.$$

We have $\operatorname{sign}(A) = \operatorname{diag}(1, -1)$ and $\operatorname{sign}(A + E) = I$. Because A is normal, (5.11) gives $\kappa_{\text{sign}}(A) = (2/(1 + \epsilon^2))\|A\|_F/\sqrt{2}$. Hence the bound (5.12) takes the form

$$\frac{2}{\sqrt{2}} \leq \frac{2}{\sqrt{2}(1 + \epsilon^2)} 2\epsilon^2 + o(\epsilon^2) = 2\sqrt{2}\epsilon^2 + o(\epsilon^2).$$

This bound is clearly incorrect. The reason is that the perturbation E causes eigenvalues to cross the imaginary axis; therefore $\operatorname{sign}(A + tE)$ does not exist for all $t \in [0, 1]$. Referring back to the analysis at the start of this section, we note that (5.7) is valid for $\|\Delta A\|_F < \|E\|_F/3$, but does not hold for $\Delta A = E$, since then $\Delta S \neq O(\|\Delta A\|)$.

Another useful characterization of the Fréchet derivative is as the limit of a matrix iteration; see Theorem 5.7.

Consider now how to estimate $\kappa_{\text{sign}}(A)$. We need to compute a norm of $B = P^{-1}(I_{n^2} - S^T \otimes S)$. For the 2-norm we can use Algorithm 3.20 (the power method). Alternatively, Algorithm 3.22 can be used to estimate the 1-norm. In both cases we need to compute $L(A, E)$, which if done via (5.9) requires solving a Sylvester equation involving N; this can be done via a matrix sign evaluation (see Section 2.4), since N is positive stable. We can compute $L^\star(X, E)$ in a similar fashion, solving a Sylvester equation of the same form. Alternatively, $L(A, E)$ can be computed using iteration (5.23) or estimated by finite differences. All these methods require $O(n^3)$ operations.

It is also of interest to understand the conditioning of the sign function for $A \approx \operatorname{sign}(A)$, which is termed the asymptotic conditioning. The next result provides useful bounds.

Theorem 5.4 (Kenney and Laub). *Let $A \in \mathbb{C}^{n\times n}$ have no pure imaginary eigenvalues and let $S = \operatorname{sign}(A)$. If $\|(A - S)S\|_2 < 1$, then*

$$\frac{\|S\|_2^2 - 1}{2(1 + \|(A - S)S\|_2)} \leq \frac{\kappa_{\text{sign}}(A)}{\|A\|_F/\|S\|_F} \leq \frac{\|S\|_2^2 + 1}{2(1 - \|(A - S)S\|_2)}. \tag{5.13}$$

In particular,

$$\frac{\|S\|_2^2 - 1}{2} \leq \kappa_{\text{sign}}(S) \leq \frac{\|S\|_2^2 + 1}{2}. \tag{5.14}$$

Proof. We need to bound $\|L_{\text{sign}}(A)\|_F = \kappa_{\text{sign}}(A)\|S\|_F/\|A\|_F$. Let $\Delta S = L_{\text{sign}}(A, \Delta A)$. Then by (5.9),

$$N\Delta S + \Delta S N = \Delta A - S\Delta A S,$$

where $N = SA = AS$. Defining $G = AS - S^2 = N - I$, we have

$$2\Delta S = \Delta A - S\Delta A S - G\Delta S - \Delta S G. \tag{5.15}$$

Taking norms, using (B.7), leads to

$$\|\Delta S\|_F \leq \frac{(\|S\|_2^2 + 1)\|\Delta A\|_F}{2(1 - \|G\|_2)},$$

which gives the upper bound.

Now let $\sigma = \|S\|_2$ and $Sv = \sigma u$, $u^*S = \sigma v^*$, where u and v are (unit-norm) left and right singular vectors, respectively. Putting $\Delta A = vu^*$ in (5.15) gives

$$2\Delta S = vu^* - Svu^*S - G\Delta S - \Delta S G = vu^* - \sigma^2 uv^* - G\Delta S - \Delta S G.$$

Hence

$$(\|S\|_2^2 - 1)\|\Delta A\|_F = (\sigma^2 - 1)\|\Delta A\|_F \le 2\|\Delta S\|_F(1 + \|G\|_2),$$

which implies the lower bound.

Setting $A = S$ in (5.13) gives (5.14). □

Theorem 5.4 has something to say about the attainable accuracy of a computed sign function. In computing $S = \operatorname{sign}(A)$ we surely cannot do better than if we computed $\operatorname{sign}(fl(S))$. But Theorem 5.4 says that relative errors in S can be magnified when we take the sign by as much as $\|S\|^2/2$, so we cannot expect a relative error in our computed sign smaller than $\|S\|^2 u/2$, whatever the method used.

5.2. Schur Method

The first method that we consider for computing $\operatorname{sign}(A)$ is expensive but has excellent numerical stability. Because the method utilizes a Schur decomposition it is not suitable for the applications in Sections 2.4 and 2.5, since those problems can be solved directly by the use of a Schur decomposition, without explicitly forming the sign function.

Let $A \in \mathbb{C}^{n\times n}$ have the Schur decomposition $A = QTQ^*$, where Q is unitary and T is upper triangular. Then $\operatorname{sign}(A) = Q\operatorname{sign}(T)Q^*$ (see Theorem 1.13 (c)). The problem therefore reduces to computing $U = \operatorname{sign}(T)$, and clearly U is upper triangular with $u_{ii} = \operatorname{sign}(t_{ii}) = \pm 1$ for all i. We will determine u_{ij} from the equation $U^2 = I$ when possible (namely, when $u_{ii} + u_{jj} \neq 0$), and from $TU = UT$ otherwise (in which case $t_{ii} \neq t_{jj}$), employing the Parlett recurrence (Algorithm 4.13) in this second case.

Algorithm 5.5 (Schur method). Given $A \in \mathbb{C}^{n\times n}$ having no pure imaginary eigenvalues, this algorithm computes $S = \operatorname{sign}(A)$ via a Schur decomposition.

1 Compute a (complex) Schur decomposition $A = QTQ^*$.
2 $u_{ii} = \operatorname{sign}(t_{ii})$, $i = 1\colon n$
3 for $j = 2\colon n$
4 for $i = j-1\colon -1\colon 1$
5
$$u_{ij} = \begin{cases} -\dfrac{\sum_{k=i+1}^{j-1} u_{ik}u_{kj}}{u_{ii}+u_{jj}}, & u_{ii}+u_{jj} \neq 0, \\ t_{ij}\dfrac{u_{ii}-u_{jj}}{t_{ii}-t_{jj}} + \dfrac{\sum_{k=i+1}^{j-1}(u_{ik}t_{kj} - t_{ik}u_{kj})}{t_{ii}-t_{jj}}, & u_{ii}+u_{jj} = 0. \end{cases}$$
6 end
7 end
8 $S = QUQ^*$

Cost: $25n^3$ flops for the Schur decomposition plus between $n^3/3$ and $2n^3/3$ flops for U and $3n^3$ flops to form S: about $28\frac{2}{3}n^3$ flops in total.

It is worth noting that the sign of an upper triangular matrix T will usually have some zero elements in the upper triangle. Indeed, suppose for some $j > i$ that t_{ii}, $t_{i+1,i+1}$, ..., t_{jj} all have the same sign, and let $T_{ij} = T(i\colon j, i\colon j)$. Then, since all the eigenvalues of T_{ij} have the same sign, the corresponding block $S(i\colon j, i\colon j)$ of $S = \operatorname{sign}(T)$ is $\pm I$. This fact could be exploited by reordering the Schur form so that the diagonal of T is grouped according to sign. Then $\operatorname{sign}(T)$ would have the form

$\begin{bmatrix} \pm I & W \\ 0 & \mp I \end{bmatrix}$, where W is computed by the Parlett recurrence. The cost of the reordering may or may not be less than the cost of (redundantly) computing zeros from the first expression for u_{ij} in Algorithm 5.5.

5.3. Newton's Method

The most widely used and best known method for computing the sign function is the Newton iteration, due to Roberts:

Newton iteration (matrix sign function):

$$X_{k+1} = \frac{1}{2}(X_k + X_k^{-1}), \qquad X_0 = A. \tag{5.16}$$

The connection of this iteration with the sign function is not immediately obvious, but in fact the iteration can be derived by applying Newton's method to the equation $X^2 = I$ (see Problem 5.8), and of course $\operatorname{sign}(A)$ is one solution of this equation (Theorem 5.1 (a)). The following theorem describes the convergence of the iteration.

Theorem 5.6 (convergence of the Newton sign iteration). *Let $A \in \mathbb{C}^{n\times n}$ have no pure imaginary eigenvalues. Then the Newton iterates X_k in* (5.16) *converge quadratically to $S = \operatorname{sign}(A)$, with*

$$\|X_{k+1} - S\| \le \frac{1}{2}\|X_k^{-1}\|\,\|X_k - S\|^2 \tag{5.17}$$

for any consistent norm. Moreover, for $k \ge 1$,

$$X_k = (I - G_0^{2^k})^{-1}(I + G_0^{2^k})S, \quad \textit{where } G_0 = (A - S)(A + S)^{-1}. \tag{5.18}$$

Proof. For $\lambda = re^{i\theta}$ we have $\lambda + \lambda^{-1} = (r + r^{-1})\cos\theta + i(r - r^{-1})\sin\theta$, and hence eigenvalues of X_k remain in their open half-plane under the mapping (5.16). Hence X_k is defined and nonsingular for all k. Moreover, $\operatorname{sign}(X_k) = \operatorname{sign}(X_0) = S$, and so $X_k + S = X_k + \operatorname{sign}(X_k)$ is also nonsingular.

Clearly the X_k are (rational) functions of A and hence, like A, commute with S. Then

$$\begin{aligned} X_{k+1} \pm S &= \frac{1}{2}\left(X_k + X_k^{-1} \pm 2S\right) \\ &= \frac{1}{2}X_k^{-1}\left(X_k^2 \pm 2X_k S + I\right) \\ &= \frac{1}{2}X_k^{-1}(X_k \pm S)^2, \end{aligned} \tag{5.19}$$

and hence

$$(X_{k+1} - S)(X_{k+1} + S)^{-1} = \left((X_k - S)(X_k + S)^{-1}\right)^2.$$

Defining $G_k = (X_k - S)(X_k + S)^{-1}$, we have $G_{k+1} = G_k^2 = \cdots = G_0^{2^{k+1}}$. Now $G_0 = (A - S)(A + S)^{-1}$ has eigenvalues $(\lambda - \operatorname{sign}(\lambda))/(\lambda + \operatorname{sign}(\lambda))$, where $\lambda \in \Lambda(A)$, all of which lie inside the unit circle since λ is not pure imaginary. Since $G_k = G_0^{2^k}$ and $\rho(G_0) < 1$, by a standard result (B.9) $G_k \to 0$ as $k \to \infty$. Hence

$$X_k = (I - G_k)^{-1}(I + G_k)S \to S \quad \text{as } k \to \infty. \tag{5.20}$$

The norm inequality (5.17), which displays the quadratic convergence, is obtained by taking norms in (5.19) with the minus sign. $\square$

Theorem 5.6 reveals quadratic convergence of the Newton iteration, but also displays in (5.18) precisely how convergence occurs: through the powers of the matrix G_0 converging to zero. Since for any matrix norm,

$$\|G_0^{2^k}\| \geq \rho(G_0^{2^k}) = \left(\max_{\lambda \in \Lambda(A)} \frac{|\lambda - \operatorname{sign}(\lambda)|}{|\lambda + \operatorname{sign}(\lambda)|} \right)^{2^k}, \tag{5.21}$$

It is clear that convergence will be slow if either $\rho(A) \gg 1$ or A has an eigenvalue close to the imaginary axis. We return to the speed of convergence in Section 5.5. For the behaviour of the iteration when it does not converge, see Problem 5.11.

The Newton iteration provides one of the rare circumstances in numerical analysis where the explicit computation of a matrix inverse is required. One way to try to remove the inverse from the formula is to approximate it by one step of Newton's method for the matrix inverse, which has the form $Y_{k+1} = Y_k(2I - BY_k)$ for computing B^{-1}; this is known as the Newton–Schulz iteration [512, 1933] (see Problem 7.8). Replacing X_k^{-1} by $X_k(2I - X_k^2)$ in (5.16) (having taken $Y_k = B = X_k$) gives

Newton–Schulz iteration:

$$X_{k+1} = \frac{1}{2}X_k(3I - X_k^2), \qquad X_0 = A. \tag{5.22}$$

This iteration is multiplication-rich and retains the quadratic convergence of Newton's method. However, it is only locally convergent, with convergence guaranteed for $\|I - A^2\| < 1$; see Theorem 5.8.

The Newton iteration also provides a way of computing the Fréchet derivative of the sign function.

Theorem 5.7 (Kenney and Laub). *Let $A \in \mathbb{C}^{n\times n}$ have no pure imaginary eigenvalues. With X_k defined by the Newton iteration* (5.16), *let*

$$Y_{k+1} = \frac{1}{2}(Y_k - X_k^{-1}Y_kX_k^{-1}), \qquad Y_0 = E. \tag{5.23}$$

Then $\lim_{k\to\infty} Y_k = L_{\mathrm{sign}}(A, E)$.

Proof. Denote by B_k the Newton sign iterates (5.16) for the matrix $B = \left[\begin{smallmatrix} A & E \\ 0 & A \end{smallmatrix}\right]$, which clearly has no pure imaginary eigenvalues. It is easy to show by induction that $B_k = \left[\begin{smallmatrix} X_k & Y_k \\ 0 & X_k \end{smallmatrix}\right]$. By Theorem 5.6 and (3.16) we have

$$B_k \to \operatorname{sign}(B) = \begin{bmatrix} \operatorname{sign}(A) & L_{\mathrm{sign}}(A, E) \\ 0 & \operatorname{sign}(A) \end{bmatrix}.$$

The result follows on equating the (1,2) blocks. $\square$

5.4. The Padé Family of Iterations

The Newton iteration is by no means the only rational matrix iteration for computing the matrix sign function. A variety of other iterations have been derived, with various aims, including to avoid matrix inversion in favour of matrix multiplication, to achieve a higher order of convergence, and to be better suited to parallel computation. Ad hoc manipulations can be used to derive new iterations, as we now indicate for the scalar case. By setting $y_k = x_k^{-1}$ in the Newton formula $x_{k+1} = (x_k + x_k^{-1})/2$, we obtain the "inverse Newton" variant

$$y_{k+1} = \frac{2y_k}{y_k^2+1}, \qquad y_0 = a, \tag{5.24}$$

which has quadratic convergence to $\mathrm{sign}(a)$. Combining two Newton steps yields $y_{k+2} = (y_k^4 + 6y_k^2 + 1)/(4y_k(y_k^2+1))$, and we can thereby define the quartically convergent iteration

$$y_{k+1} = \frac{y_k^4 + 6y_k^2 + 1}{4y_k(y_k^2+1)}, \qquad y_0 = a.$$

While a lot can be done using arguments such as these, a more systematic development is preferable. We describe an elegant Padé approximation approach, due to Kenney and Laub [343, 1991], that yields a whole table of methods containing essentially all those of current interest.

For non–pure imaginary $z \in \mathbb{C}$ we can write

$$\mathrm{sign}(z) = \frac{z}{(z^2)^{1/2}} = \frac{z}{(1-(1-z^2))^{1/2}} = \frac{z}{(1-\xi)^{1/2}}, \tag{5.25}$$

where $\xi = 1 - z^2$. Hence the task of approximating $\mathrm{sign}(z)$ leads to that of approximating

$$h(\xi) = (1-\xi)^{-1/2}, \tag{5.26}$$

where we may wish to think of ξ as having magnitude less than 1. Now h is a particular case of a hypergeometric function and hence much is known about $[\ell/m]$ Padé approximants $r_{\ell m}(\xi) = p_{\ell m}(\xi)/q_{\ell m}(\xi)$ to h, including explicit formulae for $p_{\ell m}$ and $q_{\ell m}$. (See Section 4.4.2 for the definition of Padé approximants.) Kenney and Laub's idea is to set up the family of iterations

$$x_{k+1} = f_{\ell m}(x_k) := x_k \frac{p_{\ell m}(1-x_k^2)}{q_{\ell m}(1-x_k^2)}, \qquad x_0 = a. \tag{5.27}$$

Table 5.1 shows the first nine iteration functions $f_{\ell m}$ from this family. Note that f_{11} gives Halley's method (see Problem 5.12), while f_{10} gives the Newton–Schulz iteration (5.22). The matrix versions of the iterations are defined in the obvious way:

Padé iteration:

$$X_{k+1} = X_k\, p_{\ell m}(I - X_k^2)\, q_{\ell m}(I - X_k^2)^{-1}, \qquad X_0 = A. \tag{5.28}$$

Two key questions are "what can be said about the convergence of (5.28)?" and "how should the iteration be evaluated?"

The convergence question is answered by the following theorem.

Table 5.1. *Iteration functions $f_{\ell m}$ from the Padé family* (5.27).

	$m=0$	$m=1$	$m=2$
$\ell=0$	x	$\dfrac{2x}{1+x^2}$	$\dfrac{8x}{3+6x^2-x^4}$
$\ell=1$	$\dfrac{x}{2}(3-x^2)$	$\dfrac{x(3+x^2)}{1+3x^2}$	$\dfrac{4x(1+x^2)}{1+6x^2+x^4}$
$\ell=2$	$\dfrac{x}{8}(15-10x^2+3x^4)$	$\dfrac{x}{4}\dfrac{(15+10x^2-x^4)}{1+5x^2}$	$\dfrac{x(5+10x^2+x^4)}{1+10x^2+5x^4}$

Theorem 5.8 (convergence of Padé iterations). *Let $A \in \mathbb{C}^{n\times n}$ have no pure imaginary eigenvalues. Consider the iteration* (5.28) *with $\ell+m>0$ and any subordinate matrix norm.*

(a) *For $\ell \ge m-1$, if $\|I-A^2\|<1$ then $X_k \to \mathrm{sign}(A)$ as $k\to\infty$ and $\|I-X_k^2\| < \|I-A^2\|^{(\ell+m+1)^k}$.*

(b) *For $\ell=m-1$ and $\ell=m$,*

$$(S-X_k)(S+X_k)^{-1} = \left[(S-A)(S+A)^{-1}\right]^{(\ell+m+1)^k}$$

and hence $X_k \to \mathrm{sign}(A)$ as $k\to\infty$.

Proof. See Kenney and Laub [343, 1991]. □

Theorem 5.8 shows that the iterations with $\ell=m-1$ and $\ell=m$ are globally convergent, while those with $\ell \ge m+1$ have local convergence, the convergence rate being $\ell+m+1$ in every case.

We now concentrate on the cases $\ell=m-1$ and $\ell=m$ which we call the *principal Padé iterations.* For these ℓ and m we define

$$g_r(x) \equiv g_{\ell+m+1}(x) = f_{\ell m}(x). \tag{5.29}$$

The g_r are the iteration functions from the Padé table taken in a zig-zag fashion from the main diagonal and first superdiagonal:

$$g_1(x)=x, \qquad g_2(x)=\frac{2x}{1+x^2}, \qquad g_3(x)=\frac{x(3+x^2)}{1+3x^2},$$

$$g_4(x)=\frac{4x(1+x^2)}{1+6x^2+x^4}, \quad g_5(x)=\frac{x(5+10x^2+x^4)}{1+10x^2+5x^4}, \quad g_6(x)=\frac{x(6+20x^2+6x^4)}{1+15x^2+15x^4+x^6}.$$

We know from Theorem 5.8 that the iteration $X_{k+1}=g_r(X_k)$ converges to $\mathrm{sign}(X_0)$ with order r whenever $\mathrm{sign}(X_0)$ is defined. These iterations share some interesting properties that are collected in the next theorem.

Theorem 5.9 (properties of principal Padé iterations). *The principal Padé iteration function g_r defined in* (5.29) *has the following properties.*

(a) $g_r(x) = \dfrac{(1+x)^r-(1-x)^r}{(1+x)^r+(1-x)^r}$. *In other words, $g_r(x)=p_r(x)/q_r(x)$, where $p_r(x)$ and $q_r(x)$ are, respectively, the odd and even parts of $(1+x)^r$.*

(b) $g_r(x) = \tanh(r \operatorname{arctanh}(x))$.

(c) $g_r(g_s(x)) = g_{rs}(x)$ *(the semigroup property).*

(d) g_r *has the partial fraction expansion*

$$g_r(x) = \frac{2}{r} \sum_{i=0}^{\lceil \frac{r-2}{2} \rceil}{}' \frac{x}{\sin^2\left(\frac{(2i+1)\pi}{2r}\right) + \cos^2\left(\frac{(2i+1)\pi}{2r}\right) x^2}, \tag{5.30}$$

where the prime on the summation symbol denotes that the last term in the sum is halved when r is odd.

Proof.

(a) See Kenney and Laub [343, 1991, Thm. 3.2].

(b) Recalling that $\tanh(x) = (e^x - e^{-x})/(e^x + e^{-x})$, it is easy to check that

$$\operatorname{arctanh}(x) = \frac{1}{2} \log\left(\frac{1+x}{1-x}\right).$$

Hence

$$r \operatorname{arctanh}(x) = \log\left(\frac{1+x}{1-x}\right)^{r/2}.$$

Taking the tanh of both sides gives

$$\tanh(r \operatorname{arctanh}(x)) = \frac{\left(\frac{1+x}{1-x}\right)^{r/2} - \left(\frac{1-x}{1+x}\right)^{r/2}}{\left(\frac{1+x}{1-x}\right)^{r/2} + \left(\frac{1-x}{1+x}\right)^{r/2}} = \frac{(1+x)^r - (1-x)^r}{(1+x)^r + (1-x)^r} = g_r(x).$$

(c) Using (b) we have

$$\begin{aligned} g_r(g_s(x)) &= \tanh(r \operatorname{arctanh}(\tanh(s \operatorname{arctanh}(x)))) = \tanh(rs \operatorname{arctanh}(x)) \\ &= g_{rs}(x). \end{aligned}$$

(d) The partial fraction expansion is obtained from a partial fraction expansion for the hyperbolic tangent; see Kenney and Laub [345, 1994, Thm. 3]. □

Some comments on the theorem are in order. The equality in (a) is a scalar equivalent of (b) in Theorem 5.8, and it provides an easy way to generate the g_r. Property (c) says that one rth order principal Padé iteration followed by one sth order iteration is equivalent to one rsth order iteration. Whether or not it is worth using higher order iterations therefore depends on the efficiency with which the different iterations can be evaluated. The properties in (b) and (c) are analogous to properties of the Chebyshev polynomials. Figure 5.1 confirms, for real x, that $g_r(x) = \tanh(r \operatorname{arctanh}(x))$ approximates $\operatorname{sign}(x)$ increasingly well near the origin as r increases.

Some more insight into the convergence, or nonconvergence, of the iteration $x_{k+1} = g_r(x_k)$ from (5.29) can be obtained by using Theorem 5.8 (b) to write, in polar form,

$$\rho_{k+1} e^{i\theta_{k+1}} := (s - x_{k+1})(s + x_{k+1})^{-1} := \left[(s - x_k)(s + x_k)^{-1}\right]^r = \left[\rho_k e^{i\theta_k}\right]^r,$$

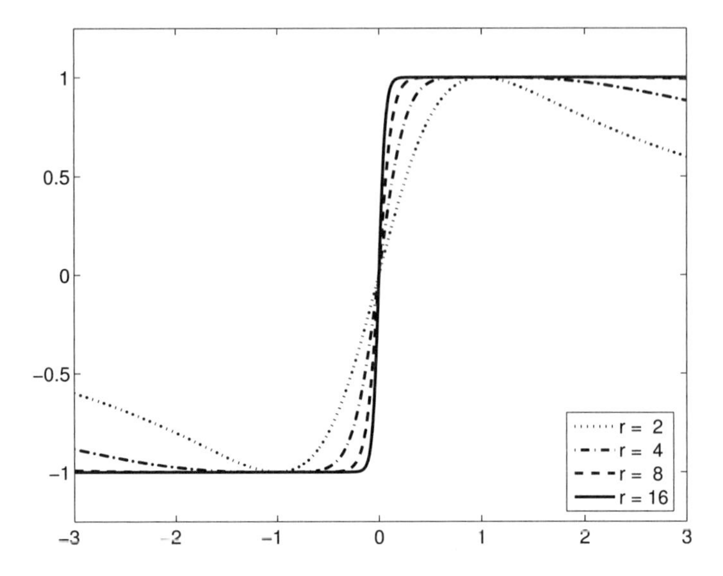

Figure 5.1. *The function* $g_r(x) = \tanh(r \operatorname{arctanh}(x))$ *for* $r = 2, 4, 8, 16$.

where $s = \operatorname{sign}(x_0)$. Hence

$$\rho_{k+1} = \rho_k^r, \qquad \theta_{k+1} = r\theta_k.$$

These relations illustrate the convergence of x_k to s for x_0 off the imaginary axis, since $\rho_0 < 1$. But they also reveal a chaotic aspect to the convergence through θ_k, which, in view of the periodicity of $e^{i\theta_k}$, can be written

$$\theta_{k+1} = r\theta_k \bmod 2\pi. \tag{5.31}$$

This recurrence can be described as a linear congruential random number generator [211, 2003, Sec. 1.2], [357, 1998, Sec. 3.2], though with a real, rather than integer, modulus. If x_0 is pure imaginary then the iteration does not converge: $\rho_k \equiv 1$, x_k remains pure imaginary for all k, and $(s - x_k)(s + x_k)^{-1}$ wanders chaotically around the circle of radius 1 centred at the origin; see also Problem 5.11.

We turn now to evaluation of the matrix iteration $X_{j+1} = g_r(X_j)$. As discussed in Section 4.4.2, several approaches are possible, based on different representations of the rational iteration function g_k. Evaluating $g_k(X_j)$ as the ratio of two polynomials may require more flops than via the partial fraction expansion (5.30). For example, evaluating g_3 from the formula $x(3 + x^2)/(1 + 3x^2)$ at an $n \times n$ matrix requires $6\frac{2}{3}n^3$ flops, whereas (5.30) can be written as

$$g_3(x) = \frac{1}{3}\left(x + \frac{8x}{1 + 3x^2}\right) \tag{5.32}$$

and evaluated in $4\frac{2}{3}n^3$ flops. An attractive feature of the partial fraction expansion (5.30) is that it comprises $\lceil \frac{r-2}{2} \rceil$ independent matrix inversions (or multiple right-hand side linear systems), which can be carried out in parallel.

5.5. Scaling the Newton Iteration

For scalar a, the Newton iteration (5.16) is

$$x_{k+1} = \frac{1}{2}(x_k + x_k^{-1}), \qquad x_0 = a, \tag{5.33}$$

which converges to $\mathrm{sign}(a) = \pm 1$ if a is not pure imaginary. This is precisely Newton's method for the square root of 1 and convergence is at a quadratic rate, as described by (5.17). Once the error is sufficiently small (in practice, less than, say, 0.5), successive errors decrease rapidly, each being approximately the square of the previous one (see (5.19)). However, initially convergence can be slow: if $|x_k| \gg 1$ then $x_{k+1} \approx x_k/2$ and the iteration is an expensive way to divide by 2! From (5.18) and (5.21) we also see that slow convergence will result when a is close to the imaginary axis. Therefore a way is needed of speeding up the initial phase of convergence in these unfavourable cases. For matrices, the same comments apply to the eigenvalues, because the Newton iteration (5.16) is effectively performing the scalar iteration (5.33) independently on each eigenvalue. However, the behaviour of the matrix iteration is not entirely determined by the eigenvalues: nonnormality of A can delay, though not prevent, convergence, as the following finite termination result shows.

Theorem 5.10 (Kenney and Laub). *For the Newton iteration* (5.16), *if* X_k *has eigenvalues* ± 1 *for some* k *then* $X_{k+p} = \mathrm{sign}(A)$ *for* $2^p \geq m$, *where* m *is the size of the largest Jordan block of* X_k *(which is no larger than the size of the largest Jordan block of* A*).*

Proof. Let X_k have the Jordan form $X_k = ZJ_kZ^{-1}$, where $J_k = D + N_k$, with $D = \mathrm{diag}(\pm 1) = \mathrm{sign}(J_k)$ and N_k strictly upper triangular. N_k has index of nilpotence m, that is, $N_k^m = 0$ but all lower powers are nonzero. We can restrict our attention to the convergence of the sequence beginning with J_k to $\mathrm{diag}(\pm 1)$, and so we can set $Z = I$. The next iterate, $X_{k+1} = D + N_{k+1}$, satisfies, in view of (5.19),

$$N_{k+1} = \frac{1}{2} X_k^{-1} N_k^2.$$

Since N_k has index of nilpotence m, N_{k+1} must have index of nilpotence $\lceil m/2 \rceil$. Applying this argument repeatedly shows that for $2^p \geq m$, N_{k+p} has index of nilpotence 1 and hence is zero, as required. That m is no larger than the order of the largest Jordan block of A follows from Theorem 1.36. □

An effective way to enhance the initial speed of convergence is to scale the iterates: prior to each iteration, X_k is replaced by $\mu_k X_k$, giving the scaled Newton iteration

Scaled Newton iteration:

$$X_{k+1} = \frac{1}{2}\left(\mu_k X_k + \mu_k^{-1} X_k^{-1}\right), \qquad X_0 = A. \tag{5.34}$$

As long as μ_k is real and positive, the sign of the iterates is preserved. Three main scalings have been proposed:

$$\text{determinantal scaling:} \quad \mu_k = |\det(X_k)|^{-1/n}, \tag{5.35}$$

$$\text{spectral scaling:} \quad \mu_k = \sqrt{\rho(X_k^{-1})/\rho(X_k)}, \tag{5.36}$$

$$\text{norm scaling:} \quad \mu_k = \sqrt{\|X_k^{-1}\|/\|X_k\|}. \tag{5.37}$$

For determinantal scaling, $|\det(\mu_k X_k)| = 1$, so that the geometric mean of the eigenvalues of $\mu_k X_k$ has magnitude 1. This scaling has the property that μ_k minimizes $d(\mu_k X_k)$, where

$$d(X) = \sum_{i=1}^{n} (\log |\lambda_i|)^2$$

and the are λ_i the eigenvalues of X. Hence determinantal scaling tends to bring the eigenvalues closer to the unit circle; see Problem 5.13.

When evaluating the determinantal scaling factor (5.35) some care is needed to avoid unnecessary overflow and underflow, especially when n is large. The quantity μ_k should be within the range of the floating point arithmetic, since its reciprocal has magnitude the geometric mean of the eigenvalues of X_k and hence lies between the moduli of the smallest and largest eigenvalues. But $\det(X_k)$ can underflow or overflow. Assuming that an LU factorization $PX_k = L_k U_k$ is computed, where U_k has diagonal elements u_{ii}, we can rewrite $\mu_k = |u_{11} \dots u_{nn}|^{-1/n}$ as $\mu_k = \exp((-1/n)\sum_{i=1}^{n} \log |u_{ii}|)$. The latter expression avoids underflow and overflow; however, cancellation in the summation can produce an inaccurate computed μ_k, so it may be desirable to use one of the summation methods from Higham [276, 2002, Chap. 4].

For spectral scaling, if $\lambda_n, \dots, \lambda_1$ are the eigenvalues of X_k ordered by increasing magnitude, then $\mu_k = |\lambda_1 \lambda_n|^{-1/2}$ and so $\mu_k X_k$ has eigenvalues of smallest and largest magnitude $|\mu_k \lambda_n| = |\lambda_n/\lambda_1|^{1/2}$ and $|\mu_k \lambda_1| = |\lambda_1/\lambda_n|^{1/2}$. If λ_1 and λ_n are real, then in the Cayley metric

$$\mathcal{C}(x, \operatorname{sign}(x)) := \begin{cases} |x-1|/|x+1|, & \operatorname{Re} x > 0, \\ |x+1|/|x-1|, & \operatorname{Re} x < 0, \end{cases}$$

$\mu_k \lambda_n$ is the same distance from $\operatorname{sign}(\lambda_n)$ as $\mu_k \lambda_1$ is from $\operatorname{sign}(\lambda_1)$, so in this case spectral scaling equalizes the extremal eigenvalue errors in the Cayley metric. The norm scaling (5.37) can be regarded as approximating the spectral scaling.

What can be said about the effectiveness of these scaling strategies? In general, all of them work well, but there are some specific advantages and disadvantages.

Spectral scaling is essentially optimal when all the eigenvalues of A are real; indeed it yields finite termination, as the following result shows.

Theorem 5.11 (Barraud). *Let the nonsingular matrix $A \in \mathbb{C}^{n\times n}$ have all real eigenvalues and let $S = \operatorname{sign}(A)$. Then, for the Newton iteration (5.34) with spectral scaling, $X_{d+p-1} = \operatorname{sign}(A)$, where d is the number of distinct eigenvalues of A and $2^p \geq m$, where m is the size of the largest Jordan block of A.*

Proof. We will need to use the following easily verified properties of the iteration function $f(x) = \frac{1}{2}(x + 1/x)$:

$$f(x) = f(1/x), \tag{5.38a}$$

$$0 \leq x_2 \leq x_1 \leq 1 \quad \text{or} \quad 1 \leq x_1 \leq x_2 \quad \Rightarrow \quad 1 \leq f(x_1) \leq f(x_2). \tag{5.38b}$$

Let the eigenvalues of $X_0 = A$, which we know to be real, be ordered $|\lambda_n| \leq \dots \leq |\lambda_1|$. Then, from (5.36), $\mu_0 = |\lambda_n \lambda_1|^{-1/2}$, and the eigenvalues of $\mu_0 X_0$ have moduli lying between $|\mu_0 \lambda_n| = |\lambda_n/\lambda_1|^{1/2}$ and $|\mu_0 \lambda_1| = |\lambda_1/\lambda_n|^{1/2}$. These values are reciprocals, and hence by (5.38a), and since the eigenvalues are real, λ_n and λ_1 are mapped to values with the same modulus. By (5.38b) these values are the eigenvalues of

X_1 of largest modulus. Hence X_1 has eigenvalues $\lambda_i^{(1)}$ satisfying $|\lambda_n^{(1)}| \le \cdots \le |\lambda_2^{(1)}| = |\lambda_1^{(1)}|$. Each subsequent iteration increases by at least 1 the number of eigenvalues with maximal modulus until, after $d-1$ iterations, X_{d-1} has eigenvalues of constant modulus. Then $\mu_{d-1}X_{d-1}$ has converged eigenvalues ± 1 (as does X_d). By Theorem 5.10, at most a further p iterations after X_{d-1} are needed to dispose of the Jordan blocks (and during these iterations $\mu_k \equiv 1$, since the eigenvalues are fixed at ± 1). □

For 1×1 matrices spectral scaling and determinantal scaling are equivalent, and both give convergence in at most two iterations (see Problem 5.14). For 2×2 matrices spectral scaling and determinantal scaling are again equivalent, and Theorem 5.11 tells us that we have convergence in at most two iterations if the eigenvalues are real. However, slightly more is true: both scalings give convergence in at most two iterations for any *real* 2×2 matrix (see Problem 5.14).

Determinantal scaling can be ineffective when there is a small group of outlying eigenvalues and the rest are nearly converged. Suppose that A has an eigenvalue 10^q ($q \ge 1$) with the rest all ± 1. Then determinantal scaling gives $\mu_k = 10^{-q/n}$, whereas spectral scaling gives $\mu_k = 10^{-q/2}$; the former quantity is close to 1 and hence the determinantally scaled iteration will behave like the unscaled iteration. Spectral scaling can be ineffective when the eigenvalues of A cluster close to the imaginary axis (see the numerical examples in Section 5.8).

All three scaling schemes are inexpensive to implement. The determinant $\det(X_k)$ can be computed at negligible cost from the LU factorization that will be used to compute X_k^{-1}. The spectral scaling parameter can be cheaply estimated by applying the power method to X_k and its inverse, again exploiting the LU factorization in the latter case. Note, however, that for a real spectrum spectral scaling increases the number of eigenvalues with maximal modulus on each iteration, which makes reliable implementation of the power method more difficult. The norm scaling is trivial to compute for the Frobenius norm, and for the 2-norm can be estimated using the power method (Algorithm 3.19).

The motivation for scaling is to reduce the length of the initial phase during which the error is reduced below 1. Should we continue to scale throughout the whole iteration? All three scaling parameters (5.35)–(5.37) converge to 1 as $X_k \to S$, so scaling does not destroy the quadratic convergence. Nor does it bring any benefit, so it is sensible to set $\mu_k \equiv 1$ once the error is sufficiently less than 1.

5.6. Terminating the Iterations

Crucial to the success of any sign iteration is an inexpensive and effective way to decide when to terminate it. We begin with a lemma that provides some bounds that help guide the choice of stopping criterion in both relative error-based and residual-based tests.

Lemma 5.12 (Kenney, Laub, Pandey, and Papadopoulos). *Let $A \in \mathbb{C}^{n\times n}$ have no pure imaginary eigenvalues, let $S = \mathrm{sign}(A)$, and let $\|\cdot\|$ be any subordinate matrix norm. If $\|S(A-S)\| = \epsilon < 1$ then*

$$\left(\frac{1-\epsilon}{2+\epsilon}\right)\|A - A^{-1}\| \le \|A - S\| \le \left(\frac{1+\epsilon}{2-\epsilon}\right)\|A - A^{-1}\| \tag{5.39}$$

and

$$\frac{\|A^2 - I\|}{\|S\|(\|A\| + \|S\|)} \le \frac{\|A - S\|}{\|S\|} \le \|A^2 - I\|. \tag{5.40}$$

The lower bound in (5.40) *always holds.*

Proof. Let $E = A - S$. Since $S^2 = I$, we have $A = S + E = (I + ES)S$. It is then straightforward to show that

$$E(2I + ES) = (A - A^{-1})(I + ES),$$

using the fact that A and S, and hence also E and S, commute. The upper bound in (5.39) is obtained by postmultiplying by $(2I + ES)^{-1}$ and taking norms, while postmultiplying by $(I + ES)^{-1}$ and taking norms gives the lower bound.

The lower bound in (5.40) is obtained by taking norms in $A^2 - I = (A - S)(A + S)$. For the upper bound, we write the last equation as $A - S = (A^2 - I)(A + S)^{-1}$ and need to bound $\|(A + S)^{-1}\|$. Since $A + S = 2S(I + \frac{1}{2}S(A - S))$, we have

$$\|(A + S)^{-1}\| = \frac{1}{2}\|S^{-1}\big(I + \tfrac{1}{2}S(A - S)\big)^{-1}\| \le \frac{\frac{1}{2}\|S^{-1}\|}{1 - \frac{1}{2}\epsilon} \le \|S\|. \qquad \square$$

Note that since the iterations of interest satisfy $\operatorname{sign}(X_k) = \operatorname{sign}(A)$, the bounds of Lemma 5.12 are applicable with A replaced by an iterate X_k.

We now describe some possible convergence criteria, using η to denote a convergence tolerance proportional to both the unit roundoff (or a larger value if full accuracy is not required) and a constant depending on the matrix dimension, n. A norm will denote any easily computable norm such as the 1-, ∞-, or Frobenius norms. We begin with the Newton iteration, describing a variety of existing criteria followed by a new one.

A natural stopping criterion, of negligible cost, is

$$\delta_{k+1} := \frac{\|X_{k+1} - X_k\|}{\|X_{k+1}\|} \le \eta. \tag{5.41}$$

As discussed in Section 4.9, this criterion is really bounding the error in X_k, rather than X_{k+1}, so it may stop one iteration too late. This drawback can be seen very clearly from (5.39): since $X_{k+1} - X_k = \frac{1}{2}(X_k^{-1} - X_k)$, (5.39) shows that $\|X_{k+1} - X_k\| \approx \|S - X_k\|$ is an increasingly good approximation as the iteration converges.

The test (5.41) could potentially never be satisfied in floating point arithmetic. The best bound for the error in the computed $\widehat{Z}_k = fl(X_k^{-1})$, which we assume to be obtained by Gaussian elimination with partial pivoting, is of the form [276, 2002, Sec. 14.3]

$$\frac{\|\widehat{Z}_k - X_k^{-1}\|}{\|X_k^{-1}\|} \le c_n u \kappa(X_k), \tag{5.42}$$

where c_n is a constant. Therefore for the computed sequence X_k, $\|X_{k+1} - X_k\| \approx \frac{1}{2}\|\widehat{Z}_k - X_k\|$ might be expected to be proportional to $\kappa(X_k)\|X_k\|u$, suggesting the test $\delta_{k+1} \le \kappa(X_k)\eta$. Close to convergence, $X_{k+1} \approx X_k \approx S = S^{-1}$ and so $\kappa(X_k) \approx \|X_{k+1}\|^2$. A test $\delta_{k+1} \le \|X_{k+1}\|^2\eta$ is also suggested by the asymptotic conditioning of the sign function, discussed at the end of Section 5.1. On the other hand, a test of the form $\delta_{k+1} \le \|X_{k+1}\|\eta$ is suggested by Byers, He, and Mehrmann [89, 1997], based on a perturbation bound for the sign function. To summarize, there are arguments for using the stopping criterion

$$\delta_{k+1} \le \|X_{k+1}\|^p \eta \tag{5.43}$$

for each of $p = 0$, 1, and 2.

A different approach is based on the bound (5.17): $\|X_{k+1} - S\| \le \frac{1}{2}\|X_k^{-1}\| \, \|X_k - S\|^2$. Since $\|X_{k+1} - X_k\| \approx \|S - X_k\|$ close to convergence, as noted above,

$$\|X_{k+1} - S\| \lesssim \frac{1}{2}\|X_k^{-1}\| \, \|X_{k+1} - X_k\|^2.$$

Hence we can expect $\|X_{k+1} - S\|/\|X_{k+1}\| \lesssim \eta$ if

$$\|X_{k+1} - X_k\| \le \left(2\eta \frac{\|X_{k+1}\|}{\|X_k^{-1}\|}\right)^{1/2}. \tag{5.44}$$

This is essentially the same test as (4.25), bearing in mind that in the latter bound $c = \|S^{-1}\|/2 \approx \|X_k^{-1}\|/2$. This bound should overcome the problem of (5.41) of stopping one iteration too late, but unlike (5.43) with $p = 1, 2$ it takes no explicit account of rounding error effects. A test of this form has been suggested by Benner and Quintana-Ortí [55, 1999]. The experiments in Section 5.8 give further insight.

For general sign iterations, intuitively appealing stopping criteria can be devised based on the fact that trace(sign(A)) is an integer, but these are of little practical use; see Problem 5.16.

The upper bound in (5.40) shows that $\|A - X_k\|/\|X_k\| \le \|X_k^2 - I\|$ and hence suggests stopping when

$$\|X_k^2 - I\| \le \eta. \tag{5.45}$$

This test is suitable for iterations that already form X_k^2, such as the Schulz iteration (5.22). Note, however, that the error in forming $fl(X_k^2 - I)$ is bounded at best by $c_n u \|X_k\|^2 \approx c_n u \|S\|^2$, so when $\|S\|$ is large it may not be possible to satisfy (5.45), and a more suitable test is then

$$\frac{\|X_k^2 - I\|}{\|X_k\|^2} \le \eta.$$

5.7. Numerical Stability of Sign Iterations

The question of the stability of sign iterations, where stability is defined in Definition 4.17, has a particularly nice answer for all the iterations of interest.

Theorem 5.13 (stability of sign iterations). *Let $S = \mathrm{sign}(A)$, where $A \in \mathbb{C}^{n\times n}$ has no pure imaginary eigenvalues. Let $X_{k+1} = g(X_k)$ be superlinearly convergent to $\mathrm{sign}(X_0)$ for all X_0 sufficiently close to S and assume that g is independent of X_0. Then the iteration is stable, and the Fréchet derivative of g at S is idempotent and is given by $L_g(S,E) = L(S,E) = \frac{1}{2}(E - SES)$, where $L(S)$ is the Fréchet derivative of the matrix sign function at S.*

Proof. Since the sign function is idempotent, stability, the idempotence of L_g, and the equality of $L_g(S)$ and $L(S)$, follow from Theorems 4.18 and 4.19. The formula for $L(S,E)$ is obtained by taking $N = I$ in Theorem 5.3. □

Theorem 5.13 says that the Fréchet derivative at S is the same for any superlinearly convergent sign iteration and that this Fréchet derivative is idempotent. Unbounded propagation of errors near the solution is therefore not possible for any such iteration. The constancy of the Fréchet derivative is not shared by iterations for all the functions in this book, as we will see in the next chapter.

Turning to limiting accuracy (see Definition 4.20), Theorem 5.13 yields $\|L_g(S,E)\| \le \frac{1}{2}(1+\|S\|^2)\|E\|$, so an estimate for the limiting accuracy of any superlinearly convergent sign iteration is $\|S\|^2 u$. Hence if, for example, $\kappa(S) = \|S\|^2 \le u^{-1/2}$, then we can hope to compute the sign function to half precision.

If S commutes with E then $L_g(S,E) = 0$, which shows that such errors E are eliminated by the iteration to first order. To compare with what convergence considerations say about E, note first that in all the sign iterations considered here the matrix whose sign is being computed appears only as the starting matrix and not within the iteration. Hence if we start the iteration at $S+E$ then the iteration converges to $\mathrm{sign}(S+E)$, for sufficiently small $\|E\|$ (so that the sign exists and any convergence conditions are satisfied). Given that S has the form (5.1), any E commuting with S has the form $Z\,\mathrm{diag}(F_{11},F_{22})Z^{-1}$, so that $\mathrm{sign}(S+E) = Z\,\mathrm{sign}(\mathrm{diag}(-I_p+F_{11}, I_q+F_{22}))Z^{-1}$. Hence there is an ϵ such that for all $\|E\| \le \epsilon$, $\mathrm{sign}(S+E) = S$. Therefore, the Fréchet derivative analysis is consistent with the convergence analysis.

Of course, to obtain a complete picture, we also need to understand the effect of rounding errors on the iteration prior to convergence. This effect is surprisingly difficult to analyze, even though the iterative methods are built purely from matrix multiplication and inversion. The underlying behaviour is, however, easy to describe. Suppose, as discussed above, that we have an iteration for $\mathrm{sign}(A)$ that does not contain A, except as the starting matrix. Errors on the $(k-1)$st iteration can be accounted for by perturbing X_k to $X_k + E_k$. If there are no further errors then (regarding $X_k + E_k$ as a new starting matrix) $\mathrm{sign}(X_k + E_k)$ will be computed. The error thus depends on the conditioning of X_k and the size of E_k. Since errors will in general occur on each iteration, the overall error will be a complicated function of $\kappa_{\mathrm{sign}}(X_k)$ and E_k for all k.

We now restrict our attention to the Newton iteration (5.16). First, we note that the iteration can be numerically unstable: the relative error is not always bounded by a modest multiple of the condition number $\kappa_{\mathrm{sign}}(A)$, as is easily shown by example (see the next section). Nevertheless, it generally performs better than might be expected, given that it inverts possibly ill conditioned matrices. We are not aware of any published rounding error analysis for the computation of $\mathrm{sign}(A)$ via the Newton iteration.

Error analyses aimed at the application of the matrix sign function to invariant subspace computation (Section 2.5) are given by Bai and Demmel [29, 1998] and Byers, He, and Mehrmann [89, 1997]. These analyses show that the matrix sign function may be more ill conditioned than the problem of evaluating the invariant subspaces corresponding to eigenvalues in the left half-plane and right half-plane. Nevertheless, they show that when Newton's method is used to evaluate the sign function the computed invariant subspaces are usually about as good as those computed by the QR algorithm. In other words, the potential instability rarely manifests itself. The analyses are complicated and we refer the reader to the two papers for details.

In cases where the matrix sign function approach to computing an invariant subspace suffers from instability, iterative refinement can be used to improve the com-

Table 5.2. *Number of iterations for scaled Newton iteration. The unnamed matrices are (quasi)-upper triangular with normal* $(0,1)$ *distributed elements in the upper triangle.*

Matrix	Scaling			
	none	determinantal	spectral	norm
Lotkin	25	9	8	9
Grcar	11	9	9	15
$A(j\colon j+1, j\colon j+1) = \begin{bmatrix} 1 & (j/n)1000 \\ -(j/n)1000 & 1 \end{bmatrix}$	24	16	19	19
$a_{jj} = 1 + 1000i(j-1)/(n-1)$	24	16	22	22
$a_{11} = 1000,\ a_{jj} \equiv 1,\ j \geq 2$	14	12	6	10
$a_{11} = 1 + 1000i,\ a_{jj} \equiv 1,\ j \geq 2$	24	22	8	19

puted subspace [29, 1998]. Iterative refinement can also be used when the sign function is used to solve algebraic Riccati equations (as described in Section 2.4) [88, 1987].

Finally, we note that all existing numerical stability analysis is for the *unscaled* Newton iteration. Our experience is that scaling tends to improve stability, not worsen it.

5.8. Numerical Experiments and Algorithm

We present some numerical experiments to illustrate the theory of the previous three sections and to give further insight into the choice of iteration, acceleration scheme, and stopping criterion. In all the tests, scaling was used as long as the relative change $\delta_k = \|X_k - X_{k-1}\|_\infty / \|X_k\|_\infty$ exceeded 10^{-2}; thereafter $\mu_k \equiv 1$ and, where relevant, μ_k is not shown in the tables.

First, we consider the effects of scaling. For a variety of matrices we ran the Newton iteration (5.34) with no scaling and with the scalings (5.35)–(5.37), with the 2-norm used for norm scaling. We recorded how many iterations are required to produce an error $\|S - X_k\|_\infty / \|S\|_\infty \leq 5 \times 10^{-14}$. The matrices are as follows:

1. The 8×8 Lotkin matrix, MATLAB's `gallery('lotkin',8)`: badly conditioned with many negative eigenvalues of small magnitude.

2. The 25×25 Grcar matrix, `gallery('grcar',25)`: a Toeplitz matrix with sensitive eigenvalues.

3. 25×25 (quasi-) upper triangular matrices with elements in the upper triangle (outside the diagonal blocks) from the normal (0,1) distribution.

Table 5.2 reports the results. The Lotkin matrix is a typical example of how scaling can greatly reduce the number of iterations. The Grcar example shows how norm scaling can perform poorly (indeed being worse than no scaling). The third matrix (real) and fourth matrix (complex) have eigenvalues on a line with real part 1 and imaginary parts between 0 and 1000. Here, spectral scaling and norm scaling are both poor. The fifth and sixth matrices, again real and complex, respectively, have eigenvalues all equal to 1 except for one large outlier, and they are bad cases for determinantal scaling.

Table 5.3 illustrates the convergence results in Theorems 5.10 and 5.11 by showing the behaviour of the Newton iteration with spectral scaling for $J(2) \in \mathbb{R}^{16\times 16}$, which

Table 5.3. *Newton iteration with spectral scaling for Jordan block* $J(2) \in \mathbb{R}^{16\times16}$.

k	$\frac{\|S-X_k\|_\infty}{\|S\|_\infty}$	δ_k	$\frac{\|X_k^2-I\|_\infty}{\|X_k\|_\infty^2}$	μ_k	(5.41)	(5.44)
1	2.5e-1	1.8e+0	3.6e-1	5.0e-1		
2	2.5e-2	2.2e-1	4.8e-2	1.0e0		
3	3.0e-4	2.5e-2	6.0e-4	1.0e0		
4	0	3.0e-4	0	1.0e0		
5	0	0	0		$\checkmark$	$\checkmark$

Table 5.4. *Newton iteration with determinantal scaling for random* $A \in \mathbb{R}^{16\times16}$ *with* $\kappa_2(A) = 10^{10}$; $\kappa_{\mathrm{sign}}(A) = 3\times10^8$, $\|S\|_F = 16$.

k	$\frac{\|S-X_k\|_\infty}{\|S\|_\infty}$	δ_k	$\frac{\|X_k^2-I\|_\infty}{\|X_k\|_\infty^2}$	μ_k	(5.41)	(5.44)
1	4.3e3	1.0e0	1.1e-1	1.0e5		
2	1.5e1	2.8e2	1.3e-1	6.8e-3		
3	1.9e0	6.3e0	5.9e-2	1.4e-1		
4	2.1e-1	1.7e0	2.1e-2	6.1e-1		
5	6.4e-2	2.3e-1	4.3e-3	9.5e-1		
6	2.0e-3	6.2e-2	1.6e-4	9.8e-1		
7	4.1e-6	2.0e-3	3.3e-7	1.0e0		
8	2.1e-9	4.1e-6	8.9e-13			
9	2.1e-9	1.1e-11	3.2e-17			$\checkmark$
10	2.1e-9	1.5e-15	3.5e-17		$\checkmark$	$\checkmark$

is a Jordan block with eigenvalue 2. Here and below the last two columns of the table indicate with a tick iterations on which the convergence conditions (5.41) and (5.44) are satisfied for the ∞-norm, with $\eta = n^{1/2}u$. In Theorem 5.11, $d = 1$ and $p = 4$, and indeed $X_{d+p-1} = X_4 = \mathrm{sign}(J(2))$. At the start of the first iteration, $\mu_0 X_0$ has eigenvalues 1, and the remaining four iterations remove the nonnormal part; it is easy to see that determinantal scaling gives exactly the same results.

Table 5.4 reports 12 iterations for a random $A \in \mathbb{R}^{16\times16}$ with $\kappa_2(A) = 10^{10}$ generated in MATLAB by `gallery('randsvd',16,1e10,3)`. Determinantal scaling was used. Note that the relative residual decreases significantly after the error has stagnated. The limiting accuracy of $\|S\|_2^2 u$ is clearly not relevant here, as the iterates do not approach S sufficiently closely.

Both these examples confirm that the relative change δ_{k+1} is a good estimate of the relative error in X_k (compare the numbers in the third column with those immediately to the northwest) until roundoff starts to dominate, but thereafter the relative error and relative change can behave quite differently.

Finally, Table 5.5 gives examples with large $\|S\|$. The matrix is of the form $A = QTQ^T$, where Q is a random orthogonal matrix and $T \in \mathbb{R}^{16\times16}$ is generated as an upper triangular matrix with normal (0,1) distributed elements and t_{ii} is replaced by $d|t_{ii}|$ for $i = 1{:}8$ and by $-d|t_{ii}|$ for $i = 9{:}16$. As d is decreased the eigenvalues of A approach the origin (and hence the imaginary axis). Determinantal scaling was used and we terminated the iteration when the relative error stopped decreasing sig-

Table 5.5. *Newton iteration with determinantal scaling for random $A \in \mathbb{R}^{16\times 16}$ with real eigenvalues parametrized by d.*

d	no. iterations	$\min_k \dfrac{\|S - X_k\|_\infty}{\|S\|_\infty}$	$\|A\|_2$	$\kappa_2(A)$	$\|S\|_2$	$\kappa_{\mathrm{sign}}(A)$
1	6	2.7e-13	6.7	4.1e3	1.3e2	4.7e3
3/4	6	4.1e-10	6.5	5.7e5	5.4e3	6.5e5
1/2	6	2.6e-6	6.2	2.6e8	3.9e5	6.5e7
1/3	3	7.8e-1	6.4	2.6e15	7.5e11	3.9e7

nificantly. This example shows that the Newton iteration can behave in a numerically unstable way: the relative error can greatly exceed $\kappa_{\mathrm{sign}}(A)u$. Note that the limiting accuracy $\|S\|_2^2 u$ provides a good estimate of the relative error for the first three values of d.

Our experience indicates that (5.44) is the most reliable termination criterion, though on badly behaved matrices such as those in Table 5.5 no one test can be relied upon to terminate at the "right moment", if at all.

Based on this and other evidence we suggest the following algorithm based on the scaled Newton iteration (5.34).

Algorithm 5.14 (Newton algorithm for matrix sign function). Given a nonsingular $A \in \mathbb{C}^{n\times n}$ with no pure imaginary eigenvalues this algorithm computes $X = \mathrm{sign}(A)$ using the scaled Newton iteration. Two tolerances are used: a tolerance tol_cgce for testing convergence and a tolerance tol_scale for deciding when to switch to the unscaled iteration.

```
1  X_0 = A; scale = true
2  for k = 1: ∞
3      Y_k = X_k^{-1}
4      if scale
5          Set μ_k to one of the scale factors (5.35)–(5.37).
6      else
7          μ_k = 1
8      end
9      X_{k+1} = 1/2 (μ_k X_k + μ_k^{-1} Y_k)
10     δ_{k+1} = ||X_{k+1} − X_k||_F / ||X_{k+1}||_F
11     if scale = true and δ_{k+1} ≤ tol_scale, scale = false, end
12     if ||X_{k+1} − X_k||_F ≤ (tol_cgce ||X_{k+1}|| / ||Y_k||)^{1/2} or
       (δ_{k+1} > δ_k/2 and scale = false)
13         goto line 16
14     end
15 end
16 X = X_{k+1}
```

Cost: $2kn^3$ flops, where k iterations are used.

The algorithm uses the unscaled Newton iteration once the relative change in the iterates is less than tol_scale. A value of tol_scale safely less than 1 is intended and the motivation is to avoid the (nonoptimal) scaling parameters interfering with the quadratic convergence once the convergence has set in. The convergence test is (5.44) combined with the requirement to stop if, in the final convergence phase, δ_k has not decreased by at least a factor 2 during the previous iteration (which is a sign that roundoff errors are starting to dominate).

We have left the choice of scale factor at line 5 open, as the best choice will depend on the class of problems considered.

5.9. Best L_∞ Approximation

Most applications of the matrix sign function involve nonnormal matrices of small to medium size. An exception is the application in lattice quantum chromodynamics (QCD) described in Section 2.7, where the action on a vector of the sign of a large, sparse, Hermitian matrix is required. For Hermitian A, approximating $\operatorname{sign}(A)$ reduces to approximating $\operatorname{sign}(x)$ at the eigenvalues of A, which is a scalar problem on the real axis. The full range of scalar approximation techniques and results can therefore be brought into play. In particular, we can use best L_∞ rational approximations. For the sign function and the interval $[-\delta_{\max}, -\delta_{\min}] \cup [\delta_{\min}, \delta_{\max}]$ an explicit formula for the best L_∞ approximation is known. It follows from a corresponding result for the inverse square root. The result is phrased in terms of elliptic functions. The Jacobi elliptic function $\operatorname{sn}(w;\kappa) = x$ is defined implicitly by the elliptic integral

$$w = \int_0^x \frac{1}{\sqrt{(1-t^2)(1-\kappa^2 t^2)}}\, dt$$

and the complete elliptic integral (for the modulus κ) is defined by

$$K = \int_0^1 \frac{1}{\sqrt{(1-t^2)(1-\kappa^2 t^2)}}\, dt.$$

Theorem 5.15 (Zolotarev, 1877).

(a) *The best L_∞ approximation $\widetilde{r}$ from $\mathcal{R}_{m-1,m}$ to $x^{-1/2}$ on the interval $[1, (\delta_{\max}/\delta_{\min})^2]$ is*

$$\widetilde{r}(x) = D\frac{\prod_{j=1}^{m-1}(x+c_{2j})}{\prod_{j=1}^{m}(x+c_{2j-1})},$$

where

$$c_j = \frac{\operatorname{sn}^2(jK/(2m);\kappa)}{1-\operatorname{sn}^2(jK/(2m);\kappa)},$$

$\kappa = (1-(\delta_{\min}/\delta_{\max})^2)^{1/2}$, *and K is the complete elliptic integral for the modulus κ. The constant D is determined by the condition*

$$\max_{x\in[1,(\delta_{\min}/\delta_{\max})^2]}(1-\sqrt{x}\,\widetilde{r}(x)) = -\min_{x\in[1,(\delta_{\min}/\delta_{\max})^2]}(1-\sqrt{x}\,\widetilde{r}(x)),$$

and the extrema occur at $x_j = \operatorname{dn}^{-2}(jK/(2m))$, $j = 0\colon 2m$, where $\operatorname{dn}^2(w;\kappa) = 1 - \kappa^2\operatorname{sn}^2(w;\kappa)$.

(b) *The best L_∞ approximation r from $\mathcal{R}_{2m-1,2m}$ to $\operatorname{sign}(x)$ on the interval $[-\delta_{\max}, -\delta_{\min}] \cup [\delta_{\min}, \delta_{\max}]$ is $r(x) = (x/\delta_{\min})\widetilde{r}((x/\delta_{\min})^2)$, where $\widetilde{r}$ is defined in* (a). □

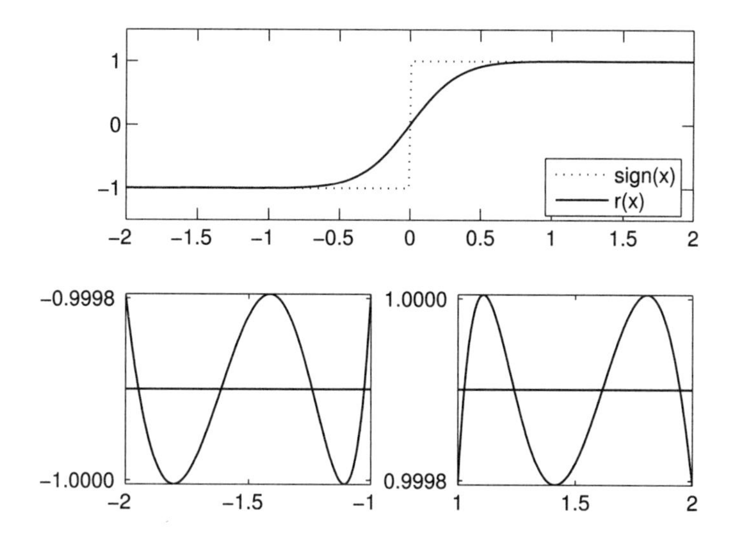

Figure 5.2. *Best L_∞ approximation $r(x)$ to* sign(x) *from $\mathcal{R}_{3,4}$ on* $[-2,-1] \cup [1,2]$. *The lower two plots show $r(x)$ in particular regions of the overall plot above.*

Figure 5.2 plots the best L_∞ approximation to $\operatorname{sign}(x)$ from $\mathcal{R}_{3,4}$ on $[-2,-1] \cup [1,2]$, and displays the characteristic equioscillation property of the error, which has maximum magnitude about 10^{-4}. In the QCD application $\delta_{\min}$ and $\delta_{\max}$ are chosen so that the spectrum of the matrix is enclosed and r is used in partial fraction form.

5.10. Notes and References

The matrix sign function was introduced by Roberts [496] in 1971 as a tool for model reduction and for solving Lyapunov and algebraic Riccati equations. He defined the sign function as a Cauchy integral and obtained the integral (5.3). Roberts also proposed the Newton iteration (5.16) for computing $\operatorname{sign}(A)$ and proposed scaling the iteration, though his scale parameters are not as effective as the ones described here.

Interest in the sign function grew steadily in the 1970s and 1980s, initially among engineers and later among numerical analysts. Kenney and Laub give a thorough survey of the matrix sign function and its history in [347, 1995].

The attractions of the concise representation $\operatorname{sign}(A) = A(A^2)^{-1/2}$ in (5.2) were pointed out by Higham [273, 1994], though the formula can be found in earlier work of Tsai, Shieh, and Yates [576, 1988].

Theorem 5.2 is from Higham, Mackey, Mackey, and Tisseur [283, 2005].

Theorems 5.3 and 5.7 are due to Kenney and Laub [342, 1991]. The expression (5.10) and upper bound (5.11) for the matrix sign function condition number are from Higham [273, 1994]. Theorem 5.4 is a refined version of a result of Kenney and Laub [342, 1991]. Another source of perturbation results for the matrix sign function is Sun [548, 1997].

The Schur method, Algorithm 5.5, is implemented in function `signm` of the Matrix Computation Toolbox [264] (on which `signm` in the Matrix Function Toolbox is based) but appears here in print for the first time.

For more on the recursions related to (5.26), and related references, see Chapter 8.

It is natural to ask how sharp the sufficient condition for convergence $\|I-A^2\| < 1$ in Theorem 5.8 (a) is for $\ell > m$ and what can be said about convergence for $\ell < m-1$. These questions are answered experimentally by Kenney and Laub [343, 1991], who give plots showing the boundaries of the regions of convergence of the scalar iterations in $\mathbb{C}$.

The principal Padé iterations for the sign function were first derived by Howland [302, 1983], though for even k his iteration functions are the inverses of those given here. Iannazzo [307, 2007] points out that these iterations can be obtained from the general König family (which goes back to Schröder [509, 1870], [510, 1992]) applied to the equation $x^2-1=0$. Parts (b)–(d) of Theorem 5.9 are from Kenney and Laub [345, 1994]. Pandey, Kenney, and Laub originally obtained the partial fraction expansion (5.30), for even k only, by applying Gaussian quadrature to an integral expression for $h(\xi)$ in (5.26) [457, 1990]. The analysis leading to (5.31) is from Kenney and Laub [345, 1994].

Theorem 5.10 is due to Kenney and Laub [344, 1992], and the triangular matrices in Table 5.2 are taken from the same paper.

Theorem 5.11 is due to Barraud [44, 1979, Sec. 4], but, perhaps because his paper is written in French, his result went unnoticed until it was presented by Kenney and Laub [344, 1992, Thm. 3.4].

Lemma 5.12 collects results from Kenney, Laub, and Papadopoulos [350, 1993] and Pandey, Kenney, and Laub [457, 1990].

The spectral scaling (5.36) and norm scaling (5.37) were first suggested by Barraud [44, 1979], while determinantal scaling (5.35) is due to Byers [88, 1987].

Kenney and Laub [344, 1992] derive a "semioptimal" scaling for the Newton iteration that requires estimates of the dominant eigenvalue (not just its modulus, i.e., the spectral radius) of X_k and of X_k^{-1}. Numerical experiments show this scaling to be generally at least as good as the other scalings we have described. Semioptimal scaling does not seem to have become popular, probably because it is more delicate to implement than the other scalings and the other scalings typically perform about as well in practice.

Theorem 5.13 on the stability of sign iterations is new. Indeed we are not aware of any previous analysis of the stability of sign iterations.

Our presentation of Zolotarev's Theorem 5.15 is based on that in van den Eshof, Frommer, Lippert, Schilling, and Van der Vorst [585, 2002] and van den Eshof [586, 2003]. In the numerical analysis literature this result seems to have been first pointed out by Kenney and Laub [347, 1995, Sec. III]. Theorem 5.15 can also be found in Achieser [1, 1956, Sec. E.27], Kennedy [338, 2004], [339, 2005], and Petrushev and Popov [470, 1987, Sec. 4.3].

A "generalized Newton sign iteration" proposed by Gardiner and Laub [205, 1986] has the form

$$X_{k+1} = \frac{1}{2}(X_k + BX_k^{-1}B), \qquad X_0 = A.$$

If B is nonsingular this is essentially the standard Newton iteration applied to $B^{-1}A$ and it converges to $B\,\mathrm{sign}(B^{-1}A)$. For singular B, convergence may or may not occur and can be at a linear rate; see Bai, Demmel, and Gu [31, 1997] and Sun and Quintana-Ortí [550, 2002]. This iteration is useful for computing invariant subspaces of matrix pencils $A - \lambda B$ (generalizing the approach in Section 2.5) and for solving generalized algebraic Riccati equations.

Problems

5.1. Show that $\operatorname{sign}(A) = A$ for any involutory matrix.

5.2. How are $\operatorname{sign}(A)$ and $\operatorname{sign}(A^{-1})$ related?

5.3. Derive the integral formula (5.3) from (5.2) by using the Cauchy integral formula (1.12).

5.4. Show that $\operatorname{sign}(A) = (2/\pi)\lim_{t\to\infty} \tan^{-1}(tA)$.

5.5. Can

$$A = \begin{bmatrix} -1 & 1 & 1/2 \\ 0 & 1 & -1 \\ 0 & 0 & 1 \end{bmatrix}$$

be the sign of some matrix?

5.6. Show that the geometric mean $A \# B$ of two Hermitian positive definite matrices A and B satisfies

$$\begin{bmatrix} 0 & A \# B \\ (A \# B)^{-1} & 0 \end{bmatrix} = \operatorname{sign}\left(\begin{bmatrix} 0 & B \\ A^{-1} & 0 \end{bmatrix}\right).$$

5.7. (Kenney and Laub [342, 1991]) Verify that for $A \in \mathbb{R}^{2\times 2}$ the matrix sign decomposition (5.5) is given as follows. If $\det(A) > 0$ and $\operatorname{trace}(A) \neq 0$ then $S = \operatorname{sign}(\operatorname{trace}(A))I$ and $N = \operatorname{sign}(\operatorname{trace}(A))A$; if $\det(A) < 0$ then

$$S = \mu\big(A - \det(A)A^{-1}\big), \qquad N = \mu\big(A^2 - \det(A)I\big),$$

where

$$\mu = \big(-\det(A - \det(A)A^{-1})\big)^{-1/2};$$

otherwise S is undefined.

5.8. Show that the Newton iteration (5.16) for the matrix sign function can be derived by applying Newton's method to the equation $X^2 = I$.

5.9. By expanding the expression $\operatorname{sign}(S + E) = (S + E)((S + E)^2)^{-1/2}$ from (5.2), show directly that the Fréchet derivative of the matrix sign function at $S = \operatorname{sign}(S)$ is given by $L(S, E) = \frac{1}{2}(E - SES)$.

5.10. Consider the scalar Newton sign iteration $x_{k+1} = \frac{1}{2}(x_k + x_k^{-1})$. Show that if $x_0 = \coth\theta_0$ then $x_k = \coth 2^k\theta_0$. Deduce a convergence result.

5.11. (Schroeder [511, 1991]) Investigate the behaviour of the Newton iteration (5.16) for scalar, pure imaginary x_0. Hint: let $x_0 = ir_0 \equiv -i\cot(\pi\theta_0)$ and work in θ coordinates.

5.12. Halley's iteration for solving $f(x) = 0$ is [201, 1985]

$$x_{k+1} = x_k - \frac{f_k/f_k'}{1 - \frac{1}{2}f_k f_k''/(f_k')^2},$$

where f_k, f_k', and f_k'' denote the values of f and its first two derivatives at x_k. Show that applying Halley's iteration to $f(x) = x^2 - 1$ yields the iteration function $f_{1,1}$ in Table 5.1.

5.13. (Byers [88, 1987]) Show that determinantal scaling $\mu = |\det(X)|^{-1/n}$ minimizes $d(\mu X)$, where

$$d(X) = \sum_{i=1}^{n} (\log|\lambda_i|)^2$$

and the λ_i are the eigenvalues of X. Show also that $d(X) = 0$ if and only if the spectrum of X lies on the unit circle and that $d(X)$ is an increasing function of $|1 - |\lambda_i||$ for each eigenvalue λ_i.

5.14. Consider the Newton iteration (5.34), with determinantal scaling (5.35) and spectral scaling (5.36). Show that with both scalings the iteration converges in at most two iterations (a) for scalars and (b) for any real 2×2 matrix.

5.15. (Higham, Mackey, Mackey, and Tisseur [283, 2005]) Suppose that $\mathrm{sign}(A) = I$ and $A^2 = I + E$, where $\|E\| < 1$, for some consistent norm. Show that

$$\|A - I\| \le \frac{\|E\|}{1 + \sqrt{1 - \|E\|}} < \|E\|.$$

How does this bound compare with the upper bound in (5.40)?

5.16. Discuss the pros and cons of terminating an iteration $X_{k+1} = g(X_k)$ for the matrix sign function with one of the tests

$$|\operatorname{trace}(X_k^2) - n| \le \eta, \tag{5.46}$$

$$|\operatorname{trace}(X_k) - \operatorname{round}(\operatorname{trace}(X_k))| \le \eta, \tag{5.47}$$

where $\mathrm{round}(x)$ denotes the nearest integer to x.

5.17. (Byers [88, 1987]) The matrix

$$W = \begin{bmatrix} A^* & G \\ F & -A \end{bmatrix}, \qquad F = F^*, \quad G = G^*,$$

arising in (2.14) in connection with the Riccati equation is Hamiltonian, that is, it satisfies the condition that JW is Hermitian, where $J = \left[\begin{smallmatrix} 0 & I_n \\ -I_n & 0 \end{smallmatrix}\right]$. Show that the Newton iteration for $\mathrm{sign}(W)$ can be written in such a way that only Hermitian matrices need to be inverted. The significance of this fact is that standard algorithms or software for Hermitian matrices can then be used, which halves the storage and computational costs compared with treating W as a general matrix.

The sign function of a square matrix can be defined in terms of a contour integral or as the result of an iterated map $Z_{r+1} = \frac{1}{2}(Z_r + Z_r^{-1})$. Application of this function enables a matrix to be decomposed into two components whose spectra lie on opposite sides of the imaginary axis.

— J. D. ROBERTS, *Linear Model Reduction and Solution of the Algebraic Riccati Equation by Use of the Sign Function* (1980)

The matrix sign function method is an elegant and, when combined with defect correction, effective numerical method for the algebraic Riccati equation.

— VOLKER MEHRMANN, *The Autonomous Linear Quadratic Control Problem: Theory and Numerical Solution* (1991)

Chapter 6
Matrix Square Root

The matrix square root is one of the most commonly occurring matrix functions, arising most frequently in the context of symmetric positive definite matrices. The key roles that the square root plays in, for example, the matrix sign function (Chapter 5), the definite generalized eigenvalue problem (page 35), the polar decomposition (Section 2.6 and Chapter 8), and the geometric mean (Section 2.10), make it a useful theoretical and computational tool. The rich variety of methods for computing the matrix square root, with their widely differing numerical stability properties, are an interesting subject of study in their own right.

We will almost exclusively be concerned with the principal square root, $A^{1/2}$. Recall from Theorem 1.29 that for $A \in \mathbb{C}^{n \times n}$ with no eigenvalues on $\mathbb{R}^-$, $A^{1/2}$ is the unique square root X of A whose spectrum lies in the open right half-plane. We will denote by $\sqrt{A}$ an arbitrary, possibly nonprincipal square root.

We note the integral representation

$$A^{1/2} = \frac{2}{\pi} A \int_0^\infty (t^2 I + A)^{-1} \, dt, \tag{6.1}$$

which is a special case of (7.1) in the next chapter. The integral can be deduced from that for the matrix sign function (see Problem 6.1).

This chapter begins with analysis of the conditioning of the matrix square root and the sensitivity of the relative residual. Then a Schur method, and a version working entirely in real arithmetic, are described. Newton's method and several variants follow, with a stability analysis revealing that the variants do not suffer the instability that vitiates the Newton iteration. After a discussion of scaling, numerical experiments are given to provide insight into the analysis. A class of coupled iterations obtained via iterations for the matrix sign function are derived and their stability proved. Linearly convergent iterations for matrices that are "almost diagonal", as well as for M-matrices, are analyzed, and a preferred iteration for Hermitian positive definite matrices is given. The issue of choosing from among the many square roots of a given matrix is addressed by considering how to compute a small-normed square root. A brief comparison of the competing methods is given. Finally, applications of involutory matrices, and some particular involutory matrices with explicit representations, are described.

6.1. Sensitivity and Conditioning

To analyze the sensitivity of the matrix square root we need to assume that we keep fixed the choice of branches of the square root function in the neighbourhoods of the eigenvalues of the matrix of interest, since otherwise the square root is not even continuous. We will concentrate on primary square roots, since a nonprimary square root belongs to a parametrized family of square roots (see Theorem 1.26) and so must be regarded as infinitely sensitive. By Theorem 3.5, the Fréchet derivative $L = L(A, E)$ of $X = f(A) = \sqrt{A}$ is the solution of $XL + LX = E$, which is the inverse of the Fréchet derivative of the function X^2. Using (B.16), we can rewrite this equation as $(I \otimes X + X^T \otimes I)\,\mathrm{vec}(L) = \mathrm{vec}(E)$, from which it can be deduced, using (3.7), that $\|L\|_F = \|(I \otimes X + X^T \otimes I)^{-1}\|_2$. Hence, by Theorem 3.1, the Frobenius norm (relative) condition number of the matrix square root at A is

$$\kappa_{\mathrm{sqrt}}(X) = \frac{\|(I \otimes X + X^T \otimes I)^{-1}\|_2 \, \|A\|_F}{\|X\|_F}, \tag{6.2}$$

where the argument of κ_{sqrt} denotes the particular square root under consideration. It follows that

$$\kappa_{\mathrm{sqrt}}(X) \geq \frac{1}{\min_{i,j=1:n} |\mu_i + \mu_j|} \frac{\|A\|_F}{\|X\|_F}, \tag{6.3}$$

where the μ_j are the eigenvalues of $X = \sqrt{A}$ (and this inequality can also be obtained from Theorem 3.14). This inequality is interesting because it reveals two distinct situations in which κ_{sqrt} must be large. The first situation is when A (and hence X) has an eigenvalue of small modulus. The second situation is when the square root is the principal square root and a real A has a pair of complex conjugate eigenvalues close to the negative real axis: $\lambda = re^{i(\pi-\epsilon)}$ $(0 < \epsilon \ll 1)$ and $\overline{\lambda}$. Then $|\lambda^{1/2} + \overline{\lambda}^{1/2}| = r^{1/2}|e^{i(\pi-\epsilon)/2} + e^{-i(\pi-\epsilon)/2}| = r^{1/2}|e^{-i\epsilon/2} - e^{i\epsilon/2}| = r^{1/2}O(\epsilon)$. In this latter case A is close to a matrix for which the principal square root is not defined.

If A is normal and X is normal (as is any primary square root of a normal A) then, either directly from (6.2) or from Corollary 3.16, we have equality in (6.3).

The formula for κ_{sqrt} allows us to identify the best conditioned square root of a Hermitian positive definite matrix. As usual, $\kappa(X) = \|X\|\,\|X^{-1}\|$.

Lemma 6.1. *If $A \in \mathbb{C}^{n\times n}$ is Hermitian positive definite and X is any primary square root of A then*

$$\kappa_{\mathrm{sqrt}}(A^{1/2}) = \frac{\|A^{-1}\|_2^{1/2}}{2} \frac{\|A\|_F}{\|A^{1/2}\|_F} \leq \kappa_{\mathrm{sqrt}}(X).$$

Moreover,

$$\frac{1}{2n^{3/2}} \kappa_F(A^{1/2}) \leq \kappa_{\mathrm{sqrt}}(A^{1/2}) \leq \frac{1}{2}\kappa_F(A^{1/2}).$$

Proof. Let A have eigenvalues $0 < \lambda_n \leq \lambda_{n-1} \leq \cdots \leq \lambda_1$. For $X = A^{1/2}$, A and X are normal so (6.3) is an equality and $\min_{i,j=1:n} |\mu_i + \mu_j| = 2\lambda_n^{1/2} = 2/\|A^{-1}\|_2^{1/2}$, which gives the expression for $\kappa_{\mathrm{sqrt}}(A^{1/2})$. Any other primary square root X has eigenvalues μ_j with moduli $\sqrt{\lambda_j}$, so the upper bound of $\kappa_{\mathrm{sqrt}}(X)$ follows from $\min_{i,j=1:n} |\mu_i + \mu_j| \leq 2\sqrt{\lambda_n}$ together with the fact that $\|X\|_F^2 = \sum_{i=1}^n \lambda_i$ is the same for all primary (and hence Hermitian) square roots of A. The upper and lower bounds for $\kappa_{\mathrm{sqrt}}(A^{1/2})$ follow using standard norm inequalities [276, 2002, Chap. 6]. □

Staying with positive definite matrices for the moment, the next result gives an elegant bound for the difference between the principal square roots of two matrices.

Theorem 6.2. *If $A, B \in \mathbb{C}^{n\times n}$ are Hermitian positive definite then for any unitarily invariant norm*

$$\|A^{1/2} - B^{1/2}\| \le \frac{1}{\lambda_{\min}(A)^{1/2} + \lambda_{\min}(B)^{1/2}} \|A - B\|,$$

where $\lambda_{\min}$ denotes the smallest eigenvalue.

Proof. This is a special case of a result of van Hemmen and Ando [591, 1980, Prop. 3.2]; see also Bhatia [62, 1987]. □

Let $\widetilde{X} = X + E$ be an approximation to a square root X of $A \in \mathbb{C}^{n\times n}$, where $\|E\| \le \epsilon\|X\|$. Then $\widetilde{X}^2 = A + XE + EX + E^2$, which leads to the relative residual bound

$$\frac{\|A - \widetilde{X}^2\|}{\|A\|} \le (2\epsilon + \epsilon^2)\alpha(X),$$

where

$$\alpha(X) = \frac{\|X\|^2}{\|A\|} = \frac{\|X\|^2}{\|X^2\|} \ge 1. \tag{6.4}$$

The quantity $\alpha(X)$ can be regarded as a condition number for the relative residual of X; if it is large then a small perturbation of X (such as $fl(X)$—the rounded square root) can have a relative residual much larger than the size of the relative perturbation. An important conclusion is that we cannot expect a numerical method to do better than provide a computed square root $\widehat{X}$ with relative residual of order $\alpha(\widehat{X})u$, where u is the unit roundoff. Where there is a choice of square root, one of minimal norm is therefore to be preferred. It is easy to show that

$$\frac{\kappa(X)}{\kappa(A)} \le \alpha(X) \le \kappa(X).$$

Thus a large value of $\alpha(X)$ implies that X is ill conditioned, and if A is well conditioned then $\alpha(X) \approx \kappa(X)$. If X is normal then $\alpha(X) = 1$ in the 2-norm.

6.2. Schur Method

Let $A \in \mathbb{C}^{n\times n}$ be nonsingular and let $f(A)$ denote any primary square root of A.

Given the Schur decomposition $A = QTQ^*$, where Q is unitary and T is upper triangular, $f(A) = Qf(T)Q^*$, so computing square roots of general A reduces to computing square roots $U = f(T)$ of upper triangular matrices, which are themselves triangular. The (i,i) and (i,j) $(j > i)$ elements of the equation $U^2 = T$ can be written

$$\begin{aligned} u_{ii}^2 &= t_{ii}, \\ (u_{ii} + u_{jj})u_{ij} &= t_{ij} - \sum_{k=i+1}^{j-1} u_{ik}u_{kj}. \end{aligned} \tag{6.5}$$

We can compute the diagonal of U and then solve for the u_{ij} either a superdiagonal at a time or a column at a time. The process cannot break down, because $0 = u_{ii} + u_{jj} = f(t_{ii}) + f(t_{jj})$ is not possible, since the t_{ii} are nonzero and f, being a primary matrix function, maps equal t_{ii} to the same square root. We obtain the following algorithm.

Algorithm 6.3 (Schur method). Given a nonsingular $A \in \mathbb{C}^{n\times n}$ this algorithm computes $X = \sqrt{A}$ via a Schur decomposition, where $\sqrt{\cdot}$ denotes any primary square root.

1 Compute a (complex) Schur decomposition $A = QTQ^*$.
2 $u_{ii} = \sqrt{t_{ii}}$, $i = 1\colon n$
3 for $j = 2\colon n$
4   for $i = j-1\colon -1\colon 1$
5         $u_{ij} = \dfrac{t_{ij} - \sum_{k=i+1}^{j-1} u_{ik}u_{kj}}{u_{ii} + u_{jj}}$
6   end
7 end
8 $X = QUQ^*$

Cost: $25n^3$ flops for the Schur decomposition plus $n^3/3$ for U and $3n^3$ to form X: $28\frac{1}{3}n^3$ flops in total.

Algorithm 6.3 generates all the primary square roots of A as different choices of sign in $u_{ii} = \sqrt{t_{ii}} \equiv \pm t_{ii}^{1/2}$ are used, subject to the restriction that $t_{ii} = t_{jj} \Rightarrow \sqrt{t_{ii}} = \sqrt{t_{jj}}$.

If A is singular with a semisimple zero eigenvalue of multiplicity k then Algorithm 6.3 can be adapted by ordering the Schur decomposition so that the zero eigenvalues are in the trailing k diagonal elements of T. Then T must have the structure

$$T = \begin{array}{c} \\ {\scriptstyle n-k} \\ {\scriptstyle k} \end{array} \!\!\begin{array}{c} \begin{array}{cc} {\scriptstyle n-k} & {\scriptstyle k} \end{array} \\ \begin{bmatrix} T_{11} & T_{12} \\ 0 & 0 \end{bmatrix} \end{array}, \tag{6.6}$$

with T_{11} nonsingular. Indeed if some element of the trailing $k \times k$ block of T were nonzero then the rank of T would exceed $n-k$, but 0 being a semisimple eigenvalue of multiplicity k implies $\operatorname{rank}(T) = n-k$. It is clear that any primary square root has the same block structure as T and that Algorithm 6.3 computes such a square root provided that we set $u_{ij} = 0$ when $i > n-k$ and $j > n-k$, which are the cases where the algorithm would otherwise incur division by zero. The behaviour of the algorithm for singular A without reordering to the form (6.6) is examined in Problem 6.5.

If A is real but has some nonreal eigenvalues then Algorithm 6.3 uses complex arithmetic. This is undesirable, because complex arithmetic is more expensive than real arithmetic and also because rounding errors may cause a computed result to be produced with nonzero imaginary part. By working with a real Schur decomposition complex arithmetic can be avoided.

Let $A \in \mathbb{R}^{n\times n}$ have the real Schur decomposition $A = QRQ^T$, where Q is orthogonal and R is upper quasi-triangular with 1×1 and 2×2 diagonal blocks. Then $f(A) = Qf(R)Q^T$, where $U = f(R)$ is upper quasi-triangular with the same block structure as R. The equation $U^2 = R$ can be written

$$\begin{aligned} U_{ii}^2 &= R_{ii}, \\ U_{ii}U_{ij} + U_{ij}U_{jj} &= R_{ij} - \sum_{k=i+1}^{j-1} U_{ik}U_{kj}. \end{aligned} \tag{6.7}$$

Once the diagonal blocks U_{ii} have been computed (6.7) provides a way to compute the remaining blocks U_{ij} a block superdiagonal or a block column at a time. The condition for the Sylvester equation (6.7) to have a unique solution U_{ij} is that U_{ii}

and $-U_{jj}$ have no eigenvalue in common (see (B.20)), and this is guaranteed for any primary square root when A is nonsingular. When neither U_{ii} nor U_{jj} is a scalar, (6.7) can solved by writing it in the form

$$(I \otimes U_{ii} + U_{jj}^T \otimes I)\,\mathrm{vec}(U_{ij}) = \mathrm{vec}\bigl(R_{ij} - \textstyle\sum_{k=i+1}^{j-1} U_{ik}U_{kj}\bigr),$$

which is a linear system $Ax = b$ of order 2 or 4 that can by solved by Gaussian elimination with partial pivoting.

We now consider the computation of $\sqrt{R_{ii}}$ for 2×2 blocks R_{ii}, which necessarily have distinct complex conjugate eigenvalues.

Lemma 6.4. *Let $A \in \mathbb{R}^{2\times 2}$ have distinct complex conjugate eigenvalues. Then A has four square roots, all primary functions of A. Two of them are real, with complex conjugate eigenvalues, and two are pure imaginary, with eigenvalues that are not complex conjugates.*

Proof. Since A has distinct eigenvalues $\theta \pm i\mu$, Theorem 1.26 shows that A has four square roots, all of them functions of A. To find them let

$$Z^{-1}AZ = \mathrm{diag}(\lambda, \overline{\lambda}) = \theta I + i\mu K, \qquad K = \begin{bmatrix} 1 & 0 \\ 0 & -1 \end{bmatrix}.$$

Then

$$A = \theta I + \mu W,$$

where $W = iZKZ^{-1}$, and since $\theta, \mu \in \mathbb{R}$ it follows that $W \in \mathbb{R}^{2\times 2}$.

If $(\alpha + i\beta)^2 = \theta + i\mu$, then the four square roots of A are given by $X = ZDZ^{-1}$, where $D = \pm\,\mathrm{diag}(\alpha+i\beta, \pm(\alpha-i\beta))$, that is, $D = \pm(\alpha I + i\beta K)$ or $D = \pm(\alpha K + i\beta I) = \pm i(\beta I - i\alpha K)$. Thus

$$X = \pm(\alpha I + \beta W),$$

that is, two real square roots with eigenvalues $\pm(\alpha + i\beta, \alpha - i\beta)$; or

$$X = \pm i(\beta I - \alpha W),$$

that is, two pure imaginary square roots with eigenvalues $\pm(\alpha + i\beta, -\alpha + i\beta)$. □

The proof of the lemma gives a way to construct $R_{ii}^{1/2}$. Writing

$$R_{ii} = \begin{bmatrix} r_{11} & r_{12} \\ r_{21} & r_{22} \end{bmatrix},$$

the eigenvalues of R_{ii} are $\theta \pm i\mu$, where

$$\theta = \frac{1}{2}(r_{11} + r_{22}), \qquad \mu = \frac{1}{2}\bigl(-(r_{11} - r_{22})^2 - 4r_{21}r_{12}\bigr)^{1/2}. \tag{6.8}$$

We now require α and β such that $(\alpha + i\beta)^2 = \theta + i\mu$. A stable way to compute them is as follows.

Algorithm 6.5. This algorithm computes the square root $\alpha+i\beta$ of $\theta+i\mu$ with $\alpha \geq 0$.

1 if $\theta = 0$ and $\mu = 0$, $\alpha = 0$, $\beta = 0$, quit, end
2 $t = \left((|\theta| + (\theta^2 + \mu^2)^{1/2})/2\right)^{1/2}$
3 if $\theta \geq 0$
4 $\quad \alpha = t$, $\beta = \mu/(2\alpha)$
5 else
6 $\quad \beta = t$, $\alpha = \mu/(2\beta)$
7 end

Finally, the real square roots of R_{ii} are obtained from

$$U_{ii} = \pm\left(\alpha I + \frac{1}{2\alpha}(R_{ii} - \theta I)\right)$$
$$= \pm \begin{bmatrix} \alpha + \frac{1}{4\alpha}(r_{11} - r_{22}) & \frac{1}{2\alpha}r_{12} \\ \frac{1}{2\alpha}r_{21} & \alpha - \frac{1}{4\alpha}(r_{11} - r_{22}) \end{bmatrix}. \tag{6.9}$$

Before giving an algorithm, we pause to summarize our theoretical findings.

Theorem 6.6. *Let $A \in \mathbb{R}^{n\times n}$ be nonsingular. If A has a real negative eigenvalue then A has no real square roots that are primary functions of A. If A has no real negative eigenvalues, then there are precisely 2^{r+c} real primary square roots of A, where r is the number of distinct real eigenvalues and c the number of distinct complex conjugate eigenvalue pairs.*

Proof. Let A have the real Schur decomposition $A = QRQ^T$. Since $f(A) = Qf(R)Q^T$, $f(A)$ is real if and only if $f(R)$ is real. If A has a real negative eigenvalue, $R_i = (r_{ii})$ say, then $f(R_i)$ is necessarily nonreal; this gives the first part of the theorem.

If A has no real negative eigenvalues, consider the 2^s primary square roots of A described in Theorem 1.26. We have $s = r + 2c$. Lemma 6.4 shows that each 2×2 block R_{ii} has two real primary square roots. Hence, of the $2^s = 2^{r+2c}$ primary square roots of A, precisely 2^{r+c} of them are real. □

Algorithm 6.7 (real Schur method). Given $A \in \mathbb{R}^{n\times n}$ with no eigenvalues on $\mathbb{R}^-$ this algorithm computes $X = \sqrt{A}$ via a real Schur decomposition, where $\sqrt{\cdot}$ denotes any real primary square root.

1 Compute a real Schur decomposition $A = QRQ^T$, where R is block $m \times m$.
2 Compute $U_{ii} = \sqrt{R_{ii}}$, $i = 1\colon m$, using (6.9) whenever R_{ii} is 2×2.
3 for $j = 2\colon m$
4 $\quad$ for $i = j-1\colon -1\colon 1$
5 $\quad\quad$ Solve $U_{ii}U_{ij} + U_{ij}U_{jj} = R_{ij} - \sum_{k=i+1}^{j-1} U_{ik}U_{kj}$ for U_{ij}.
6 $\quad$ end
7 end
8 $X = QUQ^T$

Cost: $28\frac{1}{3}n^3$ flops.

Two comments are necessary. First, the principal square root is computed if the principal square root is taken at line 2, which for 2×2 blocks means taking the positive sign in (6.9). Second, as for Algorithm 6.3, it is necessary that whenever R_{ii} and R_{jj} have the same eigenvalues, we take the same square root.

Now we consider the numerical stability of Algorithms 6.3 and 6.7. A straightforward rounding error analysis shows that the computed square root $\widehat{U}$ of T in Algorithm 6.3 satisfies (see Problem 6.6)

$$\widehat{U}^2 = T + \Delta T, \qquad |\Delta T| \le \widetilde{\gamma}_n |\widehat{U}|^2,$$

where the inequality is to be interpreted elementwise. Computation of the Schur decomposition (by the QR algorithm) is a backward stable process [224, 1996, Sec. 7.5.6], and standard error analysis leads to the overall result

$$\widehat{X}^2 = A + \Delta A, \qquad \|\Delta A\|_F \le \widetilde{\gamma}_{n^3} \|\widehat{X}\|_F^2,$$

which can be expressed as

$$\frac{\|A - \widehat{X}^2\|_F}{\|A\|_F} \le \widetilde{\gamma}_{n^3}\, \alpha_F(\widehat{X}). \tag{6.10}$$

where α is defined in (6.4). We conclude that Algorithm 6.3 has essentially optimal stability. The same conclusion holds for Algorithm 6.7, which can be shown to satisfy the same bound (6.10).

6.3. Newton's Method and Its Variants

Newton's method for solving $X^2 = A$ can be derived as follows. Let Y be an approximate solution and set $Y + E = X$, where E is to be determined. Then $A = (Y + E)^2 = Y^2 + YE + EY + E^2$. Dropping the second order term in E leads to Newton's method:

$$\left.\begin{aligned} &X_0 \text{ given,} \\ \text{Solve } X_k E_k + E_k X_k &= A - X_k^2 \\ X_{k+1} &= X_k + E_k \end{aligned}\right\} \quad k = 0, 1, 2, \ldots. \tag{6.11}$$

At each iteration a Sylvester equation must be solved for E_k. The standard way of solving Sylvester equations is via Schur decomposition of the coefficient matrices, which in this case are both X_k. But the Schur method of the previous section can compute a square root with just one Schur decomposition, so Newton's method is unduly expensive in the form (6.11). The following lemma enables us to reduce the cost. Note that E_k in (6.11) is well-defined, that is, the Sylvester equation is nonsingular, if and only if X_k and $-X_k$ have no eigenvalue in common (see (B.20)).

Lemma 6.8. *Suppose that in the Newton iteration* (6.11) X_0 *commutes with* A *and all the iterates are well-defined. Then, for all* k, X_k *commutes with* A *and* $X_{k+1} = \frac{1}{2}(X_k + X_k^{-1}A)$.

Proof. The proof is by induction. Let $Y_0 = X_0$ and $Y_k = \frac{1}{2}(X_{k-1} + X_{k-1}^{-1}A)$, $k \ge 1$. For the inductive hypothesis we take $X_k A = AX_k$ and $Y_k = X_k$, which is trivially true for $k = 0$. The matrix $F_k = \frac{1}{2}(X_k^{-1}A - X_k)$ is easily seen to satisfy $X_k F_k + F_k X_k = A - X_k^2$, so $F_k = E_k$ and $X_{k+1} = X_k + F_k = \frac{1}{2}(X_k + X_k^{-1}A) = Y_{k+1}$, which clearly commutes with A since X_k does. Hence the result follows by induction. □

The lemma shows that if X_0 is chosen to commute with A then all the X_k and E_k in (6.11) commute with A, permitting great simplification of the iteration.

The most common choice is $X_0 = A$ (or $X_0 = I$, which generates the same X_1), giving the Newton iteration

Newton iteration (matrix square root):

$$X_{k+1} = \frac{1}{2}(X_k + X_k^{-1}A), \qquad X_0 = A. \tag{6.12}$$

If A is nonsingular, standard convergence theory for Newton's method allows us to deduce quadratic convergence of (6.11) to a primary square root for X_0 sufficiently close to the square root, since the Fréchet derivative of $F(X) = X^2 - A$ is nonsingular at a primary square root. The next result shows unconditional quadratic convergence of (6.12) to the *principal* square root. Moreover, it shows that (6.12) is equivalent to the Newton sign iteration (5.16) applied to $A^{1/2}$.

Theorem 6.9 (convergence of Newton square root iteration). *Let $A \in \mathbb{C}^{n\times n}$ have no eigenvalues on $\mathbb{R}^-$. The Newton square root iterates X_k from* (6.12) *with any X_0 that commutes with A are related to the Newton sign iterates*

$$S_{k+1} = \frac{1}{2}(S_k + S_k^{-1}), \qquad S_0 = A^{-1/2}X_0$$

by $X_k \equiv A^{1/2}S_k$. Hence, provided that $A^{-1/2}X_0$ has no pure imaginary eigenvalues, the X_k are defined and X_k converges quadratically to $A^{1/2}\operatorname{sign}(A^{-1/2}X_0)$. In particular, if the spectrum of $A^{-1/2}X_0$ lies in the right half-plane then X_k converges quadratically to $A^{1/2}$ and, for any consistent norm,

$$\|X_{k+1} - A^{1/2}\| \le \frac{1}{2}\|X_k^{-1}\|\,\|X_k - A^{1/2}\|^2. \tag{6.13}$$

Proof. Note first that any matrix that commutes with A commutes with $A^{\pm 1/2}$, by Theorem 1.13 (e). We have $X_0 = A^{1/2}S_0$ and so S_0 commutes with A. Assume that $X_k = A^{1/2}S_k$ and S_k commutes with A. Then S_k commutes with $A^{1/2}$ and

$$X_{k+1} = \frac{1}{2}(A^{1/2}S_k + S_k^{-1}A^{-1/2}A) = A^{1/2}\cdot\frac{1}{2}(S_k + S_k^{-1}) = A^{1/2}S_{k+1},$$

and S_{k+1} clearly commutes with A. Hence $X_k \equiv A^{1/2}S_k$ by induction. Then, using Theorem 5.6, $\lim_{k\to\infty} X_k = A^{1/2}\lim_{k\to\infty} S_k = A^{1/2}\operatorname{sign}(S_0) = A^{1/2}\operatorname{sign}(A^{-1/2}X_0)$, and the quadratic convergence of X_k follows from that of S_k.

For the last part, if $S_0 = A^{-1/2}X_0$ has spectrum in the right half-plane then $\operatorname{sign}(S_0) = I$ and hence $X_k \to A^{1/2}$. Using the commutativity of the iterates with A, it is easy to show that

$$X_{k+1} \pm A^{1/2} = \frac{1}{2}X_k^{-1}(X_k \pm A^{1/2})^2, \tag{6.14}$$

which, with the minus sign, gives (6.13). □

We make several remarks on this result.

First, an implication of Theorem 6.9 of theoretical interest, which can also be deduced from the connection with the full Newton method, is that (6.12) converges to $A^{1/2}$ for any X_0 that commutes with A and is sufficiently close to $A^{1/2}$.

Second, it is worth noting that the sequence $\{X_k\}$ from (6.12) may be well-defined when that for the full Newton method (6.11) is not; see Problem 6.7. No analogue of the condition in Theorem 6.9 guaranteeing that the X_k are well-defined is available for (6.11).

Third, this analysis is more powerful than the Jordan-based analysis in Section 4.9.3, which also showed convergence of (6.12) to the principal square root. If X_0 is known only to commute with A then the X_k do not necessarily share the Jordan block structure of A (see Problem 1.26), and so the analysis cannot reduce to that for a single Jordan block.

Fourth, we noted in Section 5.3 that the Newton iteration for $\mathrm{sign}(A)$ requires many iterations when A has an eigenvalue close to the imaginary axis. Theorem 6.9 therefore implies that the Newton iteration (6.12) for $A^{1/2}$ will require many iterations when A has an eigenvalue close to the negative real axis, as can be seen from, using (5.18),

$$X_k = A^{1/2}(I - G_0^{2^k})^{-1}(I + G_0^{2^k}), \quad \text{where } k \geq 1 \text{ and } G_0 = (A^{1/2} - I)(A^{1/2} + I)^{-1}.$$

Fifth, when A is positive definite the convergence of (6.12) is monotonic from above in the positive semidefinite ordering; see Problem 6.9 (b).

Finally, it is interesting to consider how (6.12) behaves when X_0 does not commute with A, even though commutativity was assumed in the derivation. Lack of commutativity can cause quadratic convergence, and even convergence itself, to be lost; see Problem 6.8.

A coupled version of (6.12) can be obtained by defining $Y_k = A^{-1}X_k$. Then $X_{k+1} = \frac{1}{2}(X_k + Y_k^{-1})$ and $Y_{k+1} = A^{-1}X_{k+1} = \frac{1}{2}(Y_k + X_k^{-1})$, on using the fact that X_k commutes with A. This is the iteration of Denman and Beavers [146, 1976]:

DB iteration:

$$\begin{aligned} X_{k+1} &= \frac{1}{2}\left(X_k + Y_k^{-1}\right), & X_0 &= A, \\ Y_{k+1} &= \frac{1}{2}\left(Y_k + X_k^{-1}\right), & Y_0 &= I. \end{aligned} \tag{6.15}$$

Under the conditions of Theorem 6.9,

$$\lim_{k\to\infty} X_k = A^{1/2}, \qquad \lim_{k\to\infty} Y_k = A^{-1/2}. \tag{6.16}$$

Defining $M_k = X_kY_k$, we have $M_{k+1} = (2I + X_kY_k + Y_k^{-1}X_k^{-1})/4 = (2I + M_k + M_k^{-1})/4$. This gives the product form of the DB iteration, identified by Cheng, Higham, Kenney, and Laub [108, 2001], in which we iterate with M_k and *either*

X_k *or* Y_k:

Product form DB iteration:

$$M_{k+1} = \frac{1}{2}\left(I + \frac{M_k + M_k^{-1}}{2}\right), \qquad M_0 = A,$$
$$X_{k+1} = \frac{1}{2}X_k(I + M_k^{-1}), \qquad X_0 = A, \tag{6.17}$$
$$Y_{k+1} = \frac{1}{2}Y_k(I + M_k^{-1}), \qquad Y_0 = I.$$

Clearly, $\lim_{k\to\infty} M_k = I$ (and (6.16) continues to hold). The product form DB iteration has the advantage in efficiency over the DB iteration that it has traded one of the matrix inversions for a matrix multiplication. Another attraction of (6.17) is that a convergence test can be based on the error $\|M_k - I\|$, which is available free of charge.

Yet another variant of (6.12) can be derived by noting that

$$\begin{aligned} E_{k+1} &= \frac{1}{2}(X_{k+1}^{-1}A - X_{k+1}) \qquad (6.18)\\ &= \frac{1}{2}X_{k+1}^{-1}(A - X_{k+1}^2)\\ &= \frac{1}{2}X_{k+1}^{-1}\left(A - \frac{1}{4}(X_k + X_k^{-1}A)^2\right)\\ &= \frac{1}{2}X_{k+1}^{-1}\left(\frac{2A - X_k^2 - X_k^{-2}A^2}{4}\right)\\ &= -\frac{1}{2}X_{k+1}^{-1}\frac{(X_k - X_k^{-1}A)^2}{4}\\ &= -\frac{1}{2}X_{k+1}^{-1}E_k^2 = -\frac{1}{2}E_kX_{k+1}^{-1}E_k. \end{aligned}$$

Setting $Y_k = 2E_k$ and $Z_k = 4X_{k+1}$ we obtain the iteration

CR iteration:

$$Y_{k+1} = -Y_kZ_k^{-1}Y_k, \qquad Y_0 = I - A,$$
$$Z_{k+1} = Z_k + 2Y_{k+1}, \qquad Z_0 = 2(I + A). \tag{6.19}$$

From the derivation we have $Y_k \to 0$ and $Z_k \to 4A^{1/2}$. This iteration is derived in a different way by Meini [421, 2004]: she shows that $\frac{1}{4}A^{-1/2}$ is the constant coefficient in the inverse of the Laurent matrix polynomial $R(z) = (I-A)z^{-1}+2(I+A)+(I-A)z$ and applies cyclic reduction to an associated bi-infinite block tridiagonal matrix in order to compute the inverse of this coefficient.

A minor variation of (6.19) is worth noting. If we set $X_k = Z_k/4$ and $E_k = Y_{k+1}/2$ then (6.19) becomes

IN iteration:

$$X_{k+1} = X_k + E_k, \qquad X_0 = A,$$
$$E_{k+1} = -\frac{1}{2}E_kX_{k+1}^{-1}E_k, \qquad E_0 = \frac{1}{2}(I - A). \tag{6.20}$$

Table 6.1. *Cost per iteration of matrix square root iterations.*

Iteration	Operations	Flops
Newton, (6.12)	D	$8n^3/3$
DB, (6.15)	$2I$	$4n^3$
Product DB, (6.17)	$M+I$	$4n^3$
CR, (6.19)	$M+D$	$14n^3/3$
IN, (6.20)	$M+D$	$14n^3/3$

Here, $X_k \to A^{1/2}$ and $E_k \to 0$. This incremental form of the Newton iteration, suggested by Iannazzo [305, 2003], is of interest because it updates X_k by a correction term that is, ultimately, small and accurately computable (in an absolute sense)—this formulation therefore adheres to one of the maxims of designing stable algorithms [276, 2002, Sec. 1.18]. (The same comments apply to (6.19), in which the incremental form is slightly less clear.)

The computational cost of the Newton iteration and its variants is compared in Table 6.1, where M, I, and D denote a matrix multiplication, matrix inversion, and solution of a multiple right-hand side linear system, respectively. Clearly the Newton iteration (6.12) is the least expensive iteration.

Suppose now that A is singular with semisimple zero eigenvalues, so that a unique primary square root whose nonzero eigenvalues lie in the open right half-plane exists, by Problem 1.27. We will denote this square root by $A^{1/2}$. One difference between (6.19) and (6.20) and the other iterations is that the former do not invert A on the first step and so can potentially be used for singular A. However, the Newton iteration (6.12) is also applicable provided that we apply the first iteration formally and start with $X_1 = \frac{1}{2}(I + A)$. The next result describes the convergence behaviour of (6.12) (and hence also of (6.19) and (6.20)).

Theorem 6.10. *Let the singular matrix $A \in \mathbb{C}^{n\times n}$ have semisimple zero eigenvalues and nonzero eigenvalues lying off $\mathbb{R}^-$. The iterates X_k from the Newton iteration (6.12) started with $X_1 = \frac{1}{2}(I+A)$ are nonsingular and converge linearly to $A^{1/2}$, with*

$$\|X_k - A^{1/2}\| = O(2^{-k}). \tag{6.21}$$

Proof. We can write the Jordan canonical form of A as $A = Z\,\mathrm{diag}(J_1, 0)Z^{-1}$, where J_1 contains the Jordan blocks corresponding to the nonzero eigenvalues and hence is nonsingular. Then $X_1 = Z\,\mathrm{diag}((J_1 + I)/2, I/2)Z^{-1}$. It is easy to see that the X_k have the form $X_k = Z\,\mathrm{diag}(J_1^{(k)}, 2^{-k}I)Z^{-1}$ for all $k \ge 1$, where the $J_1^{(k)}$ are the Newton iterates for $J_1^{1/2}$. Hence, using Theorem 6.9, $X_k \to Z\,\mathrm{diag}(J_1^{1/2}, 0)Z^{-1} = A^{1/2}$ and

$$\|X_k - A^{1/2}\| \le \kappa(Z)\|\,\mathrm{diag}(J_1^{(k)} - J_1^{1/2}, 2^{-k}I)\| = O(2^{-k}). \qquad \square$$

Despite the convergence result of Theorem 6.10, it is not recommended to straightforwardly apply any of the iterations to a singular matrix; numerical instability is likely because the iterations all invert a matrix that is converging to a singular limit. It is better to iterate until the "nonsingular part" of the iterates has converged to the desired accuracy, at X_k say, and then compute the "correction step"

$$X_{k+1} = X_k^{-1}A = X_k - 2\cdot\frac{1}{2}(X_k - X_k^{-1}A). \tag{6.22}$$

This is a double-length Newton step, whose benefits for problems with a singular Fréchet derivative at the solution have been noted in the more general context of algebraic Riccati equations by Guo and Lancaster [234, 1998]. From the proof of Theorem 6.10 we see that $X_k^{-1}A = Z\,\mathrm{diag}(J_1^{(k)^{-1}} J_1, 0)Z^{-1} \approx Z\,\mathrm{diag}(J_1^{1/2}, 0)Z^{-1} = A^{1/2}$, since $J_1^{(k)} \approx J_1^{1/2}$ by assumption.

While the iterations described in this section are mathematically equivalent, they are not equivalent in finite precision arithmetic. We will see in the next section that they have quite different stability properties.

6.4. Stability and Limiting Accuracy

Standard convergence theory for Newton's method for a nonlinear system guarantees quadratic convergence to a solution provided that the Jacobian is nonsingular at the solution and the starting point is close enough to the solution. Under these conditions (sufficiently small) rounding errors in one step will be damped out in the next, close to convergence, because they are seen by Newton's method as perturbing an iterate to another point in the region of convergence. Iteration (6.11)—full Newton's method for the matrix square root—is therefore not unduly affected by rounding errors. The same cannot necessarily be said for iterations (6.12), (6.15), (6.17), (6.19), and (6.20): these iterations are equivalent to (6.11) with $X_0 = A$ only in exact arithmetic. Indeed we have already seen the instability of (6.12) in Section 4.9.4. In the presence of roundoff, the commutativity conditions in Lemma 6.8 and elsewhere that we used in deriving these iterations no longer hold and rounding errors can potentially be magnified by the iterations. Looked at another way, the Newton iteration (6.12) converges to $A^{1/2}$ along a path of matrices all of which are functions of A and hence commute with A. Rounding errors tend to perturb iterates off the path; Theorem 6.9 is then not applicable, and the local quadratic convergence of Newton's method cannot be invoked, because the equivalence of (6.12) to full Newton's is vitiated by roundoff.

In the following subsections we investigate the stability and limiting accuracy of the square root iterations using the framework of Section 4.9.4. We assume throughout that A has no eigenvalues on $\mathbb{R}^-$.

6.4.1. Newton Iteration

We consider first the Newton iteration (6.12), for which the iteration function is $g(X) = \frac{1}{2}(X + X^{-1}A)$. It is easy to show that the Fréchet derivative is given by $L_g(X, E) = \frac{1}{2}(E - X^{-1}EX^{-1}A)$. The relevant fixed point of g is $X = A^{1/2}$, for which

$$L_g(A^{1/2}, E) = \frac{1}{2}(E - A^{-1/2}EA^{1/2}). \tag{6.23}$$

The eigenvalues of $L_g(A^{1/2})$, which are most easily obtained by applying (B.17) to the Kronecker matrix form $\frac{1}{2}(I - A^{1/2^T} \otimes A^{-1/2})$, are

$$\frac{1}{2}(1 - \lambda_i^{1/2}\lambda_j^{-1/2}), \qquad i, j = 1\colon n,$$

where the λ_i are the eigenvalues of A. Hence to guarantee stability we need

$$\psi_N := \max_{i,j} \frac{1}{2}\left|1 - \lambda_i^{1/2}\lambda_j^{-1/2}\right| < 1. \tag{6.24}$$

This is a severe restriction on the matrix A. For example, if A is Hermitian positive definite then the condition is equivalent to $\kappa_2(A) < 9$, so that A must be extremely well conditioned.

The validity of the conclusion of this stability analysis is easily demonstrated. Suppose that $A = Z\Lambda Z^{-1}$, where $\Lambda = \mathrm{diag}(\lambda_i)$, and let $X_0 = ZD_0Z^{-1}$, where $D_0 = \mathrm{diag}(d_i)$. Consider a rank-1 perturbation $E_0 = \epsilon u_j v_i^* \equiv \epsilon(Ze_j)(e_i^T Z^{-1})$ to X_0, where $i \neq j$. We have $(X_0 + E_0)^{-1} = X_0^{-1} - X_0^{-1}E_0X_0^{-1}$ by the Sherman–Morrison formula (B.11). The induced perturbation in X_1 is

$$\begin{aligned} E_1 &= \frac{1}{2}(E_0 - X_0^{-1}E_0X_0^{-1}A) \\ &= \frac{1}{2}(E_0 - ZD_0^{-1}Z^{-1}E_0ZD_0^{-1}Z^{-1}A) \\ &= \frac{1}{2}(E_0 - \epsilon ZD_0^{-1}e_je_i^TD_0^{-1}\Lambda Z^{-1}) \\ &= \frac{1}{2}\Big(E_0 - \epsilon\frac{1}{d_j}Ze_je_i^T\frac{\lambda_i}{d_i}Z^{-1}\Big) \\ &= \frac{1}{2}\Big(1 - \frac{\lambda_i}{d_id_j}\Big)E_0. \end{aligned}$$

If we set $X_0 = A^{1/2}$, so that $d_i = \lambda_i^{1/2}$, then $E_1 = \frac{1}{2}(1 - \lambda_i^{1/2}\lambda_j^{-1/2})E_0$, and after k iterations we have

$$X_k + E_k = A^{1/2} + \Big[\frac{1}{2}(1 - \lambda_i^{1/2}\lambda_j^{-1/2})\Big]^k E_0.$$

This analysis shows that the Newton iteration (6.12) can diverge when started arbitrarily close to the desired square root if (6.24) is not satisfied. Of course this perturbation takes the iteration off the path of matrices that commute with A, so that Theorem 6.9 is not applicable.

Turning to the limiting accuracy, from (6.23) we have

$$\|L_g(A^{1/2}, E)\| \le \frac{1}{2}(1 + \kappa(A^{1/2}))\|E\|,$$

giving an estimate for the relative limiting accuracy of $\frac{1}{2}(1 + \kappa(A^{1/2}))u$.

6.4.2. DB Iterations

For the DB iteration (6.15) the iteration function is

$$G(X, Y) = \frac{1}{2}\begin{bmatrix} X + Y^{-1} \\ Y + X^{-1} \end{bmatrix}.$$

The Fréchet derivative of G at (X, Y) in the direction (E, F) is

$$L_g(X, Y; E, F) \equiv L_g\left(X, Y; \begin{bmatrix} E \\ F \end{bmatrix}\right) = \frac{1}{2}\begin{bmatrix} E - Y^{-1}FY^{-1} \\ F - X^{-1}EX^{-1} \end{bmatrix}.$$

Any point of the form $(X, Y) = (B, B^{-1})$ is a fixed point of G, and

$$L_g(B, B^{-1}; E, F) = \frac{1}{2}\begin{bmatrix} E - BFB \\ F - B^{-1}EB^{-1} \end{bmatrix}. \tag{6.25}$$

A straightforward computation shows that $L_g(B, B^{-1})$ is idempotent. Hence the DB iteration is stable at the fixed point $(A^{1/2}, A^{-1/2})$.

For the product form (6.17) of the DB iteration, the iteration function is

$$G(M, X) = \frac{1}{2} \begin{bmatrix} I + \frac{1}{2}(M + M^{-1}) \\ X(I + M^{-1}) \end{bmatrix}.$$

We have

$$L_g(M, X; E, F) = \frac{1}{2} \begin{bmatrix} \frac{1}{2}(E - M^{-1}EM^{-1}) \\ F(I + M^{-1}) - XM^{-1}EM^{-1} \end{bmatrix}.$$

At a fixed point (I, X),

$$L_g(I, X; E, F) = \begin{bmatrix} 0 \\ F - \frac{1}{2}XE \end{bmatrix} = \begin{bmatrix} 0 & 0 \\ -\frac{1}{2}X & I \end{bmatrix} \begin{bmatrix} E \\ F \end{bmatrix}. \tag{6.26}$$

It is easy to see that $L_g(I, X)$ is idempotent, and hence the product form of the DB iteration is stable at $(M, X) = (I, A^{1/2})$ and at $(M, Y) = (I, A^{-1/2})$.

To determine the limiting accuracy, with $B = A^{1/2}$, $\|E\| \le u\|A^{1/2}\|$, and $\|F\| \le u\|A^{-1/2}\|$, the (1,1) block of (6.25) is bounded by $\frac{1}{2}(\|A^{1/2}\| + \|A^{1/2}\|^2\|A^{-1/2}\|)u = \frac{1}{2}\|A^{1/2}\|(1 + \kappa(A^{1/2}))u$ and the (2,1) block by $\frac{1}{2}\|A^{-1/2}\|(1 + \kappa(A^{1/2}))u$. The relative limiting accuracy estimate is therefore $\frac{1}{2}(1 + \kappa(A^{1/2}))u$, as for the Newton iteration.

For the product DB iteration, by considering (6.26) with $\|E\| \le u$ and $\|F\| \le \|A^{1/2}\|u$ we obtain a relative limiting accuracy estimate of $(3/2)u$, which is independent of A.

6.4.3. CR Iteration

For the CR iteration (6.19) the iteration function is

$$G(Y, Z) = \begin{bmatrix} -YZ^{-1}Y \\ Z - 2YZ^{-1}Y \end{bmatrix}.$$

At any fixed point $(0, Z)$,

$$L_g(0, Z; E, F) = \begin{bmatrix} 0 \\ F \end{bmatrix},$$

so again $L_g(0, Z)$ is idempotent and stability of the iteration follows.

The relative limiting accuracy is trivially of order u.

6.4.4. IN Iteration

The iteration function for the IN iteration (6.20) is

$$G(X, H) = \begin{bmatrix} X + H \\ -\frac{1}{2}H(X + H)^{-1}H \end{bmatrix}.$$

It is easy to see that at the fixed point $(A^{1/2}, 0)$,

$$L_g(A^{1/2}, 0; E, F) = \begin{bmatrix} E + F \\ 0 \end{bmatrix} = \begin{bmatrix} I & I \\ 0 & 0 \end{bmatrix} \begin{bmatrix} E \\ F \end{bmatrix}.$$

Therefore $L_g(A^{1/2}, 0)$ is idempotent and the iteration is stable. The relative limiting accuracy is again trivially of order u.

Table 6.2. *Summary of stability and limiting accuracy of square root iterations.*

Iteration	Stable?	Limiting accuracy
Newton, (6.12)	Only if $\max_{i,j} \frac{1}{2}\left|1-\lambda_i^{1/2}\lambda_j^{-1/2}\right| < 1$, where $\lambda_i \in \Lambda(A)$	$\kappa(A^{1/2})u$
DB, (6.15)	Yes	$\kappa(A^{1/2})u$
Product DB, (6.17)	Yes	u
CR, (6.19)	Yes	u
IN, (6.20)	Yes	u

6.4.5. Summary

Table 6.2 summarizes what our analysis says about the stability and limiting accuracy of the iterations. The Newton iteration (6.12) is unstable at $A^{1/2}$ unless the eigenvalues λ_i of A are very closely clustered, in the sense that $(\lambda_i/\lambda_j)^{1/2}$ lies in a ball of radius 2 about $z = 1$ in the complex plane, for all i and j. The four rewritten versions of Newton's iteration, however, are all stable at $A^{1/2}$. The price to be paid for stability is a coupled iteration costing more than the original Newton iteration (see Table 6.1). The limiting accuracy of the DB iteration is essentially $\kappa(A^{1/2})u$, but the other three stable iterations have limiting accuracy of order u.

6.5. Scaling the Newton Iteration

The best way to obtain scaling parameters for the Newton iteration and its variants is to exploit connections with the matrix sign function. Theorem 6.9 implies that scaling parameters for the Newton sign iterates S_k can be translated into ones for the Newton square root iterates X_k using $X_k = A^{1/2}S_k$: thus we scale $X_k \leftarrow \mu_k X_k$ at the start of the iteration, using (for example) the formulae (5.35)–(5.37) with X_k replaced by $A^{-1/2}X_k$. The resulting parameters depend on the unknown $A^{1/2}$, but for the determinantal scaling (5.35) we can use $\det(A^{1/2}) = \det(A)^{1/2}$ to obtain

Newton iteration (scaled):

$$X_{k+1} = \frac{1}{2}(\mu_k X_k + \mu_k^{-1} X_k^{-1} A), \qquad X_0 = A, \qquad \mu_k = \left|\frac{\det(X_k)}{\det(A)^{1/2}}\right|^{-1/n}. \tag{6.27}$$

For the DB iteration, the μ_k in (6.27) can be expressed in terms of just X_k and Y_k, by using the relation $Y_k = A^{-1}X_k$. The determinantally scaled DB iteration is

DB iteration (scaled):

$$\begin{aligned} \mu_k &= \left|\frac{\det(X_k)}{\det(A)^{1/2}}\right|^{-1/n} \quad \text{or} \quad \mu_k = \left|\det(X_k)\det(Y_k)\right|^{-1/(2n)}, \\ X_{k+1} &= \frac{1}{2}\left(\mu_k X_k + \mu_k^{-1} Y_k^{-1}\right), \qquad X_0 = A, \\ Y_{k+1} &= \frac{1}{2}\left(\mu_k Y_k + \mu_k^{-1} X_k^{-1}\right), \qquad Y_0 = I. \end{aligned} \tag{6.28}$$

For the product form of the DB iteration determinantal scaling gives, using $M_k = X_k Y_k$,

Product form DB iteration (scaled):

$$\begin{aligned}
\mu_k &= \left|\det(M_k)\right|^{-1/(2n)}, \\
M_{k+1} &= \frac{1}{2}\left(I + \frac{\mu_k^2 M_k + \mu_k^{-2} M_k^{-1}}{2}\right), & M_0 &= A, \\
X_{k+1} &= \frac{1}{2}\mu_k X_k (I + \mu_k^{-2} M_k^{-1}), & X_0 &= A, \\
Y_{k+1} &= \frac{1}{2}\mu_k Y_k (I + \mu_k^{-2} M_k^{-1}), & Y_0 &= I.
\end{aligned} \tag{6.29}$$

Incorporating scaling into the IN iteration requires a little algebra. In terms of X_k, the Newton increment E_k in (6.20) is given by $E_k = \frac{1}{2}(X_k^{-1}A - X_k)$. After scaling $X_k \leftarrow \mu_k X_k$ we have

$$\begin{aligned}
\widetilde{E}_k &= \frac{1}{2}(\mu_k^{-1} X_k^{-1} A - \mu_k X_k) \\
&= \mu_k^{-1} \cdot \frac{1}{2}(X_k^{-1} A - X_k) + \frac{1}{2}\mu_k^{-1} X_k - \frac{1}{2}\mu_k X_k \\
&= \mu_k^{-1}(E_k + \frac{1}{2}X_k) - \frac{1}{2}\mu_k X_k.
\end{aligned}$$

The scaled iteration is therefore

IN iteration (scaled):

$$\begin{aligned}
X_0 &= A, \quad E_0 = \tfrac{1}{2}(I - A), \\
\mu_k &= \left|\frac{\det(X_k)}{\det(A)^{1/2}}\right|^{-1/n}, \\
\widetilde{E}_k &= \mu_k^{-1}(E_k + \tfrac{1}{2}X_k) - \tfrac{1}{2}\mu_k X_k, \\
X_{k+1} &= \mu_k X_k + \widetilde{E}_k, \\
E_{k+1} &= -\tfrac{1}{2} E_k X_{k+1}^{-1} E_k.
\end{aligned} \tag{6.30}$$

Although the formulae (6.30) appear very different from, and less elegant than, (6.27), the two iterations generate exactly the same sequence of iterates X_k.

A suitable stopping test for all these iterations is from (4.25) and (6.13),

$$\|X_{k+1} - X_k\| \le \left(\eta \frac{\|X_{k+1}\|}{\|X_k^{-1}\|}\right)^{1/2}, \tag{6.31}$$

where η is a relative error tolerance.

6.6. Numerical Experiments

We present some numerical experiments to illustrate and compare the behaviour of the methods described in this chapter. For several matrices we give the relative error $\|X - \widehat{X}\|_\infty / \|X\|_\infty$ and relative residual $\|A - \widehat{X}^2\|_\infty / \|A\|_\infty$ of the computed $\widehat{X}$. For the iterative methods $\widehat{X}$ is taken to be the first iterate after which the relative error does not decrease significantly. If the iteration is converging slowly, this is the last iterate before the error starts to increase. The iterative methods are used both without scaling and with determinantal scaling. In the latter case, scaling is used only while the relative ∞-norm change in successive iterates exceeds 10^{-2}, because after this point the quadratic convergence (if indeed it is seen) leads to rapid convergence.

The matrices are as follows.

1. $A = I + uv^T$, where $u = [1\ 2^2 \dots\ n^2]^T$, $v = [0\ 1\ 2^2 \dots\ (n-1)^2]^T$. (The principal square root is known exactly from (1.16).)

2. The 16×16 Moler matrix, MATLAB's `gallery('moler',16)`, which is symmetric positive definite. It has 15 eigenvalues of order 1 and one small eigenvalue of order 10^{-9}.

3. A moderately nonnormal 8×8 matrix formed in MATLAB as

```
n = 8;
Q = gallery('orthog',n);
A = Q*rschur(n,2e2)*Q';
```

   The function `rschur(n,mu)`, from the Matrix Computation Toolbox [264], generates an upper quasi-triangular matrix with eigenvalues $\alpha_j + i\beta_j$, $\alpha_j = -j^2/10$, $\beta_j = -j$, $j = 1{:}n/2$ and $(2j, 2j+1)$ elements `mu`.

4. A 16×16 Chebyshev–Vandermonde matrix, `gallery('chebvand',16)` in MATLAB, which has 8 complex eigenvalues with modulus of order 1 and 8 real, positive eigenvalues between 3.6 and 10^{-11}.

5. A 9×9 singular nonsymmetric matrix resulting from discretizing the Neumann problem with the usual five point operator on a regular mesh. The matrix has real nonnegative eigenvalues and a one-dimensional null space with null vector the vector of 1s.

The first matrix, for which the results are given in Table 6.3, represents an easy problem. The Newton iteration is unstable, as expected since $\psi_N > 1$ (see (6.24)): the error reaches a minimum of about $u^{1/2}$ and then grows unboundedly. The other methods all produce tiny errors and residuals. Scaling produces a useful reduction in the number of iterations.

For the Moler matrix, we can see from Table 6.4 that the Schur method performs well: the residual is consistent with (6.10), and the error is bounded by $\kappa_{\mathrm{sqrt}}(X)u$, as it should be given the size of the residual (which is also the backward error). Newton's method is unstable, as expected given the eigenvalue distribution. The convergence of the sequence of errors (not shown) for the DB, product DB, and IN iterations is more linear than quadratic, and a large number of iterations are required before convergence is reached. The DB and product DB iterations have error significantly exceeding $\kappa_{\mathrm{sqrt}}(X)u$, and are beaten for accuracy and stability by the IN iteration, which matches the Schur method. For this matrix, scaling has little effect.

For the nonnormal matrix the tables are turned and it is the IN iteration that performs badly, giving an error larger than those from the other iterative methods by a factor about 10^4; see Table 6.5. The Newton iteration performs well, which

Table 6.3. *Results for rank-1 perturbation of* I. $\alpha_\infty(X) = 1.4$, $\kappa_{\mathrm{sqrt}}(X) = 40$, $\kappa_2(A^{1/2}) = 1.7 \times 10^2$, $\psi_N = 39$.

	Iterations	Relative error	Relative residual
Schur		1.5e-15	6.5e-16
Real Schur		1.4e-15	6.7e-16
Newton, unscaled	10	8.1e-7	8.2e-7
Newton, scaled	7	7.8e-8	1.5e-7
DB, unscaled	11	1.7e-15	3.4e-15
DB, scaled	8	1.5e-15	2.9e-15
Product DB, unscaled	11	7.3e-15	3.8e-15
Product DB, scaled	8	1.2e-14	8.3e-15
IN, unscaled	11	1.5e-14	3.0e-14
IN, scaled	8	1.2e-14	2.4e-14

Table 6.4. *Results for Moler matrix.* $\alpha_\infty(X) = 1.1$, $\kappa_{\mathrm{sqrt}}(X) = 8.3 \times 10^4$, $\kappa_2(A^{1/2}) = 3.6 \times 10^5$, $\psi_N = 10^5$.

	Iterations	Relative error	Relative residual
Schur		5.3e-13	3.0e-15
Real Schur		5.3e-13	3.0e-15
Newton, unscaled	8	5.3e-4	1.0e-5
Newton, scaled	7	1.9e-2	3.8e-2
DB, unscaled	17	6.6e-10	5.1e-10
DB, scaled	13	1.1e-9	6.4e-10
Product DB, unscaled	17	1.2e-10	4.8e-11
Product DB, scaled	13	1.5e-10	5.5e-11
IN, unscaled	18	1.2e-13	1.4e-15
IN, scaled	14	8.5e-14	1.5e-15

is consistent with ψ_N being only slightly larger than 1. The residual of the Schur methods is consistent with (6.10).

The Chebyshev–Vandermonde matrix shows poor performance of the DB and product DB iterations, which are beaten by the IN iteration. The Newton iteration is wildly unstable.

These experiments confirm the predictions of the stability analysis. However, the limiting accuracy is not a reliable predictor of the relative errors. Indeed, although the product DB iteration has better limiting accuracy than the DB iteration, this is not seen in practice. The main conclusion is that no one of the DB, product DB, and IN iterations is always best. Based on these limited experiments one might choose the IN iteration when $\alpha(X) = O(1)$ and one of the DB or product DB iterations when $\alpha(X)$ is large, but this reasoning has no theoretical backing (see Problem 6.24).

Finally, for the singular Neumann matrix we applied the IN iteration (6.20) without scaling until the relative residual was smaller than 10^{-4} and then computed the double step (6.22). Six iterations of (6.20) were required, after which the relative residual was of order 10^{-5} and the relative error of order 10^{-3}. After applying (6.22) these measures both dropped to 10^{-14}, showing that the double step can be remarkably effective.

Table 6.5. *Results for nonnormal matrix.* $\alpha_\infty(X) = 1.4 \times 10^8$, $\kappa_{\mathrm{sqrt}}(X) = 5.8 \times 10^7$, $\kappa_2(A^{1/2}) = 6.2 \times 10^{10}$, $\psi_N = 1.3$.

	Iterations	Relative error	Relative residual
Schur		5.1e-10	3.6e-8
Real Schur		5.1e-10	4.3e-8
Newton, unscaled	5	9.2e-7	2.0e-3
Newton, scaled	5	8.5e-8	1.6e-5
DB, unscaled	7	2.9e-8	1.2e-5
DB, scaled	5	2.1e-7	1.6e-4
Product DB, unscaled	6	8.5e-8	8.7e-2
Product DB, scaled	5	9.8e-7	4.8e-2
IN, unscaled	5	2.8e-4	9.9e-1
IN, scaled	4	3.2e-4	1.1e0

Table 6.6. *Results for Chebyshev–Vandermonde matrix.* $\alpha_\infty(X) = 2.8$, $\kappa_{\mathrm{sqrt}}(X) = 5.2 \times 10^6$, $\kappa_2(A^{1/2}) = 1.3 \times 10^7$, $\psi_N = 3.3 \times 10^5$.

	Iterations	Relative error	Relative residual
Schur		1.0e-10	2.4e-15
Real Schur		1.9e-10	2.3e-15
Newton, unscaled	4	2.4e-2	3.4e-3
Newton, scaled	2	6.9e-1	5.9e-1
DB, unscaled	18	1.0e-5	1.4e-5
DB, scaled	11	9.1e-8	1.2e-7
Product DB, unscaled	17	1.7e-5	2.9e-5
Product DB, scaled	11	9.8e-7	7.8e-7
IN, unscaled	22	2.9e-11	5.5e-16
IN, scaled	11	4.8e-12	2.5e-14

6.7. Iterations via the Matrix Sign Function

The Newton variants described above by no means represent all possible iterations for computing the matrix square root. A wide class of iterations can be obtained from iterations for the matrix sign function, by applying a sign iteration to the matrix $\left[\begin{smallmatrix} 0 & A \\ I & 0 \end{smallmatrix}\right]$ and using (5.4), which we repeat here:

$$\operatorname{sign}\left(\begin{bmatrix} 0 & A \\ I & 0 \end{bmatrix}\right) = \begin{bmatrix} 0 & A^{1/2} \\ A^{-1/2} & 0 \end{bmatrix}. \tag{6.32}$$

The next result makes this idea precise. It assumes a specific form $g(X) = Xh(X^2)$ for the sign iteration function, in order to provide a convenient formula for the square root iteration. This form holds for all the Padé iterations, in view of (5.27), and is very natural in view of (5.2) and (5.3).

Theorem 6.11 (Higham, Mackey, Mackey, and Tisseur). *Let $A \in \mathbb{C}^{n\times n}$ have no eigenvalues on $\mathbb{R}^-$. Consider any iteration of the form $X_{k+1} = g(X_k) \equiv X_k h(X_k^2)$ that converges to* $\operatorname{sign}(X_0)$ *for $X_0 = \left[\begin{smallmatrix} 0 & A \\ I & 0 \end{smallmatrix}\right]$ with order of convergence m. Then in the coupled iteration*

$$\begin{aligned} Y_{k+1} &= Y_k h(Z_k Y_k), \qquad & Y_0 &= A, \\ Z_{k+1} &= h(Z_k Y_k) Z_k, \qquad & Z_0 &= I, \end{aligned} \tag{6.33}$$

$Y_k \to A^{1/2}$ and $Z_k \to A^{-1/2}$ as $k \to \infty$, both with order of convergence m, Y_k commutes with Z_k, and $Y_k = AZ_k$ for all k.

Proof. Observe that

$$\begin{aligned} g\left(\begin{bmatrix} 0 & Y_k \\ Z_k & 0 \end{bmatrix}\right) &= \begin{bmatrix} 0 & Y_k \\ Z_k & 0 \end{bmatrix} h\left(\begin{bmatrix} Y_k Z_k & 0 \\ 0 & Z_k Y_k \end{bmatrix}\right) \\ &= \begin{bmatrix} 0 & Y_k \\ Z_k & 0 \end{bmatrix} \begin{bmatrix} h(Y_k Z_k) & 0 \\ 0 & h(Z_k Y_k) \end{bmatrix} \\ &= \begin{bmatrix} 0 & Y_k h(Z_k Y_k) \\ Z_k h(Y_k Z_k) & 0 \end{bmatrix} \\ &= \begin{bmatrix} 0 & Y_k h(Z_k Y_k) \\ h(Z_k Y_k) Z_k & 0 \end{bmatrix} = \begin{bmatrix} 0 & Y_{k+1} \\ Z_{k+1} & 0 \end{bmatrix}, \end{aligned}$$

where the penultimate equality follows from Corollary 1.34. The initial conditions $Y_0 = A$ and $Z_0 = I$ together with (6.32) now imply that Y_k and Z_k converge to $A^{1/2}$ and $A^{-1/2}$, respectively. It is easy to see that Y_k and Z_k are polynomials in A for all k, and hence Y_k commutes with Z_k. Then $Y_k = AZ_k$ follows by induction. The order of convergence of the coupled iteration (6.33) is clearly the same as that of the sign iteration from which it arises. □

Theorem 6.11 provides an alternative derivation of the DB iteration (6.15). Take $g(X) = \frac{1}{2}(X + X^{-1}) = X \cdot \frac{1}{2}(I + X^{-2}) \equiv Xh(X^2)$. Then $Y_{k+1} = Y_k \cdot \frac{1}{2}(I + (Z_k Y_k)^{-1}) = \frac{1}{2} Y_k (I + Y_k^{-1} Z_k^{-1}) = \frac{1}{2}(Y_k + Z_k^{-1})$, and, likewise, $Z_{k+1} = \frac{1}{2}(Z_k + Y_k^{-1})$.

New iterations are obtained by applying the theorem to the Padé family (5.28):

Padé iteration:

$$\begin{aligned} Y_{k+1} &= Y_k\, p_{\ell m}(1 - Z_k Y_k)\, q_{\ell m}(1 - Z_k Y_k)^{-1}, \qquad & Y_0 &= A, \\ Z_{k+1} &= p_{\ell m}(1 - Z_k Y_k)\, q_{\ell m}(1 - Z_k Y_k)^{-1} Z_k, \qquad & Z_0 &= I. \end{aligned} \tag{6.34}$$

For $0 \le \ell, m \le 2$, $p_{\ell m}(1-x^2)$ and $q_{\ell m}(1-x^2)$ can be read off from Table 5.1. For $\ell = m-1$ and $\ell = m$, $p_{\ell m}/q_{\ell m}$ has the partial fraction form given in Theorem 5.9 (d). From Theorem 5.8 and Theorem 6.11 we deduce that $Y_k \to A^{1/2}$ and $Z_k \to A^{-1/2}$ with order $\ell + m + 1$ unconditionally if $\ell = m-1$ or $\ell = m$, or provided $\|\operatorname{diag}(I-A, I-A)\| < 1$ if $\ell \ge m+1$.

For $\ell = 1$, $m = 0$, (6.34) gives a Newton–Schulz iteration:

Newton–Schulz iteration:

$$\begin{aligned} Y_{k+1} &= \tfrac{1}{2} Y_k (3I - Z_k Y_k), & Y_0 &= A, \\ Z_{k+1} &= \tfrac{1}{2} (3I - Z_k Y_k) Z_k, & Z_0 &= I. \end{aligned} \tag{6.35}$$

A sufficient condition for convergence of this inverse-free iteration is $\|I - A\|_p < 1$, for $p = 1$, 2, or ∞. For $\ell = 0$, $m = 1$, (6.34) is

$$\begin{aligned} Y_{k+1} &= 2Y_k (Z_k Y_k + I)^{-1}, & Y_0 &= A, \\ Z_{k+1} &= 2(Z_k Y_k + I)^{-1} Z_k, & Z_0 &= I. \end{aligned} \tag{6.36}$$

This is closely related to the DB iteration (6.15) in that Y_k from (6.36) equals the inverse of X_k from (6.15) with $X_0 = A^{-1}$, and similarly for the Z_k. The DB iteration is generally preferable to (6.36) because it requires less work per iteration.

What can be said about the stability of the coupled iteration (6.33)? Since Y_k and Z_k commute, the iteration formulae can be rewritten in several ways, and the particular choice of formula turns out to be crucial to the stability. For example, (6.36) is stable, but with the rearrangement $Z_{k+1} = 2Z_k(Z_k Y_k + I)^{-1}$, stability is lost. The next theorem shows that all instances of (6.33) are stable and hence that the particular formulae (6.33) are the right choice from the point of view of stability.

Theorem 6.12 (Higham, Mackey, Mackey, and Tisseur). *Consider any iteration of the form* (6.33) *and its associated mapping*

$$G(Y, Z) = \begin{bmatrix} Y h(ZY) \\ h(ZY) Z \end{bmatrix},$$

where $X_{k+1} = g(X_k) \equiv X_k h(X_k^2)$ is any superlinearly convergent iteration for $\operatorname{sign}(X_0)$. *Then any matrix pair of the form $P = (B, B^{-1})$ is a fixed point for G, and the Fréchet derivative of G at P is given by*

$$L_g(P; E, F) = \frac{1}{2} \begin{bmatrix} E - BFB \\ F - B^{-1} E B^{-1} \end{bmatrix}. \tag{6.37}$$

The derivative map $L_g(P)$ is idempotent ($L_g(P) \circ L_g(P) = L_g(P)$) and hence iteration (6.33) *is stable.*

Proof. We know from Theorem 5.13 that $L_g(S, G) = \frac{1}{2}(G - SGS)$ if $S = \operatorname{sign}(S)$. Now $\operatorname{sign}(I) = I$ implies $I = g(I) = h(I)$. Hence (B, B^{-1}) is a fixed point for G. In view of the relations in the proof of Theorem 6.11, we have $L_g(P; E, F) = L_g(S, H)$, where

$$S = \begin{bmatrix} 0 & B \\ B^{-1} & 0 \end{bmatrix}, \qquad H = \begin{bmatrix} 0 & E \\ F & 0 \end{bmatrix}.$$

Inserting these expressions into the formula for $L_g(S, H)$ gives the result. $\square$

This result proves stability of all the Padé iterations. Note that (6.37) is identical to the expression (6.25) for the Fréchet derivative of the DB iteration, as might be expected since the inverse of the DB iteration is a member of the Padé family, as noted above.

An underlying theme in this section is that the use of commutativity is best avoided if stable iterations are to be obtained. Note that the proof of Theorem 6.11 does not draw on commutativity. Simply rearranging the "Z formula" in (6.33) to $Z_{k+1} = Z_k h(Z_k Y_k)$, which is valid mathematically, changes the Fréchet derivative, and the new iteration is generally unstable (as mentioned above for $\ell = 0$, $m = 1$) [283, 2005, Lem. 5.4]. What it *is* safe to do is to rearrange using Corollary 1.34. Thus changing the Z formula to $Z_{k+1} = Z_k h(Y_k Z_k)$ does not affect the stability of (6.33).

6.8. Special Matrices

We now consider methods for computing square roots of some special classes of matrices of practical importance, including matrices close to the identity or with "large diagonal", M-matrices, H-matrices, and Hermitian positive definite matrices. We introduce some linearly convergent iterations, all of which are matrix multiplication-based and hence easy to implement. In some applications a low accuracy approximation to a square root is sufficient and in these cases it may be enough to carry out a few iterations of one of the procedures below.

6.8.1. Binomial Iteration

The methods of this section are all related to the binomial expansion

$$(I - C)^{1/2} = \sum_{j=0}^{\infty} \binom{\frac{1}{2}}{j} (-C)^j \equiv I - \sum_{j=1}^{\infty} \alpha_j C^j, \qquad \alpha_j > 0, \tag{6.38}$$

which is valid when $\rho(C) < 1$ (see Theorem 4.7).

When $\rho(A - I) < 1$ we can approximate $A^{1/2}$ by evaluating a partial sum of (6.38) with $A \equiv I - C$. Convergence will be slow unless $\|C\| \ll 1$, but convergence acceleration techniques such as Aitken extrapolation (Aitken's Δ^2-method) [542, 2002, Sec. 5.10] can be used in an effort to obtain a better approximation from a fixed number of terms of the series.

When $\rho(A - I)$ exceeds 1 we can write

$$A = s(I - C), \tag{6.39}$$

and try to choose s so that $\rho(C) < 1$. When A has real, positive eigenvalues, $0 < \lambda_n \le \lambda_{n-1} \le \cdots \le \lambda_1$, the s that minimizes $\rho(C)$, and the corresponding minimal $\rho(C)$, are (see Problem 6.17)

$$s = (\lambda_1 + \lambda_n)/2, \qquad \rho(C) = \frac{\lambda_1 - \lambda_n}{\lambda_1 + \lambda_n} < 1. \tag{6.40}$$

An alternative choice of s, valid for all A, is $s = \operatorname{trace}(A^*A)/\operatorname{trace}(A^*)$, which minimizes $\|C\|_F$; see Problem 6.18. This choice may or may not achieve $\rho(C) < 1$.

Assume now that $\rho(C) < 1$. An iterative method can be derived by defining $(I - C)^{1/2} =: I - P$ and squaring the equation to obtain $P^2 - 2P = -C$. A natural iteration for computing P is

Binomial iteration:

$$P_{k+1} = \frac{1}{2}(C + P_k^2), \qquad P_0 = 0. \tag{6.41}$$

Our choice of name for this iteration comes from the observation that $I - P_k$ reproduces the binomial expansion (6.38) up to and including terms of order C^k. So (6.41) can be thought of as a convenient way of generating the binomial expansion.

Before analyzing the convergence of the binomial iteration we note some properties of P. From (6.38) we have

$$P = \sum_{j=1}^{\infty} \alpha_j C^j, \tag{6.42}$$

where the α_j are positive. The eigenvalues of P and C are therefore related by $\lambda_i(P) = \sum_{j=1}^{\infty} \alpha_j \lambda_i(C)^j$. Hence $\rho(P) \le \sum_{j=1}^{\infty} \alpha_j \rho(C)^j = 1 - (1 - \rho(C))^{1/2}$. Since $1 - (1 - x)^{1/2} \le x$ for $0 \le x \le 1$, we conclude that $\rho(P) \le \rho(C) < 1$. Similarly, we find that $\|P\| \le \|C\|$ for any consistent matrix norm for which $\|C\| < 1$. Finally, it is clear that of all square roots $I - Q$ of $I - C$, the principal square root is the only one for which $\rho(Q) < 1$, since $\rho(Q) < 1$ implies that the spectrum of $I - Q$ lies in the open right half-plane.

We first analyze convergence in a special case. For $A \in \mathbb{R}^{n\times n}$, let $A \ge 0$ denote that $a_{ij} \ge 0$ for all i and j. If $C \ge 0$ then the binomial iteration enjoys mononotic convergence.

Theorem 6.13. *Let $C \in \mathbb{R}^{n\times n}$ satisfy $C \ge 0$ and $\rho(C) < 1$ and write $(I - C)^{1/2} = I - P$. Then in the binomial iteration (6.41), $P_k \to P$ with*

$$0 \le P_k \le P_{k+1} \le P, \qquad k \ge 0; \tag{6.43}$$

that is, the P_k converge monotonically to P from below.

Proof. From (6.41) and (6.42) we have $0 = P_0 \le P$, $P_k \ge 0$ for all k, and $P - P_{k+1} = \frac{1}{2}(P^2 - P_k^2)$, so by induction $0 \le P_k \le P$ for all k. Moreover, $C/2 = P_1 \ge P_0 = 0$, and $P_{k+1} - P_k = \frac{1}{2}(P_k^2 - P_{k-1}^2)$, so by induction $P_k \le P_{k+1}$ for all k. The inequalities (6.43) are therefore established. Since the sequence $\{P_k\}$ is nondecreasing and bounded above, it has a limit, P_*. This limit satisfies $P_* = \frac{1}{2}(C + P_*^2)$, and so $(I - P_*)^2 = I - C$. But $\rho(P_k) \le \rho(P) < 1$ for all k by (B.13), so $\rho(P_*) < 1$. Thus $I - P_*$ is the principal square root, that is, $P_* = P$. $\square$

It is now natural to try to prove convergence for general C with $\rho(C) < 1$. Consider the scalar iteration

$$p_{k+1} = \frac{1}{2}(c + p_k^2), \qquad p_0 = 0,$$

with $(1 - c)^{1/2} = 1 - p$, $|p| \le |c| < 1$. Defining

$$q_{k+1} = \frac{1}{2}(|c| + q_k^2), \qquad q_0 = 0,$$

it follows by induction that $|p_k| \le |q_k|$ for all k. By Theorem 6.13, q_k is monotonically increasing to $1 - \sqrt{1 - |c|}$. Thus $|p_k| \le 1 - \sqrt{1 - |c|}$. Now $p_{k+1} - p = \frac{1}{2}(p_k^2 - p^2) =$

$\frac{1}{2}(p_k + p)(p_k - p)$ and so

$$\begin{aligned} |p_{k+1} - p| &\le \frac{1}{2}(|p_k| + |p|)|p_k - p| \\ &\le \frac{1}{2}(1 - (1 - |c|)^{1/2} + |c|)|p_k - p| =: \theta|p_k - p|, \end{aligned} \tag{6.44}$$

where $\theta = \frac{1}{2}(1 - (1 - |c|)^{1/2} + |c|) \le |c| < 1$. Hence $p_k \to p$ with a monotonically decreasing error. This argument can be built into a proof that in the matrix iteration (6.41) P_k converges to P if $\rho(C) < 1$. But this restriction on C is stronger than necessary! The next result describes the actual region of convergence of the binomial iteration, which is closely connected with the Mandelbrot set.

Theorem 6.14 (convergence of binomial iteration). *Let the eigenvalues of $C \in \mathbb{C}^{n\times n}$ lie in the cardioid*

$$\mathcal{D} = \{\, 2z - z^2 : z \in \mathbb{C},\ |z| < 1 \,\}. \tag{6.45}$$

Then $(I - C)^{1/2} =: I - P$ exists and in the binomial iteration (6.41), *$P_k \to P$ linearly.*

Proof. C clearly has no eigenvalues on the interval $[1, \infty)$, so $I - C$ has no eigenvalues on $\mathbb{R}^-$ and therefore $(I - C)^{1/2}$ exists.

Since P_k is a polynomial in C, the Jordan canonical form of C can be used to block-diagonalize (6.41). Hence we first analyze the behaviour of the scalar iteration $p_{k+1} = \frac{1}{2}(c + p_k^2) =: f(p_k)$, $p_0 = 0$, where $(1 - c)^{1/2} = 1 - p$ and c is an eigenvalue of C, so that $c \in \mathcal{D}$. This iteration has two fixed points: $1 \pm (1 - c)^{1/2}$. If $c \in \mathcal{D}$ then $1 - c = 1 - 2z + z^2 = (1 - z)^2$, where $|z| < 1$, and hence $(1 - c)^{1/2} = 1 - z$, so that $|f'(1 - (1 - c)^{1/2})| = |z| < 1$. Therefore the fixed point $1 - (1 - c)^{1/2}$ is attracting, while $1 + (1 - c)^{1/2} = 2 - z$ is repelling.

Change variables to $x_k = p_k/2$. Then

$$x_{k+1} = x_k^2 + \frac{c}{4} =: Q_{c/4}(x_k), \qquad x_0 = 0.$$

The quadratic map $z \to Q_a(z) = z^2 + a$ in the complex plane is much studied. The set of $a \in \mathbb{C}$ for which $Q_a^{(k)}(0)$ is bounded is the famous Mandelbrot set. Existing analysis shows that if $a \in \{\, p/2 - p^2/4 : |p| < 1 \,\}$, which is the interior of the main cardioid of the Mandelbrot set, then $Q_a^{(k)}(0)$ converges to the attracting fixed point [97, 1993, Thm. 1.3, p. 126], [149, 1989, Thm. 4.6]. We conclude that if $c \in \mathcal{D}$ then p_k converges to $1 - (1 - c)^{1/2}$.

Our analysis of the scalar iteration combines with Theorem 4.15 to show that P_k converges to a matrix P whose eigenvalues are those of $I - (I - C)^{1/2}$ and such that $(I - P)^2 = I - C$. The only P satisfying these conditions is $P = I - (I - C)^{1/2}$. Convergence is linear since this is a fixed point iteration and the derivative of the iteration function is nonzero at the eigenvalues of P. □

Figure 6.1 plots the cardioid (6.45), which is much bigger than the unit disk, reaching -3 on the negative real axis and containing points with imaginary part as large as $\pm 2.6i$. Although the binomial series expansion (6.38) does not converge for C with spectral radius greater than 1, the iteration nevertheless continues to converge when the eigenvalues of C lie outside the unit disk but within the cardioid. The monotonic decrease of the error in the eigenvalues throughout the iteration, shown in (6.44), holds only within the unit disk, however.

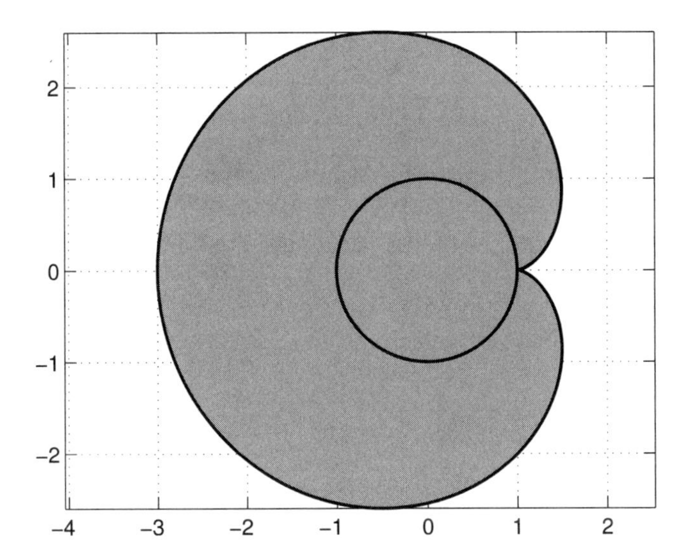

Figure 6.1. *The cardioid* (6.45)*, shaded, together with the unit circle. The binomial iteration converges for matrices* C *whose eigenvalues lie inside the cardioid.*

If $\rho(C) < 1$, there is a consistent norm such that $\|C\| < 1$ (see Section B.7). Hence when (6.41) is applicable to $I - C$, so is the Newton–Schulz iteration (6.35), which also requires only matrix multiplication but which has quadratic convergence. If medium to high accuracy is required, the Newton–Schulz iteration will be more efficient than the binomial iteration, despite requiring three matrix multiplications per iteration versus one for (6.41).

6.8.2. Modified Newton Iterations

The Newton iteration (6.11) can be written as $X_{k+1} = X_k + E_k$, where $L(X_k, E_k) = -F(X_k)$ and $L(X, E)$ is the Fréchet derivative of $F(X) = X^2 - A$ at X in the direction E. The iterations of Section 6.3 reduce the cost of Newton's method by taking an X_0 that commutes with A. Another way to reduce the cost of the method is to use approximations to the Fréchet derivative that ease solution of the Sylvester equation $L(X_k, E_k) = X_k E_k + E_k X_k = -F(X_k)$. In particular, we can freeze the Fréchet derivative at X_0 to obtain the modified Newton iteration

$$\left.\begin{aligned} \text{Solve } X_0 E_k + E_k X_0 &= A - X_k^2 \\ X_{k+1} &= X_k + E_k \end{aligned}\right\} \quad k = 0, 1, 2, \ldots, \tag{6.46}$$

and taking X_0 diagonal makes these equations easy to solve. Our first method has precisely this form, but we will derive it in a different way.

Let $A \in \mathbb{C}^{n \times n}$ have a principal square root

$$A^{1/2} = D^{1/2} + B,$$

where D is a known diagonal matrix with real, positive diagonal elements. The natural choice of D is $D = \operatorname{diag}(A)$, assuming A has positive diagonal elements. By squaring both sides of this equation we obtain the quadratic equation

$$D^{1/2} B + B D^{1/2} = A - D - B^2. \tag{6.47}$$

We can attempt to compute B by functional iteration:

Pulay iteration:

$$D^{1/2}B_{k+1} + B_{k+1}D^{1/2} = A - D - B_k^2, \qquad B_0 = 0. \tag{6.48}$$

Equation (6.48) is easily solved for B_{k+1}: $(B_{k+1})_{ij} = (A - D - B_k^2)_{ij}/(d_{ii}^{1/2} + d_{jj}^{1/2})$, $i,j = 1{:}n$. The Pulay iteration is easily shown to be a rewritten version of the modified Newton iteration (6.46) with $X_0 = D^{1/2}$; see Problem 6.16. The next result gives a sufficient condition for convergence to the principal square root.

Theorem 6.15 (convergence of Pulay iteration). *Let $A \in \mathbb{C}^{n\times n}$ have no eigenvalues on $\mathbb{R}^-$, let D be diagonal with positive diagonal entries, and let $B = A^{1/2} - D^{1/2}$. If*

$$\theta = \frac{\|B\|}{\min_i d_i^{1/2}} < \frac{2}{3} \tag{6.49}$$

for some consistent matrix norm then in the iteration (6.48), *$B_k \to A^{1/2} - D^{1/2}$ linearly.*

Proof. Let $E_k = B - B_k$ and subtract (6.48) from (6.47) to obtain

$$D^{1/2}E_{k+1} + E_{k+1}D^{1/2} = B_k^2 - B^2 = E_k^2 - BE_k - E_kB.$$

It follows that $\|E_k\| \le e_k$, where $e_{k+1} = e_k^2/(2\sigma) + (\beta/\sigma)e_k$, where $e_0 = \beta = \|B\|$ and $\sigma = \min_i d_i^{1/2}$. The change of variable $f_k = e_k/\beta$ yields, since $\theta = \beta/\sigma$,

$$f_{k+1} = \frac{\theta}{2}f_k^2 + \theta f_k, \qquad f_0 = 1. \tag{6.50}$$

Suppose $\theta < 2/3$. If $|f_k| \le 1$ then

$$|f_{k+1}| = \left|\frac{\theta}{2}f_k + \theta\right| |f_k| < |f_k|.$$

Since $f_0 = 1$ it follows by induction that the $|f_k|$ form a monotonically decreasing sequence, which, being bounded below by 0, has a limit. The iteration (6.50) has two fixed points, 0 and $2/\theta - 2 > 1$, so the limit must be 0. Hence $B_k \to B$, as required, and the convergence is clearly linear. □

Very roughly, the condition (6.49) says that $D^{1/2}$ is a reasonable approximation to $A^{1/2}$, in the sense that $|D^{1/2}| \ge \frac{3}{2}\|B\|I$.

The Pulay iteration can be simplified by taking D to be a multiple of the identity matrix. Equivalently, we can set $X_0 = I/(2\alpha)$ in (6.46). This gives the iteration

Visser iteration:

$$X_{k+1} = X_k + \alpha(A - X_k^2), \qquad X_0 = (2\alpha)^{-1}I. \tag{6.51}$$

Theorem 6.16 (convergence of Visser iteration). *Let $A \in \mathbb{C}^{n\times n}$ and $\alpha > 0$. If the eigenvalues of $I - 4\alpha^2 A$ lie in the cardioid* (6.45) *then $A^{1/2}$ exists and the iterates X_k from* (6.51) *converge linearly to $A^{1/2}$.*

Proof. The eigenvalue condition clearly implies that A has no eigenvalues on $\mathbb{R}^-$ and so $A^{1/2}$ exists. Let $\theta = 1/(2\alpha)$, $X_k = \theta Y_k$, and $\widetilde{A} = \theta^{-2}A$. Then

$$Y_{k+1} = Y_k + \frac{1}{2}(\widetilde{A} - Y_k^2), \qquad Y_0 = I.$$

This is equivalent to the binomial iteration (6.41) with $\widetilde{A} \equiv I - C$ and $Y_k = I - P_k$. Hence Theorem 6.14 shows that we have convergence of Y_k to $\widetilde{A}^{1/2}$ if the eigenvalues of $C = I - \widetilde{A}$ lie in the cardioid (6.45). In other words, $X_k \to A^{1/2}$ (since $\alpha > 0$) if the eigenvalues of $I - 4\alpha^2 A$ lie in the cardioid. □

If the eigenvalues of A are real and positive then the condition of the theorem is $0 < \alpha < \rho(A)^{-1/2}$. Convergence under these assumptions, but with equality allowed in the upper bound for α, was proved by Elsner [176, 1970].

The advantage of the Pulay iteration over the binomial iteration and the Visser iteration is that it can be applied to a matrix whose diagonal is far from constant, provided the matrix is sufficiently diagonally dominant. For example, consider the 16×16 symmetric positive definite matrix A with $a_{ii} = i^2$ and $a_{ij} = 0.1$, $i \neq j$. For the Pulay iteration (6.48) with $D = \mathrm{diag}(A)$, we have $\theta = 0.191$ in (6.49) and just 9 iterations are required to compute $A^{1/2}$ with a relative residual less than nu in IEEE double precision arithmetic. For the binomial iteration (6.41), writing $A = s(I - C)$ with s given by (6.40), $\rho(C) = 0.992$ and 313 iterations are required with the same convergence tolerance. Similarly, the Visser iteration (6.51) requires 245 iterations with $\alpha = 0.058$, which was determined by hand to approximately minimize the number of iterations. (The Newton–Schulz iteration (6.35) does not converge for this matrix.)

The stability of (6.51) under the assumptions of the theorem is easily demonstrated under the assumption that A has positive real eigenvalues. The Fréchet derivative of the iteration function $g(X) = X + \alpha(A - X^2)$ is $L_g(X, E) = E - \alpha(XE + EX)$. The eigenvalues of $L_g(A^{1/2})$ are $\mu_{ij} = 1 - \alpha(\lambda_i^{1/2} + \lambda_j^{1/2})$, $i, j = 1\!:\!n$, where λ_i denotes an eigenvalue of A. The maximum $\mu_{\max} = \max_{i,j} |\mu_{ij}|$ is obtained for $i = j$, so $\mu_{\max} = \max_i |1 - 2\alpha\lambda_i^{1/2}|$. Since $1 - 4\alpha^2\lambda_i$ lies in the cardioid (6.45), we have $1 - 4\alpha^2\lambda_i > -3$, i.e., $\alpha\lambda_i^{1/2} < 1$, so $\mu_{\max} < 1$. Hence $L_g(A^{1/2})$ is power bounded and the iteration is stable.

6.8.3. M-Matrices and H-Matrices

M-matrices arise in a variety of scientific settings, including in finite difference methods for PDEs, input-output analysis in economics, and Markov chains in stochastic processes [60, 1994].

The M-matrices are a subset of the real, square matrices with nonpositive off-diagonal elements.

Definition 6.17. *$A \in \mathbb{R}^{n\times n}$ is a nonsingular M-matrix if*

$$A = sI - B, \quad \text{where } B \geq 0 \text{ and } s > \rho(B). \tag{6.52}$$

Since $\rho(B) \geq \max_i b_{ii}$ (see Section B.11), (6.52) implies that any nonsingular M-matrix has positive diagonal entries. The representation (6.52) is not unique. We can take $s = \max_i a_{ii}$ (see Problem 6.20), which is useful for computational purposes.

An archetypal M-matrix is the tridiagonal Toeplitz matrix illustrated for $n = 3$ by

$$A = \begin{bmatrix} 2 & -1 & 0 \\ -1 & 2 & -1 \\ 0 & -1 & 2 \end{bmatrix}. \tag{6.53}$$

It is minus the second difference matrix that arises when discretizing a second derivative by a central difference, and is symmetric positive definite.

The definition of nonsingular M-matrix ensures that the eigenvalues all have positive real part and hence that the principal square root exists (see Theorem 1.29). Since this square root has spectrum in the right half-plane it is a candidate for being an M-matrix itself. In fact, the principal square root *is* an M-matrix.

Theorem 6.18 (Alefeld and Schneider). *For any nonsingular M-matrix the principal square root exists, is an M-matrix, and is the only square root that is an M-matrix.*

Proof. The first part was shown above. The analysis below of the binomial iteration (6.41) provides a constructive proof of the second part. For the third part note that any nonsingular M-matrix has spectrum in the right half-plane and that there is only one square root with this property by Theorem 1.29. □

For any nonsingular M-matrix A we can write

$$A = s(I - C), \qquad C = s^{-1}B \geq 0, \quad \rho(C) < 1, \tag{6.54}$$

and so a splitting of the form used in Section 6.8.1 automatically exists. It follows that we can use the binomial iteration (6.41) to compute $(I - C)^{1/2} =: I - P$. Since $C \geq 0$, the monotonic convergence shown in Theorem 6.13 is in effect. Moreover, since we showed in Section 6.8.1 that $P \geq 0$ and $\rho(P) \leq \rho(C) < 1$, it follows that $I - P$, and hence $A^{1/2}$, is an M-matrix (which completes the proof of Theorem 6.18).

As mentioned in Section 6.8.1, the Newton–Schulz iteration (6.35) can be used to compute $(I-C)^{1/2}$ and will generally be preferable. However, (6.41) has the advantage that it is structure-preserving: at whatever point the iteration is terminated we have an approximation $s^{1/2}(I - P_k) \approx A^{1/2}$ that is an M-matrix; by contrast, the Newton–Schulz iterates Y_k are generally not M-matrices.

The Newton iterations of Section 6.3 are applicable to M-matrices and are structure-preserving: if A is a nonsingular M-matrix and $X_0 = A$ then so are all the iterates X_k (or, in the case of the CR iteration (6.19), Z_k). This nonobvious property is shown by Meini [421, 2004] for (6.19) via the cyclic reduction derivation of the iteration; a direct proof for the Newton iteration (6.12) does not seem easy.

A numerical experiment illustrates well the differing convergence properties of the binomial and Newton–Schulz iterations. For the 8×8 instance of (6.53), with $s = \max_i a_{ii} = 2$, we find $\rho(C) = 0.94$. The binomial iteration (6.41) requires 114 iterations to produce a relative residual less than nu in IEEE double precision arithmetic, whereas the Newton–Schulz iteration requires just 9 iterations. When the residual tolerance is relaxed to 10^{-3}, the numbers of iterations are 15 and 6, respectively. If we increase the diagonal of the matrix from 2 to 4 then $\rho(C) = 0.47$ and the numbers of iterations are, respectively, 26 and 6 (tolerance nu) and 4 and 3 (tolerance 10^{-3}), so iteration (6.41) is more competitive for the more diagonally dominant matrix.

A class of matrices that includes the M-matrices is defined with the aid of the comparison matrix associated with A:

$$M(A) = (m_{ij}), \qquad m_{ij} = \begin{cases} |a_{ii}|, & i = j, \\ -|a_{ij}|, & i \neq j. \end{cases} \tag{6.55}$$

Definition 6.19. $A \in \mathbb{C}^{n \times n}$ *is a nonsingular H-matrix if $M(A)$ is a nonsingular M-matrix.*

An analogue of Theorem 6.18 holds for the subclass of the H-matrices with (real) positive diagonal entries.

Theorem 6.20 (Lin and Liu). *For any nonsingular H-matrix $A \in \mathbb{C}^{n \times n}$ with positive diagonal entries the principal square root $A^{1/2}$ exists and is the unique square root that is an H-matrix with positive diagonal entries.*

Proof. By Problem 6.20 we can write $M(A) = sI - B$, where $B \geq 0$, $s > \rho(B)$, and $s = \max_i |a_{ii}| = \max_i a_{ii}$. This means that $A = sI - \widetilde{C}$, where $|\widetilde{C}| = B$ and hence $\rho(\widetilde{C}) \leq \rho(|\widetilde{C}|) = \rho(B) < s$. It follows that the eigenvalues of A lie in the open right half-plane. Hence the principal square root $A^{1/2}$ exists. We now need to show that $A^{1/2}$ is an H-matrix with positive diagonal entries. This can be done by applying the binomial expansion (6.38) to $A = s(I - C)$, where $C = \widetilde{C}/s$ and $\rho(C) < 1$. We have $A^{1/2} = s^{1/2}(I - P)$ where, from (6.42), $\rho(P) \leq \rho(|P|) \leq \rho(|C|) = s^{-1}\rho(|\widetilde{C}|) < 1$. Since $\max_i p_{ii} \leq \rho(P) < 1$, $A^{1/2}$ has positive diagonal elements. Moreover, $M(A^{1/2}) = s^{1/2}(I - |P|)$ and $\rho(|P|) < 1$, so $A^{1/2}$ is an H-matrix.

Finally, $A^{1/2}$ is the only square root that is an H-matrix with positive diagonal entries, since it is the only square root with spectrum in the open right half-plane. □

For computing the principal square root of an H-matrix with positive diagonal elements we have the same options as for an M-matrix: the binomial iteration (6.41), the Newton iterations of Section 6.3, and the Newton–Schulz iteration (6.35). The binomial iteration no longer converges monotonically. For the behaviour of the Newton iterations, see Problem 6.25.

6.8.4. Hermitian Positive Definite Matrices

When implementing the Newton iteration variants of Section 6.3 for Hermitian positive definite A, advantage can be taken of the fact that the iterates are all Hermitian positive definite. However, our preferred iteration is based on the fact that the Hermitian positive definite square root H of A is the Hermitian polar factor of the Cholesky factor of A: if $A = R^*R$ (Cholesky factorization: R upper triangular) and $R = UH$ (polar decomposition: U unitary, H Hermitian positive definite) then $A = HU^*UH = H^2$.

Algorithm 6.21. Given a Hermitian positive definite matrix $A \in \mathbb{C}^{n \times n}$ this algorithm computes $H = A^{1/2}$.

1 $A = R^*R$ (Cholesky factorization).
2 Compute the Hermitian polar factor H of R by applying Algorithm 8.20 to R (exploiting the triangularity of R).

Cost: Up to about $15\frac{2}{3}n^3$ flops.

Algorithm 8.20, involved in step 2, is a Newton method for computing the Hermitian polar factor; see Section 8.9. Algorithm 6.21 has the advantage that it iterates on R, whose 2-norm condition number is the square root of that of A, and that it takes advantage of the excellent stability and the ready acceleration possibilities of the polar iteration. The algorithm can be extended to deal with positive *semi*definite matrices by using a Cholesky factorization with pivoting [286, 1990]. The algorithm is also easily adapted to compute the inverse square root, $A^{-1/2}$, at the same cost: after computing the unitary polar factor U by Algorithm 8.20, compute $A^{-1/2} = R^{-1}U \ (= H^{-1})$.

An alternative approach is to use the binomial iteration (6.41). If $A \in \mathbb{C}^{n\times n}$ is Hermitian positive definite then it has real, positive eigenvalues $\lambda_n \le \cdots \le \lambda_1$, and so using (6.39) and (6.40) we can write $A = s(I - C)$, where $s = (\lambda_1 + \lambda_n)/2$ and $\rho(C) = \|C\|_2 = (\kappa_2(A) - 1)(\kappa_2(A) + 1) < 1$, and this choice of s minimizes $\|C\|_2$. The convergence of (6.41) is monotonic nondecreasing in the positive semidefinite ordering, where for Hermitian matrices A and B, $A \le B$ denotes that $B - A$ is positive semidefinite (see Section B.12); this can be shown by diagonalizing the iteration and applying Theorem 6.13 to the resulting scalar iterations. However, unless A is extremely well conditioned ($\kappa_2(A) < 3$, say) the (linear) convergence of (6.41) will be very slow.

We now return to the Riccati equation $XAX = B$ discussed in Section 2.9, which generalizes the matrix square root problem. When A and B are Hermitian positive definite we can use a generalization of Algorithm 6.21 to compute the Hermitian positive definite solution. For a derivation of this algorithm see Problem 6.21. The algorithm is more efficient than direct use of the formulae in Section 2.9 or Problem 2.7 and the same comments about conditioning apply as for Algorithm 6.21.

Algorithm 6.22. Given Hermitian positive definite matrices $A, B \in \mathbb{C}^{n\times n}$ this algorithm computes the Hermitian positive definite solution of $XAX = B$.

1 $A = R^*R$, $B = S^*S$ (Cholesky factorizations).
2 Compute the unitary polar factor U of SR^* using Algorithm 8.20.
3 $X = R^{-1}U^*S$ (by solving the triangular system $RX = US$).

Cost: Up to about $19\frac{1}{3}n^3$ flops.

6.9. Computing Small-Normed Square Roots

We saw in Section 6.1 that the best a priori bound for the relative residual $\|A - \widehat{X}^2\|/\|A\|$ of the computed square root $\widehat{X}$ of $A \in \mathbb{C}^{n\times n}$ produced by any method in floating point arithmetic is of the order $\alpha(\widehat{X})u$, where $\alpha(X) = \|X\|^2/\|A\|$. Assuming that we are willing to compute any square root of A, one with minimal $\|X\|$ might therefore be preferred, on the grounds that it should give the smallest relative residual.

We therefore consider the question of determining $\min\{\, \|X\| : X^2 = A\}$. We begin with an instructive example. Consider

$$A = \begin{bmatrix} \epsilon & 1 & 0 \\ 0 & \epsilon & 0 \\ 0 & 0 & \epsilon \end{bmatrix}, \qquad 0 < \epsilon \ll 1, \tag{6.56}$$

which is in Jordan form with a 2×2 block and a 1×1 block. The two primary square roots of A are

$$X = \pm \begin{bmatrix} \epsilon^{1/2} & \frac{1}{2}\epsilon^{-1/2} & 0 \\ 0 & \epsilon^{1/2} & 0 \\ 0 & 0 & \epsilon^{1/2} \end{bmatrix},$$

and $\|X\|$ is therefore of order $\epsilon^{-1/2}$. But A also has two families of nonprimary square roots:

$$X = \pm W \begin{bmatrix} \epsilon^{1/2} & \frac{1}{2}\epsilon^{-1/2} & 0 \\ 0 & \epsilon^{1/2} & 0 \\ 0 & 0 & -\epsilon^{1/2} \end{bmatrix} W^{-1},$$

where

$$W = \begin{bmatrix} a & b & c \\ 0 & a & 0 \\ 0 & d & e \end{bmatrix},$$

with the parameters a, b, c, d, e arbitrary subject to W being nonsingular (see Theorems 1.26 and 1.25). Setting $a = e = 1$, $b = 0$, $c = -1/(2\epsilon^{1/2})$, and $d = -c$ gives the particular square roots

$$X = \pm \begin{bmatrix} \epsilon^{1/2} & 0 & 1 \\ 0 & \epsilon^{1/2} & 0 \\ 0 & 1 & -\epsilon^{1/2} \end{bmatrix}. \tag{6.57}$$

These two nontriangular square roots have $\|X\| = O(1)$, so they are much smaller normed than the primary square roots. This example shows that the square root of minimal norm may be a nonprimary square root and that the primary square roots can have norm much larger than the minimum over all square roots. Of course, if A is normal then the minimal possible value $\|X\|_2 = \|A\|_2^{1/2}$ is attained at every primary (and hence normal) square root. It is only for nonnormal matrices that there is any question over the size of $\|X\|$.

How to characterize or compute a square root of minimal norm is a difficult open question. The plausible conjecture that for a matrix with distinct, real, positive eigenvalues the principal square root will be the one of minimal norm is false; see Problem 6.22.

For the computation, Björck and Hammarling [73, 1983] consider applying general purpose constrained optimization methods. Another approach is to augment the Schur method with a heuristic designed to steer it towards a minimal norm square root. Write the equations (6.5) as

$$\begin{aligned} u_{jj} &= \pm t_{jj}^{1/2}, \\ u_{ij} &= \frac{t_{ij} - \sum_{k=i+1}^{j-1} u_{ik}u_{kj}}{u_{ii} + u_{jj}}, \qquad i = j-1\colon -1\colon 1. \end{aligned} \tag{6.58}$$

Denote the values u_{ij} resulting from the two possible choices of u_{jj} by u_{ij}^+ and u_{ij}^-. Suppose $u_{11}, \dots, u_{j-1,j-1}$ have been chosen and the first $j-1$ columns of U have been computed. The idea is to take whichever sign for u_{jj} leads to the smaller 1-norm for the jth column of U, a strategy that is analogous to one used in matrix condition number estimation [276, 2002, Chap. 15].

Algorithm 6.23 (small-normed square root). Given an upper triangular $T \in \mathbb{C}^{n \times n}$ this algorithm computes a primary square root $U = \sqrt{T}$ and attempts to minimize $\|U\|_1$.

1 $u_{11} = t_{11}^{1/2}$
2 for $j = 2\colon n$
3 Compute from (6.58) u_{ij}^+, u_{ij}^- $(i = j-1\colon -1\colon 1)$,
$s_j^+ = \sum_{i=1}^{j} |u_{ij}^+|$, and $s_j^- = \sum_{i=1}^{j} |u_{ij}^-|$.
4 if $s_j^+ \le s_j^-$ then $u_{ij} = u_{ij}^+$, $i = 1\colon j$, else $u_{ij} = u_{ij}^-$, $i = 1\colon j$.
5 end

Cost: $\frac{2}{3}n^3$ flops.

The cost of Algorithm 6.23 is twice that of a direct computation of $T^{1/2}$ with an a priori choice of the u_{ii}. This is a minor overhead in view of the overall cost of Algorithm 6.3.

For T with distinct eigenvalues, every square root of T is a candidate for computation by Algorithm 6.23. If T has some repeated eigenvalues then the algorithm will avoid nonprimary square roots, because these yield a zero divisor in (6.58) for some i and j.

It is easy to see that when T has real, positive diagonal elements Algorithm 6.23 will assign positive diagonal to U. This choice is usually, but not always, optimal, as can be confirmed by numerical example.

A weakness of Algorithm 6.23 is that it locally minimizes column norms without judging the overall matrix norm. Thus the size of the $(1, n)$ element of the square root, for example, is considered only at the last stage. The algorithm can therefore produce a square root of norm far from the minimum (as it does on the matrix in Problem 6.22, for example). Generally, though, it performs quite well.

6.10. Comparison of Methods

We offer a few observations concerning the choice of method for computing a matrix square root. For general nonsymmetric matrices, the (real) Schur method is the method of choice if it can be afforded, both because of its optimal stability and because of its ability to compute any desired primary square root (including a well conditioned one, as described in the previous section).

The iterative methods may or may not be able to beat the $28\frac{1}{3}n^3$ flops cost of the Schur method, depending on how many iterations are required, which in turn depends on the desired accuracy and the properties of the matrix. Since their flops are mostly expended in matrix multiplication and matrix inversion, these methods should run faster, relative to the Schur method, than the flop counts suggest.

For symmetric matrices the Schur method simplifies into computation of a spectral decomposition $A = QDQ^*$ $(D = \mathrm{diag}(\lambda_i)$, Q unitary) followed by evaluation of $\sqrt{A} = Q\,\mathrm{diag}(\sqrt{\lambda_i})Q^*$, which costs $10n^3$ flops. For Hermitian positive definite matrices that are not very ill conditioned, Algorithm 6.21 is an attractive alternative. Another alternative in this case is an approximation algorithm of Lu [396, 1998]. This algorithm unitarily reduces A to tridiagonal form, $A = QTQ^*$, then approximates $T^{1/2}$ using a Padé approximant of the function $\sqrt{1+x}$ expressed in linear partial fraction form. The cost of the method is $10n^3/3 + 5n^2m/2$, where m is the required degree of the Padé approximant (which is computed during the algorithm). The value of m depends on the required accuracy and the condition number of the matrix, and $m \ll n$ is possible, as shown by two well conditioned test problems in [396, 1998].

6.11. Involutory Matrices

The identity matrix has infinitely many square roots: the involutory matrices. The study of involutory matrices has been motivated by an application to cryptography. In a so-called homogeneous linear substitution cipher, encipherment is done by representing groups of characters in a message by numeric vectors and multiplying the vectors by a matrix (in modular arithmetic). If the matrix is involutory then it can be used for both encipherment and decipherment, that is, the cipher is self-reciprocal. The practical use of this idea was proposed by Hill [289, 1929], who received a US patent for a machine implementing it in 1932. Bauer [46, 2002, p. 85] notes that "the importance of Hill's invention stems from the fact that since then the value of mathematical methods in cryptology has been unchallenged." This method of encryption is nowadays mainly of pedagogical interest (see, e.g., Moler [436, 2004, Sec. 1.5]).

Potts [476, 1976] gives some constructions of symmetric involutory matrices. Levine and Nahikian [381, 1962] show that $A \in \mathbb{C}^{n\times n}$ is involutory if and only if it can be written $A = I - QP$, where $Q \in \mathbb{C}^{n\times p}$, $P \in \mathbb{C}^{p\times n}$, and $PQ = 2I$.

Some particular involutory matrices with explicit representations are worth noting. These are all nonprimary square roots of I, because the only primary square roots are $\pm I$. With $H = (1/(i+j-1))$ denoting the Hilbert matrix and

$$D = \operatorname{diag}\left((-1)^i i \binom{n+i-1}{i}\binom{n-1}{i-1}\right),$$

both $DH$and HD are involutory, as noted by Householder and Carpenter [300, 1963]. An involutory matrix with integer entries mentioned by several authors (e.g., Bellman [51, 1970, Ex. 20, p. 28]) is the upper triangular matrix U with $u_{ij} = (-1)^{i-1}\binom{j-1}{i-1}$, which is obtained from the Cholesky factor of the Pascal matrix[6] by changing the signs of alternate rows. For example, for $n = 4$,

$$U = \begin{bmatrix} 1 & 1 & 1 & 1 \\ 0 & -1 & -2 & -3 \\ 0 & 0 & 1 & 3 \\ 0 & 0 & 0 & -1 \end{bmatrix}. \tag{6.59}$$

This matrix arises in the stepsize changing mechanism in an ODE code based on backward differentiation formulae; see Shampine and Reichelt [516, 1997]. The principal submatrices $U(i\colon j, i\colon j)$ are also involutory [21, 2004]. The Cholesky factor of the Pascal matrix is also an ingredient in the involutory matrix

$$A = 2^{(1-n)/2} LDU,$$

where

$$l_{ij} = \binom{i-1}{j-1}, \quad D = \operatorname{diag}(1, -2, \dots, (-2)^{n-1}), \quad u_{ij} = \binom{n-i+1}{n-j+1},$$

with L (unit) lower triangular and U is upper triangular. For $n = 4$,

$$A = 2^{-3/2} \begin{bmatrix} 1 & 3 & 3 & 1 \\ 1 & 1 & -1 & -1 \\ 1 & -1 & -1 & 1 \\ 1 & -3 & 3 & -1 \end{bmatrix}.$$

---

[6] The Pascal matrix has (i,j) entry $\binom{i+j-2}{j-1}$; see [276, 2002, Sec. 28.4] for properties of this matrix.

This matrix is discussed by Boyd, Micchelli, Strang, and Zhou [80, 2001], who describe various applications of matrices from a more general class of "binomial matrices".

The latter three matrices are available in MATLAB as

```
gallery('invol',n)     % D*H
pascal(n,1)'
gallery('binomial',n) % Without the 2^((1-n)/2) scale factor.
```

6.12. Notes and References

The study of matrix square roots began with Cayley [99, 1858], in one of the first papers on matrix theory; he gave formulae for square roots of 2×2 and 3×3 matrices (see Problem 6.2). He later gave more details of both cases, motivated by a question of Tait of "finding the square root of a strain" [100, 1872].

In the engineering literature $A^{1/2}$ is sometimes used to denote a "square root factor" of a Hermitian positive definite matrix A: a matrix X, usually understood to be the Cholesky factor, such that $A = X^*X$. See, for example, Kailath, Sayed, and Hassibi [330, 2000, Sec. 12.1].

The formulae and inequalities of Section 6.1 for $\kappa_{\mathrm{sqrt}}(X)$ are from Higham [268, 1987]. The residual condition number $\alpha(X)$ was first defined and investigated by Björck and Hammarling [73, 1983].

Algorithm 6.3 is due to Björck and Hammarling [73, 1983] (see the quote at the end of this section). The extension to real arithmetic, Algorithm 6.7, is from Higham [268, 1987].

For a derivation of Algorithm 6.5 see Friedland [194, 1967] or Higham [276, 2002, Prob. 1.4]. For a more low level algorithm that scales to avoid overflow and underflow and is specialized to base 2 arithmetic see Kahan [327, 1987, p, 201].

The scalar version of the Newton iteration (6.12) is known as Heron's method, since it is found in the work of the Greek mathematician Heron of Alexandria (1st century AD). One step of the iteration was used by the Old Babylonians to approximate square roots (c. 2000–1600 BC) [190, 1998].

The quadratic convergence of the Newton iteration (6.12) to $A^{1/2}$ is shown by Laasonen [366, 1958] under the assumption that A has real, positive eigenvalues. Higham [267, 1986] weakens the assumption to that in Theorem 6.9: that A has no eigenvalues on $\mathbb{R}^-$. Theorem 6.9 is more general than existing results in the literature, which typically require $X_0 = A$ or X_0 a positive multiple of I. The link exploited in the theorem between Newton square root iterates and Newton sign iterates was pointed out (for $X_0 = A$) by Cheng, Higham, Kenney, and Laub [108, 2001, Thm. 4.1].

A thorough analysis of the behaviour of the full Newton method for (general) nonlinear systems in floating point arithmetic is given by Tisseur [569, 2001], who determines the limiting accuracy and limiting residual and the factors that influence them. See also Higham [276, 2002, Chap. 25].

The equivalence of the CR iteration (6.19) with Newton's method is noted by Meini [421, 2004] and investigated further by Iannazzo [305, 2003]. Our derivation of the scaled IN iteration (6.30) is from [305, 2003].

The observation that the Newton iteration (6.12) can be used for singular matrices with semisimple zero eigenvalues is new. The linear rate of convergence identified in Theorem 6.10 was obtained for the (equivalent) CR iteration by Meini [421, 2004].

The instability of Newton's iteration (6.12) was first noted by Laasonen [366, 1958]. He stated without proof that, for matrices with real, positive eigenvalues, Newton's method "if carried out indefinitely, is not stable whenever the ratio of the largest to the smallest eigenvalue of A exceeds the value 9". Analysis justifying this claim was given by Higham [267, 1986], who obtained the condition (6.24) and also proved the stability of the DB iteration. Some brief comments on the instability of (6.12) are given by Liebl [390, 1965], who computes the eigenvalues of the Fréchet derivative of the iteration function (Exercise E.10.1-18 of [453, 2000] summarizes Liebl's paper). The stability of the product form of the DB iteration and of the CR iteration is proved by Cheng, Higham, Kenney, and Laub [108, 2001] and Iannazzo [305, 2003], respectively.

The Padé iterations (6.34) with $\ell = m-1$ are derived and investigated by Higham [274, 1997], who proves the stability and instability of particular variants and shows how to scale the iterations. Theorems 6.11 and 6.12 are from Higham, Mackey, Mackey, and Tisseur [283, 2005, Thms. 4.5, 5.3]. Our proof of Theorem 6.12 is much shorter than the one in [283, 2005], which does not exploit the stability of the underlying sign iterations.

Using the binomial expansion (6.38) to compute matrix square roots is suggested by Waugh and Abel [610, 1967] in connection with the application discussed in Section 2.3 involving roots of transition matrices. The binomial expansion is also investigated by Duke [170, 1969].

The binomial iteration (6.41) is discussed for $C \geq 0$ by Alefeld and Schneider [8, 1982] and Butler, Johnson, and Wolkowicz [86, 1985], where it is used for theoretical purposes rather than suggested as a computational tool. Theorem 6.13 is essentially found in these papers. Albrecht [7, 1977] suggests the use of (6.41) for Hermitian positive definite matrices. Theorem 6.14 is new.

The iteration (6.48) was suggested for symmetric positive definite matrices by Pulay [480, 1966] (see also Bellman [51, 1970, Ex. 1, p. 334]), but no convergence result is given there. Theorem 6.15 is new. The approximation B_1 obtained after one iteration of (6.48) for symmetric positive definite A and $D = \mathrm{diag}(A)$ has been used in a stochastic differential equations application by Sharpa and Allen [517, 2000].

The iteration (6.51) can be traced back to Visser [602, 1937], who used it to show that a bounded positive definite self-adjoint operator on Hilbert space has a bounded positive definite self-adjoint square root without applying the spectral theorem. This use of the iteration can also be found in functional analysis texts, such as those of Debnath and Mikusiński [144, 1999], Riesz and Sz.-Nagy [492, 1956, Sec. 104], Schechter [502, 1971, Sec. 8.2], and Halmos [244, 1982, Prob. 121]. For a generalization of Visser's technique to pth roots see Problem 7.7. Theorem 6.16 is new. Elsner [176, 1970] obtains the stability result derived at the end of Section 6.8.2. Iteration (6.51) and its stability are also studied for symmetric positive definite matrices by Liebl [390, 1965] and Babuška, Práger, and Vitásek [25, 1966, Sec. 2.4.5].

Theorem 6.18 is an interpretation of results of Alefeld and Schneider [8, 1982], who also treat square roots of singular M-matrices. Theorem 6.20 is proved for real H-matrices by Lin and Liu [393, 2001].

Algorithm 6.21 for Hermitian positive definite matrices is due to Higham [266, 1986]. Algorithm 6.22 is suggested by Iannazzo [307, 2007].

That the matrix (6.56) has a small-normed square root (6.57) is pointed out by Björck and Hammarling [73, 1983]. Algorithm 6.23 is from Higham [268, 1987].

For more on linear substitution ciphers see Bauer [46, 2002].

Early contributions on involutory matrices are those of Cayley [99, 1858], [100, 1872] and Sylvester [556, 1882]. The number of involutory matrices over a finite field is determined by Hodges [293, 1958].

Finally, an application in which the inverse matrix square root appears is the computation of tight windows of Gabor frames [319, 2002].

Problems

6.1. Prove the integral formula (6.1) by using the integral (5.3) for the matrix sign function together with (5.4).

6.2. (Cayley [99, 1858], [100, 1872], Levinger [382, 1980]) Let $A \in \mathbb{C}^{2\times 2}$. If A has distinct complex conjugate eigenvalues then the square roots of A can be obtained explicitly from the formulae following Lemma 6.4. More generally, the following approach can be considered. By the Cayley–Hamilton theorem any 2×2 matrix X satisfies $X^2 - \mathrm{trace}(X)X + \det(X)I = 0$. Let $X^2 = A$. Then

$$A - \mathrm{trace}(X)X + \sqrt{\det(A)}\, I = 0. \tag{6.60}$$

Taking the trace gives

$$\mathrm{trace}(A) - \mathrm{trace}(X)^2 + 2\sqrt{\det(A)} = 0. \tag{6.61}$$

We can solve (6.61) for trace(X) and then solve (6.60) for X to obtain

$$X = \frac{A + \sqrt{\det(A)}\, I}{\sqrt{\mathrm{trace}(A) + 2\sqrt{\det(A)}}}.$$

Discuss the effectiveness of this procedure.

6.3. Construct a parametrized matrix $A(\epsilon)$ such that $\|A(\epsilon)^{1/2}\|^2/\|A(\epsilon)\| \to \infty$ as $\epsilon \to 0$.

6.4. Show that for any primary square root X of $A \in \mathbb{C}^{n\times n}$, range($X$) = range($A$). Show that this equality does not necessarily hold for a nonprimary square root.

6.5. Explain the behaviour of the Schur algorithm (Algorithm 6.3) when A is singular.

6.6. Show that in floating point arithmetic Algorithms 6.3 and 6.7 produce a computed square root $\widehat{U}$ of, respectively, a triangular matrix $T \in \mathbb{C}^{n\times n}$ and a quasi-triangular matrix $T \in \mathbb{R}^{n\times n}$, satisfying $\widehat{U}^2 = T + \Delta T$ with $|\Delta T| \le \widetilde{\gamma}_n|\widehat{U}|^2$ for Algorithm 6.3 and $\|\Delta T\|_F \le \widetilde{\gamma}_n\|\widehat{U}\|_F^2$ for Algorithm 6.7.

6.7. Let $A = \mathrm{diag}(a,b) \in \mathbb{C}^{2\times 2}$ with $a, b \notin \mathbb{R}^-$. The Newton iteration (6.12) converges to $A^{1/2}$ by Theorem 6.9. Show that if $a + b = -2$, the full Newton iteration (6.11) with $X_0 = A$ breaks down on the second iteration.

6.8. Investigate the behaviour of the Newton iteration (6.12) for

$$A = \begin{bmatrix} 1 & 0 \\ 0 & \mu^2 \end{bmatrix}, \qquad X_0 = \begin{bmatrix} 1 & \theta \\ 0 & \mu \end{bmatrix}, \qquad \theta \ne 0.$$

Note that this X_0 does not commute with A.

6.9. (Elsner [176, 1970]; the ideas in this problem have been generalized to algebraic Riccati equations: see Mehrmann [418, 1991, Thm. 11.3] and Lancaster and Rodman [370, 1995, Thms. 9.1.1, 9.2.1].)

(a) Let $>$ and $\geq$ denote the orderings on Hermitian matrices defined in Section B.12. Show that if $A > 0$ and $X_0 > 0$ then the iterates from the full Newton iteration (6.11) satisfy

$$0 < A^{1/2} \leq \cdots \leq X_{k+1} \leq X_k \leq \cdots \leq X_1, \qquad A \leq X_k^2, \quad k \geq 1,$$

and deduce that X_k converges monotonically to $A^{1/2}$ in the positive semidefinite ordering. Hint: first show that if $C > 0$ and $H \geq 0$ ($H > 0$) then the solution of the Sylvester equation $XC + CX = H$ satisfies $X \geq 0$ ($X > 0$).

(b) Note that X_0 is an arbitrary positive definite matrix: it need not commute with A. What can be said about the convergence of the simplified Newton iteration (6.12)?

(c) Suppose A is not Hermitian positive definite but that there is a nonsingular Z such that $Z^{-1}AZ$ and $Z^{-1}X_0Z$ are Hermitian positive definite. How can the result of (a) be applied?

6.10. Explain the behaviour of the Newton iteration (6.12) when A has an eigenvalue on $\mathbb{R}^-$.

6.11. (Iannazzo [305, 2003]) Give another derivation of the DB iteration (6.15) as follows. "Symmetrize" (6.12) by writing it as

$$X_{k+1} = \frac{1}{2}\big(X_k + A^{1/2}X_k^{-1}A^{1/2}\big), \qquad X_0 = A.$$

Heuristically, this symmetrization should improve the numerical stability. Remove the unwanted square roots by introducing a new variable $Y_k = A^{-1/2}X_kA^{-1/2}$.

6.12. In the DB iteration (6.15) it is tempting to argue that when computing Y_{k+1} the latest X iterate, X_{k+1}, should be used instead of X_k, giving $X_{k+1} = \frac{1}{2}\left(X_k + Y_k^{-1}\right)$ and $Y_{k+1} = \frac{1}{2}\left(Y_k + X_{k+1}^{-1}\right)$. (This is the same reasoning that produces the Gauss–Seidel iteration from the Jacobi iteration.) What effect does this change have on the convergence of the iteration?

6.13. From (6.25) we obtain the relation for the DB iteration function G:

$$G(A^{1/2} + E, A^{-1/2} + F) = \frac{1}{2}\begin{bmatrix} E - A^{1/2}FA^{1/2} \\ F - A^{-1/2}EA^{-1/2} \end{bmatrix} + O\left(\left\|\begin{bmatrix} E \\ F \end{bmatrix}\right\|^2\right).$$

Explain how the quadratic convergence of the DB iteration can be seen from this relation. In other words, reconcile the stability analysis with the convergence analysis.

6.14. In the coupled Padé iteration (6.34) we know that $Y_k = AZ_k$ for all k, so we can rewrite the Y_k recurrence in the uncoupled form

$$Y_{k+1} = Y_k\, p_{\ell m}(1 - A^{-1}Y_k^2)\, q_{\ell m}(1 - A^{-1}Y_k^2)^{-1}, \qquad Y_0 = A. \tag{6.62}$$

What are the pros and cons of this rearrangement? Note that for $\ell = 0$, $m = 1$, and $A \leftarrow A^{-1}$, (6.62) is the iteration

$$Y_{k+1} = 2Y_k(I + Y_kAY_k)^{-1} \tag{6.63}$$

for computing $A^{-1/2}$, which appears occasionally in the literature [518, 1991].

6.15. Discuss whether the requirement $\rho(C) < 1$ in Theorem 6.13 can be relaxed by making use of Theorem 6.14.

6.16. Show that Pulay's iteration (6.48) is equivalent to the modified Newton iteration (6.46) with $X_0 = D^{1/2}$, and specifically that $X_k = D^{1/2} + B_k$.

6.17. Show that for $A \in \mathbb{C}^{n\times n}$ with real, positive eigenvalues $0 < \lambda_n \le \lambda_{n-1} \le \cdots \le \lambda_1$, $\min_{s\in\mathbb{R}}\{\,\rho(C) : A = s(I - C)\,\}$ is attained at $s = (\lambda_1 + \lambda_n)/2$, and $\rho(C) = (\lambda_1 - \lambda_n)/(\lambda_1 + \lambda_n)$.

6.18. Show that for $A \in \mathbb{C}^{n\times n}$, $\min_{s\in\mathbb{C}}\{\,\|C\|_F : A = s(I - C)\,\}$ is attained at $s = \operatorname{trace}(A^*A)/\operatorname{trace}(A^*)$. (Cf. Theorem 4.21 (a).)

6.19. Show that Theorem 6.16 can be extended to A having a semisimple zero eigenvalue.

6.20. Show that if A is an M-matrix then in the representation (6.52) we can take $s = \max_i a_{ii}$.

6.21. Derive Algorithm 6.22 for solving $XAX = B$.

6.22. Show that for the matrix

$$T = \begin{bmatrix} 1 & -1 & 1/\epsilon^2 \\ 0 & (1+\epsilon)^2 & 1 \\ 0 & 0 & (1+2\epsilon)^2 \end{bmatrix}, \qquad 0 < \epsilon \ll 1,$$

the principal square root is not the square root of minimal norm.

6.23. (Meyer [426, 2000, Ex. 3.6.2]) Show that for all $A \in \mathbb{C}^{m\times n}$ and $B \in \mathbb{C}^{n\times m}$ the matrix

$$\begin{bmatrix} I - BA & B \\ 2A - ABA & AB - I \end{bmatrix}$$

is involutory.

6.24. (RESEARCH PROBLEM) Which of the DB, product DB, and IN iterations is to be preferred in which circumstances?

6.25. (RESEARCH PROBLEM) When the Newton iteration (6.12) is applied to a nonsingular H-matrix A with positive diagonal entries are all the iterates X_k nonsingular H-matrices with positive diagonal entries?

6.26. (RESEARCH PROBLEM) Develop algorithms for computing directly the Cholesky factors of (a) the principal square root of a Hermitian positive definite matrix and (b) the Hermitian positive definite solution X of $XAX = B$, where A and B are Hermitian positive definite.

The unexpectedly rapid convergence of Newton's method
led us to discover the simple method of the next section.

— ÅKE BJÖRCK and SVEN HAMMARLING, *A Schur Method for the Square Root of a Matrix* (1983)

Although no proof of convergence will be given,
the procedure converged rapidly in all cases examined by us.

— PÉTER PULAY, *An Iterative Method for the Determination of the Square Root of a Positive Definite Matrix* (1966)

Some simplifications in the handling of a digraphic system based on a
linear transformation would be obtained if we
designed the system so that the
enciphering matrix is its own inverse.

— ABRAHAM SINKOV, *Elementary Cryptanalysis: A Mathematical Approach* (1966)

Blackwell: *You know the algorithm for calculating the square root? ...*
It occurred to me that maybe this algorithm would work for positive definite matrices.
You take some positive definite X, add it to SX^{-1} and divide by two.
The question is: Does this converge to the square root of X. ...
In a particular example, the error at first was tremendous,
then dropped down to about .003.
Then it jumped up a bit to .02, then jumped up quite a bit to .9,
and then it exploded. Very unexpected....
It turns out that the algorithm works provided that the matrix you start with
commutes with the matrix whose square root you want.
You see, it's sort of natural because you
have to make a choice between SX^{-1} and $X^{-1}S$,
but of course if they commute it doesn't make any difference.
Of course I started out with the identity matrix and it should commute with anything.
So what happened?
MP: *You must have been having some kind of roundoff.*
Blackwell: *Exactly! If the computer had calculated exactly it would have converged.*
The problem is that the matrix the computer used didn't quite commute.

— DAVID BLACKWELL[7], in *Mathematical People: Profiles and Interviews* (1985)

[7] From [6, 1985].

Chapter 7
Matrix pth Root

X is a pth root of $A \in \mathbb{C}^{n\times n}$ if $X^p = A$. The pth root of a matrix, for $p > 2$, arises less frequently than the square root, but nevertheless is of interest both in theory and in practice. One application is in the computation of the matrix logarithm through the relation $\log A = p \log A^{1/p}$ (see Section 11.5), where p is chosen so that $A^{1/p}$ can be well approximated by a polynomial or rational function. Here, $A^{1/p}$ denotes the principal pth root, defined in the next section. Roots of transition matrices are required in some finance applications (see Section 2.3). Related to the pth root is the matrix sector function $\mathrm{sect}_p(A) = (A^p)^{-1/p}A$ discussed in Section 2.14.3.

The matrix pth root is an interesting object of study because algorithms and results for the case $p = 2$ do not always generalize easily, or in the manner that they might be expected to.

In this chapter we first give results on existence and classification of pth roots. Then we generalize the Schur method described in Chapter 6 for the square root. Newton iterations for the principal pth root and its inverse are explored, leading to a hybrid Schur–Newton algorithm. Finally, we explain how the pth root can be obtained via the matrix sign function.

Throughout the chapter p is an integer. Matrix roots A^α with α a real number can be defined (see Problem 7.2) but the methods of this chapter are applicable only when α is the reciprocal of an integer.

7.1. Theory

We summarize some theoretical results about matrix pth roots, all of which generalize results for $p = 2$ already stated.

Theorem 7.1 (classification of pth roots). *Let the nonsingular matrix $A \in \mathbb{C}^{n\times n}$ have the Jordan canonical form $Z^{-1}AZ = J = \mathrm{diag}(J_1, J_2, \dots, J_m)$, with $J_k = J_k(\lambda_k)$, and let $s \le m$ be the number of distinct eigenvalues of A. Let $L_k^{(j_k)} = L_k^{(j_k)}(\lambda_k)$, $k = 1{:}m$, denote the p pth roots given by (1.4), where $j_k \in \{1, 2, \dots, p\}$ denotes the branch of the pth root function. Then A has precisely p^s pth roots that are primary functions of A, given by*

$$X_j = Z\,\mathrm{diag}(L_1^{(j_1)}, L_2^{(j_2)}, \dots, L_m^{(j_m)})Z^{-1}, \quad j = 1{:}p^s,$$

corresponding to all possible choices of $j_1, \dots, j_m$, subject to the constraint that $j_i = j_k$ whenever $\lambda_i = \lambda_k$.

If $s < m$, A has nonprimary pth roots. They form parametrized families

$$X_j(U) = ZU\,\mathrm{diag}(L_1^{(j_1)}, L_2^{(j_2)}, \dots, L_m^{(j_m)})U^{-1}Z^{-1}, \quad j = p^s + 1{:}p^m,$$

where $j_k \in \{1, 2, \ldots, p\}$, U *is an arbitrary nonsingular matrix that commutes with* J, *and for each* j *there exist* i *and* k, *depending on* j, *such that* $\lambda_i = \lambda_k$ *while* $j_i \neq j_k$.

Proof. The proof is similar to that of Theorem 1.26; see Smith [530, 2003]. □

Theorem 7.2 (principal pth root). *Let* $A \in \mathbb{C}^{n\times n}$ *have no eigenvalues on* $\mathbb{R}^-$. *There is a unique* p*th root* X *of* A *all of whose eigenvalues lie in the segment* $\{ z : -\pi/p < \arg(z) < \pi/p \}$, *and it is a primary matrix function of* A. *We refer to* X *as the* principal pth root *of* A *and write* $X = A^{1/p}$. *If* A *is real then* $A^{1/p}$ *is real.*

Proof. The proof is entirely analogous to that of Theorem 1.29. □

Theorem 7.3 (existence of pth root). $A \in \mathbb{C}^{n\times n}$ *has a* p*th root if and only if the "ascent sequence" of integers* $d_1, d_2, \ldots$ *defined by*

$$d_i = \dim(\mathrm{null}(A^i)) - \dim(\mathrm{null}(A^{i-1}))$$

has the property that for every integer $\nu \geq 0$ *no more than one element of the sequence lies strictly between* $p\nu$ *and* $p(\nu+1)$.

Proof. See Psarrakos [479, 2002]. □

We note the integral representation

$$A^{1/p} = \frac{p\sin(\pi/p)}{\pi} A \int_0^\infty (t^p I + A)^{-1}\, dt, \tag{7.1}$$

generalizing (6.1), which holds for any real $p > 1$. For an arbitrary real power, see Problem 7.2.

Our final result shows that a pth root can be obtained from the invariant subspace of a block companion matrix associated with the matrix polynomial $\lambda^p I - A$.

Theorem 7.4 (Benner, Byers, Mehrmann, and Xu). *Let* $A \in \mathbb{C}^{n\times n}$. *If the columns of* $U = [U_1^*, \ldots, U_p^*]^* \in \mathbb{C}^{pn\times n}$ *span an invariant subspace of*

$$C = \begin{bmatrix} 0 & I & & & \\ & 0 & I & & \\ & & \ddots & \ddots & \\ & & & \ddots & I \\ A & & & & 0 \end{bmatrix} \in \mathbb{C}^{pn\times pn}, \tag{7.2}$$

that is, $CU = UY$ *for some nonsingular* $Y \in \mathbb{C}^{n\times n}$, *and* U_1 *is nonsingular, then* $X = U_2 U_1^{-1}$ *is a* p*th root of* A.

Proof. The equation $CU = UY$ yields

$$U_2 = U_1 Y, \quad \ldots, \quad U_p = U_{p-1} Y, \quad AU_1 = U_p Y.$$

Hence

$$\begin{aligned} A &= U_p Y U_1^{-1} = U_{p-1} Y^2 U_1^{-1} = \cdots = U_2 Y^{p-1} U_1^{-1} = U_1 Y^p U_1^{-1} \\ &= U_1 (U_1^{-1} U_2)^p U_1^{-1} = (U_2 U_1^{-1})^p. \quad \square \end{aligned}$$

7.2. Schur Method

Let $A \in \mathbb{R}^{n\times n}$ have the real Schur decomposition $A = QRQ^T$, where Q is orthogonal and R is upper quasi-triangular and block $m \times m$. The (real) pth root problem now reduces to computing an upper quasi-triangular U such that $U^p = R$. For $p = 2$ we were able in Section 6.2 to directly obtain a recurrence for the U_{ij}. For $p > 2$ this approach can be generalized by setting up $p-1$ coupled recurrences.

We begin by considering the cube root case: $U^3 = R$. Let $V = U^2$. In order to compute U we will need to compute V as well. The equation $UV = R$ gives

$$\begin{aligned} R_{ij} &= U_{ii}V_{ij} + U_{ij}V_{jj} + \sum_{k=i+1}^{j-1} U_{ik}V_{kj} \\ &=: U_{ii}V_{ij} + U_{ij}V_{jj} + B_{ij}^{(1)}, \end{aligned}$$

and likewise $V = U^2$ gives

$$\begin{aligned} V_{ij} &= U_{ii}U_{ij} + U_{ij}U_{jj} + \sum_{k=i+1}^{j-1} U_{ik}U_{kj} \\ &=: U_{ii}U_{ij} + U_{ij}U_{jj} + B_{ij}^{(0)}. \end{aligned}$$

Substituting the latter equation for V_{ij} into the equation for R_{ij} gives

$$R_{ij} = U_{ii}\big(U_{ii}U_{ij} + U_{ij}U_{jj} + B_{ij}^{(0)}\big) + U_{ij}V_{jj} + B_{ij}^{(1)},$$

which may be rearranged, using $V_{ii} = U_{ii}^2$, as

$$V_{ii}U_{ij} + U_{ii}U_{ij}U_{jj} + U_{ij}V_{jj} = R_{ij} - U_{ii}B_{ij}^{(0)} - B_{ij}^{(1)}.$$

This is a generalized Sylvester equation for U_{ij}. The right-hand side depends only on blocks of U and V lying to the left of and below the (i, j) blocks. Hence, just as in the square root case, we can solve for U_{ij} and V_{ij} a block column at a time or a block superdiagonal at a time.

Algorithm 7.5 (Schur method for cube root). Given a nonsingular $A \in \mathbb{R}^{n\times n}$ this algorithm computes $X = \sqrt[3]{A}$ via a real Schur decomposition, where $\sqrt[3]{\cdot}$ denotes any real primary cube root.

```
1  Compute a real Schur decomposition A = QRQ^T, where R is block m × m.
2  for j = 1: m
3      U_jj = ∛(R_jj)
4      V_jj = U_jj^2
5      for i = j − 1: −1: 1
6          B_0 = Σ_{k=i+1}^{j−1} U_ik U_kj
7          B_1 = Σ_{k=i+1}^{j−1} U_ik V_kj
8          Solve V_ii U_ij + U_ii U_ij U_jj + U_ij V_jj = R_ij − U_ii B_0 − B_1 for U_ij.
9          V_ij = U_ii U_ij + U_ij U_jj + B_0
10     end
11 end
12 X = QUQ^T
```

Cost: $25n^3$ flops for the Schur decomposition plus $2n^3/3$ for U and $3n^3$ to form X: $28\frac{2}{3}n^3$ flops in total. (The cost of forming U is dominated by that of computing $B_{ij}^{(0)}$ and $B_{ij}^{(1)}$, which is essentially that of forming U^2 and UV, respectively.)

This approach can be generalized to obtain an algorithm for pth roots. Let $U^p = R$. The idea is to define $V^{(k)} = U^{k+1}$, $k = 0\colon p-2$ and obtain recurrences for the blocks of the $V^{(k)}$ using the relations $UV^{(k)} = V^{(k+1)}$, $k = 0\colon p-3$ and $UV^{(p-2)} = R$. We simply state the algorithm; for the derivation, which involves some tedious algebra, see Smith [530, 2003].

Algorithm 7.6 (Schur method for pth root). Given a nonsingular $A \in \mathbb{R}^{n\times n}$ this algorithm computes $X = \sqrt[p]{A}$ via a real Schur decomposition, where $\sqrt[p]{\cdot}$ denotes any real primary pth root.

1 Compute a real Schur decomposition $A = QRQ^T$, where R is block $m \times m$.
2 for $j = 1\colon m$
3 $\quad U_{jj} = \sqrt[p]{R_{jj}}$
4 $\quad V_{jj}^{(k)} = U_{jj}^{k+1}$, $k = -1\colon p-2$
5 $\quad$ for $i = j-1\colon -1\colon 1$
6 $\quad\quad$ for $k = 0\colon p-2$
7 $\quad\quad\quad B_k = \sum_{\ell=i+1}^{j-1} U_{i\ell} V_{\ell j}^{(k)}$
8 $\quad\quad$ end
9 $\quad\quad$ Solve $\sum_{k=0}^{p-1} V_{ii}^{(p-2-k)} U_{ij} V_{jj}^{(k-1)} = R_{ij} - \sum_{k=0}^{p-2} V_{ii}^{(p-3-k)} B_k$ for U_{ij}.
10 $\quad\quad$ for $k = 0\colon p-2$
11 $\quad\quad\quad V_{ij}^{(k)} = \sum_{\ell=0}^{k} V_{ii}^{(k-\ell-1)} U_{ij} V_{jj}^{(\ell-1)} + \sum_{\ell=0}^{k-1} V_{ii}^{(k-2-\ell)} B_\ell$
12 $\quad\quad$ end
13 $\quad$ end
14 end
15 $X = QUQ^T$

Cost: $25n^3$ flops for the Schur decomposition plus $(p-1)n^3/3$ for U (which is essentially the cost of computing the B_k) and $3n^3$ to form X: $(28 + (p-1)/3)n^3$ flops in total.

Two details in Algorithms 7.5 and 7.6 remain to be described: how to compute $\sqrt[p]{R_{jj}}$ and how to solve the generalized Sylvester equations.

A formula for $\sqrt[p]{R_{jj}}$, for $R_{jj} \in \mathbb{R}^{2\times 2}$, can be obtained by adapting the approach used for $p = 2$ in Section 6.2. Since R_{jj} has distinct eigenvalues we can write (as in the proof of Lemma 6.4)

$$Z^{-1} R_{jj} Z = \operatorname{diag}(\lambda, \overline{\lambda}) = \theta I + i\mu K, \qquad K = \begin{bmatrix} 1 & 0 \\ 0 & -1 \end{bmatrix}.$$

Hence $R_{jj} = \theta I + \mu W$, where $W = iZKZ^{-1}$, and since $\theta, \mu \in \mathbb{R}$ it follows that $W \in \mathbb{R}^{2\times 2}$. Let $(\alpha + i\beta)^p = \theta + i\mu$. Then a pth root of R_{jj} is given by

$$U_{jj} = \alpha I + \beta W = \alpha I + \frac{\beta}{\mu}(R_{jj} - \theta I).$$

If, as is most likely, the principal pth root is required, α and β can be obtained from the polar form $\theta + i\mu = re^{i\phi}$ as $\alpha + i\beta = r^{1/p} e^{i\phi/p}$. Recall that θ and μ are given by (6.8).

The generalized Sylvester equation at line 9 of Algorithm 7.6 can be solved by applying the vec operator to obtain the linear system

$$\sum_{k=0}^{p-1}\Big(V_{jj}^{(k-1)^T}\otimes V_{ii}^{(p-2-k)}\Big)\cdot \mathrm{vec}(U_{ij}) = \mathrm{vec}\Big(R_{ij}-\sum_{k=0}^{p-2}V_{ii}^{(p-3-k)}B_k\Big),$$

which has dimension 1, 2, or 4 and can be solved by Gaussian elimination with partial pivoting. To check the nonsingularity of the coefficient matrix we note that, by (B.17), this matrix has eigenvalues

$$\sum_{k=0}^{p-1}\lambda_r^k\mu_s^{p-1-k} = \begin{cases}\dfrac{\lambda_r^p-\mu_s^p}{\lambda_r-\mu_s}, & \lambda_r\neq\mu_s,\\ p\lambda_r^{p-1}, & \lambda_r=\mu_s,\end{cases}$$

where $\lambda_r \in \Lambda(U_{jj})$ and $\mu_s \in \Lambda(U_{ii})$, with λ_r and μ_s nonzero by the assumption that A is nonsingular. Hence nonsingularity holds when $\lambda_r^p \neq \mu_s^p$ for all r and s, which requires that $\lambda_r \neq \mu_s e^{2\pi ik/p}$, $k = 1\colon p-1$. This condition is certainly satisfied if any eigenvalue appearing in two different blocks R_{ii} and R_{jj} is mapped to the same pth root in $R_{ii}^{1/p}$ and $R_{jj}^{1/p}$. Hence the algorithm will succeed when computing any primary pth root.

Algorithm 7.6 requires storage for the $2p-2$ intermediate upper quasi-triangular matrices $V^{(k)}$ and $B^{(k)}$. For large p, this storage can be a significant cost. However, if p is composite the pth root can be obtained by successively computing the roots given by the prime factors of p, which saves both storage and computation.

If A is nonsingular and has a negative real eigenvalue then the principal pth root is not defined. For odd p, Algorithm 7.6 can nevertheless compute a real, primary pth root by taking $\sqrt[p]{\lambda} = -|\lambda|^{1/p}$ when $\lambda < 0$. For even p, there is no real, primary pth root in this situation.

The stability properties of Algorithm 7.6 are entirely analogous to those for the square root case. The computed $\widehat{X}$ satisfies

$$\frac{\|A-\widehat{X}^p\|_F}{\|A\|_F} \le \widetilde{\gamma}_{n^3}\,\alpha_F(\widehat{X}), \tag{7.3}$$

where

$$\alpha(X) = \frac{\|X\|^p}{\|A\|} = \frac{\|X\|^p}{\|X^p\|} \ge 1.$$

The quantity $\alpha(X)$ can be regarded as a condition number for the relative residual of X, based on the same reasoning as used in Section 6.1.

Algorithms 7.5 and 7.6 are readily specialized to use the complex Schur decomposition, by setting all blocks to be 1×1.

7.3. Newton's Method

Given the important role Newton's method plays in iterations for the matrix sign decomposition and matrix square root, it is natural to apply it to the matrix pth root. Newton's method for the system $X^p - A = 0$ defines an increment E_k to an iterate X_k by

$$\sum_{i=1}^{p} X_k^{p-i}E_kX_k^{i-1} = A - X_k^p. \tag{7.4}$$

No $O(n^3)$ flops methods are known for solving this generalized Sylvester equation for $p > 2$ and so it is necessary to simplify the equation by using commutativity properties, just as in Lemma 6.8. When X_0 commutes with A so do all the X_k, and Newton's method can be written (see Problem 7.5)

Newton iteration (matrix pth root):

$$X_{k+1} = \frac{1}{p}\big[(p-1)X_k + X_k^{1-p}A\big], \qquad X_0A = AX_0. \tag{7.5}$$

The convergence properties of the iteration, and their dependence on X_0, are much more complicated than in the square root case.

The most obvious choice, $X_0 = A$, is unsatisfactory because no simple conditions are available that guarantee convergence to the principal pth root for general A. To illustrate, we consider the scalar case. Note that if we define $y_k = a^{-1/p}x_k$ then (7.5) reduces to

$$y_{k+1} = \frac{1}{p}\big[(p-1)y_k + y_k^{1-p}\big], \qquad y_0 = a^{-1/p}x_0, \tag{7.6}$$

which is the Newton iteration for a pth root of unity with a starting value depending on a and x_0. Figure 7.1 plots for $p = 2\colon 5$, $a = 1$, and y_0 ranging over a 400×400 grid with $\operatorname{Re} y_0, \operatorname{Im} y_0 \in [-2.5, 2.5]$, the root to which y_k from (7.6) converges, with each root denoted by a different grayscale from white (the principal root) to black. Convergence is declared if after 50 iterations the iterate is within relative distance 10^{-13} of a root; the relatively small number of points for which convergence was not observed are plotted white. For $p = 2$, the figure confirms what we already know from Chapter 6: convergence is obtained to whichever square root of unity lies nearest y_0. But for $p \geq 2$ the regions of convergence have a much more complicated structure, involving sectors with petal-like boundaries. This behaviour is well known and is illustrative of the theory of Julia sets of rational maps [468, 1992], [511, 1991]. In spite of these complications, the figures are suggestive of convergence to the principal root of unity when y_0 lies in the region

$$\{\, z : |z| \geq 1, \ -\pi/(2p) < \arg(z) < \pi/(2p) \,\},$$

which is marked by the solid line in the figure for $p \geq 3$. In fact, this *is* a region of convergence, as the next theorem, translated back to (7.5) with varying a and $x_0 = 1$ shows. We denote by $\mathbb{R}^+$ the open positive real axis.

Theorem 7.7 (convergence of scalar Newton iteration). *For all $p > 1$, the iteration*

$$x_{k+1} = \frac{1}{p}\big[(p-1)x_k + x_k^{1-p}a\big], \qquad x_0 = 1,$$

converges quadratically to $a^{1/p}$ if a belongs to the set

$$S = \{\, z \in \mathbb{C} : \operatorname{Re} z > 0 \text{ and } |z| \leq 1 \,\} \cup \mathbb{R}^+. \qquad \square \tag{7.7}$$

Proof. The rather complicated proof is given by Iannazzo [306, 2006]. $\square$

Corollary 7.8 (convergence of matrix Newton iteration). *Let $A \in \mathbb{C}^{n\times n}$ have no eigenvalues on $\mathbb{R}^-$. For all $p > 1$, the iteration (7.5) with $X_0 = I$ converges quadratically to $A^{1/p}$ if each eigenvalue λ_i of A belongs to the set S in (7.7).*

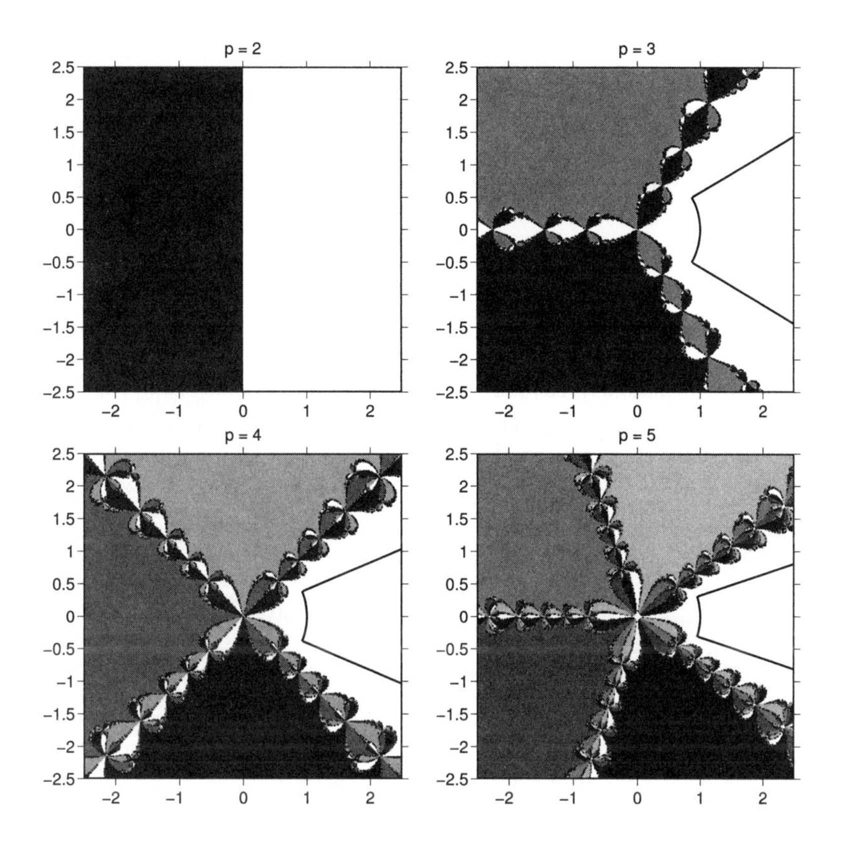

Figure 7.1. *Convergence of the Newton iteration* (7.6) *for a pth root of unity. Each point y_0 in the region is shaded according to the root to which the iteration converges, with white denoting the principal root. Equivalently, the plots are showing convergence of* (7.5) *in the scalar case with $a = 1$ and the shaded points representing x_0.*

Proof. Theorems 7.7 and 4.15 together imply that Y_k has a limit Y_* whose eigenvalues are the principal pth roots of the eigenvalues of A. The limit clearly satisfies $Y_*^p = A$, and so is a pth root of A. By Theorem 7.2, the only pth root of A having the same spectrum as Y_* is the principal pth root, and so $Y_* = A^{1/p}$. □

The restrictive condition on the eigenvalues of A in Corollary 7.8 can be overcome quite simply with some preprocessing and postprocessing and use of the identity $A^{1/p} = \big((A^{1/2})^{1/p}\big)^2$.

Algorithm 7.9 (Matrix pth root via Newton iteration). Given $A \in \mathbb{C}^{n\times n}$ having no eigenvalues on $\mathbb{R}^-$ this algorithm computes $X = A^{1/p}$ using the Newton iteration.

1 $B = A^{1/2}$
2 $C = B/\theta$, where θ is any upper bound for $\rho(B)$ (e.g., $\theta = \|B\|$)
3 Use the Newton iteration (7.5) with $X_0 = I$ to compute
$$X = \begin{cases} C^{2/p}, & p \text{ even}, \\ \big(C^{1/p}\big)^2, & p \text{ odd}. \end{cases}$$
4 $X \leftarrow \|B\|^{2/p} X$

To see why the algorithm works, note that the eigenvalues of B, and hence also C, lie in the open right half-plane, and $\rho(C) \le 1$. Therefore C satisfies the conditions of the corollary.

Unfortunately, instability vitiates the convergence properties of iteration (7.5) in practical computation—indeed we have already seen in Chapters 4 and 6 that this is the case for $p = 2$. Analysis similar to that in Section 6.4.1 (see Problem 7.12) shows that the eigenvalues of $L_g(A^{1/p})$, where $g(X) = p^{-1}[(p-1)X + X^{1-p}A]$, are of the form

$$\frac{1}{p}\left((p-1) - \sum_{r=1}^{p-1}\left(\frac{\lambda_i}{\lambda_j}\right)^{r/p}\right), \quad i,j = 1\colon n, \tag{7.8}$$

where the λ_i are the eigenvalues of A. To guarantee stability we need these eigenvalues to be less than 1 in modulus, which is a very severe restriction on A.

Inspired by Problem 6.11, we might try to obviate the instability by rewriting the iteration in the more symmetric form (using the commutativity of X_k with $A^{1/p}$)

$$\begin{aligned} X_{k+1} &= \frac{1}{p}\left((p-1)X_k + A^{1/p}X_k^{-1}A^{1/p}X_k^{-1}\dots A^{1/p}X_k^{-1}A^{1/p}\right) \\ &= \frac{1}{p}\left((p-1)X_k + (A^{1/p}X_k^{-1})^{p-1}A^{1/p}\right). \end{aligned} \tag{7.9}$$

This iteration is stable. Denoting the iteration function by g, it is straightforward to show that $L_g(A^{1/p}, E) = 0$, and stability is immediate. For practical computation, however, we need another way of rewriting the Newton iteration that does not involve $A^{1/p}$. Taking a cue from the product form DB iteration (6.17) for the square root, we define $M_k = X_k^{-p}A$. Then it is trivial to obtain, using the mutual commutativity of A, X_k, and M_k,

Coupled Newton iteration (matrix pth root):

$$\begin{aligned} X_{k+1} &= X_k\left(\frac{(p-1)I + M_k}{p}\right), & X_0 &= I, \\ M_{k+1} &= \left(\frac{(p-1)I + M_k}{p}\right)^{-p} M_k, & M_0 &= A. \end{aligned} \tag{7.10}$$

When $X_k \to A^{1/p}$, $M_k \to I$. The cost of (7.10) is $(\theta \log_2 p + 14/3)n^3$ flops per iteration, where $\theta \in [1,2]$, assuming the pth power in (7.10) is formed by binary powering (Algorithm 4.1). This will be less than the $O(pn^3)$ cost of the Schur method if p is large and not too many iterations are required.

To test for stability we write the iteration function as

$$G(X,M) = \begin{bmatrix} X\left(\dfrac{(p-1)I + M}{p}\right) \\ \left(\dfrac{(p-1)I + M}{p}\right)^{-p} M \end{bmatrix}.$$

It is easy to show that

$$L_g(X, I; E, F) = \begin{bmatrix} I & \frac{X}{p} \\ 0 & 0 \end{bmatrix}\begin{bmatrix} E \\ F \end{bmatrix}. \tag{7.11}$$

Hence $L_g(A^{1/p}, I)$ is idempotent and the iteration is stable.

7.4. Inverse Newton Method

Another Newton iteration, this time for the inverse pth root, is obtained by applying Newton's method to the system $X^{-p} - A = 0$:

Inverse Newton iteration (inverse matrix pth root):

$$X_{k+1} = \frac{1}{p}\big[(p+1)X_k - X_k^{p+1}A\big], \qquad X_0A = AX_0. \tag{7.12}$$

(Of course, (7.12) is (7.5) with $p \leftarrow -p$.) For $p = 1$ this is the well-known Newton–Schulz iteration for the matrix inverse [512, 1933] and the residuals $R_k = I - X_k^pA$ satisfy $R_{k+1} = R_k^2$ (see Problem 7.8). The next result shows that a p-term recursion for the residuals holds in general.

Theorem 7.10 (Bini, Higham, and Meini). *The residuals $R_k = I - X_k^pA$ from (7.12) satisfy*

$$R_{k+1} = \sum_{i=2}^{p+1} a_i R_k^i, \tag{7.13}$$

where the a_i are all positive and $\sum_{i=2}^{p+1} a_i = 1$. Hence if $0 < \|R_0\| < 1$ for some consistent matrix norm then $\|R_k\|$ decreases monotonically to 0 as $k \to \infty$, with $\|R_{k+1}\| < \|R_k\|^2$.

Proof. We have $X_{k+1} = p^{-1}X_k(pI + R_k)$. Since X_k commutes with A, and hence with R_k, we obtain

$$\begin{aligned} R_{k+1} &= I - \frac{1}{p^p}(I - R_k)(pI + R_k)^p \\ &= I - \frac{1}{p^p}\Big[p^pI + \sum_{i=1}^{p} b_iR_k^i - R_k^{p+1}\Big] \\ &= -\frac{1}{p^p}\left[\sum_{i=1}^{p} b_iR_k^i - R_k^{p+1}\right], \end{aligned} \tag{7.14}$$

where

$$\begin{aligned} b_i &= \binom{p}{i}p^{p-i} - \binom{p}{i-1}p^{p-i+1} \\ &= p^{p-i}\left[\binom{p}{i} - \binom{p}{i-1}p\right] \\ &= p^{p-i}\left[\frac{p!}{i!(p-i)!} - \frac{p!\,p}{(i-1)!(p-i+1)!}\right] \\ &= p^{p-i}\frac{p!}{(i-1)!(p-i)!}\left[\frac{1}{i} - \frac{p}{(p-i+1)}\right]. \end{aligned}$$

It is easy to see that $b_1 = 0$ and $b_i < 0$ for $i \ge 2$. Hence (7.13) holds, with $a_i > 0$ for all i. By setting $R_k \equiv I$ in (7.13) and (7.14) it is easy to see that $\sum_{i=2}^{p+1} a_i = 1$.

If $0 < \|R_k\| < 1$, then taking norms in (7.13) yields

$$\|R_{k+1}\| \le \sum_{i=2}^{p+1} |a_i| \, \|R_k\|^i < \|R_k\|^2 \sum_{i=2}^{p+1} |a_i| = \|R_k\|^2 < \|R_k\|.$$

Since $0 < \|R_0\| < 1$, by induction the $\|R_k\|$ form a monotonically decreasing sequence that converges to zero. □

If X_0 does not commute with A then little can be said about the behaviour of the residuals for $p \ge 2$; see Problems 7.8 and 7.9.

Theorem 7.10 does not immediately imply the convergence of X_k (except in the scalar case). We can conclude that the sequence of pth powers, $\{X_k^p\}$, is bounded, but the boundedness of $\{X_k\}$ itself does not follow when $n > 1$. If the sequence $\{X_k\}$ is indeed bounded then by writing (7.12) as $X_{k+1} - X_k = \frac{1}{p} X_k (I - X_k^p A) = \frac{1}{p} X_k R_k$ we see that $\{X_k\}$ is a Cauchy sequence and thus converges (quadratically) to a matrix X_*. This limit satisfies $I - X_*^p A = 0$ and so is *some* inverse pth root of A, but not necessarily the inverse principal pth root.

The next result proves convergence when all the eigenvalues of A lie on the real interval $(0, p+1)$ and $X_0 = I$.

Theorem 7.11 (Bini, Higham, and Meini). *Suppose that all the eigenvalues of A are real and positive. Then iteration (7.12) with $X_0 = I$ converges to $A^{-1/p}$ if $\rho(A) < p+1$. If $\rho(A) = p+1$ the iteration does not converge to the inverse of any pth root of A.*

Proof. By the same reasoning as in the proof of Corollary 7.8 it suffices to analyze the convergence of the iteration on the eigenvalues of A. We therefore consider the scalar iteration

$$x_{k+1} = \frac{1}{p}\left[(1+p)x_k - x_k^{p+1} a\right], \qquad x_0 = 1, \tag{7.15}$$

with $a > 0$. Let $y_k = a^{1/p} x_k$. Then

$$y_{k+1} = \frac{1}{p}\left[(1+p)y_k - y_k^{p+1}\right] =: f(y_k), \qquad y_0 = a^{1/p}, \tag{7.16}$$

and we need to prove that $y_k \to 1$ if $y_0 = a^{1/p} < (p+1)^{1/p}$. We consider two cases. If $y_k \in (0,1)$ then

$$y_{k+1} = y_k \left[1 + \frac{1 - y_k^p}{p}\right] > y_k.$$

Moreover, since

$$f(0) = 0, \quad f(1) = 1, \quad f'(y) = \tfrac{p+1}{p}(1 - y^p) > 0 \text{ for } y \in [0,1),$$

it follows that $f(y) < 1$ for $y \in [0,1)$. Hence $y_k < y_{k+1} < 1$ and so the y_k form a monotonically increasing sequence tending to a limit in $(0,1]$. But the only fixed points of the iteration are the roots of unity, so the limit must be 1. Now suppose $y_0 \in (1, (p+1)^{1/p})$. We have $f(1) = 1$ and $f((p+1)^{1/p}) = 0$, and $f'(y) < 0$ for $y > 1$. It follows that f maps $(1, (p+1)^{1/p})$ into $(0,1)$ and so after one iteration $y_1 \in (0,1)$ and the first case applies. The last part of the result follows from $f((p+1)^{1/p}) = 0$ and the fact that 0 is a fixed point of the iteration. □

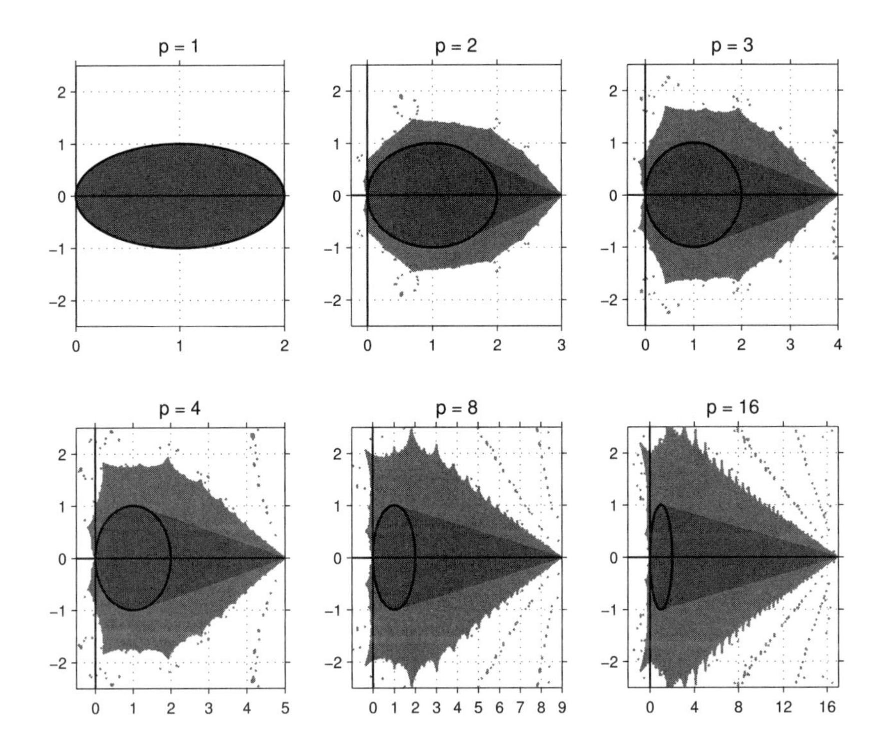

Figure 7.2. *Regions of $a \in \mathbb{C}$ for which the inverse Newton iteration (7.15) converges to $a^{-1/p}$. The dark shaded region is $E(1,p)$ in (7.17). The union of that region with the lighter shaded points is the experimentally determined region of convergence. The solid line marks the disk of radius 1, centre 1. Note the differing x-axis limits.*

The following result builds on the previous one to establish convergence when the spectrum of A lies within a much larger wedge-shaped convex region in the complex plane, the region depending on a parameter c. We denote by conv the convex hull.

Theorem 7.12 (Guo and Higham). *Let $A \in \mathbb{C}^{n\times n}$ have no eigenvalues on $\mathbb{R}^-$. For all $p \ge 1$, the iterates X_k from (7.12) with $X_0 = \frac{1}{c}I$ and $c \in \mathbb{R}^+$ converge quadratically to $A^{-1/p}$ if all the eigenvalues of A are in the set*

$$E(c,p) = \operatorname{conv}\{\, \{ z : |z - c^p| \le c^p \}, (p+1)c^p \,\} \setminus \{ 0, (p+1)c^p \}. \qquad \square \qquad (7.17)$$

The actual convergence region, determined experimentally, is shown together with $E(c,p)$ in Figure 7.2 for $c = 1$ and several values of p.

If A happens to have just one eigenvalue and we know that eigenvalue then finite convergence is obtained for an appropriate choice of c.

Lemma 7.13 (Guo and Higham). *Suppose that $A \in \mathbb{C}^{n\times n}$ has a positive eigenvalue λ of multiplicity n and that the largest Jordan block is of size q. Then for the iteration (7.12) with $X_0 = \lambda^{-1/p}I$ we have $X_m = A^{-1/p}$ for the first m such that $2^m \ge q$.*

Proof. Let A have the Jordan form $A = ZJZ^{-1}$. Then $R_0 = I - X_0^p A = Z(I - \frac{1}{\lambda}J)Z^{-1}$. Thus $R_0^q = 0$. By Theorem 7.10, $R_m = (R_0)^{2^m} h(R_0)$, where $h(R_0)$

is a polynomial in R_0. Thus $R_m = 0$ if $2^m \geq q$. That convergence is to $A^{1/p}$ follows from Theorem 7.12. □

We can now build a practical algorithm, beginning by considering stability.

A stability analysis of iteration (7.12) leads to a stability condition similar to, but even more restrictive than, (7.8). However, by introducing the matrix $M_k = X_k^p A$ the iteration can be rewritten in a coupled form analogous to (7.10), in which we now take X_0 as in Theorem 7.12:

Coupled inverse Newton iteration (inverse matrix pth root):

$$\begin{aligned} X_{k+1} &= X_k\left(\frac{(p+1)I - M_k}{p}\right), & X_0 &= \frac{1}{c}I, \\ M_{k+1} &= \left(\frac{(p+1)I - M_k}{p}\right)^p M_k, & M_0 &= \frac{1}{c^p}A. \end{aligned} \tag{7.18}$$

It is easy to show that this variant is stable (see Problem 7.14).

We begin by taking the square root twice by any iterative method. This preprocessing step brings the spectrum into the sector $\arg z \in (-\pi/4, \pi/4)$. The nearest point to the origin that is both within this sector and on the boundary of $E(c,p)$ is at a distance $c^p\sqrt{2}$. Hence the inverse Newton iteration can be applied to $B = A^{1/4}$ with $c \geq (\rho(B)/\sqrt{2})^{1/p}$. We obtain an algorithm in the same spirit as Algorithm 7.9.

Algorithm 7.14 (Matrix inverse pth root via inverse Newton iteration). Given $A \in \mathbb{C}^{n\times n}$ having no eigenvalues on $\mathbb{R}^-$ this algorithm computes $X = A^{-1/p}$ using the inverse Newton iteration.

1 $B = A^{1/4}$ (computed as two successive square roots)
2 Use the Newton iteration (7.18) with $c = (\theta/\sqrt{2})^{1/p}$, where θ is any upper bound for $\rho(B)$, to compute

$$X = \begin{cases} B^{-1/s}, & p = 4s\ , \\ (B^{-1/p})^4, & p = 4s+1 \text{ or } p = 4s+3, \\ (B^{-1/(2s+1)})^2, & p = 4s+2. \end{cases}$$

7.5. Schur–Newton Algorithm

The flop counts of Algorithms 7.9 and 7.14 are generally greater than that for the Schur method unless p is large, although the fact that the Newton iterations are rich in matrix multiplication and matrix inversion compensates somewhat on modern computers. However it is possible to combine the Schur and Newton methods to advantage. An initial Schur decomposition reveals the eigenvalues, which can be used to make a good choice of the parameter c in the inverse Newton iteration, and the subsequent computations are with triangular matrices. The algorithm can be arranged so that it computes either $A^{1/p}$ or $A^{-1/p}$, by using either (7.18) or the variant of it that is obtained by setting $Y_k = X_k^{-1}$:

Coupled inverse Newton iteration (matrix pth root):

$$\begin{aligned} Y_{k+1} &= \left(\frac{(p+1)I - M_k}{p}\right)^{-1} Y_k, \qquad Y_0 = cI, \\ M_{k+1} &= \left(\frac{(p+1)I - M_k}{p}\right)^{p} M_k, \qquad M_0 = \frac{1}{c^p}A. \end{aligned} \tag{7.19}$$

Here, $Y_k \to A^{1/p}$ under the same conditions (those in Theorem 7.12) that $X_k \to A^{-1/p}$.

We state the algorithm for real matrices, but an analogous algorithm is obtained for complex matrices by using the complex Schur decomposition.

Algorithm 7.15 (Schur–Newton algorithm for (inverse) matrix pth root). Given $A \in \mathbb{R}^{n\times n}$ with no eigenvalues on $\mathbb{R}^-$ this algorithm computes $X = A^{1/p}$ or $X = A^{-1/p}$, where $p = 2^{k_0}q$ with $k_0 \ge 0$ and q odd, using a Schur decomposition and a Newton iteration.

1 Compute a real Schur decomposition $A = QRQ^T$.
2 if $q = 1$
3 $\quad k_1 = k_0$
4 else
5 $\quad$ Choose $k_1 \ge k_0$ such that $|\lambda_1/\lambda_n|^{1/2^{k_1}} \le 2$,
$\quad$ where the eigenvalues of A are ordered $|\lambda_n| \le \cdots \le |\lambda_1|$.
6 end
7 If the λ_i are not all real and $q \ne 1$, increase k_1 as necessary so that
$\quad \arg(\lambda_i^{1/2^{k_1}}) \in (-\pi/8, \pi/8)$ for all i.
8 Compute $B = R^{1/2^{k_1}}$ by k_1 invocations of Algorithm 6.7.
9 if $q = 1$, goto line 22, end
10 Let $\mu_1 = |\lambda_1|^{1/2^{k_1}}$, $\mu_n = |\lambda_n|^{1/2^{k_1}}$.
11 if the λ_i are all real
12 $\quad$ if $\mu_1 \ne \mu_n$
13 $\quad\quad c = \left(\dfrac{\alpha^{1/q}\mu_1 - \mu_n}{(\alpha^{1/q}-1)(q+1)}\right)^{1/q}$, where $\alpha = \dfrac{\mu_1}{\mu_n}$.
14 $\quad$ else
15 $\quad\quad c = \mu_n^{1/q}$
16 $\quad$ end
17 else
18 $\quad c = \left(\dfrac{\mu_1 + \mu_n}{2}\right)^{1/q}$
19 end
20 Compute $\begin{cases} X = B^{-1/q} \text{ by (7.18)}, & \text{if } A^{-1/p} \text{ required}, \\ X = B^{1/q} \text{ by (7.19)}, & \text{if } A^{1/p} \text{ required}. \end{cases}$
21 $X \leftarrow X^{2^{k_1-k_0}}$ (repeated squaring).
22 $X \leftarrow QXQ^T$

The key idea in the algorithm is to preprocess A so that the Newton iteration converges in a small number of iterations. The algorithm begins by taking as many square roots as it can; these square roots are inexpensive since the matrix is triangular. When the eigenvalues are all real the algorithm uses a choice of c with a certain optimality property. For nonreal eigenvalues, the requirement on the arguments in line 7 ensures that the corresponding choice of c in line 18 leads to fast convergence. For more details see Guo and Higham [233, 2006], where numerical experiments show Algorithm 7.15 to perform in a numerically stable manner.

The cost of the algorithm is about

$$\Big(28 + \frac{2}{3}(k_1 + k_2) - \Big(\frac{1}{3} + \frac{k_2}{2}\Big) k_0 + \frac{k_2}{2}\log_2 p\Big) n^3 \text{ flops},$$

where we assume that k_2 iterations of (7.18) or (7.19) are needed. When $k_0 = 0$, $k_1 = 3$, and $k_2 = 4$, for example, the flop count becomes $(32\frac{2}{3} + 2\log_2 p)n^3$, while the count is always $(28 + \frac{p-1}{3})n^3$ for the Schur method. Algorithm 7.15 is slightly more expensive than the Schur method if p is small or highly composite (assuming that the Schur method is applied over the prime factors of p), but it is much less expensive than the Schur method if p is large and has a small number of prime factors.

7.6. Matrix Sign Method

Theorem 7.4 shows how to recover a pth root of A from an n-dimensional invariant subspace of the block companion matrix

$$C = \begin{bmatrix} 0 & I & & & \\ & 0 & I & & \\ & & \ddots & \ddots & \\ & & & \ddots & I \\ A & & & & 0 \end{bmatrix} \in \mathbb{C}^{pn\times pn}. \tag{7.20}$$

The relation (5.4) shows that for $p = 2$ the sign of C directly reveals $A^{1/p}$, so we might hope that $A^{1/p}$ is recoverable from $\operatorname{sign}(C)$ for all p. This is true when p is even and not a multiple of 4.

Theorem 7.16 (Bini, Higham, and Meini). *If $p = 2q$ where q is odd, then the first block column of the matrix* $\operatorname{sign}(C)$ *is given by*

$$V = \frac{1}{p}\begin{bmatrix} \gamma_0 I \\ \gamma_1 X \\ \gamma_2 X^2 \\ \vdots \\ \gamma_{p-1}X^{p-1} \end{bmatrix},$$

where $X = A^{1/p}$, $\gamma_k = \sum_{j=0}^{p-1}\omega_p^{kj}\theta_j$, $\omega_p = e^{2\pi i/p}$, and $\theta_j = -1$ for $j = \lfloor q/2\rfloor + 1\colon \lfloor q/2\rfloor + q$, $\theta_j = 1$ otherwise. If A is Hermitian positive definite then this result holds also for odd p. □

This result allows us to compute the principal pth root of A from the $(2,1)$ block of $\operatorname{sign}(C)$. For $p = 2$ we have $\theta_0 = 1$, $\theta_1 = -1$, $\gamma_0 = 0$, $\gamma_1 = 2$, and Theorem 7.16 reproduces the first block column of (5.4).

Algorithm 7.17 (Matrix pth root via matrix sign function). Given $A \in \mathbb{C}^{n\times n}$ having no eigenvalues on $\mathbb{R}^-$ this algorithm computes $X = A^{1/p}$ via the matrix sign function.

1 If p is odd
2 $\quad p \leftarrow 2p$, $A \leftarrow A^2$
3 else
4 $\quad$ if p is a multiple of 4
5 $\quad\quad$ while $p/2$ is even, $A \leftarrow A^{1/2}$, $p = p/2$, end
6 $\quad$ end
7 end
8 $S = \mathrm{sign}(C)$, where C is given in (7.20).
9 $X = \frac{p}{2\sigma} S(n+1\colon 2n, 1\colon n)$ where $\sigma = 1 + 2\sum_{j=1}^{\lfloor q/2 \rfloor} \cos(2\pi j/p)$ and $q = p/2$.

Assuming the matrix sign function is computed using one of the Newton or Padé methods of Chapter 5, the cost of Algorithm 7.17 is $O(n^3p^3)$ flops, since C is $np \times np$. The algorithm therefore appears to be rather expensive in computation and storage. However, both computation and storage can be saved by exploiting the fact that the sign iterates belong to the class of A-circulant matrices, which have the form

$$\begin{bmatrix} W_0 & W_1 & \dots & W_{p-1} \\ AW_{p-1} & W_0 & \ddots & \vdots \\ \vdots & \ddots & \ddots & W_1 \\ AW_1 & \dots & AW_{p-1} & W_0 \end{bmatrix}.$$

In particular, an A-circulant matrix can be inverted in $O(n^3 p \log p + n^2 p \log^2 p)$ flops. See Bini, Higham, and Meini [69, 2005, Sec. 7] for details.

7.7. Notes and References

In several papers Sylvester stated that the number of pth roots of an $n \times n$ matrix is p^n, sometimes mentioning exceptional cases (see [555, 1882], [557, 1883], and the quote at the end of the chapter). His count of p^n is correct if the matrix has distinct eigenvalues (see Theorem 7.1).

The representation (7.1) can be deduced from a standard identity in complex analysis, as noted by Bhatia [64, 1997, Ex. V.1.10] and Mehta [420, 1989, Sec. 5.5.5]. It can also be found in a slightly different form and for A^α, for $0 < \mathrm{Re}\,\alpha < 1$, in the functional analysis literature: see Balakrishnan [41, 1960] and Haase [236, 2006, Prop. 3.1.12].

Theorem 7.4 is from Benner, Byers, Mehrmann, and Xu [54, 2000].

Björck and Hammarling [73, 1983, Sec. 5] show how to compute a cube root using the (complex) Schur decomposition. Section 7.2 is based on M. I. Smith [530, 2003], who derives Algorithm 7.6 and proves (7.3).

The Newton iteration (7.5) is used to compute pth roots of scalars in the GNU multiple precision arithmetic library [218].

The difficulty of analyzing the convergence of the Newton iteration (7.5) in the scalar case for $p \geq 3$ was noted by Cayley [101, 1879]; see the quote at the end of this chapter.

Prior to Iannazzo's analysis the only available convergence result for the Newton iteration was for symmetric positive definite matrices [299, 1979].

The condition for stability of the Newton iteration (7.5) that the eigenvalues in (7.8) are of modulus less than 1 is obtained by M. I. Smith [530, 2003].

Algorithm 7.9 is due to Iannazzo [306, 2006], who also obtains, and proves stable, iterations (7.9), (7.10), and (7.18).

Section 7.4 is based on Bini, Higham, and Meini [69, 2005] and Guo and Higham [233, 2006], from where the results and algorithms therein are taken.

For $p = 2$, the inverse Newton iteration (7.12) is well known in the scalar case as an iteration for the inverse square root. It is employed in floating point hardware to compute the square root $a^{1/2}$ via $a^{-1/2} \times a$, since the whole computation can be done using only multiplications [117, 1999], [334, 1997]. The inverse Newton iteration is also used to compute $a^{1/p}$ in arbitrarily high precision in the MPFUN and ARPREC packages [36, 1993], [37, 2002].

The inverse Newton iteration with $p = 2$ is studied for symmetric positive definite matrices by Philippe, who proves that the residuals $R_k = I - X_k^2 A$ satisfy $R_{k+1} = \frac{3}{4}R_k^2 + \frac{1}{4}R_k^3$ [471, 1987, Prop. 2.5]. For arbitrary p, R. A. Smith [531, 1968, Thm. 5] proves quadratic convergence of (7.12) to some pth root when $\rho(R_0) < 1$, as does Lakić [368, 1998] under the assumption that A is diagonalizable, but neither determines to which root convergence is obtained.

The coupled iteration (7.18) is a special case of a family of iterations of Lakić [368, 1998], who proves stability of the whole family.

Section 7.5 is based on Guo and Higham [233, 2006]. Section 7.6 is based on Bini, Higham, and Meini [69, 2005].

Some other methods for computing the pth root are described by Bini, Higham, and Meini [69, 2005], all subject to the same restriction on p as in Theorem 7.16. One comprises numerical evaluation at the Fourier points of a transformed version of the integral (7.1), in which the integration is around the unit circle. Another computes a Wiener–Hopf factorization of the matrix Laurent polynomial $F(z) = z^{-p/2}((1+z)^p A - (1-z)^p I)$, from which $A^{1/p}$ is readily obtained. Another method expresses $A^{1/p}$ in terms of the central coefficients $H_0, \dots, H_{p/2-1}$ of the matrix Laurent series $H(z) = H_0 + \sum_{i=1}^{+\infty}(z^i + z^{-i})H_i$, where $H(z)F(z) = I$. Further work is needed to understand the behaviour of these methods in finite precision arithmetic; see Problem 7.18.

The principal pth root of a nonsingular M-matrix (see Section 6.8.3) is an M-matrix, but for $p > 3$ it need not be the only pth root that is an M-matrix; see Fiedler and Schneider [185, 1983].

Sylvester [554, 1881], [555, 1882] was interested in finding pth roots of the 2×2 identity because of a connection with a problem solved earlier by Babbage in his "calculus of functions". The problem is to find a Möbius transformation $\phi(x) = (ax + b)/(cx + d)$ such that $\phi^{(i)}(x) = x$ for a given i. This is easily seen to be equivalent to finding a matrix $A = \left[\begin{smallmatrix} a & b \\ c & d \end{smallmatrix}\right]$ for which $A^i = I_2$ (see, e.g., Mumford, Series, and Wright [442, 2002]).

A few results are available on integer pth roots of the identity matrix. Turnbull [577, 1927], [578, 1929, p. 332] observes that the antitriangular matrix A_n with $a_{ij} = (-1)^{n-j}\binom{n-j}{i-1}$ for $i + j \le n + 1$ is a cube root of I_n. For example, $A_4^3 = I_4$, where

$$A_4 = \begin{bmatrix} -1 & 1 & -1 & 1 \\ -3 & 2 & -1 & 0 \\ -3 & 1 & 0 & 0 \\ -1 & 0 & 0 & 0 \end{bmatrix}.$$

This matrix is a rotated variant of the Cholesky factor of the Pascal matrix (cf. the involutory matrix in (6.59)). Up to signs, A_n is the MATLAB matrix `pascal(n,2)`. Vaidyanathaswamy [582, 1928], [583, 1928] derives conditions on p for the existence of integer pth roots of I_n; in particular, he shows that for a given n there are only finitely many possible p. Problem 7.6 gives an integer $(n+1)$st root of I_n.

Finally, we note that Janovská and Opfer [317, 2007, Sec. 8] show how computing a pth root of a quaternion can be reduced to computing a pth root of a matrix, though the special structure of the matrix needs to be exploited to be competitive with methods working in the algebra of quaternions.

Problems

7.1. Does an $m \times m$ Jordan block with eigenvalue zero have a pth root for any p?

7.2. How should A^α be defined for real $\alpha \in [0, \infty)$?

7.3. Show that if $X^p = I_n$ then the eigenvalues of X are pth roots of unity and X is diagonalizable.

7.4. Show that the Frobenius norm (relative) condition number for the matrix pth root X of $A \in \mathbb{C}^{n\times n}$ is given by

$$\operatorname{cond}(X) = \frac{\left\| \left(\sum_{j=1}^{p} (X^{p-j})^T \otimes X^{j-1} \right)^{-1} \right\|_2 \|A\|_F}{\|X\|_F} \tag{7.21}$$

and that this condition number is finite if and only if X is a primary pth root of A.

7.5. Suppose that in the Newton iteration for a matrix pth root defined by (7.4) and $X_{k+1} = X_k + E_k$, X_0 commutes with A and all the iterates are well-defined. Show that, for all k, X_k commutes with A and $X_{k+1} = \frac{1}{p}((p-1)X_k + X_k^{1-p}A)$.

7.6. (Bambaii and Chowla [42, 1946]) Let

$$A_n = \begin{bmatrix} -1 & -1 & \dots & \dots & -1 \\ 1 & 0 & \dots & \dots & 0 \\ 0 & 1 & \ddots & & \vdots \\ \vdots & & \ddots & \ddots & \vdots \\ 0 & \dots & \dots & 1 & 0 \end{bmatrix} \in \mathbb{R}^{n\times n}.$$

Prove, or verify computationally, that $A_n^{n+1} = I_n$.

7.7. (Rice [488, 1982]) This problem develops an elementary proof of the existence of the Hermitian positive definite pth root of a Hermitian positive definite matrix $A \in \mathbb{C}^{n\times n}$. Let $\ge$ denote the Hermitian positive semidefinite ordering defined in Section B.12. First, show that if $0 \le Z \le Y \le I$ and $YZ = ZY$ then $(Y^p - Z^p)/p \le Y - Z$. Then show that we can assume $A \le I$ without loss of generality. Next, consider the iteration

$$X_{k+1} = X_k + \frac{1}{p}(A - X_k^p), \qquad X_0 = 0.$$

Show that $0 \le X_k \le X_{k+1} \le I$ for all k, and deduce that X_k converges to a pth root of A.

7.8. Show that for the Newton–Schulz iteration for the inverse of $A \in \mathbb{C}^{n\times n}$,

$$X_{k+1} = 2X_k - X_k A X_k, \qquad (7.22)$$

the residuals $R_k = I - X_k A$ and $R_k = I - AX_k$ both satisfy $R_{k+1} = R_k^2$ *for any* X_0. (Note that (7.22) is a more symmetric variant of (7.12) with $p = 1$. For (7.12) this residual recurrence is assured only for X_0 that commute with A.)

7.9. Investigate the behaviour of the inverse Newton iteration (7.12) for $p = 2$ with

$$A = \begin{bmatrix} 1 & 0 \\ 0 & \mu^{-2} \end{bmatrix}, \qquad X_0 = \begin{bmatrix} 1 & \theta \\ 0 & \mu \end{bmatrix}, \qquad \theta \neq 0.$$

(Cf. Problem 6.8.)

7.10. For the inverse Newton iteration (7.12) how does the asymptotic error constant for the residual vary with p? (Hint: evaluate a_2 in Theorem 7.10.)

7.11. Given an approximation $X \approx A^{-1/p}$, how can an approximation $Y \approx A^{1/p}$ be computed using only matrix multiplication?

7.12. (Smith [530, 2003]) Show that the eigenvalues of the Fréchet derivative $L_g(A^{1/p})$ of $g(X) = p^{-1}\big[(p-1)X + X^{1-p}A\big]$ (the Newton iteration function in (7.5)) are given by (7.8). Hint: use the expansion

$$(A+E)^{1-p} = A^{1-p} - \sum_{r=1}^{p-1} A^{r-p} E A^{-r} + O(\|E\|^2).$$

7.13. Assume that the Newton iteration (7.5) converges to a pth root B of A. By making use of the factorization

$$(p-1)x^p - px^{p-1}b + b^p = (x-b)^2 \sum_{i=0}^{p-2} (i+1)x^i b^{p-2-i},$$

obtain a constant c so that $\|X_{k+1} - B\| \le c\|X_k - B\|^2$ for all sufficiently large k.

7.14. Prove the stability of the iteration (7.18).

7.15. (Guo and Higham [233, 2006]) Let $A \in \mathbb{R}^{n\times n}$ be a stochastic matrix. Show that if A is strictly diagonally dominant (by rows—see (B.2)) then the inverse Newton iteration (7.12) with $X_0 = I$ converges to $A^{-1/p}$. The need for computing roots of stochastic matrices arises in Markov model applications; see Section 2.3. Transition matrices arising in the credit risk literature are typically strictly diagonally dominant [315, 2001], and such matrices are known to have at most one generator [127, 1972].

7.16. (Guo and Higham [233, 2006]) Let $\widetilde{X} = X + E$, with $\|E\| \le \epsilon\|X\|$, be an approximate pth root of $A \in \mathbb{C}^{n\times n}$. Obtain a bound for $\|A - \widetilde{X}^p\|$ that is sharp to first order in $\|E\|$ and hence explain why

$$\rho_A(\widetilde{X}) = \frac{\|A - \widetilde{X}^p\|}{\|\widetilde{X}\| \big\| \sum_{i=0}^{p-1} \big(\widetilde{X}^{p-1-i}\big)^T \otimes \widetilde{X}^i \big\|}$$

is a more appropriate definition of relative residual (e.g., for testing purposes) than $\|A - \widetilde{X}^p\| / \|\widetilde{X}^p\|$.

7.17. (RESEARCH PROBLEM) Determine conditions under which a stochastic matrix has a pth root that is also stochastic. How can it be computed? When is such a pth root unique? For some results on square roots of stochastic matrices and doubly stochastic matrices (those for which the matrix and its transpose are both stochastic) see Achilles and Sinkhorn [2, 1995], Iwanik and Shiflett [316, 1986], Marcus and Minc [406, 1962], and Minc [429, 1988, Sec. 5.4].

7.18. (RESEARCH PROBLEM) Investigate, and where necessary improve, the behaviour in finite precision arithmetic of Algorithm 7.17 and the other methods in Bini, Higham, and Meini [69, 2005].

7.19. (RESEARCH PROBLEM) Extend the analysis in Section 6.1 from square roots to pth roots.

It is nevertheless possible to form the powers
(positive or negative, integral or fractional) of a matrix,
and thence to arrive at the notion of a rational and integral function,
or generally of any algebraical function, of a matrix.

— ARTHUR CAYLEY, *A Memoir on the Theory of Matrices* (1858)

The solution is easy and elegant in the case of a quadric equation,
but the next succeeding case of the cubic equation
appears to present considerable difficulty.

— ARTHUR CAYLEY, *The Newton–Fourier Imaginary Problem* (1879)

In general, for a matrix of the order ω, *the number of* m^{th} *roots is* m^{ω}
and each of them is perfectly determinate.
But when the matrix is a unit-matrix or a zero-matrix ...
there are distinct genera and species of such roots,
and every species contains its own appropriate number of arbitrary constants.

— J. J. SYLVESTER, *Note on the Theory of Simultaneous Linear Differential or Difference Equations with Constant Coefficients* (1881)

Chapter 8
The Polar Decomposition

The polar decomposition is the generalization to matrices of the familiar polar representation $z = re^{i\theta}$ of a complex number. It is intimately related to the singular value decomposition (SVD), as our proof of the decomposition reveals.

Theorem 8.1 (Polar decomposition). *Let $A \in \mathbb{C}^{m\times n}$ with $m \geq n$. There exists a matrix $U \in \mathbb{C}^{m\times n}$ with orthonormal columns and a unique Hermitian positive semidefinite matrix $H \in \mathbb{C}^{n\times n}$ such that $A = UH$. The matrix H is given by $(A^*A)^{1/2}$. All possible U are given by*

$$U = P\begin{bmatrix} I_r & 0 \\ 0 & W \end{bmatrix} Q^*, \tag{8.1}$$

where $A = P\left[\begin{smallmatrix} \Sigma_r & 0 \\ 0 & 0_{m-r,n-r} \end{smallmatrix}\right]Q^*$ is an SVD, $r = \operatorname{rank}(A)$, and $W \in \mathbb{C}^{(m-r)\times(n-r)}$ is arbitrary subject to having orthonormal columns. If $\operatorname{rank}(A) = n$ then H is positive definite and U is uniquely determined.

Proof. Note first that one particular polar decomposition can be written down immediately in terms of the SVD:

$$A = P\begin{bmatrix} I_r & 0 \\ 0 & I_{m-r,n-r} \end{bmatrix} Q^* \cdot Q \begin{bmatrix} \Sigma_r & 0 \\ 0 & 0_{n-r} \end{bmatrix} Q^* \equiv UH. \tag{8.2}$$

We have $A^*A = HU^*UH = H^2$ and since H is Hermitian positive semidefinite it follows that $H = (A^*A)^{1/2}$, which uniquely defines H. If $r = n$ then H is clearly nonsingular and hence $U = AH^{-1}$ is unique.

To determine all possible U, write $A = UH$ as

$$P\begin{bmatrix} \Sigma_r & 0 \\ 0 & 0 \end{bmatrix} Q^* = UQ\begin{bmatrix} \Sigma_r & 0 \\ 0 & 0 \end{bmatrix} Q^*.$$

Premultiplying by P^* and postmultiplying by $Q\left[\begin{smallmatrix} \Sigma_r & 0 \\ 0 & I_{n-r} \end{smallmatrix}\right]^{-1}$ gives

$$\begin{bmatrix} I_r & 0 \\ 0 & 0 \end{bmatrix} = P^*UQ\begin{bmatrix} I_r & 0 \\ 0 & 0 \end{bmatrix}.$$

It follows that P^*UQ has the form $\left[\begin{smallmatrix} I_r & V \\ 0 & W \end{smallmatrix}\right]$ for some V and W. The requirement that U has orthonormal columns forces V to be zero and W to have orthonormal columns.
□

With a slight abuse of nomenclature (since a unitary matrix must be square), we will refer to U as the unitary polar factor.

When $m \le n$ we can apply Theorem 8.1 to A^* to obtain another version of the polar decomposition in which the Hermitian factor appears first: $A = \widetilde{H}\widetilde{U}$, with $\widetilde{H} \in \mathbb{C}^{m\times m}$, $\widetilde{U} \in \mathbb{C}^{m\times n}$. In fact, by constructing this factorization in an analogous way to (8.2) it is clear that $\widetilde{H} = (AA^*)^{1/2}$ and that we can take $\widetilde{U} = U$. (Equivalently, the unitary polar factor of A^* can be taken as the conjugate transpose of that of A.) When A is square and nonsingular U and $\widetilde{U}$ are unique and necessarily equal; this equality can be expressed as $A(A^*A)^{-1/2} = (AA^*)^{-1/2}A$, which is a particular case of Corollary 1.34. In fact, these two polar decomposition variants are identical (modulo any nonuniqueness of U) precisely when A is normal ($A^*A = AA^*$), as the next result shows.

Theorem 8.2. *Let $A \in \mathbb{C}^{n\times n}$. In the polar decomposition $A = UH$ the factors U and H commute if and only if A is normal.*

Proof. If $UH = HU$ then, since U is unitary, $A^*A = HU^*UH = H^2 = HUU^*H = UHHU^* = AA^*$, so A is normal.

If $A^*A = AA^*$ then $H^2 = UH^2U^*$. Taking the principal square root of both sides gives $H = UHU^*$, or $HU = UH$, as required. □

Although it is rarely discussed in the literature, the polar decomposition $A = UH$ exists also for $m < n$, with U now having orthonormal rows. However, neither U nor H is unique when $m < n$; see Problem 8.7.

The nonuniqueness of the polar decomposition for rank-deficient matrices with $m \ge n$, and for all matrices with $m < n$, can be eradicated by relaxing the orthonormality requirement on U while at the same time insisting on equality of the ranges of U^* and H. The next result provides an elegant way to obtain a unique "canonical polar decomposition" for all m and n. Although this decomposition is largely of theoretical interest, we will see in the next section that its U factor possesses a best approximation property, as does the unitary polar factor. A matrix $U \in \mathbb{C}^{m\times n}$ is a *partial isometry* if $\|Ux\|_2 = \|x\|_2$ for all $x \in \mathrm{range}(U^*)$. A more concrete characterization of a partial isometry is as a matrix whose singular values are all 0 or 1. For more details on partial isometries see Section B.6.3. We denote by H^+ the pseudoinverse of H. As the proof of the following theorem shows, the pseudoinverse is a very useful tool for analyzing the canonical polar decomposition.

Theorem 8.3 (canonical polar decomposition). *Let $A \in \mathbb{C}^{m\times n}$. There is a unique decomposition $A = UH$ with $U \in \mathbb{C}^{m\times n}$ a partial isometry, $H \in \mathbb{C}^{n\times n}$ Hermitian positive semidefinite, and* $\mathrm{range}(U^*) = \mathrm{range}(H)$. *The last condition is equivalent*[8] *to $U^*U = HH^+$. Moreover, $A^+ = H^+U^+$. The factors U and H are given by $H = (A^*A)^{1/2}$ and $U = AH^+$. Furthermore, if $A = P\left[\begin{smallmatrix}\Sigma_r & 0\\ 0 & 0_{m-r,n-r}\end{smallmatrix}\right]Q^*$ is an SVD then*

$$U = P\begin{bmatrix} I_r & 0\\ 0 & 0_{m-r,n-r}\end{bmatrix}Q^*, \qquad H = Q\begin{bmatrix}\Sigma_r & 0\\ 0 & 0_{n-r}\end{bmatrix}Q^*. \tag{8.3}$$

Proof. Note that $U^* = U^+$ for a partial isometry (see Lemma B.2). Hence the condition $U^*U = HH^+$ is equivalent to $U^+U = HH^+$, which is equivalent to $\mathrm{range}(U^*) = \mathrm{range}(H)$, since BB^+ and B^+B are the orthogonal projectors onto $\mathrm{range}(B)$ and $\mathrm{range}(B^*)$, respectively (see Problem B.10).

The formula (8.3) shows that a decomposition of the claimed form exists. However, for completeness, we prove existence without using the SVD. Note first that by

[8] In fact, $U^*U = HH^+$ implies that U is a partial isometry.

Problem B.2, H^+ is Hermitian and commutes with H. Hence, by the first and second Moore–Penrose conditions (B.3) (i) and (B.3) (ii), any product comprising p factors H and q factors H^+, in any order, equals H if $p = q+1$, H^+ if $q = p+1$, and HH^+ if $p = q$. Define $H = (A^*A)^{1/2}$ and $U = AH^+$. We first need to show that U is a partial isometry ($U^* = U^+$) and $U^*U = HH^+$. To prove the latter equality, we have

$$U^*U = H^+A^*AH^+ = H^+H^2H^+ = HH^+.$$

It remains to show that $U^* = U^+$, which involves checking the Moore–Penrose conditions. We have

$$UU^*U = AH^+ \cdot H^+A^* \cdot AH^+ = A(H^+)^2H^2H^+ = AH^+ = U,$$
$$U^*UU^* = H^+A^* \cdot AH^+ \cdot H^+A^* = H^+H^2(H^+)^2A^* = H^+A^* = U^*,$$

and UU^* and U^*U are trivially Hermitian, as required.

Now we need to verify that $A = UH$. By Problem 6.4 we have $\mathrm{range}(H) = \mathrm{range}((A^*A)^{1/2}) = \mathrm{range}(A^*A) = \mathrm{range}(A^*)$, using (B.4) for the last equality. Hence $HH^+ = A^+A$, by Problem B.10. Therefore $UH = AH^+ \cdot H = AHH^+ = AA^+A = A$.

We now show that the decomposition is unique. If $A = UH$ and $U^*U = HH^+$ then $A^*A = HU^*UH = HHH^+H = H^2$, so $H = (A^*A)^{1/2}$ is uniquely determined. Then $AH^+ = UHH^+ = UU^*U = UU^+U = U$, which determines U uniquely.

Finally, that $A^+ = H^+U^+ =: X$ is shown by verifying the Moore–Penrose conditions, of which we just illustrate the first: $AXA = UH \cdot H^+U^+ \cdot UH = UHH^+ \cdot HH^+ \cdot H = UH = A$. □

We make several comments on this theorem.

- Other ways to express the condition $\mathrm{range}(U^*) = \mathrm{range}(H)$ are as $\mathrm{null}(U) = \mathrm{null}(H)$ and as $\mathrm{range}(U) = \mathrm{range}(A)$ (or equivalently $UU^+ = AA^+$); see Problem 8.8.

- Let $m \geq n$. Then the factor U in Theorem 8.3 is the matrix U in (8.1) with $W = 0$; U in (8.3) has orthonormal columns precisely when A has full rank; and if A has full rank the polar decomposition and canonical polar decomposition are the same.

- The decompositions with $m = 1$ and $n = 1$ are

$$\begin{aligned} n = 1: \quad & a = \|a\|_2^{-1}a \cdot \|a\|_2, \\ m = 1: \quad & a^* = \|a\|_2^{-1}a^* \cdot \|a\|_2^{-1}aa^*. \end{aligned} \tag{8.4}$$

- If $A = UH$ and $A^* = VG$ are the decompositions in Theorem 8.3 then $V = U^*$, that is, A and A^* have U factors that are conjugate transposes of each other. This can be seen from

$$V = A^*\big[(AA^*)^{1/2}\big]^+ = \big[(A^*A)^{1/2}\big]^+A^* = \big(A\big[(A^*A)^{1/2}\big]^+\big)^* = U^*, \tag{8.5}$$

where the second equality follows from Corollary 1.34 with f the function $f(x) = \big(x^{1/2}\big)^D$. Here, D denotes the Drazin inverse (see Problem 1.52), which is identical to the pseudoinverse for Hermitian matrices, to which it is being applied here.

In most of the rest of this chapter we will restrict to the standard polar decomposition with $m \geq n$.

Neither polar factor is a function of A (though this can be arranged by modifying the definition of matrix function, as explained in Problem 1.53; see also Problem 8.9). The reason for including the polar decomposition in this book is that it has many intimate relations with the matrix sign function and the matrix square root, and so it is naturally treated alongside them.

A fundamental connection between the matrix sign function and the polar decomposition is obtained from Theorem 5.2 with $B = A^*$ when $A \in \mathbb{C}^{n\times n}$ is nonsingular:

$$\operatorname{sign}\left(\begin{bmatrix} 0 & A \\ A^* & 0 \end{bmatrix}\right) = \begin{bmatrix} 0 & U \\ U^* & 0 \end{bmatrix}. \tag{8.6}$$

This relation can be used to obtain an integral formula for U from the integral formula (5.3) for the sign function:

$$U = \frac{2}{\pi} A \int_0^\infty (t^2 I + A^*A)^{-1}\, dt. \tag{8.7}$$

This expression in fact holds for any $A \in \mathbb{C}^{m\times n}$ with $\operatorname{rank}(A) = n$.

The Hermitian polar factor $H = (A^*A)^{1/2}$ is also known as the matrix absolute value and written $|A|$ (which is not to be confused with the determinant of A, always written $\det(A)$ in this book). The matrix absolute value has been much studied in the linear algebra and functional analysis literatures.

The outline of this chapter is as follows. We start with some important approximation properties of the polar factors U and H. Some useful perturbation results are presented and condition numbers identified. Iterative methods of computation are treated, beginning with Newton's method and the Newton–Schulz method. We explain how any iteration for the matrix sign function leads to a corresponding iteration for the unitary polar factor and use this connection to derive a Padé family of iterations. Scaling to accelerate convergence, stopping tests for the iterations, and numerical stability are examined. The chapter finishes with an algorithm based on the Newton iteration along with some numerical illustrations.

Two general points need to be made concerning computation. First, if computational cost is not a concern then there is no better way to compute the polar decomposition than via the SVD, as in the proof of Theorem 8.1. However, this is an expensive approach, and indeed the converse strategy of computing the SVD via the polar decomposition has been suggested for parallel machines [284, 1994]. Second, we can restrict our attention to square, nonsingular matrices if we are willing to carry out a preliminary orthogonal reduction of a rectangular matrix. Let $A \in \mathbb{C}^{m\times n}$ with $m \geq n$ have the QR factorization $A = QR$, $Q \in \mathbb{C}^{m\times n}$, $R \in \mathbb{C}^{n\times n}$. If $R = UH$ is a polar decomposition then $QU \cdot H$ is a polar decomposition of A. If R is singular then the reduction can be continued to produce a

$$A = P \begin{bmatrix} R & 0 \\ 0 & 0 \end{bmatrix} Q^*, \tag{8.8}$$

where $P \in \mathbb{C}^{m\times m}$ and $Q \in \mathbb{C}^{n\times n}$ are unitary, and $R \in \mathbb{C}^{r\times r}$ is nonsingular and upper triangular [224, 1996, Sec. 5.4.2]. The polar decomposition of R can now be computed and the polar decomposition of A pieced back together.

8.1. Approximation Properties

Given that the polar decomposition is just a "compressed SVD", as the proof of Theorem 8.1 shows, it is natural to ask why the polar decomposition merits separate study. One reason is that both polar factors have best approximation properties.

Theorem 8.4 (Fan and Hoffman). *Let $A \in \mathbb{C}^{m\times n}$ $(m \geq n)$ have the polar decomposition $A = UH$. Then $\|A - U\| = \min\{ \|A - Q\| : Q^*Q = I_n \}$ for any unitarily invariant norm. The minimizer is unique for the Frobenius norm if A has full rank.*

Proof. If $Q^*Q = I_n$ then Q has singular values all equal to 1. Hence, by Theorem B.3, if A has the SVD $A = P\Sigma V^*$ then $\|A - Q\| \geq \|\Sigma - I_{m,n}\|$ for any unitarily invariant norm. Equality is achieved for $Q = U$ because, since $H = V\Sigma_1 V^*$, where $\Sigma = [\Sigma_1\ 0]^T$,

$$\|A-U\| = \|U(H-I_n)\| = \|H-I_n\| = \|V\Sigma_1V^*-I_n\| = \|\Sigma_1-I_n\| = \|\Sigma-I_{m,n}\|, \quad (8.9)$$

where we have used Problems B.7 and B.6.

To prove the uniqueness result for the Frobenius norm we prove the result again from scratch, thereby illustrating a proof technique applicable to related problems.

Let $Q \in \mathbb{C}^{m\times n}$ be any matrix satisfying $Q^*Q = I_n$ and set $E = Q - U$. Then $I_n = Q^*Q = E^*U + U^*E + E^*E + I_n$, so that

$$E^*U + U^*E + E^*E = 0. \quad (8.10)$$

Hence

$$\begin{aligned}(A - Q)^*(A - Q) &= (A - U)^*(A - U) - (A - U)^*E - E^*(A - U) + E^*E \\ &= (A - U)^*(A - U) - A^*E - E^*A. \quad (8.11)\end{aligned}$$

Taking the trace in (8.11) gives $\|A - Q\|_F^2 = \|A - U\|_F^2 - \operatorname{trace}(A^*E + E^*A)$, and

$$\begin{aligned}\operatorname{trace}(A^*E + E^*A) &= \operatorname{trace}(HU^*E + E^*UH) = \operatorname{trace}(H(U^*E + E^*U)) \\ &= -\operatorname{trace}(HE^*E) = -\operatorname{trace}(EHE^*) \leq 0,\end{aligned}$$

where we have used (8.10). Hence $\|A - Q\|_F^2 \geq \|A - U\|_F^2$, with equality only if $\operatorname{trace}(EHE^*) = 0$, and if A has full rank then H is nonsingular and this last condition implies $E = 0$, giving the uniqueness condition. □

For $m < n$ we can apply Theorem 8.4 to A^* to deduce that if $A^* = \widehat{U}\widehat{H}$ is a polar decomposition then $\|A - \widehat{U}^*\| = \min\{ \|A - Q\| : Q^*Q = I_m \}$ for any unitarily invariant norm. The next result, concerning best approximation by partial isometries, holds for arbitrary m and n and reduces to Theorem 8.4 when $\operatorname{rank}(A) = n$.

Theorem 8.5 (Laszkiewicz and Ziętak). *Let $A \in \mathbb{C}^{m\times n}$ have the canonical polar decomposition $A = UH$. Then U solves*

$$\min\{ \|A - Q\| : Q \in \mathbb{C}^{m\times n} \text{ is a partial isometry with } \operatorname{range}(Q) = \operatorname{range}(A) \}$$

for any unitarily invariant norm.

Proof. Let $Q \in \mathbb{C}^{m\times n}$ be any partial isometry satisfying $\mathrm{range}(Q) = \mathrm{range}(A)$, and let A have the SVD $A = P\left[\begin{smallmatrix} \Sigma_r & 0 \\ 0 & 0_{m-r,n-r} \end{smallmatrix}\right] V^*$, where $r = \mathrm{rank}(A)$. Since $\mathrm{rank}(Q) = \mathrm{rank}(A)$, Theorem B.3 implies that

$$\|A - Q\| \geq \left\| \begin{bmatrix} \Sigma_r - I_r & 0 \\ 0 & 0_{m-r,n-r} \end{bmatrix} \right\|$$

for any unitarily invariant norm. Equality is achieved for $Q = U$ (which satisfies the range condition by Problem 8.8) because

$$\begin{aligned} \|A - U\| &= \left\| P \begin{bmatrix} \Sigma_r & 0 \\ 0 & 0_{m-r,n-r} \end{bmatrix} V^* - P \begin{bmatrix} I_r & 0 \\ 0 & 0_{m-r,n-r} \end{bmatrix} V^* \right\| \\ &= \left\| \begin{bmatrix} \Sigma_r - I_r & 0 \\ 0 & 0_{m-r,n-r} \end{bmatrix} \right\|. \qquad \square \end{aligned}$$

Theorem 8.4 says that the polar factor U is the nearest matrix to A with orthonormal columns. Hence the polar decomposition provides an optimal way of orthogonalizing a matrix. Applications are discussed in Section 2.6. For square matrices U is the nearest unitary matrix to A. A related approximation problem is that of finding the nearest unitary matrix with determinant 1—the nearest "rotation matrix". This problem is a special case of one of the Procrustes problems discussed in Section 2.6, for which we now derive solutions.

Theorem 8.6 (orthogonal Procrustes problems). *Let $A, B \in \mathbb{C}^{m\times n}$.*

(a) *Any solution to the orthogonal Procrustes problem*

$$\min\{\, \|A - BW\|_F : W \in \mathbb{C}^{n\times n},\ W^*W = I \,\} \tag{8.12}$$

is a unitary polar factor of B^*A. There is a unique solution if B^*A is nonsingular.

(b) *For real A and B the rotation variant*

$$\min\{\, \|A - BW\|_F : W \in \mathbb{R}^{n\times n},\ W^*W = I,\ \det(W) = 1 \,\} \tag{8.13}$$

has solution $W = PVQ^*$, where $V = \mathrm{diag}(1,\dots,1,\det(PQ^*))$ and $B^*A = P\Sigma Q^*$ is an SVD. The solution is unique if (i) $\det(PQ^*) = 1$ *and $\sigma_{n-1} \neq 0$ or* (ii) $\det(PQ^*) = -1$ *and $\sigma_{n-1} > \sigma_n$, where $\Sigma = \mathrm{diag}(\sigma_i)$.*

Proof. Since

$$\|A - BW\|_F^2 = \|A\|_F^2 - \mathrm{trace}(A^*BW + W^*B^*A) + \|B\|_F^2, \tag{8.14}$$

the problem reduces to maximizing $\mathrm{trace}(A^*BW + W^*B^*A) = 2\,\mathrm{Re}\,\mathrm{trace}(W^*B^*A)$. The result follows from Problem 8.13. $\square$

Before explaining the approximation properties of the Hermitian polar factor we note the solution to the problem of finding the nearest Hermitian matrix. For $A \in \mathbb{C}^{n\times n}$, we introduce the notation

$$A_H = \frac{1}{2}(A + A^*), \quad A_K = \frac{1}{2}(A - A^*), \qquad A \equiv A_H + A_K.$$

A_H is called the Hermitian part of A and A_K the skew-Hermitian part.

Theorem 8.7 (Fan and Hoffman). *For $A \in \mathbb{C}^{n\times n}$ and any unitarily invariant norm, $\|A - A_H\| = \min\{ \|A - X\| : X = X^* \}$. The solution is unique for the Frobenius norm.*

Proof. For any Hermitian Y,

$$\begin{aligned} \|A - A_H\| = \|A_K\| &= \frac{1}{2}\|(A - Y) + (Y^* - A^*)\| \\ &\le \frac{1}{2}\|A - Y\| + \frac{1}{2}\|(Y - A)^*\| \\ &= \|A - Y\|, \end{aligned}$$

using the fact that $\|A\| = \|A^*\|$ for any unitarily invariant norm. The uniqueness for the Frobenius norm is a consequence of the strict convexity of the norm. □

While A_H is the nearest Hermitian matrix to A, the nearest positive semidefinite matrix involves the Hermitian polar factor of A_H.

Theorem 8.8. *Let $A \in \mathbb{C}^{n\times n}$ and let A_H have the polar decomposition $A_H = UH$. Then $X = (A_H + H)/2$ is the unique solution to*

$$\min\{ \|A - X\|_F : X \textit{ is Hermitian positive semidefinite} \}.$$

Proof. Let X be any Hermitian positive semidefinite matrix. From the fact that $\|S + K\|_F^2 = \|S\|_F^2 + \|K\|_F^2$ if $S = S^*$ and $K = -K^*$, we have

$$\|A - X\|_F^2 = \|A_H - X\|_F^2 + \|A_K\|_F^2,$$

so the problem reduces to that of approximating A_H. Let $A_H = Q\Lambda Q^*$ be a spectral decomposition ($Q^*Q = I$, $\Lambda = \mathrm{diag}(\lambda_i)$) and let $Y = Q^*XQ$. Then

$$\|A_H - X\|_F^2 = \|\Lambda - Y\|_F^2 = \sum_{i\ne j} y_{ij}^2 + \sum_i |\lambda_i - y_{ii}|^2 \ge \sum_{\lambda_i<0} |\lambda_i - y_{ii}|^2 \ge \sum_{\lambda_i<0} \lambda_i^2,$$

since $y_{ii} \ge 0$ because Y is positive semidefinite. This lower bound is uniquely attained when $Y = \mathrm{diag}(d_i) \equiv \mathrm{diag}(\max(\lambda_i, 0))$, for which $X = Q\,\mathrm{diag}(d_i)Q^*$. The representation $X = (A_H + H)/2$ follows, since $H = Q\,\mathrm{diag}(|\lambda_i|)Q^*$. □

If A is Hermitian then Theorem 8.8 says that the nearest Hermitian positive semidefinite matrix is obtained by replacing every eigenvalue λ_i of A by $\max(\lambda_i, 0)$ and leaving the eigenvectors unchanged. If, instead, λ_i is replaced by $|\lambda_i|$ then H itself is obtained, and $\|A - H\|_F$ is at most twice the Frobenius norm distance to the nearest Hermitian positive semidefinite matrix. These ideas have been used in Newton's method in optimization to modify indefinite Hessian matrices to make them definite [215, 1981, Sec. 4.4.2], [449, 1999, Sec. 6.3],

8.2. Sensitivity and Conditioning

We now consider the sensitivity of the polar factors to perturbations in the matrix. Here, and in the rest of this chapter, we denote the ith largest singular value of A by $\sigma_i = \sigma_i(A)$.

Theorem 8.9. *Let $A \in \mathbb{C}^{m\times n}$ $(m \ge n)$ have the polar decomposition $A = UH$ and let $A + \Delta A$ have the polar decomposition $A + \Delta A = (U + \Delta U)(H + \Delta H)$. Then*

$$\|\Delta H\|_F \le \sqrt{2}\|\Delta A\|_F.$$

If A has full rank and $\|\Delta A\|_2 < \sigma_n$ then U and $U + \Delta U$ are uniquely defined and

$$\|\Delta U\|_F \le \theta\|\Delta A\|_F + O(\|\Delta A\|_F^2),$$

where the value of θ is given in the table

	$A, \Delta A \in \mathbb{R}^{m\times n}$	$A, \Delta A \in \mathbb{C}^{m\times n}$
$m = n$	$2/(\sigma_n + \sigma_{n-1})$	σ_n^{-1}
$m > n$	σ_n^{-1}	σ_n^{-1}

and in all cases the bound is attainable to first order in $\|\Delta A\|_F$.

Proof. For the bound on $\|\Delta H\|_F$, see Problem 8.15. We outline the proofs for $m = n$, which are from Kenney and Laub [342, 1991, Thms. 2.2, 2.3]. To first order we have

$$\Delta A = \Delta U H + U\Delta H, \tag{8.15}$$

$$0 = \Delta U^* U + U^*\Delta U. \tag{8.16}$$

Taking the conjugate transpose of (8.15) and premultiplying and postmultiplying by U gives

$$U\Delta A^* U = UH\Delta U^* U + U\Delta H = -UHU^*\Delta U + U\Delta H,$$

where we have used (8.16). Subtracting this equation from (8.15) gives

$$UHU^*\Delta U + \Delta UH = \Delta A - U\Delta A^* U.$$

Inserting the expressions for U and H in terms of the SVD $A = P\Sigma Q^*$ we have

$$P\Sigma P^*\Delta U + \Delta UQ\Sigma Q^* = \Delta A - PQ^*\Delta A^* PQ^*,$$

or

$$\Sigma W + W\Sigma = E - E^*, \qquad W = P^*\Delta UQ, \quad E = P^*\Delta AQ.$$

Hence

$$w_{ij} = \frac{e_{ij} - \bar{e}_{ji}}{\sigma_i + \sigma_j},$$

so that

$$|w_{ij}|^2 \le \frac{|e_{ij}|^2 + |e_{ji}|^2}{2\sigma_n^2}.$$

Thus $\|\Delta U\|_F = \|W\|_F \le \|E\|_F/\sigma_n = \|\Delta A\|_F/\sigma_n$. This bound is attained for $\Delta A = P(ie_n e_n^T)Q^*$.

For real A and ΔA the same argument gives $\Sigma W + W\Sigma = E - E^T$, with real W and E given by $W = P^T\Delta UQ$ and $E = P^T\Delta AQ$. We have

$$w_{ij} = \frac{e_{ij} - e_{ji}}{\sigma_i + \sigma_j},$$

and $w_{ii} \equiv 0$. For $i \neq j$,

$$|w_{ij}|^2 \leq \frac{2(|e_{ij}|^2 + |e_{ji}|^2)}{(\sigma_n + \sigma_{n-1})^2}.$$

Hence $\|\Delta U\|_F = \|W\|_F \leq 2\|E\|_F/(\sigma_n + \sigma_{n-1}) = 2\|\Delta A\|_F/(\sigma_n + \sigma_{n-1})$. The bound is attained when $\Delta A = P(e_{n-1}e_n^T - e_n e_{n-1}^T)Q^T$.

For the bounds for $m > n$, see Chaitin-Chatelin and Gratton [103, 2000]. □

We can define condition numbers of the polar factors using (3.2) and (3.4), where f maps A to U or H and we take the Frobenius norm. We can conclude from the theorem that $\kappa_H \leq \sqrt{2}$, where κ_H is the absolute condition number for H. The actual value of κ_H is $\sqrt{2}(1+\kappa_2(A)^2)^{1/2}/(1+\kappa_2(A)) \in [1, \sqrt{2}]$, as shown by Chaitin-Chatelin and Gratton [103, 2000]. For U, $\kappa_U = \theta\|A\|_F/\|U\|_F$ is a relative condition number. Interestingly, for square A, κ_U depends very much on whether the data are real or complex, and in the complex case it is essentially the condition number with respect to inversion, $\kappa(A)$. For rectangular A, this dichotomy between real and complex data does not hold; see Problem 8.17. To summarize, H is always perfectly conditioned, but the condition of U depends on the smallest one or two singular values of A.

A number of variations on the bounds for ΔU are available. Particularly elegant are the following nonasymptotic bounds. Problem 8.16 asks for a proof of the first bound for $n = 1$.

Theorem 8.10. *Let $A, \widetilde{A} \in \mathbb{C}^{n\times n}$ be nonsingular, with smallest singular values σ_n and $\widetilde{\sigma}_n$, respectively. The unitary polar factors U of A and $\widetilde{U}$ of $\widetilde{A}$ satisfy*

$$\|U - \widetilde{U}\| \leq \frac{2}{\sigma_n + \widetilde{\sigma}_n}\|A - \widetilde{A}\|$$

for any unitarily invariant norm.

Proof. See R.-C. Li [386, 1995, Thm. 1]. □

Note that the bound of the theorem is attained whenever A and $\widetilde{A}$ are unitary. The bound remains valid in the Frobenius norm for $A, \widetilde{A} \in \mathbb{C}^{m\times n}$ of rank n [389, 2002, Thm. 2.4].

Theorem 8.11. *Let $A, \widetilde{A} \in \mathbb{R}^{n\times n}$ be nonsingular, with ith largest singular values σ_i and $\widetilde{\sigma}_i$, respectively. If $\|A - \widetilde{A}\|_F < \sigma_n + \widetilde{\sigma}_n$ then the unitary polar factors U of A and $\widetilde{U}$ of $\widetilde{A}$ satisfy*

$$\|U - \widetilde{U}\|_F \leq \frac{4}{\sigma_{n-1} + \sigma_n + \widetilde{\sigma}_{n-1} + \widetilde{\sigma}_n}\|A - \widetilde{A}\|_F.$$

Proof. See W. Li and Sun [389, 2002, Thm. 3.4]. □

A perturbation result for the orthogonal Procrustes problem (8.13) with the constraint $\det(W) = 1$ is given by Söderkvist [533, 1993].

8.3. Newton's Method

Newton's method for the unitary polar factor of a square matrix A can be derived by applying Newton's method to the equation $X^*X = I$ (see Problem 8.18).

Newton iteration (polar decomposition):

$$X_{k+1} = \frac{1}{2}(X_k + X_k^{-*}), \qquad X_0 = A. \tag{8.17}$$

Theorem 8.12. *Let $A \in \mathbb{C}^{n\times n}$ be nonsingular. Then the Newton iterates X_k in (8.17) converge quadratically to the unitary polar factor U of A, with*

$$\|X_{k+1} - U\| \le \frac{1}{2}\|X_k^{-1}\|\,\|X_k - U\|^2. \tag{8.18}$$

Proof. Let $A = P\Sigma Q^*$ be an SVD, so that $U = PQ^*$. It is easy to show by induction that $X_k = PD_kQ^*$, where $D_{k+1} = \frac{1}{2}(D_k + D_k^{-1})$, $D_0 = \Sigma$. This latter iteration is the Newton iteration (5.16) for $\operatorname{sign}(\Sigma) = I$, so $D_k \to I$ quadratically by Theorem 5.6. Finally, (8.18) is obtained from $X_{k+1} - U = \frac{1}{2}(X_k - U)X_k^{-1}(X_k - U)$. □

A variant of (8.17) is

Newton iteration variant:

$$Y_{k+1} = 2Y_k(I + Y_k^*Y_k)^{-1}, \qquad Y_0 = A, \tag{8.19}$$

for which $Y_k \equiv X_k^{-*}$, $k \ge 1$ (see Problem 8.19) and $\lim_{k\to\infty} Y_k = \lim_{k\to\infty} X_k^{-*} = U^{-*} = U$. Note that this iteration is applicable to rectangular A.

As for the Newton sign iteration in Section 5.3, one step of the Schulz iteration can be used to remove the matrix inverse from (8.17), giving

Newton–Schulz iteration:

$$X_{k+1} = \frac{1}{2}X_k(3I - X_k^*X_k), \qquad X_0 = A. \tag{8.20}$$

This iteration is quadratically convergent to U if $0 < \sigma_i < \sqrt{3}$ for every singular value σ_i of A (see Problem 8.20). Again, the iteration is applicable to rectangular A.

Iterations (8.19) and (8.20) are both members of the Padé family (8.22).

8.4. Obtaining Iterations via the Matrix Sign Function

An alternative way to derive the three iterations of the previous section is by using the relation (8.6) in conjunction with appropriate iterations for the matrix sign function. Specifically, the idea is to apply an iteration for the matrix sign function to the block 2×2 matrix $\left[\begin{smallmatrix} 0 & A \\ A^* & 0 \end{smallmatrix}\right]$, check that the iterates have the structure $\left[\begin{smallmatrix} 0 & X_k \\ X_k^* & 0 \end{smallmatrix}\right]$, and then read off an iteration for the unitary polar factor from the (1,2) blocks. This derivation is valid only for square A, although the resulting iterations may work for rectangular A. An alternative connection that is valid for rectangular A is explained in the next result.

Theorem 8.13 (Higham, Mackey, Mackey, and Tisseur). *Let $A \in \mathbb{C}^{m\times n}$ be of rank n and have the polar decomposition $A = UH$. Let g be any matrix function of the form $g(X) = X h(X^2)$ such that the iteration $X_{k+1} = g(X_k)$ converges to* $\operatorname{sign}(X_0)$ *for $X_0 = H$ with order of convergence m. Assume that g has the property that $g(X)^* = g(X^*)$. Then the iteration*

$$Y_{k+1} = Y_k h(Y_k^* Y_k), \qquad Y_0 = A \tag{8.21}$$

converges to U with order of convergence m.

Proof. Let $X_{k+1} = g(X_k)$ with $X_0 = H$, so that $\lim_{k\to\infty} X_k = \operatorname{sign}(H) = I$. We claim that $X_k^* = X_k$ and $Y_k = UX_k$ for all k. These equalities are trivially true for $k = 0$. Assuming that they are true for k, we have

$$X_{k+1}^* = g(X_k)^* = g(X_k^*) = g(X_k) = X_{k+1}$$

and

$$Y_{k+1} = UX_k h(X_k^* U^* U X_k) = UX_k h(X_k^2) = UX_{k+1}.$$

The claim follows by induction. Hence $\lim_{k\to\infty} Y_k = U \lim_{k\to\infty} X_k = U$. The order of convergence is readily seen to be m. □

This result is analogous to Theorem 6.11 (recall that the assumed form of g is justified at the start of Section 6.7). Another way to prove the result is to use the SVD to show that the convergence of (8.21) is equivalent to the convergence of the scalar sign iteration $x_{k+1} = g(x_k)$ with starting values the singular values of A.

To illustrate the theorem, we write the Newton sign iteration (5.16) as $X_{k+1} = \frac{1}{2}(X_k + X_k^{-1}) = X_k \cdot \frac{1}{2}(I + (X_k^2)^{-1})$. Then, for square matrices, the theorem yields $Y_{k+1} = Y_k \cdot \frac{1}{2}(I + (Y_k^* Y_k)^{-1}) = \frac{1}{2} Y_k (I + Y_k^{-1} Y_k^{-*}) = \frac{1}{2}(Y_k + Y_k^{-*})$. More straightforwardly, starting with the Newton–Schulz iteration (5.22), $X_{k+1} = \frac{1}{2} X_k (3I - X_k^2)$, we immediately obtain (8.20).

8.5. The Padé Family of Methods

Theorem 8.13 enables us to translate the Padé family of iterations (5.28) for the matrix sign function into a family of iterations for the unitary polar factor[9]:

Padé iteration:

$$X_{k+1} = X_k p_{\ell m}(I - X_k^* X_k) q_{\ell m}(I - X_k^* X_k)^{-1}, \qquad X_0 = A \in \mathbb{C}^{s\times n}. \tag{8.22}$$

Recall that $p_{\ell m}(\xi)/q_{\ell m}(\xi)$ is the $[\ell/m]$ Padé approximant to $h(\xi) = (1-\xi)^{-1/2}$ and that some of the iteration functions $f_{\ell m}(x) = x p_{\ell m}(1-x^2)/q_{\ell m}(1-x^2)$ are given in Table 5.1. (Of course, the derivation of the Padé iterations could be done directly for the polar decomposition by using the analogue of (5.25): $e^{i\theta} = z(1 - (1 - |z|^2))^{-1/2}$). In particular, $\ell = 0$, $m = 1$ gives (8.19), while $\ell = m = 1$ gives the Halley iteration

$$X_{k+1} = X_k (3I + X_k^* X_k)(I + 3X_k^* X_k)^{-1} = \frac{1}{3} X_k \big[I + 8\big(I + 3X_k^* X_k\big)^{-1}\big], \qquad X_0 = A,$$

in which the second expression is the more efficient to evaluate (cf. (5.32)). The convergence of the Padé iterations is described in the next result.

[9]In this section only we denote the row dimension of A by s in order to avoid confusion with the Padé denominator degree.

Corollary 8.14. *Let $A \in \mathbb{C}^{s\times n}$ be of rank n and have the polar decomposition $A = UH$. Consider the iteration* (8.22) *with $\ell+m > 0$ and any subordinate matrix norm.*

(a) *For $\ell \ge m-1$, if $\|I - A^*A\| < 1$ then $X_k \to U$ as $k \to \infty$ and $\|I - X_k^*X_k\| < \|I - A^*A\|^{(\ell+m+1)^k}$.*

(b) *For $\ell = m-1$ and $\ell = m$,*

$$(I - H_k)(I + H_k)^{-1} = \left[(I - H)(I + H)^{-1}\right]^{(\ell+m+1)^k},$$

where $X_k = UH_k$ is a polar decomposition, and hence $X_k \to U$ as $k \to \infty$ with rate of convergence $\ell + m + 1$.

Proof. Substituting $X_k = UH_k$ in (8.22) yields $H_{k+1} = H_k p_{\ell m}(I - H_k^2) q_{\ell m}(I - H_k^2)^{-1}$, $H_0 = H$. The result therefore follows by applying Theorem 5.8 to H. □

The partial fraction expansion (5.30) can be used to rewrite (8.22). For example, the iterations with $\ell = m-1$ have the partial fraction form (after a little manipulation of (5.30))

$$X_{k+1} = \frac{1}{m} X_k \sum_{i=1}^{m} \frac{1}{\xi_i} (X_k^* X_k + \alpha_i^2 I)^{-1}, \quad X_0 = A \in \mathbb{C}^{s\times n}, \tag{8.23}$$

where

$$\xi_i = \frac{1}{2}\left(1 + \cos\frac{(2i-1)\pi}{2m}\right), \quad \alpha_i^2 = \frac{1}{\xi_i} - 1, \quad i = 1\colon m.$$

For $s = 1$ this formula is just (8.19). The following lemma proves a somewhat unexpected property of these iterations: after the first iteration all iterates have residual less than 1.

Lemma 8.15. *Let $A \in \mathbb{C}^{s\times n}$ be of rank n and have the polar decomposition $A = UH$. The iterates from* (8.22) *with $\ell = m-1$ satisfy $\|X_k^*X_k - I\|_2 < 1$ for $k \ge 1$.*

Proof. By using the SVDs of A and X_k the inequality can be reduced to the corresponding inequality for the singular values of X_k, which satisfy the scalar iteration $x_{k+1} = x_k p_r(1 - x_k^2)/q_r(1 - x_k^2) =: g_r(x_k)$. Applying Theorem 5.9(a) with $r = \ell + m + 1 = 2m$ it suffices to note that (a) $0 < g_r(x)$ for $x > 0$ and (b) $g_r(x) < 1$ for all x since r is even. □

The lemma is useful because for the iterations with $\ell > m$ we need $\|I - A^*A\| < 1$ to ensure convergence (Corollary 8.14). An example is the Newton–Schulz iteration (8.20) (for which $0 < \|A\|_2 < \sqrt{3}$ is in fact sufficient for convergence, as noted in Section 8.3). The lemma opens the possibility of carrying out at least one step of one of the iterations with $\ell = m - 1$ and then switching to (8.20) or some other multiplication-rich iteration, safe in the knowledge that this second iteration will converge.

These convergence results for the Padé iterations can be generalized to rank-deficient A. As we have already observed, it suffices to consider the convergence of the scalar iteration on the singular values of A. Zero singular values are fixed points of the iteration and for the nonzero singular values the convergence to 1 is determined by the results above. So for $\ell = m - 1$ and $\ell = m$ we are guaranteed convergence to the factor U in the *canonical* polar decomposition, while for $\ell > m$ convergence to U is assured if $|1 - \sigma_i^2| < 1$ for all nonzero singular values σ_i of A. However, these results are mainly of theoretical interest because in practice rounding errors will usually perturb the zero singular values, which will then converge to 1; the limit matrix will then be of full rank and therefore a factor U in the polar decomposition.

8.6. Scaling the Newton Iteration

Scaling iterations for the polar decomposition in order to speed up the convergence is desirable for the same reasons as for the matrix sign function (see Section 5.5). An analogous theory holds, but the results are stronger—essentially because the iterations can be reduced to uncoupled scalar iterations via unitary transformations. Consider the Newton iteration (8.17) and its scaled form, for nonsingular $A \in \mathbb{C}^{n\times n}$ and μ_k real and positive,

Scaled Newton iteration:

$$X_{k+1} = \frac{1}{2}\big(\mu_k X_k + \mu_k^{-1} X_k^{-*}\big), \qquad X_0 = A. \tag{8.24}$$

We consider three scalings:

$$\text{optimal scaling:} \quad \mu_k^{\text{opt}} = \big(\sigma_1(X_k)\sigma_n(X_k)\big)^{-1/2}, \tag{8.25}$$

$$1,\infty\text{-norm scaling:} \quad \mu_k^{1,\infty} = \left(\frac{\|X_k^{-1}\|_1\|X_k^{-1}\|_\infty}{\|X_k\|_1\|X_k\|_\infty}\right)^{1/4}, \tag{8.26}$$

$$\text{Frobenius norm scaling:} \quad \mu_k^F = \left(\frac{\|X_k^{-1}\|_F}{\|X_k\|_F}\right)^{1/2}. \tag{8.27}$$

It is easy to show that the iterates from (8.24) satisfy

$$\|X_{k+1} - U\|_2 \le \theta(\mu_k)^2 \|X_{k+1} + U\|_2, \qquad \theta(\mu_k) = \max_{i=1:n}\left|\frac{\mu_k\sigma_i(X_k) - 1}{\mu_k\sigma_i(X_k) + 1}\right|.$$

The quantity (8.25) minimizes $\theta(\mu_k)$ over all μ_k, and so is in this sense optimal. To analyze optimal scaling it suffices to analyze the convergence of the singular values of X_k to 1 (see the proof of Theorem 8.12). Write $\sigma_i^{(k)} = \sigma_i(X_k)$. Observe that this scaling makes the smallest and largest singular values of X_k reciprocals of each other:

$$\sigma_1(\mu_k X_k) = \sqrt{\frac{\sigma_1^{(k)}}{\sigma_n^{(k)}}}, \qquad \sigma_n(\mu_k X_k) = \sqrt{\frac{\sigma_n^{(k)}}{\sigma_1^{(k)}}}. \tag{8.28}$$

These reciprocal values then map to the same value by (5.38a), which is the largest singular value of X_1 by (5.38b). Hence the singular values of X_{k+1} satisfy

$$1 \le \sigma_n^{(k+1)} \le \cdots \le \sigma_2^{(k+1)} = \sigma_1^{(k+1)} = \frac{1}{2}\left(\sqrt{\frac{\sigma_1^{(k)}}{\sigma_n^{(k)}}} + \sqrt{\frac{\sigma_n^{(k)}}{\sigma_1^{(k)}}}\right). \tag{8.29}$$

It follows that

$$\kappa_2(X_{k+1}) \le \frac{1}{2}\left(\kappa_2(X_k)^{1/2} + \frac{1}{\kappa_2(X_k)^{1/2}}\right) \le \kappa_2(X_k)^{1/2}.$$

By comparison, $\mu_k = 1$ and if we assume for simplicity that $\sigma_n^{(k)} = 1$ then $\kappa_2(X_{k+1}) = \frac{1}{2}(\sigma_1^{(k)} + 1/\sigma_1^{(k)}) \approx \frac{1}{2}\kappa_2(X_k)$, which shows a much less rapid reduction in large values of $\kappa_2(X_k)$ for the unscaled iteration. A particularly interesting feature of optimal scaling is that it ensures finite termination of the iteration.

Theorem 8.16 (Kenney and Laub). *For the scaled Newton iteration* (8.24) *with optimal scaling* (8.25), $X_d = U$, *where* d *is the number of distinct singular values of* A.

Proof. The argument just before the theorem shows that the multiplicity of the largest singular value of X_k increases by one on every iteration. Hence at the end of the dth iteration X_{d-1} has all its singular values equal. The next scaling maps all the singular values to 1 (i.e., $\mu_{d-1}X_{d-1} = U$), and so $X_d = U$. □

In general we will not know that d in Theorem 8.16 is small, so the theorem is of little practical use. However, it is possible to predict the number of iterations accurately and cheaply via just scalar computations. By (8.29), the extremal singular values of X_1 satisfy $1 \le \sigma_n^{(1)} \le \sigma_1^{(1)}$. The largest singular value of $\mu_1 X_1$ is then $\left(\sigma_1^{(1)}/\sigma_n^{(1)}\right)^{1/2} \le \left(\sigma_1^{(1)}\right)^{1/2}$, and so from the properties in (5.38) of the map $f(x) = \frac{1}{2}(x + 1/x)$ it follows that the singular values of X_2 are

$$1 \le \sigma_n^{(2)} \le \cdots \le \sigma_1^{(2)} = \frac{1}{2}\left(\sqrt{\frac{\sigma_1^{(1)}}{\sigma_n^{(1)}}} + \sqrt{\frac{\sigma_n^{(1)}}{\sigma_1^{(1)}}}\right) \le \frac{1}{2}\left(\sqrt{\sigma_1^{(1)}} + \frac{1}{\sqrt{\sigma_1^{(1)}}}\right) = f\left(\sqrt{\sigma_1^{(1)}}\right).$$

Therefore, with $g^{(i)}$ denoting the i-fold composition of g,

$$\|U - X_k\|_2 \le g^{(k)}\left(\sigma_1^{(0)}/\sigma_n^{(0)}\right) - 1, \qquad g(x) = \frac{1}{2}(\sqrt{x} + 1/\sqrt{x}). \tag{8.30}$$

Hence, given only knowledge of the extremal singular values of A we can bound the number of iterations required by the Newton iteration with optimal scaling. For the unscaled iteration a similar procedure can be employed to obtain the error *exactly*; see Problem 8.22.

Figure 8.1 plots a bound on the number of iterations for optimal scaling for a range of $\kappa_2(A) = \sigma_1/\sigma_n$. The bound is the smallest value of k for which the bound in (8.30) is at most $u = 2^{-53}$. The figure shows, for example, that if $\kappa_2(A) \le 10^{13}$ then seven iterations suffice.

The optimal scaling (8.25) requires the extremal singular values of X_k. While these can be approximated by the power method on A and A^{-1}, directly computable scale parameters are perhaps more attractive. The $1,\infty$-norm scaling (8.26) is within a constant factor of μ_k^{opt}: in view of the inequalities $\|A\|_2 \le \sqrt{\|A\|_1\|A\|_\infty} \le \sqrt{n}\|A\|_2$, we have

$$\frac{1}{n^{1/4}}\mu_k^{\text{opt}} \le \mu_k^{1,\infty} \le n^{1/4}\mu_k^{\text{opt}}.$$

The Frobenius norm scaling (8.27) has the property that it minimizes $\|X_{k+1}\|_F$ over all μ_k; see Problem 8.23.

Scaling can also be applied to the Padé iterations (8.22). Indeed because of the relations in Theorem 5.9, for the principal Padé iterations the same scalings $X_k \leftarrow \mu_k X_k$ with μ_k given by (8.25), (8.26), or (8.27) are applicable. Three caveats must be borne in mind, however. First, higher order iterations permit less frequent scaling than the Newton iteration (relative to the total number of iterations), so the benefits

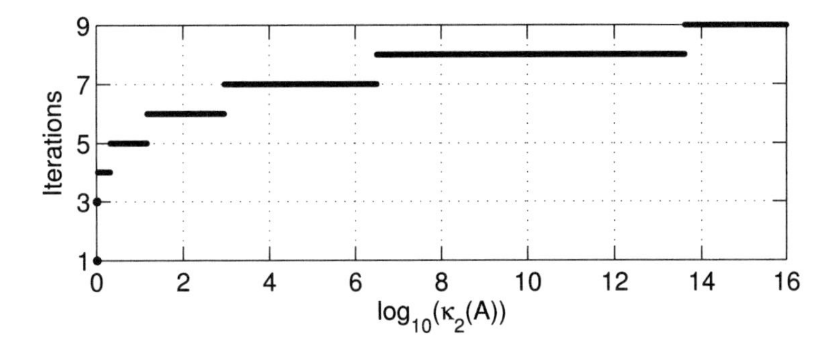

Figure 8.1. *Bounds on number of iterations for Newton iteration with optimal scaling for* $1 \le \kappa_2(A) \le 10^{16}$.

of scaling are reduced. Second, the scaling parameters require the computation of X_k^{-1}, which does not appear in the Padé iterations, so this is an added cost. Third, practical experience shows that scaling degrades the numerical stability of the Padé iterations [285, 1994], so there is a tradeoff between speed and stability. For more on numerical stability, see Section 8.8.

8.7. Terminating the Iterations

We begin with two results that are analogues of Lemma 5.12 for the matrix sign function.

Lemma 8.17. *Let* $A \in \mathbb{C}^{m\times n}$ $(m \ge n)$ *have the polar decomposition* $A = UH$. *Then*

$$\frac{\|A^*A - I\|}{1+\sigma_1(A)} \le \|A - U\| \le \frac{\|A^*A - I\|}{1+\sigma_n(A)},$$

for any unitarily invariant norm.

Proof. It is straightforward to show that $A^*A - I = (A-U)^*(A+U)$. Taking norms and using (B.7) gives the lower bound. Since $A+U = U(H+I)$ we have, from the previous relation,

$$(A-U)^*U = (A^*A - I)(H+I)^{-1}.$$

Hence

$$\|A - U\| = \|(A-U)^*U\| \le \|A^*A - I\|\,\|(H+I)^{-1}\|_2 \le \frac{\|A^*A - I\|}{1+\sigma_n(A)},$$

since the eigenvalues of H are the singular values of A. $\square$

This result shows that the two measures of orthonormality $\|A^*A - I\|$ and $\|A - U\|$ are essentially equivalent, in that they have the same order of magnitude if $\|A\| < 1/2$ (say). Hence the residual $\|X_k^*X_k - I\|$ of an iterate is essentially the same as the error $\|U - X_k\|$. This result is useful for the Padé iterations, which form $X_k^*X_k$.

The next result bounds the distance $\|A - U\|$ in terms of the Newton correction $\frac{1}{2}(A - A^{-*})$.

Lemma 8.18. *Let $A \in \mathbb{C}^{n\times n}$ be nonsingular and have the polar decomposition $A = UH$. If $\|A-U\|_2 = \epsilon < 1$ then for any unitarily invariant norm*

$$\left(\frac{1-\epsilon}{2+\epsilon}\right)\|A-A^{-*}\| \le \|A-U\| \le \left(\frac{1+\epsilon}{2-\epsilon}\right)\|A-A^{-*}\|.$$

Proof. Let $E = A - U$. It is straightforward to show that

$$E = (A - A^{-*})(I + E^*U)(2I + E^*U)^{-1}.$$

Since $\|E^*U\|_2 = \|E\|_2 = \epsilon < 1$, taking norms and using (B.7) yields

$$\|E\| \le \|A - A^{-*}\| \left(\frac{1+\epsilon}{2-\epsilon}\right).$$

The lower bound for E is obtained by taking norms in

$$A - A^{-*} = E(2I + E^*U)(I + E^*U)^{-1}. \qquad \square$$

Note that if all the singular values of A are at least 1 (as is the case for the scaled or unscaled Newton iterates X_k for $k \ge 1$) then $\|A - U\| \le \|A - A^{-*}\|$ for any unitarily invariant norm, with no restriction on $\|A - U\|$. This follows from the inequality $\sigma_i - 1 \le \sigma_i - \sigma_i^{-1}$ for each singular value of A and the characterization of $\|A\|$ as a symmetric gauge function of the singular values.

Let us consider how to build a termination criterion for the (unscaled) Newton iteration. We can use similar reasoning as we used to derive the test (5.44) for the matrix sign function. Lemma 8.18 suggests that once convergence has set in, $\|X_k - U\| \approx \frac{1}{2}\|X_k - X_k^{-*}\|$. However, having computed X_k^{-*} we might as well compute the next iterate, X_{k+1}, so it is the error in X_{k+1} that we wish to bound in terms of $\|X_k - X_k^{-*}\|$. In view of (8.18) we expect that

$$\|X_{k+1} - U\|_F \lesssim \frac{1}{2}\|X_k^{-1}\|_F \frac{1}{4}\|X_k - X_k^{-*}\|_F^2 = \frac{1}{2}\|X_k^{-1}\|_F \|X_{k+1} - X_k\|_F^2.$$

This suggests that we accept X_{k+1} when

$$\|X_{k+1} - X_k\|_F \le \left(\frac{2\eta\|X_{k+1}\|_F}{\|X_k^{-1}\|_F}\right)^{1/2},$$

where η is the desired relative error tolerance. Since X_k is converging to a unitary matrix, close to convergence this test is effectively

$$\|X_{k+1} - X_k\|_F \le (2\eta)^{1/2}. \tag{8.31}$$

Modulo the constant 2, this is precisely the test (4.25) when $X_k \approx U$, since $c = n^{1/2}/2$ therein. This test with an extra factor $n^{1/2}$ on the right-hand side is also recommended by Kiełbasiński and Ziętak [352, 2003, App. B]. Such a test should avoid the weakness of a test of the form $\delta_k \le \eta$ that it tends to terminate the iteration too late (see the discussion in Section 4.9.2). In particular, (8.31) allows termination after the first iteration, which will be required in orthogonalization applications where the starting matrix can be very close to being orthogonal (as in [601, 2006], for example).

8.8. Numerical Stability and Choice of H

What can be said about the computed polar factors obtained using any of the methods described in this chapter? Let $A \in \mathbb{C}^{m\times n}$ with $m \geq n$. The best that can be expected in floating point arithmetic is that the computed matrices $\widehat{U}$ and $\widehat{H}$ satisfy

$$\widehat{U} = V + \Delta U, \qquad V^*V = I, \quad \|\Delta U\| \leq \epsilon\|V\|, \tag{8.32a}$$

$$\widehat{H} = K + \Delta H, \qquad \widehat{H}^* = \widehat{H}, \quad \|\Delta H\| \leq \epsilon\|K\|, \tag{8.32b}$$

$$VK = A + \Delta A, \qquad \|\Delta A\| \leq \epsilon\|A\|, \tag{8.32c}$$

where K is Hermitian positive semidefinite (and necessarily positive definite for the 2-norm, by (8.32c), if A has full rank and $\kappa_2(A) < 1/\epsilon$) and ϵ is a small multiple of the unit roundoff u. These conditions say that $\widehat{U}$ and $\widehat{H}$ are relatively close to the true polar factors of a matrix near to A. If the polar decomposition is computed via the SVD then these ideal conditions hold with ϵ a low degree polynomial in m and n, which can be shown using the backward error analysis for the SVD [224, 1996, Sec. 5.5.8].

We turn now to iterations for computing the unitary polar factor. The next result describes their stability.

Theorem 8.19 (stability of iterations for the unitary polar factor). *Let the nonsingular matrix $A \in \mathbb{C}^{n\times n}$ have the polar decomposition $A = UH$. Let $X_{k+1} = g(X_k)$ be superlinearly convergent to the unitary polar factor of X_0 for all X_0 sufficiently close to U and assume that g is independent of X_0. Then the iteration is stable, and the Fréchet derivative of g at U is idempotent and is given by $L_g(U,E) = L(U,E) = \frac{1}{2}(E - UE^*U)$, where $L(U)$ is the Fréchet derivative of the map $A \to A(A^*A)^{-1/2}$ at U.*

Proof. Since the map is idempotent, stability, the idempotence of L_g, and the equality of $L_g(U)$ and $L(U)$, follow from Theorems 4.18 and 4.19. To find $L(U,E)$ it suffices to find $L_g(U,E)$ for the Newton iteration, $g(X) = \frac{1}{2}(X + X^{-*})$. From

$$\begin{aligned} g(U+E) &= \frac{1}{2}\big(U + E + \big(U^{-1} - U^{-1}EU^{-1} + O(\|E\|^2)\big)^*\big) \\ &= \frac{1}{2}\big(U + E + U - UE^*U + O(\|E\|^2)\big) \end{aligned}$$

we have $L(U,E) = L_g(U,E) = \frac{1}{2}(E - UE^*U)$. $\square$

Theorem 8.19 shows that, as for the matrix sign iterations, all the iterations in this book for the unitary polar factor are stable. The polar iterations generally have better behaviour than the sign iterations because the limit matrix is unitary (and hence of unit 2-norm) rather than just idempotent (and hence of unrestricted norm).

Since these iterations compute only U, a key question is how to obtain H. Given a computed $\widehat{U}$ it is natural to take $\widetilde{H} = \widehat{U}^*A$. This matrix will in general not be Hermitian, so we will replace it by the nearest Hermitian matrix (see Theorem 8.7):

$$\widehat{H} = \frac{(\widehat{U}^*A)^* + \widehat{U}^*A}{2}. \tag{8.33}$$

(Ideally, we would take the nearest Hermitian positive semidefinite matrix, but this would be tantamount to computing the polar decomposition of $\widehat{H}$, in view of Theorem 8.8). If $\widehat{U}$ satisfies (8.32a), then $\widehat{U}^*\widehat{U} = I + O(\epsilon)$ and

$$\|A - \widehat{U}\widehat{H}\|_F = \frac{1}{2}\|(\widehat{U}^*A)^* - \widehat{U}^*A\|_F + O(\epsilon).$$

Hence the quantity

$$\beta(\widehat{U}) = \frac{\|(\widehat{U}^*A)^* - \widehat{U}^*A\|_F}{2\|A\|_F} \tag{8.34}$$

is a good approximation to the relative residual $\|A - \widehat{U}\widehat{H}\|_F/\|A\|_F$. Importantly, $\beta(\widehat{U})$ can be computed without an extra matrix multiplication.

We now summarize how to measure a posteriori the quality of a computed polar decomposition $A \approx \widehat{U}\widehat{H}$ obtained via an iterative method for $\widehat{U}$ and (8.33). The quantities $\|\widehat{U}^*\widehat{U} - I\|_F$ and $\|A - \widehat{U}\widehat{H}\|_F/\|A\|_F$ should be of order the convergence tolerance and $\widehat{H}$ (which is guaranteed to be Hermitian) should be positive semidefinite. The latter condition can be tested by computing the smallest eigenvalue or attempting a Cholesky decomposition (with pivoting). If $\|\widehat{U}^*\widehat{U} - I\|_F$ is small then the relative residual can be safely approximated by $\beta(\widehat{U})$.

The ultimate question regarding any iteration for the unitary polar factor is "what can be guaranteed a priori about the computed $\widehat{U}$?" In other words, is there an a priori forward or backward error bound that takes account of all the rounding errors in the iteration as well as the truncation error due to terminating an iterative process? Such results are rare for any matrix iteration, but Kiełbasiński and Ziętak [352, 2003] have done a detailed analysis for the Newton iteration with the $1,\infty$-norm scaling (8.26). Under the assumptions that matrix inverses are computed in a mixed backward–forward stable way and that $\mu_k^{1,\infty}$ is never too much smaller than μ_k^{opt} they show that the computed factors are backward stable. The assumption on the computed inverses is not always satisfied when the inverse is computed by Gaussian elimination with partial pivoting [276, 2002, Sec. 14.1], but it appears to be necessary in order to push through the already very complicated analysis. Experiments and further analysis are given by Kiełbasiński, Zieliński, and Ziętak [351, 2006].

8.9. Algorithm

We give an algorithm based on the Newton iteration (8.24) with the $1,\infty$-norm scaling (8.26).

Algorithm 8.20 (Newton algorithm for polar decomposition). Given $A \in \mathbb{C}^{m\times n}$ of rank n this algorithm computes the polar decomposition $A = UH$ using the Newton iteration. Two tolerances are used: a tolerance tol_cgce for testing convergence and a tolerance tol_scale for deciding when to switch to the unscaled iteration.

1 if $m > n$
2   compute a QR factorization $A = QR$ ($R \in \mathbb{C}^{n\times n}$)
3    $A := R$
4 end
5 $X_0 = A$; scale = true
6 for $k = 1{:}\infty$

```
7       Y_k = X_k^{-1}    % If m > n, exploit triangularity of X_0.
8       if scale
9          μ_k = (||Y_k||_1 ||Y_k||_∞ / (||X_k||_1 ||X_k||_∞))^{1/4}
10      else
11         μ_k = 1
12      end
13      X_{k+1} = 1/2 (μ_k X_k + μ_k^{-1} Y_k^*)
14      δ_{k+1} = ||X_{k+1} - X_k||_F / ||X_{k+1}||_F
15      if scale = true and δ_{k+1} ≤ tol_scale, scale = false, end
16      if ||X_{k+1} - X_k||_F ≤ (tol_cgce)^{1/2} or (δ_{k+1} > δ_k/2 and scale = false)
17         goto line 20
18      end
19   end
20   U = X_{k+1}
21   H_1 = U^* A
22   H = 1/2 (H_1 + H_1^*)
23   if m > n, U = QU, end
```

Cost: For $m = n$: $2(k+1)n^3$ flops, where k iterations are used. For $m > n$: $6mn^2 + (2k - 3\frac{1}{3})n^3$ flops (assuming Q is kept in factored form).

The strategy for switching to the unscaled iteration, and the design of the convergence test, are exactly as for Algorithm 5.14.

Possible refinements to the algorithm include:

(a) Doing an initial complete orthogonal decomposition (8.8) instead of a QR factorization when the matrix is not known to be of full rank.

(b) Switching to the matrix multiplication-rich Newton–Schulz iteration once $\|I - X_k^* X_k\| \le \theta$, for some $\theta < 1$. The computation of $X_k^* X_k$ in this test can be avoided by applying a matrix norm estimator to $I - X_k^* X_k$ [276, 2002, Chap. 15] (see Higham and Schreiber [286, 1990]).

We now give some numerical examples to illustrate the theory, obtained with a MATLAB implementation of Algorithm 8.20. We took tol_cgce $= n^{1/2}u$ and tol_scale $= 10^{-2}$. We used three real test matrices:

1. a nearly orthogonal matrix, `orth(gallery('moler',16))+ones(16)*1e-3`,

2. a scalar multiple of an involutory matrix, `gallery('binomial',16)`, and

3. the Frank matrix, `gallery('frank',16)`.

We report in Tables 8.1–8.3 various statistics for each iteration, including $\delta_{k+1} = \|X_{k+1} - X_k\|_F / \|X_{k+1}\|_F$ and $\beta(\widehat{U})$ in (8.34). The line across each table follows the iteration number for which the algorithm detected convergence. In order to compare the stopping criterion in the algorithm with the criterion $\delta_{k+1} \le$ tol_cgce, we continued the iteration until the latter test was satisfied, and those iterations appear after the line.

Several points are worth noting. First, the convergence test is reliable and for these matrices saves one iteration over the simple criterion $\delta_{k+1} \le$ tol_cgce. Second, the number of iterations increases with $\kappa_2(A)$, as expected from the theory, but even for the most ill conditioned matrix only seven iterations are required. Third, the Frank matrix is well conditioned with respect to U but very ill conditioned with respect

Table 8.1. *Results for nearly orthogonal matrix, $n = 16$. $\kappa_U = 1.0$, $\kappa_2(A) = 1.0$, $\beta(\widehat{U}) = 3.3 \times 10^{-16}$, $\|\widehat{U}^*\widehat{U} - I\|_F = 1.4 \times 10^{-15}$, $\|A - \widehat{U}\widehat{H}\|_F/\|A\|_F = 5.1 \times 10^{-16}$, $\lambda_{\min}(\widehat{H}) = 9.9 \times 10^{-1}$.*

k	$\dfrac{\|X_k - U\|_F}{\|U\|_F}$	$\|X_k^*X_k - I\|_F$	δ_{k+1}	$\|X_k\|_F$	μ_k
1	1.3e-5	1.1e-4	2.9e-3	4.0e+0	
2	3.3e-10	2.6e-9	1.3e-5	4.0e+0	
3	8.1e-16	1.5e-15	3.3e-10	4.0e+0	
4	8.2e-16	1.4e-15	2.7e-16	4.0e+0	

Table 8.2. *Results for binomial matrix, $n = 16$. $\kappa_U = 1.7 \times 10^3$, $\kappa_2(A) = 4.7 \times 10^3$, $\beta(\widehat{U}) = 3.5 \times 10^{-16}$, $\|\widehat{U}^*\widehat{U} - I\|_F = 1.4 \times 10^{-15}$, $\|A - \widehat{U}\widehat{H}\|_F/\|A\|_F = 4.2 \times 10^{-16}$, $\lambda_{\min}(\widehat{H}) = 2.6$.*

k	$\dfrac{\|X_k - U\|_F}{\|U\|_F}$	$\|X_k^*X_k - I\|_F$	δ_{k+1}	$\|X_k\|_F$	μ_k
1	1.7e+1	2.4e+3	2.6e+2	7.0e+1	5.5e-3
2	1.4e+0	2.2e+1	7.0e+0	9.2e+0	1.8e-1
3	1.3e-1	1.1e+0	1.2e+0	4.5e+0	5.9e-1
4	2.6e-3	2.1e-2	1.3e-1	4.0e+0	9.2e-1
5	1.4e-6	1.1e-5	2.6e-3	4.0e+0	
6	1.3e-12	1.0e-11	1.4e-6	4.0e+0	
7	3.3e-14	1.2e-15	1.3e-12	4.0e+0	
8	3.3e-14	1.4e-15	2.3e-16	4.0e+0	

Table 8.3. *Results for Frank matrix, $n = 16$. $\kappa_U = 5.2 \times 10^1$, $\kappa_2(A) = 2.3 \times 10^{14}$, $\beta(\widehat{U}) = 2.5 \times 10^{-16}$, $\|\widehat{U}^*\widehat{U} - I\|_F = 1.1 \times 10^{-15}$, $\|A - \widehat{U}\widehat{H}\|_F/\|A\|_F = 3.7 \times 10^{-16}$, $\lambda_{\min}(\widehat{H}) = 3.5 \times 10^{-13}$.*

k	$\dfrac{\|X_k - U\|_F}{\|U\|_F}$	$\|X_k^*X_k - I\|_F$	δ_{k+1}	$\|X_k\|_F$	μ_k
1	2.9e+6	8.4e+13	1.0e+0	1.1e+7	1.8e+5
2	1.9e+0	4.5e+1	1.1e+6	1.0e+1	1.1e-6
3	2.7e-1	2.5e+0	1.4e+0	4.9e+0	4.2e-1
4	9.5e-3	7.7e-2	2.6e-1	4.0e+0	8.3e-1
5	3.9e-5	3.1e-4	9.5e-3	4.0e+0	
6	2.5e-9	2.0e-8	3.9e-5	4.0e+0	
7	3.3e-15	9.6e-16	2.5e-9	4.0e+0	
8	3.3e-15	1.1e-15	1.7e-16	4.0e+0	

to inversion, since it has just one small singular value. Since the Newton iteration inverts A on the first step, we might expect instability. Yet the algorithm performs in a backward stable fashion, as it does for all three matrices. Finally, if optimal scaling is used then the numbers of iterations are unchanged for the first and last matrices and one less for the second, emphasizing the effectiveness of the $1,\infty$-norm scaling.

Experience from these and many other experiments (see, e.g., [625, 1995]) suggests that

- Algorithm 8.20 with tol_cgce $= nu$ requires at most about 8 iterations in IEEE double precision arithmetic for matrices A not too close to being rank-deficient (say, $\kappa_2(A) \le 10^{14}$) and at most 1 or 2 more iterations for matrices numerically rank-deficient;

- as long as A is not numerically rank-deficient, the algorithm produces computed $\widehat{U}$ and $\widehat{H}$ with $\|\widehat{U}^*\widehat{U} - I\|_F$ and $\|A - \widehat{U}\widehat{H}\|_F/\|A\|_F$ of order tol_cgce and $\widehat{H}$ positive definite.

Thus the algorithm is remarkably stable, quick to converge, and robust—much more so than the Newton sign iteration with any scaling (cf. Table 5.2, for example). And its flop count is less than that for computation of the polar factors via the SVD (see Problem 8.24).

8.10. Notes and References

The polar decomposition was introduced by Autonne in 1902 [24]; hence it is much older than the matrix sign function and almost as old as the SVD (which was derived by Beltrami in 1873). For a detailed history of the polar decomposition see Horn and Johnson [296, 1991, Sec. 3.0].

Some applications of the polar decomposition can be found in Section 2.6. Another application is to the construction of block Householder transformations (block reflectors), which have the form $H = I_m - 2Z(Z^*Z)^+Z^*$, where $Z \in \mathbb{C}^{m\times n}$ $(m \ge n)$. The desired task is to construct H so that for a given $E \in \mathbb{C}^{m\times n}$, $HE = [F^* \; 0]^*$, where $F \in \mathbb{C}^{n\times n}$. The stable algorithms given by Schreiber and Parlett [508, 1988] need to compute the unitary polar factor of one or more blocks of a matrix whose columns form an orthonormal basis for range(E).

The terminology "canonical polar decomposition" is ours; unfortunately, there is no standard name for this decomposition, which, confusingly, is often simply called "the polar decomposition" in the literature. The term "generalized polar decomposition" has been used by some authors, but this term is best reserved for decompositions defined in terms of an underlying scalar product, with U belonging to the corresponding matrix automorphism group, or for appropriate decompositions in a Lie group setting; see the references cited at the end of Section 14.1.1. Some authors use the same name "polar decomposition" for the decomposition in Theorem 8.1 and that in Theorem 8.3 without any restriction on the ranges of U and H, and regard them as one decomposition. We feel this is inappropriate. What we call the polar decomposition is defined only for $m \ge n$ and is distinguished by providing the nearest matrix with orthonormal columns. The canonical polar decomposition is defined for any m and n and is nonunique without the range condition, but with this condition the factors U and H are both rank-deficient if A is. We are dealing with two different decompositions.

Theorem 8.3 goes back at least to von Neumann [603, 1923, Satz 7], but it has been rediscovered by several authors. The result is most easily found in the literature on generalized inverses—for example in Penrose's classic paper [469, 1955] and in Ben-Israel and Greville [52, 2003, Thm. 7, p. 220]. For square A, the result appears in an exercise of Halmos [243, 1974, p. 171].

A "refined" polar decomposition $A = UPD \in \mathbb{C}^{m\times n}$ $(m \geq n)$ is investigated by Eirola [175, 2000]. Here, U has orthonormal columns, P is Hermitian positive semidefinite with unit diagonal (and so is a correlation matrix), and D is diagonal with nonnegative diagonal elements. Eirola shows that the decomposition always exists (the proof is nontrivial) and considers uniqueness, computation, and an application.

An analytic polar decomposition $A(t) = U(t)H(t)$ can be defined for a real matrix $A(t)$ whose elements are analytic functions of a real variable t. The factors are required to be analytic, $U(t)$ orthogonal, and $H(t)$ symmetric but not definite. Existence of the analytic polar decomposition is discussed by Mehrmann and Rath [419, 1993, Sec. 3.1].

The integral formula (8.7) is due to Higham [273, 1994].

The case $m = n$ in Theorem 8.4 is proved by Fan and Hoffman [181, 1955]; for $m > n$ the result is stated without proof by Rao [484, 1980] and a proof is given by Laszkiewicz and Ziętak [372, 2006].

Theorem 8.5 is equivalent to a result of Laszkiewicz and Ziętak [372, 2006], in which the constraint on Q is $\text{rank}(Q) = \text{rank}(A)$, and was obtained for the Frobenius norm by Sun and Chen [549, 1989]. For a related result with no constraints on the partial isometry and that generalizes earlier results of Maher [404, 1989] and Wu [618, 1986] see Problem 8.12.

Theorem 8.7 is from Fan and Hoffman [181, 1955]. Theorem 8.8 is from Higham [269, 1988]. The best approximation property of $(A_H + H)/2$ identified in Theorem 8.8 remains true in any unitarily invariant norm if A is normal, as shown by Bhatia and Kittaneh [66, 1992]. For general matrices and the 2-norm, the theory of best Hermitian positive semidefinite approximation is quite different and does not involve the polar decomposition; see Halmos [242, 1972] and Higham [269, 1988].

The orthogonal Procrustes problem (8.12) was first solved by Green [229, 1952] (for full rank A and B) and Schönemann [507, 1966] (with no restrictions on A and B). The rotation variant (8.13) was posed by Wahba [605, 1965], who explains that it arises in the determination of the attitude of a satellite. Unlike the problem of finding the nearest matrix with orthonormal columns, the solution to the orthogonal Procrustes problem is not the same for all unitarily invariant norms, as shown by Mathias [410, 1993].

Given that the QR factorization $A = QR$ can be more cheaply computed than the polar decomposition $A = UH$, for the purposes of orthogonalization it is natural to ask whether Q can be used in place of U; in other words, is $\|A-Q\|$ close to $\|A-U\|$? Two bounds provide some insight, both holding for $A \in \mathbb{C}^{m\times n}$ of rank n under the assumption that R has positive diagonal elements. Chandrasekaran and Ipsen [105, 1994] show that $\|A - Q\|_F \leq 5\sqrt{n}\|A - U\|_2$, assuming that A has columns of unit 2-norm. Sun [547, 1995] proves that if $\|A^*A - I\|_2 < 1$ then

$$\|A - Q\|_F \leq \frac{1 + \|A\|_2}{\sqrt{2}(1 - \|A^*A - I\|_2)}\|A - U\|_F.$$

The latter bound is the sharper for $\|A^*A - I\|_2 < 1/2$.

The use of the complete orthogonal decomposition as a "preprocessor" before applying an iterative method was suggested by Higham and Schreiber [286, 1990].

Theorem 8.9 has an interesting history. A version of the theorem was first developed by Higham [266, 1986], with a larger constant in the bound for ΔU and the sharpness of the bounds not established. Barrlund [45, 1990] and Kenney and Laub [342, 1991, Thm. 2.3] were the first to recognize the differing sensitivity of U for real and complex data. Unbeknown to the numerical analysts, in the functional analysis literature the bound for ΔH had already been obtained via the analysis of the Lipschitz continuity of the absolute value map. Araki and Yamagami [16, 1981] showed that $\| |A| - |B| \|_F \le \sqrt{2} \|A - B\|_F$ and that the constant $\sqrt{2}$ is as small as possible; see Problem 8.15 for an elegant proof using ideas of Kittaneh. For some history of this topic and further bounds on the norm of $|A| - |B|$ see Bhatia [63, 1994, Sec. 5], [64, 1997, Sec. X.2].

R.-C. Li [387, 1997] gives a different style of perturbation result for U in which the perturbation is expressed in multiplicative form ($A \to XAY$); it shows that the change in U depends only on how close X and Y are to the identity and not on the condition of A. In subsequent work Li [388, 2005] obtains a perturbation bound for H for the case where A is graded, that is, $A = BD$ where B is well conditioned and the scaling matrix D (usually diagonal) can be very ill conditioned; the bounds show that the smaller elements of H can be much less sensitive to perturbations in A than the bound in Theorem 8.9 suggests. Li also investigates how to compute H accurately, suggesting the use of the SVD computed by the one-sided Jacobi algorithm; interestingly, it proves important to compute H from U^*A and not directly from the SVD factors.

Some authors have carried out perturbation analysis for the canonical polar decomposition under the assumption that $\operatorname{rank}(A) = \operatorname{rank}(A + \Delta A)$; see, for example, R.-C. Li [385, 1993] and W. Li and Sun [389, 2002].

The polar decomposition cannot in general be computed in a finite number of arithmetic operations and radicals, as shown by George and Ikramov [212, 1996], [213, 1997]. For a companion matrix the polar decomposition *is* finitely computable and Van Den Driessche and Wimmer [584, 1996] provide explicit formulae; formulae for the block companion matrix case are given by Kalogeropoulos and Psarrakos [332, 2005].

The Newton–Schulz iteration (8.20) is used in Mathematica's NDSolve function within the projected integration method that solves matrix ODEs with orthogonal solutions; see Sofroniou and Spaletta [534, 2002].

Theorem 8.13 is essentially a special case of a result of Higham, Mackey, Mackey, and Tisseur [283, 2005, Thm. 4.6] that applies to the generalized polar decomposition referred to above.

Early papers on iterations for computing the unitary polar factor are those by Björck and Bowie [72, 1971], Kovarik [361, 1970], and Leipnik [379, 1971]. Each of these papers (none of which cites the others) develops families of polynomial iterations—essentially the $[\ell/0]$ Padé iterations.

The Newton iteration (8.17) was popularized as a general tool for computing the polar decomposition by Higham [266, 1986]. It had earlier been used in the aerospace application mentioned in Section 2.6 to orthogonalize the 3×3 direction cosine matrix; see Bar-Itzhack, Meyer, and Fuhrmann [43, 1976].

Gander [202, 1990] (see also Laszkiewicz and Ziętak [372, 2006]) obtains conditions on h for iterations of the form (8.21) to have a particular order of convergence and also considers applying such iterations to rank deficient matrices.

The particular Padé iteration (8.23) in partial fraction form was derived from the corresponding sign iteration by Higham [273, 1994]. It was developed into a practical algorithm for parallel computers by Higham and Papadimitriou [285, 1994], who obtain order of magnitude speedups over computing the polar factors via the SVD on one particular virtual shared memory MIMD computer.

Lemma 8.15 is due to Higham [273, 1994].

The optimal scaling (8.25) and its approximations (8.26) and (8.27) were suggested by Higham [265, 1985], [266, 1986]. All three scalings are analyzed in detail by Kenney and Laub [344, 1992]. The Frobenius norm scaling is also analyzed by Dubrulle [169, 1999]; see Problem 8.23.

Theorem 8.16 is due to Kenney and Laub [344, 1992], who also derive (8.30).

Lemmas 8.17 and 8.18 are from Higham [273, 1994]. Theorem 8.19 is new.

All the globally convergent iterations described in this chapter involve matrix inverses or the solution of multiple right-hand side linear systems. Problem 8.26 describes how the Newton iteration variant (8.19) can be implemented in an inversion-free form. The idea behind this implementation is due to Zha and Zhang [622, 1996], who consider Hermitian matrices and apply the idea to subspace iteration and to iterations for the matrix sign function.

A drawback of the algorithms described in this chapter is that they cannot take advantage of a known polar decomposition of a matrix close to A. Thus knowing a polar decomposition $\widetilde{A} = \widetilde{U}\widetilde{H}$ is of no help when computing the polar decomposition of A, however small the norm or rank of $A - \widetilde{A}$. Indeed the iterations all require A as the starting matrix. The same comment applies to the iterations for the matrix sign function and matrix roots. The trace maximization algorithm outlined in Problem 8.25 can take advantage of a "nearby" polar decomposition but, unfortunately, the basic algorithm is not sufficiently efficient to be of practical use. An SVD updating technique of Davies and Smith [138, 2004] could potentially be used to update the polar decomposition. See Problem 8.27.

Problems

8.1. Find all polar decompositions of the Jordan block $J_m(0) \in \mathbb{C}^{m\times m}$.

8.2. (Uhlig [581, 1981]) Verify that for nonsingular $A \in \mathbb{R}^{2\times 2}$ the polar factors are

$$U = \gamma\big(A + |\det(A)|A^{-T}\big), \qquad H = \gamma\big(A^TA + |\det(A)|I\big),$$

where

$$\gamma = \big|\det\big(A + |\det(A)|\,A^{-T}\big)\big|^{-1/2}.$$

8.3. What are the polar and canonical polar decompositions of $0 \in \mathbb{C}^{m\times n}$?

8.4. (Higham, Mackey, Mackey, and Tisseur [283, 2005, Thm. 4.7]) If $Q \in \mathbb{C}^{n\times n}$ is unitary and $-1 \notin \Lambda(Q)$ what is the polar decomposition of $I + Q$? How might we compute $Q^{1/2}$ iteratively using (8.17)?

8.5. (Moakher [434, 2002]) Show that if $Q_1, Q_2 \in \mathbb{C}^{n\times n}$ are unitary and $-1 \notin \Lambda(Q_1^*Q_2)$ then then the unitary polar factor of $Q_1 + Q_2$ is $Q_1(Q_1^*Q_2)^{1/2}$.

8.6. Let $A \in \mathbb{C}^{m\times n}$. Show that if $H = (A^*A)^{1/2}$ then $\mathrm{null}(H) = \mathrm{null}(A)$.

8.7. For $A \in \mathbb{C}^{m\times n}$ with $m < n$ investigate the existence and uniqueness of polar decompositions $A = UH$ with $U \in \mathbb{C}^{m\times n}$ having orthonormal rows and $H \in \mathbb{C}^{n\times n}$ Hermitian positive semidefinite.

8.8. Show that the condition $\text{range}(U^*) = \text{range}(H)$ in Theorem 8.3 is equivalent to $\text{range}(U) = \text{range}(A)$.

8.9. Let $A \in \mathbb{C}^{n\times n}$ be normal and nonsingular. Show that the polar factors U and H may be expressed as functions of A.

8.10. Give a proof of Theorem 8.4 from first principles for the 2-norm.

8.11. Let $A, B \in \mathbb{C}^{m\times n}$. Show that $\min\{\, \|A - BW\|_F : W \in \mathbb{C}^{n\times n},\ W^*W = I\,\}$ and $\min\{\, \|B^*A - W\|_F : W \in \mathbb{C}^{n\times n},\ W^*W = I\,\}$ are attained at the same matrix W. Thus the orthogonal Procrustes problem problem reduces to the nearest unitary matrix problem.

8.12. (Laszkiewicz and Ziętak [372, 2006]) Show that for $A \in \mathbb{C}^{m\times n}$ and any unitarily invariant norm

$$\min\{\, \|A - Q\| : Q \in \mathbb{C}^{m\times n} \text{ is a partial isometry}\,\} = \max_{i=1:\min(m,n)} \max(\sigma_i, |1 - \sigma_i|),$$

where the σ_i are the singular values of A, and determine a maximizer.

8.13. (a) Show that for $A \in \mathbb{C}^{n\times n}$, $\text{Re}\,\text{trace}(W^*A)$ is maximized over all unitary $W \in \mathbb{C}^{n\times n}$ if and only if W is a unitary polar factor of A (or equivalently, W^*A is Hermitian positive semidefinite). Deduce that the maximizer is unique if A is nonsingular.
(b) Show that for $A \in \mathbb{R}^{n\times n}$, $\text{Re}\,\text{trace}(W^*A)$ is maximized over all orthogonal $W \in \mathbb{R}^{n\times n}$ with $\det(W) = 1$ by $W = P\,\text{diag}(1, \ldots, 1, \det(PQ^*))Q^*$, where $A = P\Sigma Q^*$ is an SVD. Show that the maximizer is unique if (i) $\det(PQ^*) = 1$ and $\sigma_{n-1} \neq 0$ or (ii) $\det(PQ^*) = -1$ and $\sigma_{n-1} > \sigma_n$, where $\Sigma = \text{diag}(\sigma_i)$.

8.14. (Beattie and Smith [49, 1992]) Solve the following weighted variation of the nearest matrix with orthonormal columns problem in Theorem 8.4:

$$\min\{\, \|M^{1/2}(A - Q)\|_F : Q^*MQ = I_n\,\},$$

where $A \in \mathbb{C}^{m\times n}$ $(m \geq n)$ and $M \in \mathbb{C}^{n\times n}$ is Hermitian positive definite. This problem arises in structural identification.

8.15. (Bhatia [63, 1994], Bhatia [64, 1997, Sec. VII.5], Kittaneh [353, 1985], Kittaneh [354, 1986]) Show that for all $A, B \in \mathbb{C}^{m\times n}$

$$\|\,|A| - |B|\,\|_F \leq \sqrt{2}\|A - B\|_F \tag{8.35}$$

by following these steps.

(i) Let f be any function satisfying

$$|f(z) - f(w)| \leq c|z - w| \quad \text{for all } z, w \in \mathbb{C}.$$

Show that for all normal $A \in \mathbb{C}^{n\times n}$ and all $X \in \mathbb{C}^{n\times n}$,

$$\|f(A)X - Xf(A)\|_F \leq c\|AX - XA\|_F.$$

(ii) By applying the previous part to $A \leftarrow \left[\begin{smallmatrix} A & 0 \\ 0 & B \end{smallmatrix}\right]$ and $X \leftarrow \left[\begin{smallmatrix} 0 & X \\ 0 & 0 \end{smallmatrix}\right]$ ("") deduce that

$$\|f(A)X - Xf(B)\|_F \leq c\|AX - XB\|_F$$

for all normal $A, B \in \mathbb{C}^{n\times n}$ and all $X \in \mathbb{C}^{n\times n}$.

(iii) Show that if A and B are both normal then

$$\||A| - |B|\|_F \le \|A - B\|_F.$$

(iv) By applying the previous part to $A \leftarrow \left[\begin{smallmatrix} 0 & A \\ A^* & 0 \end{smallmatrix}\right]$ and $B \leftarrow \left[\begin{smallmatrix} 0 & B \\ B^* & 0 \end{smallmatrix}\right]$ deduce that for all $A, B \in \mathbb{C}^{m\times n}$

$$\||A| - |B|\|_F^2 + \||A^*| - |B^*|\|_F^2 \le 2\|A - B\|_F^2.$$

8.16. (Scalar version of Theorem 8.10. This bound is not readily found in texts on complex analysis.) Show if $z_1 = r_1 e^{i\theta_1}$ and $z_2 = r_2 e^{i\theta_2}$ are complex numbers in polar form then

$$|e^{i\theta_1} - e^{i\theta_2}| \le 2\frac{|z_1 - z_2|}{r_1 + r_2}.$$

8.17. (Li [386, 1995]) Let

$$A = \begin{bmatrix} 1 & 0 \\ 0 & \epsilon \\ 0 & 0 \end{bmatrix}, \qquad \widetilde{A} = \begin{bmatrix} 1 & 0 \\ 0 & \epsilon \\ 0 & \delta \end{bmatrix},$$

where $\epsilon \gg \delta > 0$. Show that the difference $\|U - \widetilde{U}\|_F$ between the respective polar factors is of order δ/ϵ, showing that the sensitivity of U depends on $1/\sigma_n$ and not on $1/(\sigma_n + \sigma_{n-1})$ as it would if A were square.

8.18. Show that the Newton iteration (8.17) can be derived by applying Newton's method to the equation $X^*X = I$.

8.19. Prove that for nonsingular $A \in \mathbb{C}^{n\times n}$ the iterates X_k from (8.17) and Y_k from (8.19) are related by $Y_k = X_k^{-*}$ for $k \ge 1$.

8.20. Let $A \in \mathbb{C}^{m\times n}$ of rank n have the polar decomposition $A = UH$. Show that in the Newton–Schulz iteration (8.20), $X_k \to U$ quadratically as $k \to \infty$ if $\|A\|_2 < \sqrt{3}$, and that

$$\|X_{k+1} - U\|_2 \le \frac{1}{2}\|X_k + 2U\|_2 \, \|X_k - U\|_2^2. \tag{8.36}$$

Show also that $R_k = I - X_k^* X_k$ satisfies $R_{k+1} = \frac{3}{4}R_k^2 + \frac{1}{4}R_k^3$.

8.21. The Newton iteration (8.17) and the Newton–Schulz iteration (8.20) are both quadratically convergent. The convergence of the Newton iteration is described by (8.18) and that of Newton–Schulz by (8.36). What do the error constants in the error relations for these two iterations imply about their relative speeds of convergence?

8.22. Show for the unscaled Newton iteration (8.17) that $\|U - X_k\|_2 = f^{(k-1)}(\sigma_1^{(1)}) - 1$, where $f(x) = \frac{1}{2}(x + 1/x)$ and $\sigma_i^{(k)} = \sigma_i(X_k)$. [Cf. (8.30) for the optimally scaled iteration.]

8.23. (Dubrulle [169, 1999]) (a) Show that the Frobenius norm scaling (8.27) minimizes $\|X_{k+1}\|_F$ over all μ_k.

(b) Show that the $1,\infty$-norm scaling (8.26) minimizes a bound on $\|X_{k+1}\|_1 \|X_{k+1}\|_\infty$.

8.24. Compare the flop count of Algorithm 8.20 with that for computation of the polar factors via the SVD.

8.25. We know from Problem 8.13 that the polar factors U and H of $A \in \mathbb{R}^{n\times n}$ satisfy $\max\{\operatorname{trace}(W^*A) : W^*W = I\} = \operatorname{trace}(U^*A) = \operatorname{trace}(H)$. Thus premultiplying A by U^* both symmetrizes it and maximizes the trace. This suggests developing an algorithm for computing the polar decomposition that iteratively maximizes the trace by premultiplying A by a succession of suitably chosen orthogonal matrices.

(a) Show that if $a_{ij} \neq a_{ji}$ then a Givens rotation $G = \left[\begin{smallmatrix} \cos\theta & \sin\theta \\ -\sin\theta & \cos\theta \end{smallmatrix}\right]$ exists such that $G^T \left[\begin{smallmatrix} a_{ii} & a_{ij} \\ a_{ji} & a_{jj} \end{smallmatrix}\right]$ is symmetric and has maximal trace over all θ.

(b) Show that if A is symmetric but indefinite then a Householder transformation $G = I - 2vv^T/(v^Tv)$ can be chosen so that $\operatorname{trace}(GA) > \operatorname{trace}(A)$.

(c) Develop an algorithm for computing the polar decomposition based on a combination of the transformations in (a) and (b).

8.26. (a) Let $A \in \mathbb{C}^{m\times n}$ and let

$$\begin{array}{c} \\ n \\ m \end{array}\!\!\begin{array}{c} n \\ \begin{bmatrix} I \\ A \end{bmatrix} \end{array} = \begin{array}{c} \\ n \\ m \end{array}\!\!\begin{array}{c} n \\ \begin{bmatrix} Q_1 \\ Q_2 \end{bmatrix} \end{array} R$$

be a QR factorization. The polar decomposition of this matrix is

$$\begin{array}{c} \\ n \\ m \end{array}\!\!\begin{array}{c} n \\ \begin{bmatrix} I \\ A \end{bmatrix} \end{array} = U(I + A^*A)^{1/2}.$$

Since the columns of Q and U span the same space,

$$\begin{bmatrix} Q_1 \\ Q_2 \end{bmatrix} = UM = \begin{bmatrix} I \\ A \end{bmatrix} (I + A^*A)^{-1/2} \cdot M$$

for some unitary M. By examining the blocks of this equation, obtain an expression for $Q_2Q_1^*$ and hence show that the Newton iteration variant (8.19), which we rewrite as

$$X_{k+1} = 2X_k(I + X_k^*X_k)^{-1}, \qquad X_0 = A, \tag{8.37}$$

can be written as

$$\begin{array}{c} \\ n \\ m \end{array}\!\!\begin{array}{c} n \\ \begin{bmatrix} I \\ X_k \end{bmatrix} \end{array} = \begin{array}{c} \\ n \\ m \end{array}\!\!\begin{array}{c} n \\ \begin{bmatrix} Q_1^{(k)} \\ Q_2^{(k)} \end{bmatrix} \end{array} R_k \quad \text{(QR factorization)}, \tag{8.38a}$$

$$X_{k+1} = 2Q_2^{(k)}Q_1^{(k)*}. \tag{8.38b}$$

This is an inverse-free implementation of the iteration.

(b) Evaluate and compare the operation counts for (8.37) and (8.38).

(c) How can iteration (8.38) be scaled?

8.27. (RESEARCH PROBLEM) Develop ways to efficiently update the polar decomposition after a small-normed or low rank perturbation.

Chapter 9
Schur–Parlett Algorithm

We have seen in earlier chapters that reliable algorithms for computing the matrix sign function and matrix roots can be constructed by using the Schur decomposition to reduce the problem to the triangular case and then exploiting particular properties of the functions. We now develop a general purpose algorithm for computing $f(A)$ via the Schur decomposition. We assume that the reader has read the preliminary discussion in Section 4.6.

Our algorithm for computing $f(A)$ consists of several stages. The Schur decomposition $A = QTQ^*$ is computed; T is reordered and blocked to produce another triangular matrix $\widetilde{T}$ with the property that distinct diagonal blocks have "sufficiently distinct" eigenvalues and the eigenvalues within each diagonal block are "close"; the diagonal blocks $f(\widetilde{T}_{ii})$ are computed; the rest of $f(\widetilde{T})$ is obtained using the block form of the Parlett recurrence; and finally the unitary similarity transformations from the Schur decomposition and the reordering are reapplied. We consider first, in Section 9.1, the evaluation of f on the atomic triangular blocks $\widetilde{T}_{ii}$, for which we use a Taylor series expansion. "Atomic" refers to the fact that these blocks cannot be further reduced. This approach is mainly intended for functions whose Taylor series have an infinite radius of convergence, such as the trigonometric and hyperbolic functions, but for some other functions this step can be adapted or replaced by another technique, as we will see in later chapters. In Section 9.2 we analyze the use of the block form of Parlett's recurrence. Based on the conflicting requirements of these two stages we describe a Schur reordering and blocking strategy in Section 9.3. The overall algorithm is summarized in Section 9.4, where its performance on some test matrices is illustrated. The relevance of several preprocessing techniques is discussed in Section 9.5.

9.1. Evaluating Functions of the Atomic Blocks

In this section $T \in \mathbb{C}^{n\times n}$ represents an atomic block of the reordered Schur form. It is therefore an upper triangular matrix whose eigenvalues are "close". Given an arbitrary function f, we need a method for evaluating $f(T)$ efficiently and accurately. A natural approach is to expand f in a Taylor series about the mean of the eigenvalues of T. Write

$$T = \sigma I + M, \qquad \sigma = \operatorname{trace}(T)/n, \tag{9.1}$$

which defines M as T shifted by the mean of its eigenvalues. If f has a Taylor series representation

$$f(\sigma + z) = \sum_{k=0}^{\infty} \frac{f^{(k)}(\sigma)}{k!} z^k \tag{9.2}$$

for z in an open disk containing $\Lambda(T-\sigma I)$, then, by Theorem 4.7,

$$f(T)=\sum_{k=0}^{\infty}\frac{f^{(k)}(\sigma)}{k!}M^k. \tag{9.3}$$

If T has just one eigenvalue, so that $t_{ii}\equiv\sigma$, then M is strictly upper triangular and hence is nilpotent with $M^n=0$; the series (9.3) is then finite. More generally, if the eigenvalues of T are sufficiently close, then the powers of M can be expected to decay quickly after the $(n-1)$st, and so a suitable truncation of (9.3) should yield good accuracy. This notion is made precise in the following lemma, in which M is represented by $M=D+N$, with D diagonal and N strictly upper triangular and hence nilpotent with $M^n=0$.

Lemma 9.1 (Davies and Higham). *Let $D\in\mathbb{C}^{n\times n}$ be diagonal with $|D|\le\delta I$ and let $N\in\mathbb{C}^{n\times n}$ be strictly upper triangular. Then*

$$|(D+N)^k|\le\sum_{i=0}^{\min(k,n-1)}\binom{k}{i}\delta^{k-i}|N|^i$$

and the same inequality holds with the absolute values replaced by any matrix norm subordinate to an absolute vector norm.

Proof. The bound follows from

$$|(D+N)^k|\le(|D|+|N|)^k\le(\delta I+|N|)^k,$$

followed by a binomial expansion of the last term. Since $|N|^{n-1}=0$ we can drop the terms involving $|N|^i$ for $i\ge n-1$. The corresponding bound for matrix norms is obtained by taking norms in the binomial expansion of $(D+N)^k$ and using (B.6). □

If $\delta<1$ and $\delta\ll\|N\|$ in Lemma 9.1, then, for $k\ge n-1$,

$$\|(D+N)^k\|=O(\delta^{k+1-n}\|N\|^{n-1}),$$

and hence the powers of $D+N$ decay rapidly after the $(n-1)$st, irrespective of N.

This analysis shows that as long as the scalar multipliers $f^{(k)}(\sigma)/k!$ in (9.3) are not too large we should be able to truncate the series (9.3) soon after the $(n-1)$st term (and possibly much earlier if M is small).

We need a reliable criterion for deciding when to truncate the Taylor series. When summing a series whose terms decrease monotonically it is safe to stop as soon as a term is smaller than the desired error. Unfortunately, our matrix Taylor series can exhibit very nonmonotonic convergence. Indeed, when $n=2$, $M=T-\sigma I$ always has the form

$$M=\begin{bmatrix}\epsilon & \alpha\\ 0 & -\epsilon\end{bmatrix}, \tag{9.4}$$

and its powers are

$$M^{2k}=\begin{bmatrix}\epsilon^{2k} & 0\\ 0 & \epsilon^{2k}\end{bmatrix},\quad M^{2k+1}=\begin{bmatrix}\epsilon^{2k+1} & \alpha\epsilon^{2k}\\ 0 & -\epsilon^{2k+1}\end{bmatrix}.$$

For $|\epsilon|<1$, $\|M^k\|\to0$ as $k\to\infty$, but $\|M^{2k+1}\|\gg\|M^{2k}\|$ for $\alpha\gg1$. The next theorem shows that this phenomenon of the "disappearing nonnormal part" is connected with the fact that f can map distinct λ_i into the same value.

Theorem 9.2 (Davies and Higham). *Let $D \in \mathbb{C}^{n\times n}$ be diagonal with distinct eigenvalues $\lambda_1,\dots,\lambda_p$ $(1 \le p \le n)$ of multiplicity $k_1,\dots,k_p$, respectively, and let the values $f^{(j)}(\lambda_i)$, $j = 0\colon k_i - 1$, $i = 1\colon p$, be defined. Then $f(D+N) = f(D)$ for all strictly triangular $N \in \mathbb{C}^{n\times n}$ if and only if $f(D) = f(\lambda_1)I$ and*

$$f^{(j)}(\lambda_i) = 0, \quad j = 1\colon k_i - 1, \quad i = 1\colon p. \tag{9.5}$$

Note that (9.5) is vacuous when $k_i = 1$.

Proof. See Problem 9.1. □

Applying Theorem 9.2 to the function $f(x) = x^k$ we obtain the following corollary.

Corollary 9.3. *Let $D \in \mathbb{C}^{n\times n}$ be a nonzero diagonal matrix and let $k \ge 2$. Then $(D+N)^k = D^k$ for all strictly triangular matrices $N \in \mathbb{C}^{n\times n}$ if and only if*

$$D = \beta \operatorname{diag}(e^{2k_1\pi i/k}, e^{2k_2\pi i/k}, \dots, e^{2k_n\pi i/k}),$$

where $\beta \ne 0$, $k_i \in \{0,1,\dots,k-1\}$ and the k_i are distinct (and hence $k \ge n$).

Proof. By Theorem 9.2, all the diagonal elements of D must be kth roots of the same number, β^k say. The condition (9.5) implies that any repeated diagonal element d_{ii} must satisfy $f'(d_{ii}) = kd_{ii}^{k-1} = 0$, which implies $d_{ii} = 0$ and hence $D = 0$; therefore D has distinct diagonal elements. □

As a check, we note that the diagonal of M in (9.4) is of the form in the corollary for even powers k. The corollary shows that this phenomenon of very nonmonotonic convergence of the Taylor series can occur when the eigenvalues are a constant multiple of kth roots of unity. As is well known, the computed approximations to multiple eigenvalues occurring in a single Jordan block tend to have this distribution. We will see in the experiment of Section 9.4 that this eigenvalue distribution also causes problems in finding a good blocking.

We now develop a strict bound for the truncation error of the Taylor series, which we will use to decide when to terminate the series. We apply Theorem 4.8 with $A := \sigma I + M$, $\alpha := \sigma$, M from (9.1), and the Frobenius norm, and so we need to be able to bound $\max_{0\le t\le 1} \|M^s f^{(s)}(\sigma I + tM)\|_F$. We will bound it by the product of the norms, noting that the term M^s is needed anyway if we form the next term of the series. To bound $\max_{0\le t\le 1} \|f^{(s)}(\sigma I + tM)\|_F$ we can use Theorem 4.28 to show that

$$\max_{0\le t\le 1} \|f^{(s)}(\sigma I + tM)\|_F \le \max_{0\le r\le n-1} \frac{\omega_{s+r}}{r!} \|(I - |N|)^{-1}\|_F, \tag{9.6}$$

where $\omega_{s+r} = \sup_{z\in\Omega} |f^{(s+r)}(z)|$ and N is the strictly upper triangular part of M. By using (9.6) in (4.8) we can therefore bound the truncation error. Approximating the Frobenius norm by the ∞-norm, the term $\|(I - |N|)^{-1}\|_\infty$ can be evaluated in just $O(n^2)$ flops, since $I - |N|$ is an M-matrix: we solve the triangular system $(I - |N|)y = e$, where $e = [1,1,\dots,1]^T$, and then $\|y\|_\infty = \|(I - |N|)^{-1}\|_\infty$ [276, 2002, Sec. 8.3].

We now state our algorithm for evaluating a function of an atomic block via the Taylor series.

Algorithm 9.4 (evaluate function of atomic block). Given a triangular matrix $T \in \mathbb{C}^{n\times n}$ whose eigenvalues $\lambda_1,\dots,\lambda_n$ are "close," a function f having the Taylor series

(9.2) for z in an open disk containing $\lambda_i - \sigma$, $i = 1\colon n$, where $\sigma = n^{-1}\sum_{i=1}^n \lambda_i$, and the ability to evaluate derivatives of f, this algorithm computes $F = f(T)$ using a truncated Taylor series.

1 $\sigma = n^{-1}\sum_{i=1}^n \lambda_i$, $M = T - \sigma I$, tol $= u$
2 $\mu = \|y\|_\infty$, where y solves $(I - |N|)y = e$ and N is the strictly
upper triangular part of T.
3 $F_0 = f(\sigma)I_n$
4 $P = M$
5 for $s = 1\colon\infty$
6 $\quad F_s = F_{s-1} + f^{(s)}(\sigma)P$
7 $\quad P = PM/(s+1)$
8 $\quad$ if $\|F_s - F_{s-1}\|_F \le \text{tol}\|F_s\|_F$
$\quad\quad$ % Successive terms are close so check the truncation error bound.
9 $\quad\quad$ Estimate or bound $\Delta = \max_{0\le r\le n-1} \omega_{s+r}/r!$, where
$\quad\quad$ $\omega_{s+r} = \sup_{z\in\Omega} |f^{(s+r)}(z)|$, with Ω a closed convex set containing $\Lambda(T)$.
10 $\quad\quad$ if $\mu\Delta\|P\|_F \le \text{tol}\|F_s\|_F$, quit, end
11 $\quad$ end
12 end

Unless we are able to exploit particular properties of f, we can in practice take $\omega_{s+r} = \max\{\, |f^{(s+r)}(\lambda_i)| : \lambda_i \in \Lambda(T)\,\}$.

Algorithm 9.4 costs $O(n^4)$ flops, since even if T has constant diagonal, so that M is nilpotent with $M^n = 0$, the algorithm may need to form the first $n-1$ powers of M. Although we usually insist on $O(n^3)$ flops algorithms in numerical linear algebra, this higher order operation count is mitigated by two factors. First, n here is the size of a block, and in most cases the blocks will be of much smaller dimension than the original matrix. Second, M is an upper triangular matrix, so forming all the powers $M^2, \dots, M^{n-1}$ costs $n^4/3$ flops—a factor 6 less than the flop count for multiplying full matrices.

Since in our overall $f(A)$ algorithm we are not able to impose a fixed bound on the spread $\max_{i,j} |t_{ii} - t_{jj}|$ of the diagonal of T, Algorithm 9.4 is suitable in its stated form only for functions that have a Taylor series with an infinite radius of convergence, such as exp, cos, sin, cosh, and sinh.

We now turn to the effects of rounding errors on Algorithm 9.4. Ignoring truncation errors, standard error analysis shows that the best possible forward error bound is of the form

$$|F - \widehat{F}| \le \widetilde{\gamma}_n \sum_{k=0}^{\infty} \frac{|f^{(k)}(\sigma)|}{k!} |M|^k.$$

If there is heavy cancellation in the sum (9.3) then a large relative error $\|F - \widehat{F}\|/\|F\|$ is possible. This danger is well known, particularly in the case of the matrix exponential (see Chapter 10). A mitigating factor here is that our matrix T is chosen to have eigenvalues that are clustered, which tends to limit the amount of cancellation in the sum. However, for sufficiently far from normal T, damaging cancellation can take place. For general functions there is little we can do to improve the accuracy; for particular f we can of course apply alternative methods, as illustrated in Chapters 10 and 11.

9.2. Evaluating the Upper Triangular Part of $f(T)$

Let T denote the triangular Schur factor of A, which we assume to have been reordered and blocked so that T_{ii} and T_{jj} have no eigenvalue in common for $i \neq j$. To evaluate the upper triangular part of $F = f(T)$ we use the block form (4.19) of Parlett's recurrence, which we rewrite here as

$$T_{ii}F_{ij} - F_{ij}T_{jj} = F_{ii}T_{ij} - T_{ij}F_{jj} + \sum_{k=i+1}^{j-1} (F_{ik}T_{kj} - T_{ik}F_{kj}). \tag{9.7}$$

This Sylvester equation is nonsingular and it is easy to see that F_{ij} can be computed a column at a time from first to last, with each column obtained as the solution of a triangular system. Of particular concern is the propagation of errors in the recurrence. These errors are of two sources: errors in the evaluation of the diagonal blocks F_{ii}, and rounding errors in the formation and solution of (9.7). To gain insight into both types of error we consider the residual of the computed solution $\widehat{F}$:

$$T\widehat{F} - \widehat{F}T =: R, \tag{9.8}$$

where R_{ij} is the residual from the solution of the Sylvester equation (9.7). Although it is possible to obtain precise bounds on R, these are not important to our argument. Writing $\widehat{F} = F + \Delta F$, on subtracting $TF - FT = 0$ from (9.8) we obtain

$$T\Delta F - \Delta F T = R.$$

As for the original equation $TF - FT = 0$, this equation uniquely determines the off-diagonal blocks ΔF in terms of the diagonal blocks. Equating (i, j) blocks yields

$$\begin{aligned} T_{ii}\Delta F_{ij} - \Delta F_{ij}T_{jj} &= R_{ij} + \Delta F_{ii}T_{ij} - T_{ij}\Delta F_{jj} + \sum_{k=i+1}^{j-1} (\Delta F_{ik}T_{kj} - T_{ik}\Delta F_{kj}) \\ &=: B_{ij}, \end{aligned} \tag{9.9}$$

and these equations can be solved to determine ΔF_{ij} a block superdiagonal at a time.

It is straightforward to show that

$$\|\Delta F_{ij}\|_F \leq \operatorname{sep}(T_{ii}, T_{jj})^{-1} \|B_{ij}\|_F, \tag{9.10}$$

where sep is the *separation* of T_{ii} and T_{jj}:

$$\operatorname{sep}(T_{ii}, T_{jj}) = \min_{X \neq 0} \frac{\|T_{ii}X - XT_{jj}\|_F}{\|X\|_F}.$$

It follows that rounding errors introduced during the stage at which F_{ij} is computed (i.e., represented by R_{ij}) can lead to an error ΔF_{ij} of norm proportional to $\operatorname{sep}(T_{ii}, T_{jj})^{-1}\|R_{ij}\|$. Moreover, earlier errors (represented by the ΔF_{ij} terms on the right-hand side of (9.9)) can be magnified by a factor $\operatorname{sep}(T_{ii}, T_{jj})^{-1}$. It is also clear from (9.9) that even if $\operatorname{sep}(T_{ii}, T_{jj})^{-1}$ is not large, serious growth of errors in the recurrence (9.9) is possible if some off-diagonal blocks T_{ij} are large.

To minimize the bounds (9.10) for all i and j we need the blocks T_{ii} to be as well separated as possible in the sense of sep. However, trying to maximize the separations between the diagonal blocks T_{ii} tends to produce larger blocks with less tightly clustered eigenvalues, which increases the difficulty of evaluating $f(T_{ii})$, so any strategy for reordering the Schur form is necessarily a compromise.

Computing $\operatorname{sep}(T_{ii}, T_{jj})$ exactly when both blocks are $m \times m$ costs $O(m^4)$ flops, while condition estimation techniques allow an estimate to be computed at the cost of solving a few Sylvester equations, that is, in $O(m^3)$ flops [87, 1984], [270, 1988], [324, 1992]. It is unclear how to develop a reordering and blocking strategy for producing "large seps" at reasonable cost; in particular, it is unclear how to define "large." Indeed the maximal separations are likely to be connected with the conditioning of $f(T)$, but little or nothing is known about any such connections. More generally, how to characterize matrices for which the condition number of f is large is not well understood, even for the matrix exponential (see Section 10.2). Recalling the equivalence mentioned in Section 4.7 between block diagonalization and the use of the Parlett recurrence, a result of Gu [232, 1995] provides further indication of the difficulty of maximizing the seps: he shows that, given a constant τ, finding a similarity transformation with condition number bounded by τ that block diagonalizes (with at least two diagonal blocks) a triangular matrix is NP-hard.

In the next section we will adopt a reordering and blocking strategy that bounds the right-hand side of the approximation

$$\operatorname{sep}(T_{ii}, T_{jj})^{-1} \approx \frac{1}{\min\{\, |\lambda - \mu| : \lambda \in \Lambda(T_{ii}),\ \mu \in \Lambda(T_{jj})\,\}}$$

by the reciprocal of a given tolerance. The right-hand side is a lower bound for the left that can be arbitrarily weak, but it is a reasonable approximation for matrices not too far from being normal.

It is natural to look for ways of improving the accuracy of the computed $\widehat{F}$ from the Parlett recurrence. One candidate is fixed precision iterative refinement of the systems (9.7). However, these systems are essentially triangular, and standard error analysis shows that the backward error is already small componentwise [276, 2002, Thm. 8.5]; fixed precision iterative refinement therefore cannot help. The only possibility is to use extended precision when solving the systems.

9.3. Reordering and Blocking the Schur Form

We wish to reorder the upper triangular Schur factor T into a partitioned upper triangular matrix $\widetilde{T} = U^*TU = (\widetilde{T}_{ij})$, where U is unitary and two conditions hold:

1. *separation between blocks*:

$$\min\{\, |\lambda - \mu| : \lambda \in \Lambda(\widetilde{T}_{ii}),\ \mu \in \Lambda(\widetilde{T}_{jj}),\ i \neq j\,\} > \delta, \tag{9.11}$$

2. *separation within blocks*: for every block $\widetilde{T}_{ii}$ with dimension bigger than 1, for every $\lambda \in \Lambda(\widetilde{T}_{ii})$ there is a $\mu \in \Lambda(\widetilde{T}_{ii})$ with $\mu \neq \lambda$ such that $|\lambda - \mu| \le \delta$.

Here, $\delta > 0$ is a blocking parameter. The second property implies that for $\widetilde{T}_{ii} \in \mathbb{R}^{m\times m}$ $(m > 1)$

$$\max\{\, |\lambda - \mu| : \lambda, \mu \in \Lambda(\widetilde{T}_{ii}),\ \lambda \neq \mu\,\} \le (m-1)\delta,$$

and this bound is attained when, for example, $\Lambda(\widetilde{T}_{ii}) = \{\delta, 2\delta, \dots, m\delta\}$.

The following algorithm is the first step in obtaining such an ordering. It can be interpreted as finding the connected components of the graph on the eigenvalues of T in which there is an edge between two nodes if the corresponding eigenvalues are a distance at most δ apart.

Algorithm 9.5 (block pattern). Given a triangular matrix $T \in \mathbb{C}^{n\times n}$ with eigenvalues $\lambda_i \equiv t_{ii}$ and a blocking parameter $\delta > 0$, this algorithm produces a block pattern, defined by an integer vector q, for the block Parlett recurrence: the eigenvalue λ_i is assigned to the set S_{q_i}, and it satisfies the conditions that $\min\{|\lambda_i - \lambda_j| \colon \lambda_i \in S_p, \lambda_j \in S_q, p \neq q\} > \delta$ and, for each set S_i with more than one element, every element of S_i is within distance at most δ from some other element in the set. For each set S_q, all the eigenvalues in S_q are intended to appear together in an upper triangular block $\widetilde{T}_{ii}$ of $\widetilde{T} = U^*TU$.

```
1  p = 1
2  Initialize the S_p to empty sets.
3  for i = 1: n
4      if λ_i ∉ S_q for all 1 ≤ q < p
5        Assign λ_i to S_p.
6        p = p + 1
7      end
8      for j = i + 1: n
9          Denote by S_{q_i} the set that contains λ_i.
10         if λ_j ∉ S_{q_i}
11           if |λ_i − λ_j| ≤ δ
12             if λ_j ∉ S_k for all 1 ≤ k < p
13               Assign λ_j to S_{q_i}.
14             else
15               Move the elements of S_{max(q_i,q_j)} to S_{min(q_i,q_j)}.
16               Reduce by 1 the indices of sets S_q for q > max(q_i, q_j).
17               p = p − 1
18             end
19           end
20         end
21     end
22 end
```

Algorithm 9.5 provides a mapping from each eigenvalue λ_i of T to an integer q_i such that the set S_{q_i} contains λ_i. Our remaining problem is equivalent to finding a method for swapping adjacent elements in q to obtain a confluent permutation q'. A confluent permutation of n integers, $q_1, \dots, q_n$, is a permutation such that any repeated integers q_i are next to each other. For example, there are 3! confluent permutations of $(1, 2, 1, 3, 2, 1)$, which include $(1, 1, 1, 3, 2, 2)$ and $(3, 2, 2, 1, 1, 1)$. Ideally we would like a confluent permutation that requires a minimal number of swaps to transform q to q'. Ng [448, 1984] notes that finding such a permutation is an NP-complete problem. He proves that the minimum number of swaps required to obtain a given confluent permutation is bounded above by $\frac{n^2}{2}(1 - \frac{1}{k})$, where k is the number of distinct q_i, and that this bound is attainable [448, 1984, Thm. A.1]. In practice, since the QR algorithm tends to order the eigenvalues by absolute value in the Schur form, complicated strategies for determining a confluent permutation are not needed. The following method works well in practice: find the average index of the integers in q and then order the integers in q' by ascending average index. If we take our example $(1, 2, 1, 3, 2, 1)$ and let g_k denote the average index of the integer k, we see that $g_1 = (1 + 3 + 6)/3 = 3\frac{1}{3}$, $g_2 = (2 + 5)/2 = 3\frac{1}{2}$, and $g_3 = 4$. Therefore we try to obtain the confluent permutation $q' = (1, 1, 1, 2, 2, 3)$ by a sequence of swaps of adjacent elements:

$$\begin{aligned} q = (1,2,1,3,2,1) &\to (1,1,2,3,2,1) \to (1,1,2,3,1,2) \\ &\to (1,1,2,1,3,2) \to (1,1,1,2,3,2) \to (1,1,1,2,2,3) = q'. \end{aligned} \tag{9.12}$$

Swapping two adjacent diagonal elements of T requires $20n$ flops, plus another $20n$ flops to update the Schur vectors, so the cost of the swapping is $40n$ times the number of swaps. The total cost is usually small compared with the overall cost of the algorithm.

9.4. Schur–Parlett Algorithm for $f(A)$

The complete algorithm for computing $f(A)$ is as follows.

Algorithm 9.6 (Schur–Parlett algorithm). Given $A \in \mathbb{C}^{n\times n}$, a function f analytic on a closed convex set Ω whose interior contains the eigenvalues of A, and the ability to evaluate derivatives of f, this algorithm computes $F = f(A)$.

```
1  Compute the Schur decomposition A = QTQ* (Q unitary, T upper triangular).
2  If T is diagonal, F = f(T), goto line 12, end
3  Using Algorithm 9.5 with blocking parameter δ = 0.1,
   assign each eigenvalue λ_i to a set S_{q_i}.
4  Choose a confluent permutation q' of q ordered by average index.
5  Reorder T according to q' and update Q.
   % Now A = QTQ* is our reordered Schur decomposition, with block m × m T.
6  for i = 1:m
7      Use Algorithm 9.4 to evaluate F_ii = f(T_ii).
8      for j = i − 1: −1: 1
9          Solve the Sylvester equation (9.7) for F_ij.
10     end
11 end
12 F = QFQ*
```

The cost of Algorithm 9.6 depends greatly on the eigenvalue distribution of A, and is roughly between $28n^3$ flops and $n^4/3$ flops. Note that Q, and hence F, can be kept in factored form, with a significant computational saving. This is appropriate if F needs just to be applied to a few vectors, for example.

We have set the blocking parameter $\delta = 0.1$, which our experiments indicate is as good a default choice as any. The optimal choice of δ in terms of cost or accuracy is problem-dependent.

Algorithm 9.6 has a property noted as being desirable by Parlett and Ng [462, 1985]: it acts simply on simple cases. Specifically, if A is normal, so that the Schur decomposition is $A = QDQ^*$ with D diagonal, the algorithm simply evaluates $f(A) = Qf(D)Q^*$. At another extreme, if A has just one eigenvalue of multiplicity n, then the algorithm works with a single block, $T_{11} \equiv T$, and evaluates $f(T_{11})$ via its Taylor series expanded about the eigenvalue.

Another attraction of Algorithm 9.6 is that it allows a function of the form $f(A) = \sum_i f_i(A)$ (e.g., $f(A) = \sin A + \cos A$) to be computed with less work than is required to compute each $f_i(A)$ separately, since the Schur decomposition and its reordering need only be computed once.

Reordering the Schur form is a nontrivial subject, the state of the art of which is described by Bai and Demmel [28, 1993]. The algorithm described therein uses unitary similarities and effects a sequence of swaps of adjacent blocks as in (9.12). The algorithm has guaranteed backward stability and, for swapping only 1×1 blocks as we are here, always succeeds. LAPACK routine `xTREXC` implements this algorithm and is called by the higher level Schur reordering routine `xTRSEN` [12, 1999]; the MATLAB function `ordschur` calls `xTRSEN`.

Algorithm 9.6 was developed by Davies and Higham [135, 2003] and is implemented by the MATLAB function `funm`.

For real matrices, it might seem that by using the real Schur decomposition in the first step of Algorithm 9.6 it would be possible to work entirely in real arithmetic. However, the algorithm's strategy of placing eigenvalues that are not close in different blocks requires splitting complex conjugate pairs of eigenvalues having large imaginary parts, forcing complex arithmetic, so the algorithm does not in general lend itself to exploitation of the real Schur form. However, if A is real and normal then the real Schur decomposition is block diagonal, no reordering is necessary, and Algorithm 9.6 can be reduced to computation of the Schur form and evaluation of f on the diagonal blocks. For the 2×2 diagonal blocks Problem 9.2 provides appropriate formulae.

We focus for a moment on some negative aspects of the algorithm revealed by numerical experiments in [135, 2003]. The algorithm can be unstable, in the sense that the normwise relative error can greatly exceed $\mathrm{cond}_{\mathrm{rel}}(f, A)u$. Changing the blocking parameter δ (say from 0.1 to 0.2) may produce a different blocking that cures the instability. However, instability can be present for all choices of δ. Moreover, instability can be present for all nontrivial blockings (i.e., any blocking with more than one block)—some of which it might not be possible to generate by an appropriate choice of δ in the algorithm. The latter point indicates a fundamental weakness of the Parlett recurrence.

To illustrate the typical behaviour of Algorithm 9.6 we present a numerical experiment with f the exponential function. We took 71 test matrices, which include some from MATLAB (in particular, from the `gallery` function), some from the Matrix Computation Toolbox [264], and test matrices from the e^A literature; most matrices are 10×10, with a few having smaller dimension. We evaluated the normwise relative errors of the computed matrices from a modified version `funm_mod` of MATLAB 7.6's `funm` (when invoked as `funm(A,@exp)` the modified version uses Algorithm 9.4 for the diagonal blocks), where the "exact" e^A is obtained at 100 digit precision using MATLAB's Symbolic Math Toolbox.

Figure 9.1 displays the relative errors together with a solid line representing $\mathrm{cond}_{\mathrm{rel}}(\exp, A)u$, where $\mathrm{cond}_{\mathrm{rel}}(\exp, A)$ is computed using Algorithm 3.17 with Algorithm 10.27, and the results are sorted by decreasing condition number; the norm is the Frobenius norm. For `funm` to perform in a forward stable manner its error should lie not too far above this line on the graph; note that we must accept some dependence of the error on n.

The errors are mostly very satisfactory, but with two exceptions. The first exception is the MATLAB matrix `gallery('chebspec',10)`. This matrix is similar to a Jordan block of size 10 with eigenvalue 0 (and hence is nilpotent), modulo the rounding errors in its construction. The computed eigenvalues lie roughly on a circle with centre 0 and radius 0.2; this is the most difficult distribution for Algorithm 9.6 to handle. With the default $\delta = 0.1$, `funm_mod` chooses a blocking with eight 1×1 blocks and one 2×2 block. This leads to an error $\approx 10^{-7}$, which greatly exceeds

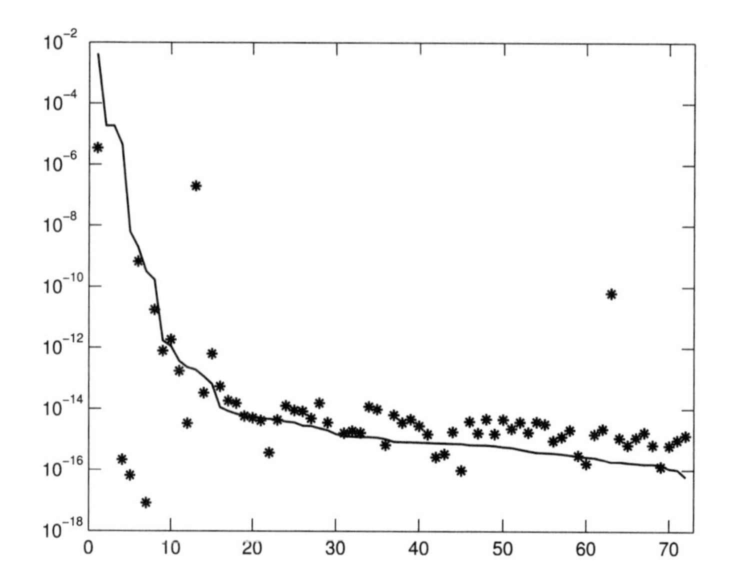

Figure 9.1. *Normwise relative errors for* `funm_mod` (*) *and* $\mathrm{cond}_{\mathrm{rel}}(\exp, A)u$ (*solid line*).

$\mathrm{cond}_{\mathrm{rel}}(\exp, A)u \approx 10^{-13}$. However, increasing δ to 0.2 produces just one 10×10 block, which after using 20 terms of the Taylor series leads to an error $\approx 10^{-13}$. The other exceptional matrix is `gallery('forsythe',10)`, which is a Jordan block of size 10 with eigenvalue 0 except for a (10,1) entry of $u^{1/2}$. The computed eigenvalue distribution and the behaviour for $\delta = 0.1$ and $\delta = 0.2$ are similar to that for the chebspec matrix.

The main properties of Algorithm 9.6 can be summarized as follows.

1. The algorithm requires $O(n^3)$ flops unless close or repeated eigenvalues force a large block T_{ii} to be chosen, in which case the operation count can be up to $n^4/3$ flops.

2. The algorithm needs to evaluate derivatives of the function when there are blocks of dimension greater than 1.

3. In practice the algorithm usually performs in a forward stable manner. However, the error can be greater than the condition of the problem warrants; if this behaviour is detected then a reasonable strategy is to recompute the blocking with a larger δ.

Points 1 and 2 are the prices to be paid for catering for general functions and nonnormal matrices with possibly repeated eigenvalues.

9.5. Preprocessing

In an attempt to improve the accuracy of Algorithm 9.6 we might try to preprocess the data before applying a particular stage of the algorithm, using one or more of the techniques discussed in Section 4.10.

Translation has no effect on our algorithm. Algorithm 9.4 for evaluating the Taylor series already translates the diagonal blocks, and further translations before applying the Parlett recurrence are easily seen to have no effect, because (9.7) is invariant under translations $T \to T - \alpha I$ and $F \to F - \beta I$.

A diagonal similarity transformation could be applied at any stage of the algorithm and then undone later. For example, such a transformation could be used in conjunction with Parlett's recurrence in order to make $U := D^{-1}TD$ less nonnormal than T and to increase the separations between diagonal blocks. In fact, by choosing D of the form $D = \mathrm{diag}(\theta^{n-1}, \dots, 1)$ we can make U arbitrarily close to diagonal form. Unfortunately, no practical benefit is gained: Parlett's recurrence involves solving triangular systems and the substitution algorithm is invariant under diagonal scalings (at least, as long as they involve only powers of the machine base). Similar comments apply to the evaluation of the Taylor series in Algorithm 9.4.

It may be beneficial to apply balancing at the outset, prior to computing the Schur decomposition, particularly when we are dealing with badly scaled matrices.

9.6. Notes and References

This chapter is based on Davies and Higham [135, 2003].

The use of the Taylor series expansion (9.3) was suggested by Stewart [438, 2003, Method 18] for the matrix exponential and investigated for general f by Kågström [323, 1977].

For more on the separation of two matrices see Golub and Van Loan [224, 1996, Sec. 7.2.4], Stewart [536, 1973], and Varah [599, 1979].

Parlett's recurrence was used by Kågström in his thesis [323, 1977]. There are three main differences between Kågström's approach and that of this chapter. First, he uses an initial block diagonalization, carried out with the method of Kågström and Ruhe [325, 1980], whereas we compute a Schur decomposition and reorder the triangular form. Second, Kågström uses the scalar rather than the block form of the Parlett recurrence and when t_{ii} and t_{jj} are sufficiently close he uses the explicit formula for f_{ij} in Theorem 4.11 with derivatives substituted for the divided differences. Finally, we use a combination of Taylor series and the Parlett recurrence, whereas Kågström investigates the separate use of these two tools upon his block diagonal form.

Problems

9.1. Prove Theorem 9.2.

9.2. Let $A \in \mathbb{R}^{2\times 2}$ have distinct complex conjugate eigenvalues $\lambda = \theta + i\mu$ and $\overline{\lambda}$ and let $f(\lambda) = \alpha + i\beta$. Show that $f(A) = \alpha I + \beta\mu^{-1}(A - \theta I)$. If A is normal how does this formula simplify?

Chapter 10
Matrix Exponential

The matrix exponential is by far the most studied matrix function. The interest in it stems from its key role in the solution of differential equations, as explained in Chapter 2. Depending on the application, the problem may be to compute e^A for a given A, to compute e^{At} for a fixed A and many t, or to apply e^A or e^{At} to a vector (cf. (2.3)); the precise task affects the choice of method.

Many methods have been proposed for computing e^A, typically based on one of the formulae summarized in Table 10.1. Most of them are of little practical interest when numerical stability, computational cost, and range of applicability are taken into account. In this chapter we make no attempt to survey the range of existing methods but instead restrict to a few of proven practical value.

A broad assortment of methods is skilfully classified and analyzed in the classic "Nineteen dubious ways" paper of Moler and Van Loan [437, 1978], reprinted with an update in [438, 2003]. The conclusion of the paper is that there are three candidates for best method. One of these—employing methods for the numerical solution of ODEs—is outside the scope of this book, and in fact the converse approach of using the matrix exponential in the solution of differential equations has received increasing attention in recent years (see Section 2.1.1). The others—the scaling and squaring method and the use of the Schur form—are treated here in some detail. The scaling and squaring method has become the most widely used, not least because it is the method implemented in MATLAB.

This chapter begins with a summary of some basic properties, followed by results on the Fréchet derivative and the conditioning of the e^A problem. The scaling and squaring method based on underlying Padé approximation is then described in some detail. Three approaches based on the Schur decomposition are outlined. The behaviour of the scaling and squaring method and two versions of the Schur method of the previous chapter is illustrated on a variety of test problems. Several approaches for approximating the Fréchet derivative and estimating its norm are explained. A final section treats miscellaneous topics: best rational L_∞ approximation, essentially nonnegative matrices, preprocessing, and the ψ functions that we first saw in Section 2.1.

10.1. Basic Properties

The matrix exponential is defined for $A \in \mathbb{C}^{n\times n}$ by

$$e^A = I + A + \frac{A^2}{2!} + \frac{A^3}{3!} + \cdots. \tag{10.1}$$

We know from Theorem 4.7 that the series has infinite radius of convergence. Hence we can differentiate term by term to obtain $\frac{d}{dt}e^{At} = Ae^{At} = e^{At}A$.

Table 10.1. *Some formulae for e^A.*

Power series	Limit	Scaling and squaring
$I + A + \frac{A^2}{2!} + \frac{A^3}{3!} + \cdots$	$\lim_{s\to\infty} (I + A/s)^s$	$(e^{A/2^s})^{2^s}$
Cauchy integral	**Jordan form**	**Interpolation**
$\frac{1}{2\pi i}\int_\Gamma e^z (zI - A)^{-1}\, dz$	$Z \operatorname{diag}(e^{J_k}) Z^{-1}$	$\sum_{i=1}^{n} f[\lambda_1,\ldots,\lambda_i] \prod_{j=1}^{i-1}(A - \lambda_j I)$
Differential system	**Schur form**	**Padé approximation**
$Y'(t) = AY(t),\ Y(0) = I$	$Q \operatorname{diag}(e^T) Q^*$	$p_{km}(A) q_{km}(A)^{-1}$

Another representation is

$$e^A = \lim_{s\to\infty} (I + A/s)^s. \tag{10.2}$$

This formula is the limit of the first order Taylor expansion of A/s raised to the power $s \in \mathbb{Z}$. More generally, we can take the limit as $r \to \infty$ or $s \to \infty$ of r terms of the Taylor expansion of A/s raised to the power s, thereby generalizing both (10.1) and (10.2). The next result shows that this general formula yields e^A and it also provides an error bound for finite r and s.

Theorem 10.1 (Suzuki). *For $A \in \mathbb{C}^{n\times n}$, let*

$$T_{r,s} = \left[\sum_{i=0}^{r} \frac{1}{i!}\left(\frac{A}{s}\right)^i \right]^s. \tag{10.3}$$

Then for any consistent matrix norm

$$\|e^A - T_{r,s}\| \le \frac{\|A\|^{r+1}}{s^r (r+1)!} e^{\|A\|} \tag{10.4}$$

and $\lim_{r\to\infty} T_{r,s}(A) = \lim_{s\to\infty} T_{r,s}(A) = e^A$.

Proof. Let $T = \sum_{i=0}^{r} \frac{1}{i!}(A/s)^i$ and $B = e^{A/s}$. Then, since B and T commute,

$$e^A - T_{r,s} = B^s - T^s = (B - T)(B^{s-1} + B^{s-2}T + \cdots + T^{s-1}).$$

Hence

$$\|e^A - T_{r,s}\| \le \|B - T\|\ s \max_{i=0:s-1} \|B\|^i \|T\|^{s-i-1}.$$

Now $\|T\| \le \sum_{i=0}^{r} \frac{1}{i!}(\|A\|/s)^i \le e^{\|A\|/s}$, and $\|B\|$ satisfies the same bound, so

$$\|e^A - T_{r,s}\| \le s\, \|e^{A/s} - T\|\, e^{\frac{s-1}{s}\|A\|}.$$

But, by Theorem 4.8,

$$\|e^{A/s} - T\| \le \frac{1}{(r+1)!}\left(\frac{\|A\|}{s}\right)^{r+1} e^{\|A\|/s}.$$

This yields (10.4) and the limits are immediate. $\square$

Although $e^{A+B} \neq e^A e^B$ in general, the equality holds when A and B commute, as the following well-known result shows.

Theorem 10.2. *For $A, B \in \mathbb{C}^{n\times n}$, $e^{(A+B)t} = e^{At}e^{Bt}$ for all t if and only if $AB = BA$.*

Proof. If $AB = BA$ then all the terms in the power series expansions of $e^{(A+B)t}$ and $e^{At}e^{Bt}$ commute and so these matrices are equal for the same reasons as in the scalar case. If $e^{(A+B)t} = e^{At}e^{Bt}$ for all t then equating coefficients of t^2 in the power series expansions of both sides yields $(AB + BA)/2 = AB$ or $AB = BA$. $\square$

The commutativity of A and B is not necessary for $e^{A+B} = e^A e^B$ to hold, as the example $A = \left[\begin{smallmatrix} 0 & 0 \\ 0 & 2\pi i \end{smallmatrix}\right]$, $B = \left[\begin{smallmatrix} 0 & 1 \\ 0 & 2\pi i \end{smallmatrix}\right]$ shows ($e^{A+B} = e^A = e^B = I$). But if A and B have algebraic entries then their commutativity is necessary for $e^{A+B} = e^A e^B = e^B e^A$ to hold. An algebraic number is defined by the property that it is a root of a polynomial with rational (or equivalently, integer) coefficients.

Theorem 10.3 (Wermuth). *Let $A \in \mathbb{C}^{n\times n}$ and $B \in \mathbb{C}^{n\times n}$ have algebraic elements and let $n \ge 2$. Then $e^A e^B = e^B e^A$ if and only if $AB = BA$.*

Proof. The "if" part is trivial. For the "only if", note that Lindemann's theorem on the transcendence of π implies that no two eigenvalues of A differ by a nonzero integer multiple of $2\pi i$, since the eigenvalues of A, being the roots of a polynomial with algebraic coefficients, are themselves algebraic. Hence A is a primary logarithm of e^A, since the nonprimary logarithms (if any) are characterized by two copies of a repeated eigenvalue being mapped to different logarithms, which must differ by a nonzero integer multiple of $2\pi i$ (see Theorem 1.28). Thus A is a polynomial in e^A, and likewise B is a polynomial in e^B. Since e^A and e^B commute, A and B commute. $\square$

The *Lie bracket*, or *commutator*, is defined by $[A, B] = AB - BA$. It can be shown [296, 1991, Prob. 6.5.35] that if A and B commute with $[A, B]$ then

$$e^A e^B = e^{A+B+\frac{1}{2}[A,B]} = e^{A+B} e^{\frac{1}{2}[A,B]} = e^{\frac{1}{2}[A,B]} e^{A+B}.$$

This relation can be seen as the first two terms in the following general result connecting $e^{tA}e^{tB}$ and $e^{t(A+B)}$.

Theorem 10.4 (Baker–Campbell–Hausdorff formula). *For $A, B \in \mathbb{C}^{n\times n}$ we have*

$$e^{tA}e^{tB} = \exp\Big(t(A+B) + \frac{t^2}{2}[A, B] + \frac{t^3}{12}([A, [A, B]] - [B, [A, B]]) + O(t^4)\Big), \quad (10.5)$$

where the $O(t^4)$ term is expressible as an infinite series of iterated commutators of A and B. $\square$

Note that the third order term in (10.5) can be written in other ways, such as $(t^3/12)([[A,B],B]+[[B,A],A])$. A kind of dual to (10.5) is an infinite product formula of Zassenhaus.

Theorem 10.5 (Zassenhaus formula). *For $A, B \in \mathbb{C}^{n\times n}$ we have*

$$e^{t(A+B)} = e^{tA}e^{tB}e^{C_2t^2}e^{C_3t^3}\dots, \tag{10.6a}$$

$$C_2 = -[A,B]/2, \quad C_3 = [B,[A,B]]/3 + [A,[A,B]]/6, \quad \dots . \quad \square \tag{10.6b}$$

Yet another variant of relations between $e^{tA}e^{tB}$ and $e^{t(A+B)}$ is $e^{t(A+B)} = e^{tA}e^{tB} + \sum_{i=2}^{\infty} E_i t^i$, for which Richmond [489, 1989] gives recurrences for the E_i. In a different vein, the Strang splitting breaks e^{tB} into its square root factors and thereby provides a second order accurate approximation: $e^{t(A+B)} = e^{tB/2}e^{tA}e^{tB/2} + O(t^3)$.

Another way to relate the exponential of a sum to a related product of exponentials is via the limit in the next result.

Theorem 10.6 (Suzuki). *For $A_1, \dots, A_p \in \mathbb{C}^{n\times n}$ and any consistent matrix norm,*

$$\|e^{A_1+\cdots+A_p} - \left(e^{A_1/m}\dots e^{A_p/m}\right)^m\| \le \frac{2}{m}\left(\sum_{j=1}^{p}\|A_j\|\right)^2 \exp\left(\frac{m+2}{m}\sum_{j=1}^{p}\|A_j\|\right). \tag{10.7}$$

Hence

$$e^{A_1+\cdots+A_p} = \lim_{m\to\infty}\left(e^{A_1/m}\dots e^{A_p/m}\right)^m. \tag{10.8}$$

Proof. Let $G = e^{(A_1+\cdots+A_p)/m}$ and $H = e^{A_1/m}\dots e^{A_p/m}$. Then we need to bound $\|G^m - H^m\|$. Using Lemma B.4 and Theorem 10.10 we have

$$\begin{aligned}\|G^m - H^m\| &\le \|G-H\|\left(\|G\|^{m-1} + \|G\|^{m-2}\|H\| + \cdots + \|H\|^{m-1}\right)\\ &\le m\|G-H\|\, e^{\frac{m-1}{m}\sum_{j=1}^{p}\|A_j\|}.\end{aligned}$$

Now

$$\begin{aligned}\|G-H\| \le \|H\|\,\|GH^{-1} - I\| &\le \|H\|\left(e^{\frac{2}{m}\sum_{j=1}^{p}\|A_j\|} - \left(1 + \frac{2}{m}\sum_{j=1}^{p}\|A_j\|\right)\right)\\ &\le \|H\|\frac{1}{2}\left(\frac{2}{m}\sum_{j=1}^{p}\|A_j\|\right)^2 e^{\frac{2}{m}\sum_{j=1}^{p}\|A_j\|},\end{aligned}$$

where for the second inequality we used $\|GH^{-1}\| \le \|G\|\|H^{-1}\| \le e^{\frac{2}{m}\sum_{j=1}^{p}\|A_j\|}$ and the fact that the first two terms of the expansion of $GH^{-1} - I$ are zero and for the third inequality we used a Taylor series with remainder (Theorem 4.8). Hence

$$\|G^m - H^m\| \le m\, e^{\frac{1}{m}\sum_{j=1}^{p}\|A_j\|}\,\frac{2}{m^2}\left(\sum_{j=1}^{p}\|A_j\|\right)^2 e^{\frac{2}{m}\sum_{j=1}^{p}\|A_j\|}\, e^{\frac{m-1}{m}\sum_{j=1}^{p}\|A_j\|},$$

as required. $\square$

Corollary 10.7 (Trotter product formula). *For $A, B \in \mathbb{C}^{n\times n}$,*

$$e^{A+B} = \lim_{m\to\infty}\left(e^{A/m}e^{B/m}\right)^m. \quad \square \tag{10.9}$$

The following result provides a better error bound for $p = 2$ than (10.7) in the sense that the bound is small when the commutator is small.

Theorem 10.8. *For $A, B \in \mathbb{C}^{n\times n}$,*

$$\|e^{A+B} - \bigl(e^{A/m}e^{B/m}\bigr)^m\|_2 \le \frac{\|[A,B]\|_2}{2m}\, e^{\|A\|_2+\|B\|_2}.$$

Proof. See Moler and Van Loan [438, 2003, App. 2]. □

The commutativity requirement in Theorem 10.2 can be removed if the usual summation and multiplication operators are replaced by their Kronecker counterparts (see Section B.13 for the definitions and basic properties of the Kronecker product $\otimes$ and Kronecker sum $\oplus$).

Theorem 10.9. *Let $A \in \mathbb{C}^{n\times n}$ and $B \in \mathbb{C}^{m\times m}$. Then $e^{A\otimes I} = e^A \otimes I$, $e^{I\otimes B} = I \otimes e^B$, and $e^{A\oplus B} = e^A \otimes e^B$.*

Proof. The first two relations are easily obtained from the power series (10.1) and also follow from Theorem 1.13 (h), (i). The proof of the second relation is analogous. Since $A\otimes I$ and $I\otimes B$ commute, $e^{A\oplus B} = e^{A\otimes I + I\otimes B} = e^{A\otimes I}e^{I\otimes B} = (e^A\otimes I)(I\otimes e^B) = e^A \otimes e^B$. □

Bounds for the norm of the matrix exponential are of great interest. We state just three of the many bounds available in the literature. We need the spectral abscissa,

$$\alpha(A) = \max\{\, \mathrm{Re}\,\lambda_i : \lambda_i \in \Lambda(A)\,\}, \tag{10.10}$$

and the logarithmic norm [546, 1975]

$$\mu(A) = \lambda_{\max}\bigl(\tfrac{1}{2}(A + A^*)\bigr). \tag{10.11}$$

Theorem 10.10. *For $A \in \mathbb{C}^{n\times n}$ and any subordinate matrix norm,*

$$e^{-\|A\|} \le \|e^A\| \le e^{\|A\|}.$$

The upper bound is attained in the ∞-norm if A is a nonnegative multiple of a stochastic matrix.

Proof. The upper bound is immediate on taking norms in the power series for e^A. Since $I = e^A e^{-A}$ we have $1 \le \|e^A\|\,\|e^{-A}\| \le \|e^A\|\,e^{\|A\|}$, which gives the lower bound. For the last part, write $A = cP$, where $c \ge 0$ and P is stochastic. Then $Af = cPf = cf$, where $f = [1,1,\dots,1]^T$, so $c = \|A\|_\infty$. Hence, since $e^A \ge 0$, $\|e^A\|_\infty = \|e^A f\|_\infty = \|(I + A + A^2/2! + \cdots)f\|_\infty = \|(1 + c + c^2/2! + \cdots)f\|_\infty = e^c = e^{\|A\|_\infty}$. □

Theorem 10.11. *For $A \in \mathbb{C}^{n\times n}$ and any unitarily invariant norm,*

$$\|e^A\| \le \|e^{(A+A^*)/2}\|.$$

Proof. See Bhatia [64, 1997, Thm. IX.3.1]. □

For the 2-norm, Theorem 10.11 says that $\|e^A\|_2 \le e^{\mu(A)}$, a bound due to Dahlquist [128, 1958].

Theorem 10.12. *Let $A \in \mathbb{C}^{n\times n}$ have the Schur decomposition $A = Q(D+N)Q^*$, where D is diagonal and N is strictly upper triangular. Then*

$$e^{\alpha(A)} \le \|e^A\|_2 \le e^{\alpha(A)} \sum_{k=0}^{n-1} \frac{\|N\|_2^k}{k!}. \tag{10.12}$$

Proof. See Van Loan [594, 1977]. □

A Frobenius norm analogue to (10.12) can be obtained from Theorem 4.28.

The 2-norm bounds in Theorems 10.11 and 10.12 are equalities when A is normal.

10.2. Conditioning

In this section we investigate the sensitivity of the matrix exponential to perturbations.

The matrix exponential satisfies the identity (see Problem 10.1)

$$e^{(A+E)t} = e^{At} + \int_0^t e^{A(t-s)} E e^{(A+E)s}\, ds. \tag{10.13}$$

Using this expression to substitute for $e^{(A+E)s}$ inside the integral yields

$$e^{(A+E)t} = e^{At} + \int_0^t e^{A(t-s)} E e^{As}\, ds + O(\|E\|^2). \tag{10.14}$$

Hence, from the definition (3.6), the Fréchet derivative of the exponential at A in the direction E is given by

Fréchet derivative of matrix exponential:

$$L(A,E) = \int_0^1 e^{A(1-s)} E e^{As}\, ds. \tag{10.15}$$

This formula simplifies to $L(A,E) = Ee^A = e^A E$ when A commutes with E. Another representation of the Fréchet derivative is, from the Taylor series for $e^{A+E} - e^A$,

$$L(A,E) = E + \frac{AE+EA}{2!} + \frac{A^2E + AEA + EA^2}{3!} + \cdots,$$

but the integral formula is generally easier to work with.

The following result provides three explicit expressions for $\mathrm{vec}(L(A,E))$. Define the sinch function by

$$\mathrm{sinch}(x) = \begin{cases} \sinh(x)/x, & x \ne 0, \\ 1, & x = 0. \end{cases} \tag{10.16}$$

Note that sinch can be written in terms of the more common sinc function—defined by $\mathrm{sinc}(x) = \sin(x)/x$ if $x \ne 0$ or 1 otherwise—as $\mathrm{sinch}(x) = \mathrm{sinc}(ix)$.

Theorem 10.13 (Kronecker representation of Fréchet derivative). *For $A \in \mathbb{C}^{n\times n}$, $\mathrm{vec}(L(A,E)) = K(A)\,\mathrm{vec}(E)$, where $K(A) \in \mathbb{C}^{n^2\times n^2}$ has the representations*

$$K(A) = \begin{cases} (I \otimes e^A)\,\psi_1\big(A^T \oplus (-A)\big), & (10.17\text{a}) \\ (e^{A^T/2} \otimes e^{A/2})\,\mathrm{sinch}\big(\tfrac{1}{2}[A^T \oplus (-A)]\big), & (10.17\text{b}) \\ \dfrac{1}{2}(e^{A^T} \oplus e^A)\,\tau\big(\tfrac{1}{2}[A^T \oplus (-A)]\big), & (10.17\text{c}) \end{cases}$$

where $\psi_1(x) = (e^x - 1)/x$ *and* $\tau(x) = \tanh(x)/x$. *The third expression is valid if* $\frac{1}{2}\|A^T \oplus (-A)\| < \pi/2$ *for some consistent matrix norm.*

Proof. Applying the vec operator to (10.15) gives

$$\begin{aligned}\operatorname{vec}(L(A,E)) &= \int_0^1 \big(e^{A^T s} \otimes e^{A(1-s)}\big) \operatorname{vec}(E)\, ds \\ &= (I \otimes e^A) \int_0^1 \big(e^{A^T s} \otimes e^{-As}\big)\, ds \cdot \operatorname{vec}(E) \\ &= (I \otimes e^A) \int_0^1 e^{[A^T \oplus (-A)]s}\, ds \cdot \operatorname{vec}(E) \qquad (10.18)\end{aligned}$$

by Theorem 10.9. Now

$$\int_0^1 e^{xs}\, ds = 1 + \frac{x}{2!} + \frac{x^2}{3!} + \cdots = \frac{e^x - 1}{x} = \psi_1(x) \qquad (10.19)$$

$$= \begin{cases} \dfrac{e^{x/2}(e^{x/2} - e^{-x/2})}{x} = e^{x/2}\,\dfrac{\sinh(x/2)}{x/2}, \\ \dfrac{e^{x/2}}{x}(e^{x/2} + e^{-x/2})\dfrac{e^{x/2} - e^{-x/2}}{e^{x/2} + e^{-x/2}} = \dfrac{e^x + 1}{2}\,\dfrac{\tanh(x/2)}{x/2}. \end{cases} \qquad (10.20)$$

The first formula in (10.17) follows from (10.18) and (10.19). The second formula follows from (10.20) and by using Theorem 10.9:

$$\begin{aligned}K(A) &= (I \otimes e^A)\, e^{\frac{1}{2}[A^T \oplus (-A)]} \operatorname{sinch}\big(\tfrac{1}{2}[A^T \oplus (-A)]\big) \\ &= (I \otimes e^A)(e^{A^T/2} \otimes e^{-A/2}) \operatorname{sinch}\big(\tfrac{1}{2}[A^T \oplus (-A)]\big) \\ &= (e^{A^T/2} \otimes e^{A/2}) \operatorname{sinch}\big(\tfrac{1}{2}[A^T \oplus (-A)]\big).\end{aligned}$$

Similarly, for the third formula,

$$\begin{aligned}K(A) &= (I \otimes e^A)\, \tfrac{1}{2}(e^{A^T \oplus (-A)} + I)\,\tau\big(\tfrac{1}{2}[A^T \oplus (-A)]\big) \\ &= (I \otimes e^A)\, \tfrac{1}{2}(e^{A^T} \otimes e^{-A} + I)\,\tau\big(\tfrac{1}{2}[A^T \oplus (-A)]\big) \\ &= \tfrac{1}{2}(e^{A^T} \otimes I + I \otimes e^A)\,\tau\big(\tfrac{1}{2}[A^T \oplus (-A)]\big) \\ &= \tfrac{1}{2}(e^{A^T} \oplus e^A)\,\tau\big(\tfrac{1}{2}[A^T \oplus (-A)]\big).\end{aligned}$$

The norm restriction is due to $\tanh(x)$ having poles with $|x| = \pi/2$. □

Each of the three expressions (10.17) is of potential interest for computational purposes, as we will see in Section 10.6.

Note that the matrix $A^T \oplus (-A)$ has eigenvalues $\lambda_i(A) - \lambda_j(A)$, $i, j = 1{:}n$ (see Section B.13), so it is singular with at least n zero eigenvalues. This does not affect the formulae (10.17) since ϕ, τ, sinch, and their derivatives are defined from their power series expansions at the origin.

The relative condition number of the exponential is (see Theorem 3.1)

$$\kappa_{\exp}(A) = \frac{\|L(A)\|\,\|A\|}{\|e^A\|}.$$

From Theorem 10.13 we obtain explicit expressions for $\|L(A)\|_F$.

Corollary 10.14 (norm of Fréchet derivative). *For $A \in \mathbb{C}^{n\times n}$,*

$$\|L(A)\|_F = \begin{cases} \big\|(I\otimes e^A)\psi_1(A^T \oplus (-A))\big\|_2, \\ \big\|(e^{A^T/2}\otimes e^{A/2})\,\mathrm{sinch}\big(\tfrac{1}{2}[A^T\oplus(-A)]\big)\big\|_2, \\ \big\|\tfrac{1}{2}(e^{A^T/2}\oplus e^{A/2})\,\tau\big(\tfrac{1}{2}[A^T\oplus(-A)]\big)\big\|_2. \end{cases} \tag{10.21}$$

Proof. The formulae follow immediately from the theorem on using the fact that $\|\mathrm{vec}(B)\|_2 = \|B\|_F$. □

The next lemma gives upper and lower bounds for $\kappa_{\exp}(A)$.

Lemma 10.15. *For $A \in \mathbb{C}^{n\times n}$ we have, for any subordinate matrix norm,*

$$\|A\| \le \kappa_{\exp}(A) \le \frac{e^{\|A\|}\|A\|}{\|e^A\|}. \tag{10.22}$$

Proof. From (10.15) we have

$$\|L(A,E)\| \le \|E\|\int_0^1 e^{\|A\|(1-s)}e^{\|A\|s}\,ds = \|E\|\int_0^1 e^{\|A\|}\,ds = \|E\|e^{\|A\|},$$

so that $\|L(A)\| \le e^{\|A\|}$. Also, $\|L(A)\| \ge \|L(A,I)\| = \|\int_0^1 e^A\,ds\| = \|e^A\|$, and the result follows. □

An important class of matrices that has minimal condition number is the normal matrices.

Theorem 10.16 (Van Loan). *If $A \in \mathbb{C}^{n\times n}$ is normal then in the 2-norm, $\kappa_{\exp}(A) = \|A\|_2$.*

Proof. We need a slight variation of the upper bound in (10.22). First, note that for a normal matrix B, $\|e^B\|_2 = e^{\alpha(B)}$. From (10.15) we have

$$\begin{aligned}\|L(A,E)\|_2 &\le \|E\|_2\int_0^1 e^{\alpha(A)(1-s)}e^{\alpha(A)s}\,ds \\ &= \|E\|_2\int_0^1 e^{\alpha(A)}\,ds = e^{\alpha(A)}\|E\|_2 = \|e^A\|_2\|E\|_2.\end{aligned}$$

Hence $\|L(A)\|_2 \le \|e^A\|_2$, which implies the result in view of the lower bound in (10.22). □

A second class of matrices with perfect conditioning comprises nonnegative scalar multiples of stochastic matrices.

Theorem 10.17 (Melloy and Bennett). *Let $A \in \mathbb{R}^{n\times n}$ be a nonnegative scalar multiple of a stochastic matrix. Then in the ∞-norm, $\kappa_{\exp}(A) = \|A\|_\infty$.*

Proof. From Theorem 10.10 we have $\|e^A\|_\infty = e^{\|A\|_\infty}$. Hence the result follows from (10.22) with the ∞-norm. □

For a third class of perfectly conditioned matrices, see Theorem 10.30 below.

Unfortunately, no useful characterization of matrices for which $\kappa_{\exp}$ is large is known; see Problem 10.15.

The next result is useful because it shows that argument reduction $A \leftarrow A - \mu I$ reduces the condition number if it reduces the norm. Since $e^{A-\mu I} = e^{-\mu}e^A$, argument reduction is trivial to incorporate in any algorithm.

Theorem 10.18 (Parks). *If μ is a scalar such that $\|A-\mu I\| < \|A\|$ then $\kappa_{\exp}(A-\mu I) < \kappa_{\exp}(A)$.*

Proof. For any E,

$$\begin{aligned}\|L(A-\mu I,E)\| &= \left\|\int_0^1 e^{(A-\mu I)(1-s)}Ee^{(A-\mu I)s}\,ds\right\| \\ &= \left\|e^{-\mu I}\int_0^1 e^{A(1-s)}Ee^{As}\,ds\right\| = e^{-\mathrm{Re}\,\mu}\,\|L(A,E)\|.\end{aligned}$$

Hence $\|L(A-\mu I)\| = e^{-\mathrm{Re}\,\mu}\|L(A)\|$. Thus if $\|A-\mu I\| < \|A\|$ then

$$\begin{aligned}\kappa_{\exp}(A-\mu I) &= \frac{\|L(A-\mu I)\|\,\|A-\mu I\|}{\|e^{A-\mu I}\|} = \frac{\|L(A-\mu I)\|\,\|A-\mu I\|}{\|e^{A}\|e^{-\mathrm{Re}\,\mu}} \\ &< \frac{\|L(A)\|\,\|A\|}{\|e^A\|} = \kappa_{\exp}(A). \qquad \square\end{aligned}$$

10.3. Scaling and Squaring Method

The scaling and squaring method exploits the relation $e^A = (e^{A/\sigma})^\sigma$, for $A \in \mathbb{C}^{n\times n}$ and $\sigma \in \mathbb{C}$, together with the fact that e^A can be well approximated by a Taylor or Padé approximant near the origin, that is, for small $\|A\|$. The idea is to choose σ an integral power of 2, $\sigma = 2^s$ say, so that A/σ has norm of order 1; approximate $e^{A/2^s} \approx r(A/2^s)$, where r is a Taylor or Padé approximant to the exponential; and then take $e^A \approx r(A/2^s)^{2^s}$, where the approximation is formed by s repeated squarings.

Padé approximants are preferred to Taylor series approximations in this context because they provide a given accuracy with lower computational cost [438, 2003, Sec. 3]. Recall from Section 4.4.2 that the $[k/m]$ Padé approximant $r_{km}(x) = p_{km}(x)/q_{km}(x)$ to the exponential is defined by the properties that p and q are polynomials of degrees at most k and m, respectively, and that $e^x - r_{km}(x) = O(x^{k+m+1})$. These Padé approximants are known explicitly for all k and m:

$$p_{km}(x) = \sum_{j=0}^{k}\frac{(k+m-j)!\,k!}{(k+m)!\,(k-j)!}\frac{x^j}{j!}, \quad q_{km}(x) = \sum_{j=0}^{m}\frac{(k+m-j)!\,m!}{(k+m)!\,(m-j)!}\frac{(-x)^j}{j!}. \tag{10.23}$$

Note that $p_{km}(x) = q_{mk}(-x)$, which reflects the property $1/e^x = e^{-x}$ of the exponential function. Later we will exploit the fact that $p_{mm}(x)$ and $q_{mm}(x)$ approximate $e^{x/2}$ and $e^{-x/2}$, respectively, though they do so *much* less accurately than $r_{mm} = p_{mm}/q_{mm}$ approximates e^x. That r_{km} satisfies the definition of Padé approximant is demonstrated by the error expression

$$e^x - r_{km}(x) = (-1)^m\frac{k!m!}{(k+m)!(k+m+1)!}x^{k+m+1} + O(x^{k+m+2}). \tag{10.24}$$

We also have the exact error expression, for $A \in \mathbb{C}^{n\times n}$ [438, 2003, App. A], [501, 1973],

$$e^A - r_{km}(A) = \frac{(-1)^m}{(k+m)!}A^{k+m+1}q_{km}(A)^{-1}\int_0^1 e^{tA}(1-t)^k t^m\,dt. \tag{10.25}$$

Diagonal approximants ($k = m$) are preferred, since r_{km} with $k \neq m$ is less accurate than r_{jj}, where $j = \max(k, m)$, but r_{jj} can be evaluated at a matrix argument at the same cost. Moreover, the diagonal approximants have the property that if the eigenvalues of A lie in the open left half-plane then the eigenvalues of $r_{mm}(A)$ have modulus less than 1, which is important in applications to differential equations [600, 2000, Chap. 8]. We will write the diagonal approximants as $r_m(x) = p_m(x)/q_m(x)$.

Our aim is to choose s, in the initial scaling $A \leftarrow A/2^s$, and the degree m of the Padé approximant, so that the exponential is computed with backward error bounded by the unit roundoff and with minimal cost. In bounding the backward error we assume exact arithmetic and examine solely the effects of the approximation errors in the Padé approximant.

The choice of s will be based on $\|A\|$, where the norm can be any consistent matrix norm. Our initial aim is therefore to bound the backward error in terms of $\|2^{-s}A\|$ and then to determine, for each degree m, the maximum $\|2^{-s}A\|$ for which r_m can be guaranteed to deliver the desired backward error. Let

$$e^{-A} r_m(A) = I + G = e^H, \tag{10.26}$$

where we assume that $\|G\| < 1$, so that $H = \log(I + G)$ is guaranteed to exist. (Here, log denotes the principal logarithm.) It is easy to show that $\|H\| \leq -\log(1 - \|G\|)$ (see Problem 10.8). Now G is clearly a function of A hence so is H, and therefore H commutes with A. It follows that

$$r_m(A) = e^A e^H = e^{A+H}.$$

Now we replace A by $A/2^s$, where s is a nonnegative integer, and raise both sides of this equation to the power 2^s, to obtain

$$r_m(A/2^s)^{2^s} = e^{A+E},$$

where $E = 2^s H$ satisfies

$$\|E\| \leq -2^s \log(1 - \|G\|)$$

and G satisfies (10.26) with A replaced by $2^{-s}A$. We summarize our findings in the following theorem.

Theorem 10.19. *Let the diagonal Padé approximant r_m satisfy*

$$e^{-2^{-s}A} r_m(2^{-s}A) = I + G, \tag{10.27}$$

where $\|G\| < 1$ and the norm is any consistent matrix norm. Then

$$r_m(2^{-s}A)^{2^s} = e^{A+E},$$

where E commutes with A and

$$\frac{\|E\|}{\|A\|} \leq \frac{-\log(1 - \|G\|)}{\|2^{-s}A\|}. \qquad \square \tag{10.28}$$

Theorem 10.19 is a backward error result: it interprets the truncation errors in the Padé approximant as equivalent to a perturbation in the original matrix A. (Note that the result holds for any rational approximation r_m, as we have not yet used specific properties of a Padé approximant.) The advantage of this backward error viewpoint is that it takes into account the effect of the squaring phase on the error in the Padé approximant and, compared with a forward error bound, avoids the need to consider the conditioning of the problem.

Our task now is to bound the norm of G in (10.27) in terms of $\|2^{-s}A\|$. One way to proceed is to assume an upper bound on $\|A\|$ and use the error formula (10.25) to obtain an explicit bound on G, or at least one that is easy to evaluate. This approach, which is used by Moler and Van Loan [438, 2003] and is illustrated in Problem 10.10, is mathematically elegant but does not yield the best possible bounds and hence does not lead to the best algorithm. We will use a bound on $\|G\|$ that makes no a priori assumption on $\|A\|$ and is as sharp as possible. The tradeoff is that the bound is hard to evaluate, but this is a minor inconvenience because the evaluation need only be done during the design of the algorithm.

Define the function

$$\rho(x) = e^{-x} r_m(x) - 1.$$

In view of the Padé approximation property (10.24), ρ has a power series expansion

$$\rho(x) = \sum_{i=2m+1}^{\infty} c_i x^i, \tag{10.29}$$

and this series will converge absolutely for $|x| < \min\{\, |t| : q_m(t) = 0 \,\} =: \nu_m$. Hence

$$\|G\| = \|\rho(2^{-s}A)\| \le \sum_{i=2m+1}^{\infty} |c_i|\theta^i =: f(\theta), \tag{10.30}$$

where $\theta := \|2^{-s}A\| < \nu_m$. It is clear that if A is a general matrix and only $\|A\|$ is known then (10.30) provides the smallest possible bound on $\|G\|$. The corresponding bound of Moler and Van Loan [438, 2003, Lem. 4] is easily seen to be less sharp, and a refined analysis of Dieci and Papini [157, 2000, Sec. 2], which bounds a different error, is also weaker when adapted to bound $\|G\|$.

Combining (10.30) with (10.28) we have

$$\frac{\|E\|}{\|A\|} \le \frac{-\log(1 - f(\theta))}{\theta}. \tag{10.31}$$

Evaluation of $f(\theta)$ in (10.30) would be easy if the coefficients c_i were one-signed, for then we would have $f(\theta) = |\rho(\theta)|$. Experimentally, the c_i are one-signed for some, but not all, m. Using MATLAB's Symbolic Math Toolbox we evaluated $f(\theta)$, and hence the bound (10.31), in 250 decimal digit arithmetic, summing the first 150 terms of the series, where the c_i in (10.29) are obtained symbolically. For $m = 1\colon 21$ we used a zero-finder to determine the largest value of θ, denoted by θ_m, such that the backward error bound (10.31) does not exceed $u = 2^{-53} \approx 1.1 \times 10^{-16}$. The results are shown to two significant figures in Table 10.2.

The second row of the table shows the values of ν_m, and we see that $\theta_m < \nu_m$ in each case, confirming that the bound (10.30) is valid. The inequalities $\theta_m < \nu_m$ also confirm the important fact that $q_m(A)$ is nonsingular for $\|A\| \le \theta_m$ (which is in any case implicitly enforced by our analysis).

Next we need to determine the cost of evaluating $r_m(A)$. Because of the relation $q_m(x) = p_m(-x)$ between the numerator and denominator polynomials, an efficient scheme can be based on explicitly computing the even powers of A, forming p_m and

Table 10.2. *Maximal values θ_m of $\|2^{-s}A\|$ such that the backward error bound* (10.31) *does not exceed $u = 2^{-53}$, values of $\nu_m = \min\{\,|x| : q_m(x) = 0\}$, and upper bound ξ_m for $\|q_m(A)^{-1}\|$.*

m	1	2	3	4	5	6	7	8	9	10
θ_m	3.7e-8	5.3e-4	1.5e-2	8.5e-2	2.5e-1	5.4e-1	9.5e-1	1.5e0	2.1e0	2.8e0
ν_m	2.0e0	3.5e0	4.6e0	6.0e0	7.3e0	8.7e0	9.9e0	1.1e1	1.3e1	1.4e1
ξ_m	1.0e0	1.0e0	1.0e0	1.0e0	1.1e0	1.3e0	1.6e0	2.1e0	3.0e0	4.3e0

m	11	12	13	14	15	16	17	18	19	20	21
θ_m	3.6e0	4.5e0	5.4e0	6.3e0	7.3e0	8.4e0	9.4e0	1.1e1	1.2e1	1.3e1	1.4e1
ν_m	1.5e1	1.7e1	1.8e1	1.9e1	2.1e1	2.2e1	2.3e1	2.5e1	2.6e1	2.7e1	2.8e1
ξ_m	6.6e0	1.0e1	1.7e1	3.0e1	5.3e1	9.8e1	1.9e2	3.8e2	8.3e2	2.0e3	6.2e3

q_m, and then solving the matrix equation $q_m r_m = p_m$. If $p_m(x) = \sum_{i=0}^{m} b_i x^i$, we have, for the even degree case,

$$\begin{aligned} p_{2m}(A) &= b_{2m}A^{2m} + \cdots + b_2A^2 + b_0I + A(b_{2m-1}A^{2m-2} + \cdots + b_3A^2 + b_1I) \\ &=: U + V, \end{aligned} \tag{10.32}$$

which can be evaluated with $m+1$ matrix multiplications by forming A^2, A^4, ..., A^{2m}. Then

$$q_{2m}(A) = U - V$$

is available at no extra cost. For odd degrees,

$$\begin{aligned} p_{2m+1}(A) &= A(b_{2m+1}A^{2m} + \cdots + b_3A^2 + b_1I) + b_{2m}A^{2m} + \cdots + b_2A^2 + b_0I \\ &=: U + V, \end{aligned} \tag{10.33}$$

so p_{2m+1} and $q_{2m+1} = -U + V$ can be evaluated at exactly the same cost as p_{2m} and q_{2m}. However, for $m \ge 12$ this scheme can be improved upon. For example, we can write

$$\begin{aligned} p_{12}(A) &= A^6(b_{12}A^6 + b_{10}A^4 + b_8A^2 + b_6I) + b_4A^4 + b_2A^2 + b_0I \\ &\quad + A\big[A^6(b_{11}A^4 + b_9A^2 + b_7I) + b_5A^4 + b_3A^2 + b_1I\big] \\ &=: U + V, \end{aligned} \tag{10.34}$$

and $q_{12}(A) = U - V$. Thus p_{12} and q_{12} can be evaluated in just six matrix multiplications (for A^2, A^4, A^6, and three additional multiplications). For $m = 13$ an analogous formula holds, with the outer multiplication by A transferred to the U term. Similar formulae hold for $m \ge 14$. Table 10.3 summarizes the number of matrix multiplications required to evaluate p_m and q_m, which we denote by π_m, for $m = 1\colon 21$.

The information in Tables 10.2 and 10.3 enables us to determine the optimal algorithm when $\|A\| \ge \theta_{21}$. From Table 10.3, we see that the choice is between $m = 1, 2, 3, 5, 7, 9, 13, 17$ and 21 (there is no reason to use $m = 6$, for example, since the cost of evaluating the more accurate q_7 is the same as the cost of evaluating q_6). Increasing from one of these values of m to the next requires an extra matrix multiplication to evaluate r_m, but this is offset by the larger allowed $\theta_m = \|2^{-s}A\|$ if θ_m jumps by more than a factor 2, since decreasing s by 1 saves one multiplication in

Table 10.3. *Number of matrix multiplications, π_m, required to evaluate $p_m(A)$ and $q_m(A)$, and measure of overall cost C_m in* (10.35).

m	1	2	3	4	5	6	7	8	9	10
π_m	0	1	2	3	3	4	4	5	5	6
C_m	25	12	8.1	6.6	5.0	4.9	4.1	4.4	3.9	4.5

m	11	12	13	14	15	16	17	18	19	20	21
π_m	6	6	6	7	7	7	7	8	8	8	8
C_m	4.2	3.8	3.6	4.3	4.1	3.9	3.8	4.6	4.5	4.3	4.2

the final squaring stage. Table 10.2 therefore shows that $m = 13$ is the best choice. Another way to arrive at this conclusion is to observe that the cost of the algorithm in matrix multiplications is, since $s = \lceil \log_2 \|A\|/\theta_m \rceil$ if $\|A\| \geq \theta_m$ and $s = 0$ otherwise,

$$\pi_m + s = \pi_m + \max\left(\lceil \log_2 \|A\| - \log_2 \theta_m \rceil, 0\right).$$

(We ignore the required matrix equation solution, which is common to all m.) We wish to determine which m minimizes this quantity. For $\|A\| \geq \theta_m$ we can remove the max and ignore the $\|A\|$ term, which is essentially a constant shift, so we minimize

$$C_m = \pi_m - \log_2 \theta_m. \tag{10.35}$$

The C_m values are shown in the second line of Table 10.3. Again, $m = 13$ is optimal. We repeated the computations with $u = 2^{-24} \approx 6.0 \times 10^{-8}$, which is the unit roundoff in IEEE single precision arithmetic, and $u = 2^{-105} \approx 2.5 \times 10^{-32}$, which corresponds to quadruple precision arithmetic; the optimal m are now $m = 7$ ($\theta_7 = 3.9$) and $m = 17$ ($\theta_{17} = 3.3$), respectively.

Now we consider the effects of rounding errors on the evaluation of $r_m(A)$. We immediately rule out $m = 1$ and $m = 2$ because r_1 and r_2 can suffer from loss of significance in floating point arithmetic. For example, r_1 requires $\|A\|$ to be of order 10^{-8} after scaling, and then the expression $r_1(A) = (I + A/2)(I - A/2)^{-1}$ loses about half the significant digits in A in double precision arithmetic; yet if the original A has norm of order at least 1 then all the significant digits of some of the elements of A should contribute to the result. Applying Theorem 4.5 to $p_m(A)$, where $\|A\|_1 \leq \theta_m$, and noting that p_m has all positive coefficients, we deduce that

$$\begin{aligned}
\|p_m(A) - \widehat{p}_m(A)\|_1 &\leq \widetilde{\gamma}_{mn}\, p_m(\|A\|_1) \\
&\approx \widetilde{\gamma}_{mn}\, e^{\|A\|_1/2} \\
&\leq \widetilde{\gamma}_{mn} \|e^{A/2}\|_1\, e^{\|A\|_1} \\
&\approx \widetilde{\gamma}_{mn} \|p_m(A)\|_1\, e^{\|A\|_1} \leq \widetilde{\gamma}_{mn} \|p_m(A)\|_1\, e^{\theta_m}.
\end{aligned}$$

Hence the relative error is bounded approximately by $\widetilde{\gamma}_{mn} e^{\theta_m}$, which is a satisfactory bound given the values of θ_m in Table 10.2. Replacing A by $-A$ in the latter bound we obtain

$$\|q_m(A) - \widehat{q}_m(A)\|_1 \lesssim \widetilde{\gamma}_{mn} \|q_m(A)\|_1\, e^{\theta_m}.$$

In summary, the errors in the evaluation of p_m and q_m are nicely bounded.

To obtain r_m we solve a multiple right-hand side linear system with $q_m(A)$ as coefficient matrix, so to be sure that this system is solved accurately we need to check

Table 10.4. *Coefficients $b(0\colon m)$ in numerator $p_m(x) = \sum_{i=0}^m b_i x^i$ of Padé approximant $r_m(x)$ to e^x, normalized so that $b(m) = 1$.*

m	$b(0\colon m)$
3	[120, 60, 12, 1]
5	[30240, 15120, 3360, 420, 30, 1]
7	[17297280, 8648640, 1995840, 277200, 25200, 1512, 56, 1]
9	[17643225600, 8821612800, 2075673600, 302702400, 30270240, 2162160, 110880, 3960, 90, 1]
13	[64764752532480000, 32382376266240000, 7771770303897600, 1187353796428800, 129060195264000, 10559470521600, 670442572800, 33522128640, 1323241920, 40840800, 960960, 16380, 182, 1]

that $q_m(A)$ is well conditioned. It is possible to obtain a priori bounds for $\|q_m(A)^{-1}\|$ under assumptions such as (for any subordinate matrix norm) $\|A\| \le 1/2$ [438, 2003, Lem. 2], $\|A\| \le 1$ [606, 1977, Thm. 1], or $q_m(-\|A\|) < 2$ [157, 2000, Lem. 2.1] (see Problem 10.9), but these assumptions are not satisfied for all the m and $\|A\|$ of interest to us. Therefore we take a similar approach to the way we derived the constants θ_m. With $\|A\| \le \theta_m$ and by writing

$$q_m(A) = e^{-A/2}\big(I + e^{A/2}q_m(A) - I\big) =: e^{-A/2}(I+F),$$

we have, if $\|F\| < 1$,

$$\|q_m(A)^{-1}\| \le \|e^{A/2}\|\,\|(I+F)^{-1}\| \le \frac{e^{\theta_m/2}}{1-\|F\|}.$$

We can expand $e^{x/2}q_m(x) - 1 = \sum_{i=2}^\infty d_i x^i$, from which $\|F\| \le \sum_{i=2}^\infty |d_i|\theta_m^i$ follows. Our overall bound is

$$\|q_m(A)^{-1}\| \le \frac{e^{\theta_m/2}}{1-\sum_{i=2}^\infty |d_i|\theta_m^i}.$$

By determining the d_i symbolically and summing the first 150 terms of the sum in 250 decimal digit arithmetic, we obtained the bounds in the last row of Table 10.2, which confirm that q_m is very well conditioned for m up to about 13 when $\|A\| \le \theta_m$.

The overall algorithm is as follows. It first checks whether $\|A\| \le \theta_m$ for $m \in \{3, 5, 7, 9, 13\}$ and, if so, evaluates r_m for the smallest such m. Otherwise it uses the scaling and squaring method with $m = 13$.

Algorithm 10.20 (scaling and squaring algorithm). This algorithm evaluates the matrix exponential $X = e^A$ of $A \in \mathbb{C}^{n\times n}$ using the scaling and squaring method. It uses the constants θ_m given in Table 10.2 and the Padé coefficients in Table 10.4. The algorithm is intended for IEEE double precision arithmetic.

```
1  for m = [3 5 7 9]
2      if ||A||_1 ≤ θ_m
          % Form r_m(A) = [m/m] Padé approximant to A.
3         Evaluate U and V using (10.33) and solve (−U + V)X = U + V.
4         quit
```

5 end
6 end
7 $A \leftarrow A/2^s$ with $s \geq 0$ a minimal integer such that $\|A/2^s\|_1 \leq \theta_{13}$ (i.e., $s = \lceil \log_2(\|A\|_1/\theta_{13}) \rceil$).
8 % Form [13/13] Padé approximant to e^A.
9 $A_2 = A^2$, $A_4 = A_2^2$, $A_6 = A_2 A_4$
10 $U = A\big[A_6(b_{13}A_6 + b_{11}A_4 + b_9 A_2) + b_7 A_6 + b_5 A_4 + b_3 A_2 + b_1 I\big]$
11 $V = A_6(b_{12}A_6 + b_{10}A_4 + b_8 A_2) + b_6 A_6 + b_4 A_4 + b_2 A_2 + b_0 I$
12 Solve $(-U + V)r_{13} = U + V$ for r_{13}.
13 $X = {r_{13}}^{2^s}$ by repeated squaring.

Cost: $\big(\pi_m + \lceil \log_2(\|A\|_1/\theta_m) \rceil\big)M + D$, where m is the degree of Padé approximant used and π_m is tabulated in Table 10.3. (M and D are defined at the start of Chapter 4.)

It is readily checked that the sequences $\theta_{13}^{2k} b_{2k}$ and $\theta_{13}^{2k+1} b_{2k+1}$ are approximately monotonically decreasing with k, and hence the ordering given in Algorithm 10.20 for evaluating U and V takes the terms in approximately increasing order of norm. This ordering is certainly preferable when A has nonnegative elements, and since there cannot be much cancellation in the sums it cannot be a bad ordering [276, 2002, Chap. 4].

The part of the algorithm most sensitive to rounding errors is the final scaling phase. The following general result, in which we are thinking of B as the Padé approximant, shows why.

Theorem 10.21. *For $B \in \mathbb{R}^{n\times n}$ let $\widehat{X} = fl(B^{2^k})$ be computed by repeated squaring. Then, for the 1-, ∞-, and Frobenius norms,*

$$\|B^{2^k} - \widehat{X}\| \leq (2^k - 1)nu\|B\|^2 \cdot \|B^2\|\,\|B^4\| \dots \|B^{2^{k-1}}\| + O(u^2). \tag{10.36}$$

Proof. See Problem 10.11. □

If B^{2^k} were computed by repeated multiplication, the upper bound in (10.36) would be $2^k nu\|B\|^{2^k} + O(u^2)$ [276, 2002, Lem. 3.6]. This bound is much weaker than (10.36), so repeated squaring can be expected to be more accurate as well as more efficient than repeated multiplication. To see that (10.36) can nevertheless be unsatisfactory, we rewrite it as the relative error bound

$$\frac{\|B^{2^k} - \widehat{X}\|}{\|B^{2^k}\|} \leq \mu(2^k - 1)nu + O(u^2), \tag{10.37}$$

where

$$\mu = \frac{\|B\|^2\,\|B^2\|\,\|B^4\| \dots \|B^{2^{k-1}}\|}{\|B^{2^k}\|} \geq 1. \tag{10.38}$$

The ratio μ can be arbitrarily large, because cancellation can cause an intermediate power B^{2^j} ($j < k$) to be much larger than the final power B^{2^k}. In other words, the powers can display the *hump phenomenon*, illustrated in Figure 10.1 for the matrix

$$A = \begin{bmatrix} -0.97 & 25 \\ 0 & -0.3 \end{bmatrix}. \tag{10.39}$$

We see that while the powers ultimately decay to zero (since $\rho(A) = 0.97 < 1$), initially they increase in norm, producing a hump in the plot. The hump can be arbitrarily high relative to the starting point $\|A\|$. Moreover, there may not be a single hump, and indeed scalloping behaviour is observed for some matrices. Analysis of the hump for 2×2 matrices, and various bounds on matrix powers, can be found in [276, 2002, Chap. 18] and the references therein.

Another way of viewing this discussion is through the curve $\|e^{At}\|$, shown for the matrix (10.39) in Figure 10.2. This curve, too, can be hump-shaped. Recall that we are using the relation $e^A = (e^{A/\sigma})^\sigma$, and if $t = 1/\sigma$ falls under a hump but $t = 1$ is beyond it, then $\|e^A\| \ll \|e^{A/\sigma}\|^\sigma$. The connection between powers and exponentials is that if we set $B = e^{A/r}$ then the values $\|B^j\|$ are the points on the curve $\|e^{At}\|$ for $t = 1/r, 2/r, \ldots$.

The bound (10.37) also contains a factor $2^k - 1$. If $\|A\| > \theta_{13}$ then $2^k \approx \|A\|$ and so the overall relative error bound contains a term of the form $\mu\|A\|nu$. However, since $\kappa_{\exp}(A) \ge \|A\|$ (see Lemma 10.15), a factor $\|A\|$ is not troubling.

Our conclusion is that the overall effect of rounding errors in the final squaring stage may be large relative to the computed exponential $\widehat{X}$, and so $\widehat{X}$ may have large relative error. This may or may not indicate instability of the algorithm, depending on the conditioning of the e^A problem at the matrix A. Since little is known about the size of the condition number $\kappa_{\exp}$ for nonnormal A, no clear general conclusions can be drawn about the stability of the algorithm (see Problem 10.16).

In the special case where A is normal the scaling and squaring method is guaranteed to be forward stable. For normal matrices there is no hump because $\|A^k\|_2 = \|A\|_2^k$, so $\mu = 1$ in (10.38), and $2^k \approx \|A\|_2 = \kappa_{\exp}(A)$ by Theorem 10.16. Therefore the squaring phase is innocuous and the error in the computed exponential is consistent with the conditioning of the problem. Another case in which the scaling and squaring method is forward stable is when $a_{ij} \ge 0$ for $i \ne j$, as shown by Arioli, Codenotti, and Fassino [17, 1996]. The reason is that the exponential of such a matrix is nonnegative (see Section 10.7.2) and multiplying nonnegative matrices is a stable procedure since there can be no cancellation.

Finally, we note that the scaling and squaring method has a weakness when applied to block triangular matrices. Suppose $A = \left[\begin{smallmatrix} A_{11} & A_{12} \\ 0 & A_{22} \end{smallmatrix}\right]$. Then

$$e^A = \begin{bmatrix} e^{A_{11}} & \int_0^1 e^{A_{11}(1-s)} A_{12} e^{A_{22}s} \, ds \\ 0 & e^{A_{22}} \end{bmatrix} \tag{10.40}$$

(see Problem 10.12). The linear dependence of the (1,2) block of e^A on A_{12} suggests that the accuracy of the corresponding block of a Padé approximant should not unduly be affected by $\|A_{12}\|$ and hence that in the scaling and squaring method only the norms of A_{11} and A_{22} should influence the amount of scaling (specified by s in Algorithm 10.20). But since s depends on the norm of A as a whole, when $\|A_{12}\| \gg \max(\|A_{11}\|, \|A_{22}\|)$ the diagonal blocks are overscaled with regard to the computation of $e^{A_{11}}$ and $e^{A_{22}}$, and this may have a harmful effect on the accuracy of the computed exponential (cf. the discussion on page 245 about rejecting degrees $m = 1, 2$). In fact, the block triangular case merits special treatment. If the spectra of A_{11} and A_{22} are well separated then it is best to compute $e^{A_{11}}$ and $e^{A_{22}}$ individually and obtain F_{12} from the block Parlett recurrence by solving a Sylvester equation of the form (9.7). In general, analysis of Dieci and Papini [157, 2000] suggests that if the scaling and squaring method is used with s determined so that $2^{-s}\|A_{11}\|$ and

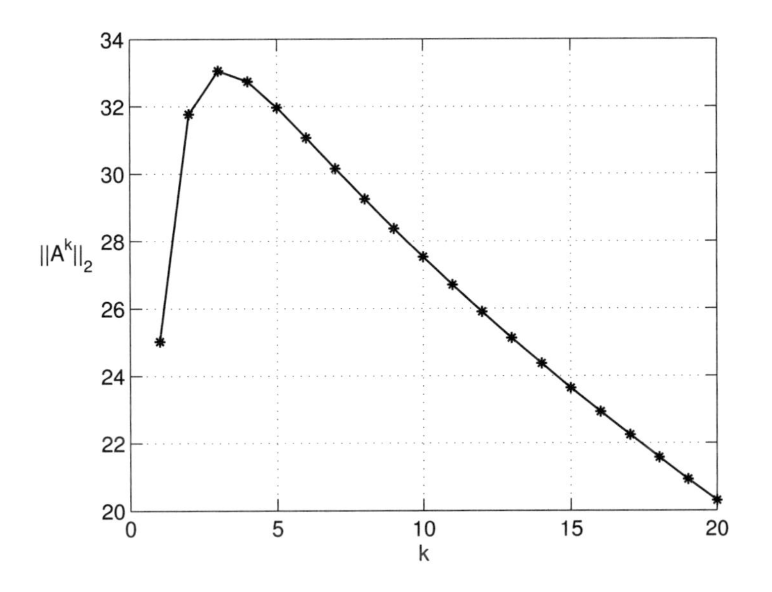

Figure 10.1. *2-norms of first* 20 *powers of* A *in* (10.39).

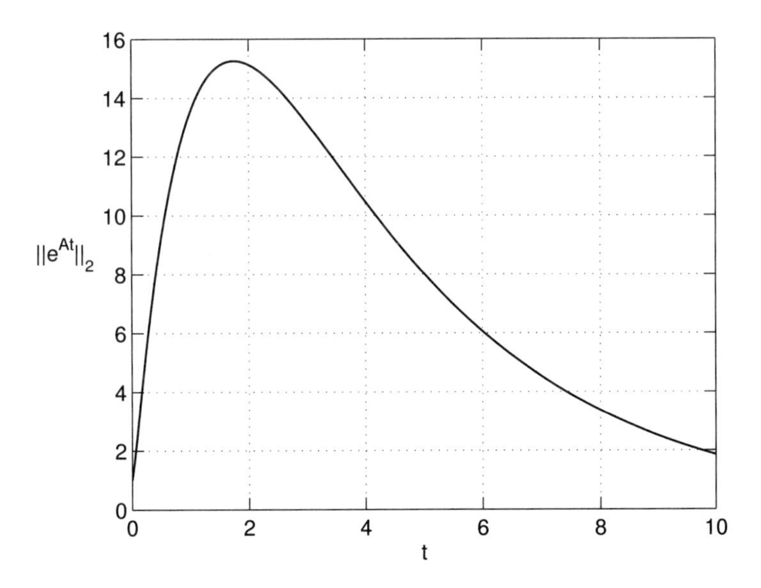

Figure 10.2. *2-norm of* $\exp(At)$ *for* A *in* (10.39).

$2^{-s}\|A_{22}\|$ are appropriately bounded, without consideration of $\|A_{12}\|$, then an accurate approximation to e^A will still be obtained.

10.4. Schur Algorithms

Another general purpose way to compute e^A is to employ a Schur decomposition $A = QTQ^*$, thereby reducing the problem to that of computing the exponential of a triangular matrix. In this section we describe the three main contenders for how to compute e^T.

10.4.1. Newton Divided Difference Interpolation

Parlett and Ng [462, 1985] develop an algorithm based on the Schur decomposition and the Parlett recurrence that is tailored to the exponential in both its blocking (a two-level scheme) and how the diagonal blocks are exponentiated. We will concentrate on the latter aspect.

To exponentiate an atomic block $T_{ii} \in \mathbb{C}^{m\times m}$, Parlett and Ng use the Newton divided difference form of the Hermite interpolating polynomial

$$p(t) = \sum_{i=1}^{m} c_i \prod_{j=1}^{i-1} (t - \lambda_j), \tag{10.41}$$

where $c_i = f[\lambda_1, \lambda_2, \ldots, \lambda_i]$ with $\lambda_i \equiv t_{ii}$ and $f(t) = e^t$. The cost of evaluating $p(T_{ii})$ is $O(m^4)$ flops.

For general functions f (or a set of given function values) divided differences are computed using the standard recurrence (B.24). However, the recurrence can produce inaccurate results in floating point arithmetic [276, 2002, Sec. 5.3]. This can be seen from the first order divided difference $f[\lambda_k, \lambda_{k+1}] = (f(\lambda_{k+1}) - f(\lambda_k))/(\lambda_{k+1} - \lambda_k)$ $(\lambda_k \neq \lambda_{k+1})$, in which for λ_k close to λ_{k+1} the subtraction in the numerator will suffer cancellation and the resulting error will be magnified by a small denominator. Obtaining accurate divided differences is important because the matrix product terms in (10.41) can vary greatly in norm. For a particular f that is given in functional form rather than simply as function values $f(\lambda_i)$, we would hope to be able to obtain the divided differences more accurately by exploiting properties of f. The next result offers one way to do this because it shows that evaluating f at a certain bidiagonal matrix yields the divided differences in the first row.

Theorem 10.22 (Opitz). *The divided difference $f[\lambda_1, \lambda_2, \ldots, \lambda_m]$ is the $(1, m)$ element of $f(Z)$, where*

$$Z = \begin{bmatrix} \lambda_1 & 1 & & & \\ & \lambda_2 & 1 & & \\ & & \ddots & \ddots & \\ & & & \ddots & 1 \\ & & & & \lambda_m \end{bmatrix}. \qquad \square$$

For $m = 2$ the theorem is just (4.16), while for $\lambda_1 = \lambda_2 = \cdots = \lambda_m$ it reproduces the Jordan block formula (1.4), in view of (B.27).

McCurdy, Ng, and Parlett [416, 1984] investigate in detail the accurate evaluation of the divided differences of the exponential. They derive a hybrid algorithm that uses the standard recurrence when it is safe to do so (i.e., when the denominator is not too small) and otherwise uses the Taylor series of the exponential in conjunction with Theorem 10.22 in a sophisticated way that employs scaling and squaring. They also discuss the use of matrix argument reduction, which has some benefits for their algorithm since it can reduce the size of the imaginary parts of the eigenvalues. Note that if $J_k(\lambda)$ is a Jordan block then $\exp(J_k(\lambda)) = \exp(J_k(\lambda - 2\pi i j))$ for any integer j (since in (1.4) the values of e^t and its derivatives are unchanged by the shift $\lambda \to \lambda - 2\pi i j$). Hence for an arbitrary A, each eigenvalue can be shifted by an integer multiple of $2\pi i$ so that its imaginary part is of modulus at most π without changing e^A. If A is in triangular form then a technique developed by Ng [448, 1984] based on the Parlett recurrence can be used to carry out matrix argument reduction. Li [384, 1988] shows how when A is real (10.41) can be evaluated in mainly real arithmetic, with complex arithmetic confined to the computation of the divided differences.

Unfortunately, no precise statement of an overall algorithm for e^A is contained in the above references and thorough numerical tests are lacking therein.

10.4.2. Schur–Fréchet Algorithm

Kenney and Laub [348, 1998] develop an algorithm based on Theorem 4.12, which shows that if $T = \begin{bmatrix} T_{11} & T_{12} \\ 0 & T_{22} \end{bmatrix}$ then the (1,2) block of e^T is the (1,2) block of the Fréchet derivative $L\big(\mathrm{diag}(T_{11}, T_{22}), \begin{bmatrix} 0 & T_{12} \\ 0 & 0 \end{bmatrix}\big)$. They evaluate the Fréchet derivative using Algorithm 10.27 described in Section 10.6.2 below. For an overall algorithm, $e^{T_{11}}$ and $e^{T_{22}}$ are recursively blocked and evaluated in the same way. The rather intricate details can be found in [348, 1998].

10.4.3. Schur–Parlett Algorithm

A simple modification of Algorithm 9.6 (Schur–Parlett) to call Algorithm 10.20 (scaling and squaring) on the diagonal blocks specializes it to the matrix exponential. However, it is worth evaluating the exponential of any 2×2 diagonal blocks $\begin{bmatrix} \lambda_1 & t_{12} \\ 0 & \lambda_2 \end{bmatrix}$ from an explicit formula. The usual formula (4.16) for the $(1,2)$ element in the nonconfluent case suffers from cancellation when $\lambda_1 \approx \lambda_2$. But we can rewrite this formula as

$$\begin{aligned} t_{12}\,\frac{e^{\lambda_2} - e^{\lambda_1}}{\lambda_2 - \lambda_1} &= t_{12}\,e^{(\lambda_1+\lambda_2)/2}\,\frac{e^{(\lambda_2-\lambda_1)/2} - e^{(\lambda_1-\lambda_2)/2}}{\lambda_2 - \lambda_1} \\ &= t_{12}\,e^{(\lambda_1+\lambda_2)/2}\,\frac{\sinh\big((\lambda_2 - \lambda_1)/2\big)}{(\lambda_2 - \lambda_1)/2}. \end{aligned}$$

This leads to the formula, valid for all λ_1 and λ_2,

$$\exp\left(\begin{bmatrix} \lambda_1 & t_{12} \\ 0 & \lambda_2 \end{bmatrix}\right) = \begin{bmatrix} e^{\lambda_1} & t_{12}\,e^{(\lambda_1+\lambda_2)/2}\,\mathrm{sinch}\big((\lambda_1 - \lambda_2)/2\big) \\ 0 & e^{\lambda_2} \end{bmatrix}. \tag{10.42}$$

This will provide an accurate result as long as an accurate implementation of sinch is available.

Algorithm 10.23 (Schur–Parlett algorithm for matrix exponential). Given $A \in \mathbb{C}^{n \times n}$ this algorithm computes $F = e^A$ via a Schur decomposition.

1 Execute Algorithm 9.6 with line 7 replaced by
"7 Evaluate $F_{ii} = e^{T_{ii}}$ directly if T_{ii} is 1×1, by (10.42) if T_{ii} is 2×2,
or else by Algorithm 10.20 (scaling and squaring algorithm)."

The evaluation of the exponential of the atomic diagonal blocks by Algorithm 10.20 brings the greater efficiency of Padé approximation compared with Taylor approximation to the exponential and also exploits scaling and squaring (though the latter could of course be used in conjunction with a Taylor series).

Compared with Algorithm 10.20 applied to the whole (triangular) matrix, the likelihood of overscaling is reduced because the algorithm is being applied only to the (usually small-dimensioned) diagonal blocks (and not to the 1×1 or 2×2 diagonal blocks).

Algorithm 10.23 is our preferred alternative to Algorithm 10.20. It has the advantage over the methods of the previous two subsections of simplicity and greater efficiency, and its potential instabilities are more clear.

10.5. Numerical Experiment

We describe an experiment that compares the accuracy of four methods for computing e^A. The matrices (mostly 10×10) and most of the details are exactly the same as in the experiment of Section 9.4. The methods are as follows.

1. MATLAB's `expm`, which implements Algorithm 10.20.

2. The modified version `funm_mod` of MATLAB's `funm` tested in Section 9.4, which exponentiates the diagonal blocks in the Schur form using Taylor series.

3. MATLAB 7.6's `funm`, which implements Algorithm 10.23 when it is invoked as `funm(A,@exp)`.

4. MATLAB's `expmdemo1`: a function that implements the scaling and squaring method with $m = 6$ and $\|2^{-s}A\|_\infty \le 0.5$ as the scaling criterion. This is an M-file version of the `expm` function that was used in MATLAB 7 (R14SP3) and earlier versions.

Figure 10.3 shows the normwise relative errors $\|\widehat{X} - e^A\|_F / \|e^A\|_F$ of the computed $\widehat{X}$. Figure 10.4 presents the same data in the form of a performance profile: for a given α on the x-axis, the y coordinate of the corresponding point on the curve is the probability that the method in question has an error within a factor α of the smallest error over all the methods on the given test set. Both plots are needed to understand the results: the performance profile reveals the typical performance, while Figure 10.3 highlights the extreme cases.[10] For more on performance profiles see Dolan and Moré [161, 2002] and Higham and Higham [263, 2005, Sec. 22.4].

Several observations can be made.

- `expm` is the most accurate and reliable code overall.

- `funm` and `funm_mod` perform very similarly.

[10] To be more precise, a performance profile shows the existence and extent of unusually large errors for a method if the x-axis is extended far enough to the right, but we have limited to $x \in [1, 15]$ for readability.

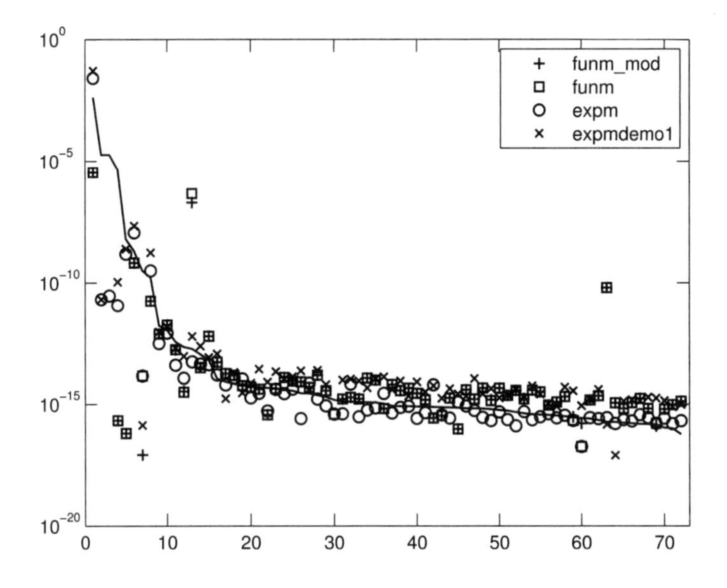

Figure 10.3. *Normwise relative errors for MATLAB's* `funm`, `expm`, `expmdemo1`, *and* `funm_mod`; *the solid line is* $\kappa_{\exp}(A)u$.

- `expmdemo1` is the least reliable of the four codes. This is mainly due to its suboptimal choice of m and s; `expm` usually takes a larger s and hence requires fewer squarings.

10.6. Evaluating the Fréchet Derivative and Its Norm

In some applications it is desired to compute not only e^A but also the Fréchet derivative, in order to obtain sensitivity information or to apply an optimization algorithm requiring derivatives. Two problems are of interest: approximating $L(A,E)$ for a given E and computing or estimating $\|L(A)\|$. We will begin by discussing several different approaches to the former problem and return to the norm problem at the end of the section.

The relation (3.16) gives

$$\exp\left(\begin{bmatrix} A & E \\ 0 & A \end{bmatrix}\right) = \begin{bmatrix} e^A & L(A,E) \\ 0 & e^A \end{bmatrix}. \tag{10.43}$$

Hence the Fréchet derivative $L(A,E)$ can be obtained by applying any existing method for the exponential to the above block upper triangular $2n\times 2n$ matrix and reading off the (1,2) block. Of course it may be possible to take advantage of the block triangular form in the method. This approach has the major advantage of simplicity but is likely to be too expensive if n is large.

It is interesting to note a duality between the matrix exponential and its Fréchet derivative. Either one can be used to compute the other: compare (10.43) with Section 10.4.2.

Some other techniques for evaluating the Frechét derivative exploit scaling and squaring. Applying the chain rule (Theorem 3.4) to the identity $e^A = (e^{A/2})^2$ gives,

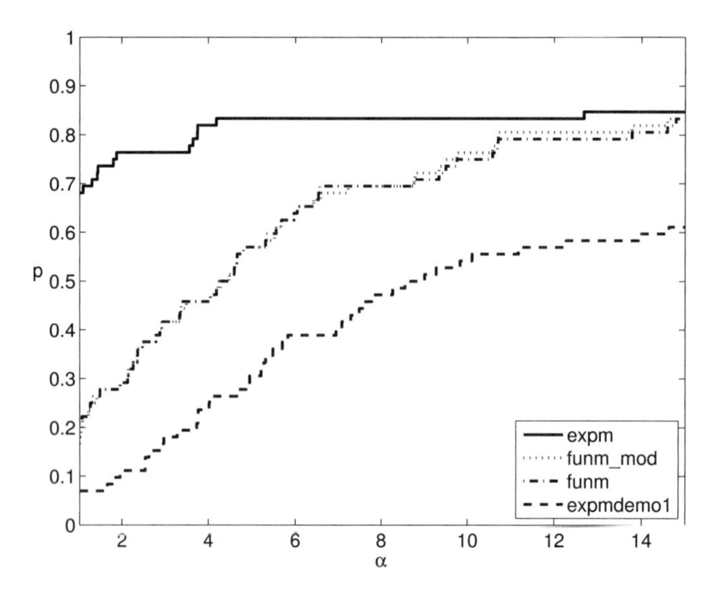

Figure 10.4. *Same data as in Figure* 10.3 *presented as a performance profile.*

on recalling that $L_{x^2}(A,E) = AE + EA$ (see (3.10)),

$$\begin{aligned} L_{\exp}(A,E) &= L_{x^2}\big(e^{A/2}, L_{\exp}(A/2,E/2)\big) \\ &= e^{A/2}L_{\exp}(A/2,E/2) + L_{\exp}(A/2,E/2)e^{A/2}. \end{aligned}$$

This relation yields the following recurrence for computing $L_0 = L_{\exp}(A,E)$:

$$L_s = L_{\exp}(2^{-s}A, 2^{-s}E), \tag{10.44a}$$

$$L_{i-1} = e^{2^{-i}A}L_i + L_i e^{2^{-i}A}, \qquad i = s\colon -1\colon 1. \tag{10.44b}$$

The advantage of the recurrence is that it reduces the problem of approximating the Fréchet derivative for A to that of approximating the Fréchet derivative for $2^{-s}A$, and s can be chosen to make $\|2^{-s}A\|$ small enough that the latter problem can be solved accurately and efficiently. In the recurrence the relation $e^{2^{-i}A} = \big(e^{2^{-(i+1)}A}\big)^2$ can be exploited to save computational effort, although repeated squaring could worsen the effects of rounding error. Another attraction of the recurrence is that it is possible to intertwine the computation of $L(A,E)$ with that of e^A by the scaling and squaring method and thereby compute both matrices with less effort than is required to compute each separately; however, the requirements on s for the approximation of $e^{2^{-s}A}$ and $L_{2^{-s}A}$ may differ, as we will see below. We now describe two approaches based on the above recurrence.

10.6.1. Quadrature

One way to approximate $L(A,E)$ is to apply quadrature to the integral representation (10.15). For example, we can apply the m-times repeated trapezium rule

$$\int_0^1 f(t)\,dt \approx \frac{1}{m}\big(\tfrac{1}{2}f_0 + f_1 + f_2 + \cdots + f_{m-1} + \tfrac{1}{2}f_m\big), \qquad f_i := f(i/m)$$

to obtain

$$L(A,E) \approx \frac{1}{m}\Big(\frac{1}{2}e^{A}E + \sum_{i=1}^{m-1} e^{A(1-i/m)}Ee^{Ai/m} + \frac{1}{2}Ee^{A}\Big).$$

This formula is expensive to evaluate, because it requires $e^{A/m}$, $e^{2A/m}$, ..., $e^{(m-1)A/m}$. However, when $m = 2^s$ a much more efficient evaluation is possible, for any integration rule.

Theorem 10.24. *Consider a quadrature formula* $\int_0^1 f(t)\,dt \approx \sum_{i=1}^p w_i f(t_i) = R_1(f)$ *and denote the result of applying its* m*-times repeated form to* (10.15) *by* $R_m(A,E)$, *so that* $R_1(A,E) = \sum_{i=1}^p w_i e^{A(1-t_i)}Ee^{At_i}$. *If*

$$Q_s = R_1(2^{-s}A, 2^{-s}E), \tag{10.45a}$$

$$Q_{i-1} = e^{2^{-i}A}Q_i + Q_i e^{2^{-i}A}, \quad i = s\colon -1\colon 1, \tag{10.45b}$$

then $Q_0 = R_{2^s}(A,E)$.

Proof. We have

$$R_1(A/2, E/2) = \frac{1}{2}\sum_{i=1}^p w_i e^{A(1-t_i)/2}Ee^{At_i/2}.$$

Now

$$\begin{aligned} e^{A/2}R_1(A/2,E/2) + R_1(A/2,E/2)e^{A/2} &= \frac{1}{2}\sum_{i=1}^p w_i\big(e^{A(1-t_i/2)}Ee^{At_i/2} \\ &\qquad + e^{A(1-t_i)/2}Ee^{A(1+t_i)/2}\big) \\ &= R_2(A,E), \end{aligned} \tag{10.46}$$

since $R_2(f) = \frac{1}{2}\sum_{i=1}^p w_i\big(f(t_i/2) + f(1/2 + t_i/2)\big)$. Repeated use of the relation (10.46) yields the result. □

Note that the approximation Q_0 can be derived by using $L_{\exp} \approx R_1$ in (10.44). The extra information in Theorem 10.24 is that the resulting approximation is precisely R_{2^s}.

Two key questions are how large the error $R_{2^s}(A,E) - L(A,E)$ is and which quadrature rule to choose. The next result helps in these regards. We define $\widetilde{\mu}(A) = \max\big(0, \frac{1}{2}(\mu(A) + \mu(-A))\big)$, where μ is the logarithmic norm (10.11), and note that both $\widetilde{\mu}(A)$ and $\mu(A)$ are bounded above by $\|A\|_2$.

Theorem 10.25 (error bounds). *Assume that* $c := \|A\|_2^2\, e^{\widetilde{\mu}(A)/m}/(6m^2) < 1$. *Then for the repeated trapezium rule* $R_{T,m}$, *the repeated Simpson rule* $R_{S,m}$, *and any unitarily invariant norm,*

$$\|L(A,E) - R_{T,m}(A,E)\| \le \frac{2c}{1-c}\|L(A)\|\,\|E\|, \tag{10.47}$$

$$\|L(A,E) - R_{S,m}(A,E)\| \le \frac{e^{\widetilde{\mu}(A)/m}}{180(1-c)}\|m^{-1}A\|_2^4\,\|L(A)\|\,\|E\|. \tag{10.48}$$

Proof. See Mathias [408, 1992, Cor. 2.7]. □

If, for example, $\|A\|_2/m \le 1/2$ then $c < 0.07$ and $\|L(A,E) - R_{T,m}(A,E)\| \le 0.15\|L(A)\|\,\|E\|$.

In order that s not be unduly large, the quadrature approach is restricted to providing low accuracy approximations to $L(A,E)$.

Note that (10.47) and (10.48) are not relative errors bounds, because $\|L(A)\|$ rather than $\|L(A,E)\|$ appears on the right-hand side, and $\|L(A,E)\| \ll \|L(A)\|\,\|E\|$ is possible. However, by maximizing these bounds over all E we obtain $\|L(A) - R_{T,m}(A)\| \le \theta\|L(A)\|$, where $\theta = 2c/(1-c)$, and similarly for $R_{S,m}$. Also note that (10.47) readily yields

$$(1-\theta)\|L(A)\| \le \|R_{T,m}(A)\| \le (1+\theta)\|L(A)\|,$$

and similarly for (10.48).

In order to guarantee that $c < 1$, given only $\|A\|_2$, we need to assume that $\|A\|_2/m < 1.28$. If we intertwine the computation of $L(A)$ with Algorithm 10.20 then since $\theta_{13} \approx 5.37$ exceeds 1.28 by more than a factor 4, we may be forced to scale by an extra factor 8 in the scaling and squaring method, resulting in 3 extra squarings and possible loss of accuracy. Hence evaluation of $L(A,E)$ by quadrature does not mesh particularly well with the scaling and squaring method.

Mathias [408, 1992] shows that if we scale so that $\|A\|_2/2^s \in (1/4, 1/2]$ (for which $c = 0.069$ and $2c/(1-c) = 0.15$) then Simpson's rule $R_{S,2^{s}-1}$ is more accurate than the trapezium rule $T_{S,2^s}$ both in theory (from Theorem 10.25) and in practice, while having the same cost.

Algorithm 10.26 (low accuracy approximation to Fréchet derivative of e^A). Given $A \in \mathbb{C}^{n\times n}$ this algorithm approximates the Fréchet derivative $L = L(A,E)$ of the matrix exponential via the repeated Simpson rule, aiming to provide a result with norm of the correct order of magnitude.

1 $B = A/2^s$ with $s \ge 0$ a minimal integer such that $\|A/2^s\|_1 \le 1/2$
2 $X = e^B$
3 $\widetilde{X} = e^{B/2}$
4 $Q_s = 2^{-s}(XE + 4\widetilde{X}E\widetilde{X} + EX)/6$
5 for $i = s\colon -1\colon 1$
6   if $i < s$, $X = e^{2^{-i}A}$, end
7     $Q_{i-1} = XQ_i + Q_iX$
8 end
9 $L = Q_0$

Cost: $(4+2s)M$ and $s+1$ matrix exponentials (or 1 exponential and $(4+3s)M$ if repeated squaring is used at line 6).

10.6.2. The Kronecker Formulae

Another way to approximate $L(A,E)$ is by using the Kronecker formulae (10.17). Consider the formula

$$\mathrm{vec}(L(A,E)) = \frac{1}{2}(e^{A^T} \oplus e^A)\,\tau\big(\tfrac{1}{2}[A^T \oplus (-A)]\big)\,\mathrm{vec}(E), \qquad (10.49)$$

where $\tau(x) = \tanh(x)/x$ and $\|\frac{1}{2}[A^T \oplus (-A)]\| < \pi/2$ is assumed. Using (B.16), we have

$$L(A,E) = \frac{1}{2}(Ye^A + e^A Y), \qquad \mathrm{vec}(Y) = \tau\big(\tfrac{1}{2}[A^T \oplus (-A)]\big)\,\mathrm{vec}(E).$$

Note that $\frac{1}{2}[A^T \oplus (-A)]\,\mathrm{vec}(E) = \frac{1}{2}(A^T \otimes I - I \otimes A)\,\mathrm{vec}(E) = \frac{1}{2}\,\mathrm{vec}(EA - AE)$. Hence if $r(x)$ is a rational approximation to $\tau(x)$ in which both numerator and denominator are factored into linear factors then $r\big(\frac{1}{2}[A^T \oplus (-A)]\big)\,\mathrm{vec}(E)$ can be evaluated at the cost of solving a sequence of Sylvester equations containing "Sylvester products" on the right-hand side. To be specific, let

$$\tau(x) \approx r_m(x) = \prod_{i=1}^{m} (x/\beta_i - 1)^{-1}(x/\alpha_i - 1) \tag{10.50}$$

be a rational approximation. Then we can approximate $L(A,E)$ using the following "Sylvester cascade":

$$\begin{gathered} G_0 = E, \\ \left(I + \frac{A}{\beta_i}\right) G_i + G_i\left(I - \frac{A}{\beta_i}\right) = \left(I + \frac{A}{\alpha_i}\right) G_{i-1} + G_{i-1}\left(I - \frac{A}{\alpha_i}\right), \quad i = 1\colon m, \\ L(A,E) \approx \frac{1}{2}\big(G_m e^A + e^A G_m\big). \end{gathered}$$

A natural choice for r_m is a Padé approximant. Padé approximants to τ can be obtained by truncating the continued fraction expansion

$$\tau(x) = \tanh(x)/x = 1 + \cfrac{1}{1 + \cfrac{x^2/(1\cdot 3)}{1 + \cfrac{x^2/(3\cdot 5)}{1 + \cdots + \cfrac{x^2/((2k-1)\cdot(2k+1))}{1 + \cdots}}}}. \tag{10.51}$$

Kenney and Laub [348, 1998] show that for $\|A\| < \pi/2$ we have $\|\tau(A) - r_m(A)\| \le g(\|A\|)$, where $g(x) = \tau(ix) - r_m(ix)$ and the norm is any consistent matrix norm. Defining $C = \frac{1}{2}[A^T \oplus (-A)]$ and noting that $\|C\|_p \le \|A\|_p$ for $1 \le p \le \infty$, it follows that if $\|A\|_p \le 1$ then $\|\tau(C) - r_8(C)\|_p \le g(1) = 1.44 \times 10^{-16}$, and the condition $\|C\|_p \le \pi/2$ for (10.49) to be valid is then satisfied. The condition $\|A\|_p \le 1$ can be arranged by scaling $A \leftarrow 2^{-s}A$ and using the recurrence (10.44) to undo the effects of the scaling. The [8/8] Padé approximant is given by

$$r_8(x) = \frac{p_8(x)}{q_8(x)} = \frac{x^8 + 990x^6 + 135135x^4 + 4729725x^2 + 34459425}{45x^8 + 13860x^6 + 945945x^4 + 16216200x^2 + 34459425}$$

and the zeros of p_8 and q_8 are given in Table 10.5. In choosing how far to scale A we also need to consider the conditioning of the Sylvester cascade. This boils down to ensuring that the upper bound in

$$\prod_{i=1}^{m} \|(C/\beta_i - I)^{-1}\|_p \|C/\alpha_i - I\|_p \le \prod_{i=1}^{m} \frac{1 + \|C/\alpha_i\|_p}{1 - \|C/\beta_i\|_p} \le \prod_{i=1}^{m} \frac{1 + \|A/\alpha_i\|_p}{1 - \|A/\beta_i\|_p}$$

is not too large. For $m = 8$ and the 1-norm this bound is at most 55.2 if $\|A\|_p \le 1$, which is quite acceptable.

The ideas above are collected in the following algorithm.

Table 10.5. *Zeros α_j of numerator p_8 and β_j of denominator q_8 of [8/8] Padé approximant r_8 to $\tau(x) = \tanh(x)/x$, shown to 5 significant digits.*

α_j	β_j
$\pm 3.1416e0i$	$\pm 1.5708e0i$
$\pm 6.2900e0i$	$\pm 4.7125e0i$
$\pm 1.0281e1i$	$\pm 7.9752e0i$
$\pm 2.8894e1i$	$\pm 1.4823e1i$

Algorithm 10.27 (Fréchet derivative of matrix exponential). Given $A \in \mathbb{C}^{n\times n}$ this algorithm evaluates the Fréchet derivative $L = L(A, E)$ of the matrix exponential via (10.49), using scaling and squaring and the [8/8] Padé approximant to $\tau = \tanh(x)/x$. It uses the constants in Table 10.5. The algorithm is intended for IEEE double precision arithmetic.

1 $B = A/2^s$ with $s \geq 0$ a minimal integer such that $\|A/2^s\|_1 \leq 1$.
2 $G_0 = 2^{-s}E$
3 for $i = 1\colon 8$
4   Solve for G_i the Sylvester equation $(I + B/\beta_i)G_i + G_i(I - B/\beta_i) =$
      $(I + B/\alpha_i)G_{i-1} + G_{i-1}(I - B/\alpha_i)$.
5 end
6 $X = e^B$
7 $L_s = (G_8X + XG_8)/2$
8 for $i = s\colon -1\colon 1$
9   if $i < s$, $X = e^{2^{-i}A}$, end
10     $L_{i-1} = XL_i + L_iX$
11 end
12 $L = L_0$

Cost: $(18 + 2s)M$ and s matrix exponentials (or 1 exponential and $(17 + 3s)M$ if repeated squaring is used at line 9), and the solution of 8 Sylvester equations.

In practice we would use an initial Schur decomposition of A to reduce the cost, making use of Problem 3.2; this is omitted from the algorithm statement to avoid clutter.

Similar algorithms could be developed for the other two formulae in (10.17).

10.6.3. Computing and Estimating the Norm

The Fréchet derivative norm $\|L(A)\|_F$ is given by the explicit formulae in Corollary 10.14, which involve the 2-norms of $n^2 \times n^2$ matrices. If n is small it is practical to compute the matrix directly from its formula, for example by Algorithm 9.6, provided a means for evaluating the relevant scalar functions and their derivatives is available; for ψ_1 see Section 10.7.4. However, direct evaluation in this way is prone to overflow. For example, if A has a negative eigenvalue λ_i with $|\lambda_i| \gg |\lambda_j| \geq 1$ for some $j \neq i$, then $\psi_1(A^T \oplus (-A))$ has an eigenvalue approximately $\psi_1(-\lambda_i) = \psi_1(|\lambda_i|)$ and so $\|\psi_1(A^T \oplus (-A))\|$ will be large and could overflow, even though e^A may be of modest norm.

An alternative approach, which avoids unnecessary overflow, is to apply Algorithm 3.17, using Algorithm 10.26 or 10.27 to evaluate $L(X, E)$ for the n^2 specified choice of E. For large n, this is prohibitively expensive and the condition number must be estimated rather than exactly computed. One possibility is to use the power method, Algorithm 3.20, in conjunction with Algorithm 10.26 or Algorithm 10.27. A good starting matrix is $Z_0 = XX^* + X^*X$, where $X = e^A$, the reason being that $X = L(A, I)$ and the adjoint step of the power method gives $L^\star(A, X) = L(A^*, X) = \int_0^1 e^{A^*(1-s)} e^A e^{A^* s}\, ds \approx (XX^* + X^*X)/2$, using a trapezium rule approximation. Hence Z_0 approximates at relatively small cost the result of a power method iteration.

The following is our preferred approach.

Algorithm 10.28 (condition number estimate for matrix exponential). Given $A \in \mathbb{C}^{n\times n}$ this algorithm produces an estimate γ of the 1-norm condition number $\kappa_{\exp}(A)$ of the matrix exponential; more precisely, γ is a lower bound for a quantity in the interval $[n^{-1}\kappa_{\exp}(A), n\kappa_{\exp}(A)]$.

1 Invoke Algorithm 3.22 to produce an estimate γ of $\|L(A)\|_1$, using Algorithm 10.27 to evaluate $L(A, E)$ and $L^\star(A, E) = L(A^*, E)$.
2 $\gamma \leftarrow \gamma \|A\|_1 / \|e^A\|_1$

Cost: About 4–5 invocations of Algorithm 10.27.

For efficiency, A would in practice be reduced to Schur form at the start of Algorithm 3.17 and Algorithm 10.28.

We ran Algorithm 10.28 on the same set of matrices used in Section 10.5. We also ran a modified version of Algorithm 10.28 that estimates $L(A, E)$ by the finite difference (3.22) with increment (3.23). We compared the estimates with a 1-norm analogue of Algorithm 3.17 in which each norm on line 7 is the 1-norm and $L(A, E)$ is evaluated by Algorithm 10.27. The three estimates were mostly of the same order of magnitude, the only exceptions being for extremely ill conditioned problems where the estimate via finite differences was occasionally several orders of magnitude too large (a relatively harmless overestimation).

10.7. Miscellany

In this section we discuss some miscellaneous methods, techniques, and special classes of matrices, as well as a variation of the e^A problem.

10.7.1. Hermitian Matrices and Best L_∞ Approximation

If A is Hermitian (or, more generally, normal) then approximating e^A is equivalent to approximating the exponential function on the spectrum of A. Best rational L_∞ approximations r_m from $\mathcal{R}_{m,m}$ to e^x on $(-\infty, 0]$ (or, equivalently, to e^{-x} on $[0, \infty)$) have been determined for $m \le 14$ by Cody, Meinardus, and Varga [110, 1969] and to high accuracy by Carpenter, Ruttan, and Varga [98, 1984] (see Section 4.4 for the definition of $\mathcal{R}_{m,m}$). Define the L_∞ error $\mathcal{E}_m = \sup_{x\in(-\infty,0]} |r_m(x) - e^x|$. It is known that $\lim_{m\to\infty} \mathcal{E}_m^{1/m} = 1/9.289025\ldots$ [98, 1984] (see Trefethen, Weideman, and Schmelzer [574, 2006] or Finch [187, 2003, Sec. 4.5] for the interesting history of this result and references), which implies that $\mathcal{E}_m \approx 10^{-m}$.

The approximations r_m are valid on the whole negative real axis, whereas Padé approximations are accurate only near the origin. In applications from differential equations in which e^{Ah} is required, with h a timestep, these best L_∞ approximations allow the use of much larger timesteps than Padé approximants [200, 1992].

For evaluating $r_m(A)$ the partial fraction form is preferred. The poles have been observed in practice to be distinct and to come in complex conjugate pairs; see [200, 1992], where values of the poles are tabulated for degrees 10 and 14.

Lu [395, 1998] suggests shifting by the smallest eigenvalue μ and evaluating $e^A \approx e^\mu r_m(A-\mu I)$, using unitary reduction to tridiagonal form to reduce the evaluation of r_m to the tridiagonal case and then employing the partial fraction expansion of r_m.

10.7.2. Essentially Nonnegative Matrices

A matrix $A \in \mathbb{R}^{n\times n}$ is *essentially nonnegative* if $a_{ij} \geq 0$ for $i \neq j$. This class of matrices merits special consideration because of its occurrence in important applications such as Markov chains (see Section 2.3) and the numerical solution of PDEs [600, 2000, Sec. 8.2]. It is worth noting that the set $\mathbb{Z}^{n\times n}$ arising in the theory of nonnegative matrices comprises the negatives of the essentially nonnegative matrices [60, 1994, Chap. 6].

Interestingly, the essentially nonnegative matrices are precisely those for which e^{At} is nonnegative for all $t \geq 0$.

Theorem 10.29. *Let $A \in \mathbb{R}^{n\times n}$. Then $e^{At} \geq 0$ for all $t \geq 0$ if and only if A is essentially nonnegative.*

Proof. Suppose $e^{At} \geq 0$ for all $t \geq 0$. Taking $t \geq 0$ sufficiently small in the expansion $e^{At} = I + At + O(t^2)$ shows that the off-diagonal elements of A must be nonnegative.

Conversely, suppose A is essentially nonnegative. We can write $A = D+N$, where $D = \mathrm{diag}(A)$ and $N \geq 0$, which can be rewritten as

$$A = \theta I + (D - \theta I + N) =: \theta I + B, \qquad \theta = \min_i a_{ii}. \tag{10.52}$$

Then $B \geq 0$ and $e^{At} = e^{\theta t}e^{Bt} \geq 0$ for $t \geq 0$, since the exponential of a matrix with nonnegative elements is clearly nonnegative. □

The formula $e^A = e^\theta e^B$ with θ, B as in (10.52) is used in the uniformization method in Markov chain analysis [523, 2007], [524, 1999, Sec. 2], [540, 1994, Chap. 8], in which $e^B v$ ($v \geq 0$) is approximated using a truncated Taylor series expansion of e^B. The Taylor series for e^B contains only nonnegative terms, whereas that for e^A contains terms of both sign and so can suffer from cancellation. Note that in the case of A being an intensity matrix the shift θ is optimal in the sense of Corollary 4.22.

Recall from Section 2.3 that an intensity matrix is an essentially nonnegative matrix with zero row sums, and that the exponential of such a matrix is stochastic and so has unit ∞-norm. An intensity matrix is perfectly conditioned for the matrix exponential.

Theorem 10.30. *If $A \in \mathbb{R}^{n\times n}$ is an intensity matrix then in the ∞-norm, $\kappa_{\exp}(A) = \|A\|_\infty$.*

Proof. For $s \in [0,1]$, sA is also an intensity matrix. Hence on taking ∞-norms in (10.15) we obtain

$$\|L(A,E)\|_\infty \leq \int_0^1 \|e^{A(1-s)}\|_\infty \|E\|_\infty \|e^{As}\|_\infty \, ds$$

$$= \int_0^1 \|E\|_\infty \, ds = \|E\|_\infty = \|e^A\|_\infty \|E\|_\infty.$$

Thus $\kappa_{\exp}(A) \le \|A\|_\infty$, and this is an equality in view of (10.22). □

10.7.3. Preprocessing

Balancing, and argument reduction via a single shift$A \leftarrow A - \mu I$, can be used as preprocessing steps in any of the algorithms described in this chapter. Theorem 10.18 in conjunction with Theorem 4.21 (a) suggests that $\mu = n^{-1}\,\mathrm{trace}(A)$ is an excellent choice of shift. However, this shift is not always appropriate. If A has one or more eigenvalues with large negative real part then μ can be large and negative and $A - \mu I$ can have an eigenvalue with large positive real part, causing $e^{A-\mu I}$ to overflow.

Preprocessing is potentially particularly helpful to Algorithm 10.20 (the scaling and squaring method), since it can reduce the norm and hence directly reduce the cost of the method. Our own numerical experience suggests that preprocessing typically has little effect on the cost or accuracy of Algorithm 10.20, and since it could even worsen the accuracy we do not recommend its automatic use.

10.7.4. The ψ Functions

The functions $\psi_k(z) = \sum_{j=0}^{\infty} z^j/(j+k)!$ are closely related to the exponential:

$$\psi_0(z) = e^z, \quad \psi_1(z) = \frac{e^z - 1}{z}, \quad \psi_2(z) = \frac{e^z - 1 - z}{z^2}, \quad \psi_3(z) = \frac{e^z - 1 - z - \frac{1}{2}z^2}{z^3}, \ldots.$$

As explained in Section 2.1, these functions play a fundamental role in exponential integrators for ordinary differential equations, and we saw earlier in this chapter that ψ_1 arises in one of the representations (10.17) for the Fréchet derivative of the matrix exponential. An integral representation is

$$\psi_k(z) = \frac{1}{(k-1)!} \int_0^1 e^{(1-t)z} t^{k-1} \, dt \tag{10.53}$$

and the ψ_k satisfy the recurrence relation

$$\psi_k(z) = z\psi_{k+1}(z) + \frac{1}{k!}. \tag{10.54}$$

From the last two equations it follows that ψ_{k+1} is the divided difference

$$\psi_{k+1}(z) = \psi_k[z, 0].$$

Even for scalar arguments it is a nontrivial task to evaluate the ψ_k accurately. For example, $\psi_1(z)$ suffers badly from cancellation if evaluated as $(e^z - 1)/z$ for $|z| \ll 1$, though this formula can be reworked in a nonobvious way to avoid the cancellation [276, 2002, Sec. 1.14.1] (or the function $\mathrm{expm1}(x) := e^x - 1$, available as `expm1` in MATLAB, can be invoked [562, 1992]).

Padé approximants to the ψ functions are known explicitly. The following theorem includes (10.23) and (10.24) (with $k = m$ therein) as special cases.

Theorem 10.31. *The diagonal $[m/m]$ Padé approximant $r_m(z) = p_m(z)/q_m(z)$ to $\psi_k(z)$ is given by*

$$p_m(z) = \frac{m!}{(2m+k)!} \sum_{i=0}^{m} \left(\sum_{j=0}^{i} \frac{(2m+k-j)!(-1)^j}{j!(m-j)!(k+i-j)!} \right) z^i,$$

$$q_m(z) = \frac{m!}{(2m+k)!} \sum_{i=0}^{m} \frac{(2m+k-i)!}{i!(m-i)!} (-z)^i,$$

and

$$\psi_k(z) - r_m(z) = \frac{(-1)^m m!(m+k)!}{(2m+k)!(2m+k+1)!} z^{2m+1} + O(z^{2m+2}).$$

Proof. See Skaflestad and Wright [527, 2006]. Magnus and Wynn [401, 1975] had earlier obtained the $[k/m]$ Padé approximants to ψ_0. □

With the aid of these Padé approximants, evaluation of ψ_j, $j = 0\colon k$, can be done via an analogue of the scaling and squaring method for the exponential. The squaring phase makes use (with $\alpha = \beta = 1$) of the identity [527, 2006], for $\alpha, \beta \in \mathbb{R}$,

$$\psi_k\big((\alpha+\beta)z\big) = \frac{1}{(\alpha+\beta)^k} \left(\beta^k \psi_0(\alpha z)\psi_k(\beta z) + \sum_{j=1}^{k} \frac{\alpha^j \beta^{k-j}}{(k-j)!} \psi_j(\alpha z) \right). \tag{10.55}$$

A basic Padé-based scaling and squaring algorithm is proposed by Hochbruck, Lubich, and Selhofer [292, 1998] for $\psi_1(A)$, while Koikari [360, 2007] and Skaflestad and Wright [527, 2006] develop more sophisticated versions, analogous to Algorithm 10.20, for evaluating the sequence $\psi_0(A), \ldots, \psi_k(A)$.

As a special case of (10.40) (or directly, by using $Bf(B) = f(B)B$), we have

$$B = \begin{bmatrix} A & I \\ 0 & 0 \end{bmatrix} \quad \Rightarrow \quad e^B = \begin{bmatrix} e^A & \psi_1(A) \\ 0 & I \end{bmatrix}, \tag{10.56}$$

so the evaluation of ψ_1 can be reduced to that of the exponential. The practical use of (10.56) for Hermitian A is explored by Lu [397, 2003]. This relation can be extended to provide a sequence of higher order ψ functions from a single exponential evaluation; see Sidje [522, 1998, Thm. 1] and Minchev [430, 2004].

10.8. Notes and References

An early exposition of the matrix exponential and its use in solving differential equations is given by Frazer, Duncan, and Collar [193, 1938].

Moler and Van Loan's classic paper [437, 1978], [438, 2003] is highly cited, with over 500 citations in the ISI Citation Index. Golub and Van Loan [224, 1996, Sec. 11.3] give an overview of the scaling and squaring method and of perturbation theory for the matrix exponential. Van Loan [592, 1975] is a good source for fundamental theoretical results concerning the matrix exponential.

Theorem 10.1 is due to Suzuki [552, 1976] and Theorem 10.3 to Wermuth [613, 1989].

For more details on the Baker–Campbell–Hausdorff formula (10.5) see Hall [241, 2003, Chap. 3] or Varadarajan [598, 1974, Thm. 2.15.4], and for its history see Weiss and Maradudin [612, 1962]. The domain of convergence of the formula is discussed by Blanes and Casas [74, 2004]. A simple way to compute the coefficients in the formula, involving a matrix $\log(e^F e^G)$ where F and G are certain strictly upper bidiagonal matrices, is given by Reinsch [486, 2000].

For more on the Zassenhaus formula (10.6) see Magnus [403, 1954] or Wilcox [615, 1967], and see Scholz and Weyrauch [506, 2006] for a simple way to compute the coefficients C_i.

The Strang splitting is due to Strang [544, 1968]. For a survey of splitting methods see McLachlan and Quispel [417, 2002].

Theorem 10.6 is from Suzuki [552, 1976]. The formula (10.8) is widely used in physics, where it is known as the Suzuki–Trotter formula. For an example of the use of the formula see Bai, Chen, Scalettar, and Yamazaki [26, 2007]. The Trotter product formula in Corollary 10.7 is from Trotter [575, 1959].

Other bounds on $\|e^A\|$ can be found in Van Loan [594, 1977] and Kågström [322, 1977]. For various bounds on $\sup_{t\geq 0}\|e^{tA}\|$ see Trefethen and Embree [573, 2005].

The representation (10.15) for the Fréchet derivative appears in Karplus and Schwinger [335, 1948], and Najfeld and Havel [445, 1995] state that this is the earliest appearance.

Theorem 10.13 is a mixture of relations from Kenney and Laub [348, 1998] and Najfeld and Havel [445, 1995]. It is possible to "unvec" the relations (10.17), but the resulting expressions (see [580, 1995] or [241, 2003, Thm. 3.5] for (10.17a)) are less useful for computational purposes.

Theorem 10.16 is from Van Loan [594, 1977]. Theorem 10.17 is due to Melloy and Bennett [422, 1993], whose paper seems to have hitherto been overlooked. Theorem 10.18 is from Parks [458, 1994].

For more on the theory and computation of the Fréchet derivative (and higher derivatives) of the exponential see Najfeld and Havel [445, 1995].

The formulae (10.23) and (10.24) are due to Padé, and proofs are given by Gautschi [207, 1997, Thm. 5.5.1]. A proof of the "$\rho(r_{mm}(A)) < 1$" property can be found in [207, 1997, p. 308].

It is known that the poles of the $[k/m]$ Padé approximant $r_{km}(x) = p_{km}(x)/q_{km}(x)$, that is, the zeros of q_{km}, are distinct, though they are complex; see Zakian [621, 1975, Lem. 4.1]. Thus r_{km} can be expressed in linear partial fraction form, as is done by Gallopoulos and Saad [200, 1992], for example.

An early reference to the scaling and squaring method is Lawson [376, 1967], and Ward [606, 1977] attributes the method to Lawson.

The evaluation schemes (10.32) and (10.33) were suggested by Van Loan [593, 1977].

Algorithm 10.20 and the supporting analysis are due to Higham [278, 2005]. Section 10.3 is based on [278].

The analysis of Moler and Van Loan [438, 2003, Sec. 3, App. A] proceeds as in Problem 10.10 and makes the assumption that $\|2^{-s}A\| \leq 1/2$. The backward error bound is used to determine a suitable Padé degree m. Based on this analysis, in MATLAB 7 (R14SP3) and earlier the `expm` function used an implementation of the scaling and squaring method with $m = 6$ and $\|2^{-s}A\|_\infty \leq 1/2$ as the scaling criterion (and Sidje [522, 1998] uses the same parameters in his function `padm`). Version 7.2 (R2006a) of MATLAB introduced a new `expm` that implements Algorithm 10.20. This new version of `expm` is more efficient than the old, as the experiment of Section 10.5 indicates, and typically more accurate, as a side effect of the reduced scaling and hence fewer squarings. Mathematica's MatrixExp function has used Algorithm 10.20 for matrices of machine numbers since Version 5.1.

Ward [606, 1977] uses $m = 8$ and $\|2^{-s}A\|_1 \le 1$ in his implementation of the scaling and squaring method and preprocesses with argument reduction and balancing, as discussed in Section 10.7.3. He gives a rounding error analysis that leads to an a posteriori forward error bound. This analysis is similar to that given here, but for the squaring phase it uses a recurrence based on computed quantities in place of the a prior bound of Theorem 10.21.

Najfeld and Havel [445, 1995] suggest a variation of the scaling and squaring method that uses Padé approximations to the function

$$\begin{aligned}\tau(x) &= x\coth(x) = x(e^{2x}+1)(e^{2x}-1)^{-1} \\ &= 1 + \cfrac{x^2}{3 + \cfrac{x^2}{5 + \cfrac{x^2}{7 + \cdots}}},\end{aligned} \tag{10.57}$$

from which the exponential can be recovered via $e^{2x} = (\tau(x)+x)/(\tau(x)-x)$. The continued fraction expansion provides an easy way of generating the Padé approximants to τ. Higham [278, 2005] shows that the algorithm suggested by Najfeld and Havel is essentially a variation of the standard scaling and squaring method with direct Padé approximation, but with weaker guarantees concerning its behaviour both in exact arithmetic (because a backward error result is lacking) and in floating point arithmetic (because a possibly ill conditioned matrix must be inverted).

The overscaling phenomenon discussed at the end of Section 10.3 was first pointed out by Kenney and Laub [348, 1998] and Dieci and Papini [157, 2000].

Equation (10.40) shows that the (1,2) block of the exponential of a block triangular matrix can be expressed as an integral. Van Loan [595, 1978] uses this observation in the reverse direction to show how certain integrals involving the matrix exponential can be obtained by evaluating the exponential of an appropriate block triangular matrix.

A conditioning analysis of the exponential taking account of the structure of block 2×2 triangular matrices is given by Dieci and Papini [158, 2001].

Theorem 10.22 is due to Opitz [452, 1964] and was rediscovered by McCurdy: see [416, 1984]. The result is also proved and discussed by de Boor [143, 2005].

The use of quadrature for evaluating $L(A, E)$ was investigated by Kenney and Laub [340, 1989], who concentrate on the repeated trapezium rule, for which they obtain the recurrence (10.45) and a weaker error bound than (10.47).

Section 10.6.2 is based on Kenney and Laub [348, 1998], wherein the focus is not on computing the Fréchet derivative of the exponential per se, but on using the Fréchet derivative within the Schur–Fréchet algorithm (see Section 10.4.2). The continued fraction expansion (10.51) can be found in Baker [38, 1975, p. 65]. The starting matrix $Z_0 = XX^* + X^*X$ for the power method mentioned in Section 10.6.3 was suggested by Kenney and Laub [340, 1989].

Theorem 10.29 can be found in Bellman [51, 1970, p. 176] and in Varga [600, 2000, Sec. 8.2]. Theorem 10.30 is new.

While MATLAB, Maple, and Mathematica all have functions to compute the matrix exponential, the only freely available software we are aware of is Sidje's Expokit package of MATLAB and Fortran codes [522, 1998].

Problems

10.1. Prove the identity (10.13).

10.2. Give a proof of $AB = BA \Rightarrow e^{(A+B)t} = e^{At}e^{Bt}$ for all t (the "if" part of Theorem 10.2) that does not use the power series for the exponential.

10.3. Show that for any $A, B \in \mathbb{C}^{n\times n}$,

$$\|e^A - e^B\| \le \|A - B\|\, e^{\max(\|A\|,\|B\|)}.$$

10.4. For $A, B \in \mathbb{C}^{n\times n}$ show that $\det(e^{A+B}) = \det(e^A e^B)$ (even though $e^{A+B} \neq e^A e^B$ in general).

10.5. Show that for $A \in \mathbb{C}^{n\times n}$ and any subordinate matrix norm, $\kappa_{\exp}(A) \le e^{2\|A\|}\|A\|$.

10.6. Let $A \in \mathbb{C}^{n\times n}$ be normal. Theorem 10.16 says that the absolute condition number $\mathrm{cond}_{\mathrm{abs}}(\exp, X) = \|e^A\|_2$, but Corollary 3.16 says that $\mathrm{cond}_{\mathrm{abs}}(\exp, A) = \max_{\lambda,\mu\in\Lambda(A)} |f[\lambda,\mu]|$, where $f = \exp$. Reconcile these two formulae.

10.7. Determine when the Fréchet derivative $L(A)$ of the exponential is nonsingular.

10.8. Show that if $\|G\| < 1$ then

$$\|\log(I+G)\| \le -\log(1-\|G\|) \le \frac{\|G\|}{1-\|G\|} \tag{10.58}$$

while if $\|G\| < 2/3$ then

$$\|G\|\left(1 - \frac{\|G\|}{2(1-\|G\|)}\right) \le \|\log(I+G)\|,$$

where the norm is any consistent matrix norm.

10.9. (Dieci and Papini [157, 2000], Moler and Van Loan [438, 2003, Lem. 2]) Show that if the denominator $q_m(A)$ of the $[m/m]$ Padé approximant to e^A satisfies $q_m(-\|A\|) < 2$ then it is nonsingular and

$$\|q_m(A)^{-1}\| \le \frac{1}{2 - q_m(-\|A\|)}.$$

The norm here is any subordinate matrix norm. Show further that if $\|A\| < \log 4 = 1.39$ then $\|q_m(A)^{-1}\| \le 1/(2 - e^{\|A\|/2})$, and hence that $\|q_m(A)^{-1}\| \le 3$ for $\|A\| \le 1$.

10.10. Let $r_m = p_m/q_m$ be the $[m/m]$ Padé approximant to the exponential. Show that $G = e^{-A}r_m(A) - I$ satisfies

$$\|G\| \le \frac{q_m(\|A\|)}{2 - q_m(-\|A\|)}\left|e^{\|A\|} - r_m(\|A\|)\right| \tag{10.59}$$

for any subordinate matrix norm, assuming that $\|G\| < 1$ and $q_m(-\|A\|) < 2$.

10.11. Prove Theorem 10.21, using the basic result that for the 1-, ∞-, and Frobenius norms [276, 2002, Sec. 3.5], $\|AB - fl(AB)\| \le \gamma_n\|A\|\,\|B\|$.

10.12. Derive the formula (10.40) for the exponential of a block 2×2 block triangular matrix.

10.13. Prove Rodrigues' formula [407, 1999, p. 291] for the exponential of a skew-symmetric matrix $A \in \mathbb{R}^{3\times 3}$:

$$e^A = I + \frac{\sin\theta}{\theta}A + \frac{1-\cos\theta}{\theta^2}A^2, \tag{10.60}$$

where $\theta = \sqrt{\|A\|_F^2/2}$. (For generalizations of this formula to higher dimensions see Gallier and Xu [199, 2002].)

10.14. Given that e^A is the matrix

$$\begin{bmatrix} \cos t & 0 & 0 & 0 & 0 & 0 & 0 & -\sin t \\ 0 & \cos t & 0 & 0 & 0 & 0 & -\sin t & 0 \\ 0 & 0 & \cos t & 0 & 0 & -\sin t & 0 & 0 \\ 0 & 0 & 0 & \cos t & -\sin t & 0 & 0 & 0 \\ 0 & 0 & 0 & \sin t & \cos t & 0 & 0 & 0 \\ 0 & 0 & \sin t & 0 & 0 & \cos t & 0 & 0 \\ 0 & \sin t & 0 & 0 & 0 & 0 & \cos t & 0 \\ \sin t & 0 & 0 & 0 & 0 & 0 & 0 & \cos t \end{bmatrix},$$

what is A?

10.15. (RESEARCH PROBLEM) Obtain conditions that are necessary or sufficient—and ideally both—for $\kappa_{\exp}(A)$ to be large.

10.16. (RESEARCH PROBLEM) Prove the stability or otherwise of the scaling and squaring algorithm, by relating rounding errors in the square root phase to the conditioning of the e^A problem.

The problem of expressing $e^{At}e^{Bt}$ in the form e^{Ct},
where A and B do not commute,
has important ramifications not only in the theory of Lie groups and algebras,
but also in modern quantum field theory.

— RICHARD BELLMAN, *Introduction to Matrix Analysis* (1970)

Calculating the exponential of a matrix can be quite complicated,
requiring generalized eigenvectors or
Jordan Canonical Form if you want explicit formulas.
Although, compared to nonlinear problems, this is still "simple,"
if you try to put this into practice,
you will come to understand the expletive that
"hell is undiagonalizable matrices."

— JOHN H. HUBBARD and BEVERLY H. WEST, *Differential Equations:*
A Dynamical Systems Approach. Higher Dimensional Systems (1995)

The [error of the] best rational approximation to e^{-x} on $[0, +\infty)$ exhibits
geometric convergence to zero.
It is this geometric convergence which has fascinated so many researchers.

— A. J. CARPENTER, A. RUTTAN, and R. S. VARGA, *Extended Numerical Computations*
on the "1/9" Conjecture in Rational Approximation Theory (1984)

In this survey we try to describe all the methods that appear to be practical,
classify them into five broad categories,
and assess their relative effectiveness.

— CLEVE B. MOLER and CHARLES F. VAN LOAN,
Nineteen Dubious Ways to Compute the Exponential of a Matrix (1978)

[The] availability of `expm(A)` *in early versions of MATLAB*
quite possibly contributed to
the system's technical and commercial success.

...

Scaling and squaring has emerged as the
least dubious of the original nineteen ways,
but we still do not understand the method completely.

— CLEVE B. MOLER and CHARLES F. VAN LOAN,
Nineteen Dubious Ways to Compute the Exponential of a Matrix,
Twenty-Five Years Later (2003)

Chapter 11
Matrix Logarithm

A logarithm of $A \in \mathbb{C}^{n\times n}$ is any matrix X such that $e^X = A$. As we saw in Theorem 1.27, any nonsingular A has infinitely many logarithms. In this chapter $A \in \mathbb{C}^{n\times n}$ is assumed to have no eigenvalues on $\mathbb{R}^-$ and "log" always denotes the principal logarithm, which we recall from Theorem 1.31 is the unique logarithm whose spectrum lies in the strip $\{ z : -\pi < \mathrm{Im}(z) < \pi \}$.

The importance of the matrix logarithm can be ascribed to it being the inverse function of the matrix exponential and this intimate relationship leads to close connections between the theory and computational methods for the two functions.

This chapter is organized as follows. We begin by developing some basic properties of the logarithm, including conditions under which the product formula $\log(BC) = \log(B) + \log(C)$ holds. Then we consider the Fréchet derivative and conditioning. The Mercator and Gregory series expansions are derived and various properties of the diagonal Padé approximants to the logarithm are explained. Two versions of the inverse scaling and squaring method are developed in some detail, one using the Schur form and the other working with full matrices. A Schur–Parlett algorithm employing inverse scaling and squaring on the diagonal blocks together with a special formula for 2×2 blocks is then derived. A numerical experiment comparing four different methods is then presented. Finally, an algorithm for evaluating the Fréchet derivative is described.

11.1. Basic Properties

We begin with an integral expression for the logarithm.

Theorem 11.1 (Richter). *For $A \in \mathbb{C}^{n\times n}$ with no eigenvalues on $\mathbb{R}^-$,*

$$\log(A) = \int_0^1 (A - I)\big[t(A - I) + I\big]^{-1} dt. \tag{11.1}$$

Proof. It suffices to prove the result for diagonalizable A, by Theorem 1.20, and hence it suffices to show that $\log x = \int_0^1 (x-1)\big[t(x-1)+1\big]^{-1} dt$ for $x \in \mathbb{C}$ lying off $\mathbb{R}^-$; this latter equality is immediate. $\square$

Now we turn to useful identities satisfied by the logarithm. Because of the multivalued nature of the logarithm it is not generally the case that $\log(e^A) = A$, though a sufficient condition for the equality is derived in Problem 1.39. To understand the scalar case of this equality we use the *unwinding number* of $z \in \mathbb{C}$ defined by

$$\mathcal{U}(z) = \frac{z - \log(e^z)}{2\pi i}. \tag{11.2}$$

It is easy to show that

$$\mathcal{U}(z) = \left\lceil \frac{\operatorname{Im} z - \pi}{2\pi} \right\rceil \in \mathbb{Z}.$$

The equation

$$\log(e^z) = z - 2\pi i \mathcal{U}(z) \tag{11.3}$$

completely describes the relationship between $\log(e^z)$ and z, in particular showing that $\log(e^z) = z$ for $\operatorname{Im} z \in (-\pi, \pi]$.

The next result gives an important identity.

Theorem 11.2. *For $A \in \mathbb{C}^{n\times n}$ with no eigenvalues on $\mathbb{R}^-$ and $\alpha \in [-1,1]$ we have $\log(A^\alpha) = \alpha \log(A)$. In particular, $\log(A^{-1}) = -\log(A)$ and $\log(A^{1/2}) = \frac{1}{2}\log(A)$.*

Proof. Consider first the scalar case. By definition, $a^\alpha = e^{\alpha \log a}$. Hence, using (11.3), $\log a^\alpha = \log(e^{\alpha \log a}) = \alpha \log a - 2\pi i \mathcal{U}(\alpha \log a)$. But $\mathcal{U}(\alpha \log a) = 0$ for $a \notin \mathbb{R}^-$ and $\alpha \in [-1,1]$. Thus the identity holds for scalar A and follows for general A by Theorem 1.20. (For the last part, cf. Problem 1.34.) □

We now turn to the question of when $\log(BC) = \log(B) + \log(C)$. First, consider the scalar case. Even here, the equality may fail. Let $b = c = e^{(\pi-\epsilon)i}$ for ϵ small and positive. Then

$$\log bc = -2\epsilon i \neq (\pi - \epsilon)i + (\pi - \epsilon)i = \log b + \log c. \tag{11.4}$$

Understanding of this behaviour is aided by using the unwinding number. We will denote by $\arg z \in (-\pi, \pi]$ the principal value of the argument of $z \in \mathbb{C}$. For $z_1, z_2 \in \mathbb{C}$ we have $z_1 z_2 = e^{\log z_1} e^{\log z_2} = e^{\log z_1 + \log z_2}$ and hence, by (11.3),

$$\log(z_1 z_2) = \log\left(e^{\log z_1 + \log z_2}\right) = \log z_1 + \log z_2 - 2\pi i \mathcal{U}(\log z_1 + \log z_2). \tag{11.5}$$

Note that since $\operatorname{Im} \log z \in (-\pi, \pi]$, we have $\mathcal{U}(\log z_1 + \log z_2) \in \{-1, 0, 1\}$. Now $\mathcal{U}(\log z_1 + \log z_2) = 0$ precisely when $|\arg z_1 + \arg z_2| < \pi$, so this is the condition for $\log(z_1 z_2) = \log z_1 + \log z_2$ to hold. As a check, the inequality in (11.4) is confirmed by the fact that $\mathcal{U}(\log b + \log c) = \mathcal{U}(2\pi - 2\epsilon) = 1$.

The next theorem generalizes this result from scalars to matrices. Recall from Corollary 1.41 that if X and Y commute then for each eigenvalue λ_j of X there is an eigenvalue μ_j of Y such that $\lambda_j + \mu_j$ is an eigenvalue of $X + Y$. We will call μ_j the eigenvalue corresponding to λ_j.

Theorem 11.3. *Suppose $B, C \in \mathbb{C}^{n\times n}$ both have no eigenvalues on $\mathbb{R}^-$ and that $BC = CB$. If for every eigenvalue λ_j of B and the corresponding eigenvalue μ_j of C,*

$$|\arg \lambda_j + \arg \mu_j| < \pi, \tag{11.6}$$

then $\log(BC) = \log(B) + \log(C)$.

Proof. Note first that $\log(B)$, $\log(C)$, and $\log(BC)$ are all defined—by (11.6) in the latter case. The matrices $\log(B)$ and $\log(C)$ commute, since B and C do. By Theorem 10.2 we therefore have

$$e^{\log(B)+\log(C)} = e^{\log(B)} e^{\log(C)} = BC.$$

Thus $\log(B) + \log(C)$ is *some* logarithm of BC. The imaginary part of the jth eigenvalue of $\log(B) + \log(C)$ is

$$\operatorname{Im}(\log \lambda_j + \log \mu_j) = \arg \lambda_j + \arg \mu_j \in (-\pi, \pi)$$

by (11.6), so $\log(B) + \log(C)$ is the *principal* logarithm of BC. □

The following variant of Theorem 11.3 may be easier to apply in some situations. We need to define the open half-plane associated with $z = \rho e^{i\theta}$, which is the set of complex numbers $w = \zeta e^{i\phi}$ such that $-\pi/2 < \phi - \theta < \pi/2$. Figure 11.1 gives a pictorial representation of condition (b) of the result.

Theorem 11.4 (Cheng, Higham, Kenney, and Laub). *Let $B, C \in \mathbb{C}^{n\times n}$. Suppose that $A = BC$ has no eigenvalues on $\mathbb{R}^-$ and*

(a) $BC = CB$,

(b) *every eigenvalue of B lies in the open half-plane of the corresponding eigenvalue of $A^{1/2}$ (or, equivalently, the same condition holds for C).*

Then $\log(A) = \log(B) + \log(C)$.

Proof. As noted above, since B and C commute there is a correspondence between the eigenvalues a, b, and c (in some ordering) of A, B, and C: $a = bc$. Express these eigenvalues in polar form as

$$a = \alpha e^{i\theta}, \quad b = \beta e^{i\phi}, \quad c = \gamma e^{i\psi}, \qquad -\pi < \theta, \phi, \psi \le \pi.$$

Since A has no eigenvalues on $\mathbb{R}^-$,

$$-\pi < \theta < \pi. \tag{11.7}$$

The eigenvalues of B lie in the open half-planes of the corresponding eigenvalues of $A^{1/2}$, that is,

$$-\frac{\pi}{2} < \phi - \frac{\theta}{2} < \frac{\pi}{2}. \tag{11.8}$$

From (11.7) and (11.8) it follows that $-\pi < \phi < \pi$. Now $a = bc$ implies $\theta = \phi + \psi + 2k\pi$ for some integer k. Subtracting one half (11.7) from (11.8) gives $-\pi < \phi - \theta < \pi$, that is, $-\pi < \psi + 2k\pi < \pi$, and since $-\pi < \psi \le \pi$ we must have $k = 0$, so that $-\pi < \psi < \pi$ and $\theta = \phi + \psi$. Thus, by (11.7), $\phi + \psi \in (-\pi, \pi)$ and so the result holds by Theorem 11.3. □

In the terminology of Theorem 11.4, the reason for the behaviour in (11.4) is that b and c are equal to a nonprincipal square root of $a = bc$ and hence are not in the half-plane of $a^{1/2}$.

A special case of Theorems 11.2–11.4 is the relation $\log(A) = 2\log(A^{1/2})$, which is the basis of the inverse scaling and squaring method described in Section 11.5.

Finally, it is worth noting that the difference between $\log(XY)$ and $\log(X) + \log(Y)$ is described for X and Y sufficiently close to I by setting $t = 1$, $A = \log(X)$, and $B = \log(Y)$ in the Baker–Campbell–Hausdorff formula (10.5) and taking the log of both sides, giving

$$\log(XY) = \log(X) + \log(Y) + [\log(X), \log(Y)]/2 + \text{higher order terms}.$$

The "error", however, is expressed in terms of $\log(X)$ and $\log(Y)$ rather than X and Y.

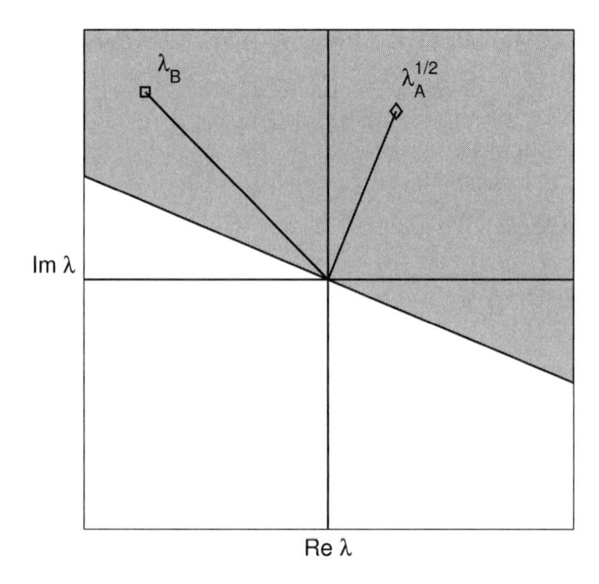

Figure 11.1. *Illustration of condition* (b) *of Theorem* 11.4, *which requires every eigenvalue of* B (λ_B) *to lie in the open half-plane (shaded) of the corresponding eigenvalue of* $A^{1/2}$ ($\lambda_A^{1/2}$).

11.2. Conditioning

From Theorem 3.5 we know that the Fréchet derivative of the logarithm is the inverse of the Fréchet derivative of the exponential, so that $L_{\exp}\bigl(\log(A), L_{\log}(A, E)\bigr) = E$. From (10.15) we have $L_{\exp}(A, E) = \int_0^1 e^{A(1-s)} E e^{As}\, ds$, and hence $L(A, E) = L_{\log}(A, E)$ satisfies $E = \int_0^1 e^{\log(A)(1-s)} L(A, E) e^{\log(A)s}\, ds$, that is (using Theorem 11.2),

$$E = \int_0^1 A^{1-s} L(A, E) A^s\, ds. \tag{11.9}$$

A different representation of the Fréchet derivative can be obtained from Theorem 11.1 (see Problem 11.2):

$$L(A, E) = \int_0^1 \bigl(t(A - I) + I\bigr)^{-1} E \bigl(t(A - I) + I\bigr)^{-1} dt. \tag{11.10}$$

On setting $E = I$ we obtain $L(A, I) = \int_0^1 \bigl(t(A - I) + I\bigr)^{-2} dt = -(A - I)^{-1}\bigl[(t(A - I) + I)^{-1}\bigr]_0^1 = A^{-1}$. Hence $\|L(A)\| \ge \|A^{-1}\|$ and thus

$$\kappa_{\log}(A) \ge \frac{\kappa(A)}{\|\log(A)\|}.$$

(Note that taking norms in (11.9) gives the weaker bound $\|L(A)\| \ge \|A\|^{-1}$.) It is reasonable to guess that equality will be achieved in this lower bound for any normal matrix. However, this is not the case. From the general lower bound in Theorem 3.14 we have

$$\|L(A)\| \ge \max_{\lambda,\mu \in \Lambda(A)} |f[\lambda, \mu]| = \max\left(\max_{\lambda \in \Lambda(A)} \frac{1}{|\lambda|}, \max_{\substack{\lambda,\mu \in \Lambda(A) \\ \lambda \ne \mu}} \frac{|\log \lambda - \log \mu|}{|\lambda - \mu|} \right). \tag{11.11}$$

For a normal matrix and the Frobenius norm this inequality is an equality (Corollary 3.16). But even for normal A, the right-hand side generally exceeds $\max_{\lambda\in\Lambda(A)} |\lambda|^{-1}$. For Hermitian positive definite matrices we do, however, have the equality $\|L(A)\|_F = \max_{\lambda\in\Lambda(A)} |\lambda|^{-1} = \|A^{-1}\|_2$.

The lower bound (11.11) is large in two situations: if A has an eigenvalue of small modulus or if A has a pair of complex eigenvalues lying close to, but on opposite sides of, the negative real axis; in both cases A is close to a matrix for which the principal logarithm is not defined.

11.3. Series Expansions

One way to obtain the matrix logarithm is via series expansions, of which there are several variants. We begin with scalars. By integrating $(1+t)^{-1} = 1-t+t^2-t^3+\cdots$ between 0 and x we obtain Mercator's series (1668),

$$\log(1+x) = x - \frac{x^2}{2} + \frac{x^3}{3} - \frac{x^4}{4} + \cdots, \tag{11.12}$$

which converges for $|x| < 1$ (and in fact at every x on the unit circle except $x = -1$). Replacing x by $-x$ we have

$$\log(1-x) = -x - \frac{x^2}{2} - \frac{x^3}{3} - \frac{x^4}{4} - \cdots$$

and subtracting from the first equation gives

$$\log\frac{1+x}{1-x} = 2\left(x + \frac{x^3}{3} + \frac{x^5}{5} + \cdots\right). \tag{11.13}$$

This is Gregory's series (1668). Writing

$$y := \frac{1-x}{1+x} \quad\Longleftrightarrow\quad x = \frac{1-y}{1+y}$$

we have

$$\log y = -2\sum_{k=0}^{\infty} \frac{1}{2k+1}\left(\frac{1-y}{1+y}\right)^{2k+1}. \tag{11.14}$$

This series converges when $|1-y|/|1+y| < 1$, that is, $\operatorname{Re} y > 0$.

For matrices the corresponding series are

$$\log(I+A) = A - \frac{A^2}{2} + \frac{A^3}{3} - \frac{A^4}{4} + \cdots, \quad \rho(A) < 1 \tag{11.15}$$

and

$$\log(A) = -2\sum_{k=0}^{\infty} \frac{1}{2k+1}\left((I-A)(I+A)^{-1}\right)^{2k+1}, \quad \min_i \operatorname{Re}\lambda_i(A) > 0. \tag{11.16}$$

The Gregory series (11.16) has a much larger region of convergence than (11.15), and in particular converges for any Hermitian positive definite A. However, convergence can be slow if A is not close to I.

11.4. Padé Approximation

The diagonal Padé approximants r_m to $\log(1+x)$ have two useful explicit representations. First is the continued fraction expansion

$$r_m(x) = \cfrac{c_1 x}{1 + \cfrac{c_2 x}{1 + \cfrac{c_3 x}{1 + \cdots + \cfrac{c_{2m-2} x}{1 + \cfrac{c_{2m-1} x}{1 + c_{2m} x}}}}}, \tag{11.17a}$$

$$c_1 = 1, \qquad c_{2j} = \frac{j}{2(2j-1)}, \quad c_{2j+1} = \frac{j}{2(2j+1)}, \qquad j = 1, 2, \ldots, \tag{11.17b}$$

which is the truncation of an infinite continued fraction for $\log(1+x)$. Second is the partial fraction form

$$r_m(x) = \sum_{j=1}^{m} \frac{\alpha_j^{(m)} x}{1 + \beta_j^{(m)} x}, \tag{11.18}$$

where the $\alpha_j^{(m)}$ are the weights and the $\beta_j^{(m)}$ the nodes of the m-point rule on $[0,1]$. Codes for computing these weights and nodes, which the theory of Gaussian quadrature guarantees are real, are given in [142, 1984, App. 2], [206, 1994], [208, 2004, Sec. 3.1], [478, 2007, Sec. 4.6].

An interesting connection is that $r_m(A-I)$ is produced by applying the m-point Gauss–Legendre quadrature rule to (11.1) [154, 1996, Thm. 4.3].

The logarithm satisfies $\log(1+x) = -\log(1/(1+x))$. The diagonal Padé approximant satisfies the corresponding identity.

Lemma 11.5 (Kenney and Laub). *The diagonal Padé approximant r_m to $\log(1+x)$ satisfies $r_m(x) = -r_m(-x/(1+x))$.* □

Proof. The result is a special case of a more general invariance result applying to an arbitrary function under an origin-preserving linear fractional transformation of its argument [38, 1975, Thm. 9.1], [39, 1996, Thm. 1.5.2]. □

The next result states the important fact that the error in matrix Padé approximation is bounded by the error in scalar Padé approximation at the norm of the matrix.

Theorem 11.6 (Kenney and Laub). *For $\|X\| < 1$ and any subordinate matrix norm,*

$$\|r_m(X) - \log(I+X)\| \le \big|r_m(-\|X\|) - \log(1-\|X\|)\big|. \quad \square \tag{11.19}$$

The bound (11.19) is sharp in that it is exact for nonpositive scalar X and hence for nonpositive diagonal X in any p-norm.

Several possibilities exist for evaluating $r_m(X) = q_m(X)^{-1} p_m(X)$:

- Horner's method applied to p_m and q_m (Algorithm 4.2), followed by solution of $q_m r_m = p_m$;
- the Paterson–Stockmeyer method applied to p_m and q_m (see Section 4.4.3), followed by solution of $q_m r_m = p_m$;

- the continued fraction (11.17) evaluated top-down (Algorithm 4.9) or bottom-up (Algorithm 4.10);
- the partial fraction expansion (11.18).

The choice between these methods is a compromise between computational cost and numerical stability. For the first two approaches it is crucial to establish that the denominator polynomial is not too ill conditioned. The next result provides a bound.

Lemma 11.7 (Kenney and Laub). *For $\|X\| < 1$ and any subordinate matrix norm, the denominator q_m of the diagonal Padé approximant r_m satisfies*

$$\kappa(q_m(X)) \le \frac{q_m(\|X\|)}{q_m(-\|X\|)}. \qquad \square \tag{11.20}$$

In the detailed analysis of Higham [275, 2001] the partial fraction method emerges as the best overall method for evaluating r_m. Its cost is m solutions of multiple right-hand side linear systems. Its numerical stability is governed by the condition of the linear systems that are solved, and for a stable linear system solver the normwise relative error will be bounded approximately by $d(m,n)u\phi_m$, where

$$\phi_m = \max_j \kappa\big(I + \beta_j^{(m)} X\big) \le \max_j \frac{1 + |\beta_j^{(m)}|\|X\|}{1 - |\beta_j^{(m)}|\|X\|}. \tag{11.21}$$

Since $\beta_j^{(m)} \in (0,1)$, ϕ_m is guaranteed to be of order 1 provided that $\|X\|$ is not too close to 1. An advantage for parallel computation is that the m terms in (11.18) can be evaluated in parallel.

So far, we have concentrated on Padé approximants to $\log(1+x)$. Motivated by the Gregory series (11.13) we might also consider approximating $\log\big((1+x)/(1-x)\big)$. It turns out that these two sets of Padé approximants are closely related.

Lemma 11.8. *The Padé approximants $r_m(x)$ to $\log(1+x)$ and $s_m(x)$ to $\log\big((1+x)/(1-x)\big)$ are related by $r_m(x) = s_m(x/(x+2))$.*

Proof. The proof is very similar to that of Lemma 11.5. $\square$

Analogously to Theorem 11.6, the error in s_m satisfies the bound

$$\big\|s_m(X) - \log\big((I+X)(I-X)^{-1}\big)\big\| \le \big|s_m(\|X\|) - \log\big((1+\|X\|)/(1-\|X\|)\big)\big|,$$

provided that $\|X\| < 1$. This inequality is obtained in the same way as (11.19) from a more general result in [341, 1989, Cor. 4] holding for a class of hypergeometric functions.

11.5. Inverse Scaling and Squaring Method

One of the best methods for computing the matrix logarithm is a generalization of the method that Briggs used in the 17th century to compute tables of logarithms of numbers. Briggs knew that for positive real numbers a and b, $\log(ab) = \log a + \log b$ and hence that $\log a = 2 \log a^{1/2}$. Using this equation repeatedly he obtained $\log a = 2^k \log a^{1/2^k}$. Briggs also knew that $\log(1+x) \approx x$ is a good approximation for small x.

He used this approximation with $a^{1/2^k} = 1+x$ and a sufficiently large k, and multiplied by the scale factor needed to obtain logarithms to base 10. His approximation was therefore $\log_{10} a \approx 2^k \cdot \log_{10} e \cdot (a^{1/2^k} - 1)$.

For a matrix A with no eigenvalues on $\mathbb{R}^-$, the analogue of Briggs' idea is to use the identity, from Theorem 11.2,

$$\log(A) = 2^k \log\bigl(A^{1/2^k}\bigr). \tag{11.22}$$

The integer k is to be chosen so that $\log(A^{1/2^k})$ is easy to approximate. We will use a Padé approximant of $\log(1+x)$, so we require $A^{1/2^k}$ sufficiently close to the identity, which holds for large enough k since $\lim_{k\to\infty} A^{1/2^k} = I$. This method is called the inverse scaling and squaring method by analogy with the scaling and squaring method for the matrix exponential.

At this point it is worth pointing out a danger in implementing Briggs' original method in floating point arithmetic. Suppose we are working in IEEE double precision arithmetic and wish to obtain the logarithm to full accuracy of about 16 significant decimal digits. Then we will need $|a^{1/2^k} - 1| \approx 10^{-8}$ (as can be seen from Theorem 4.8). The subtraction $a^{1/2^k} - 1$ must therefore suffer massive cancellation, with about half of the significant digits in $a^{1/2^k}$ being lost. This was not a problem for Briggs, who calculated to 30 decimal digits in order to produce a table of 14-digit logarithms. And the problem can be avoided by using higher order approximations to $\log a^{1/2^k}$. However, in the matrix context a value of k sufficiently large for the matrix as a whole may be too large for some particular submatrices and could lead to damaging subtractive cancellation. This is an analogue of the overscaling phenomenon discussed at the end of Section 10.3 in the context of the scaling and squaring method for the exponential.

The design of an algorithm depends on what kinds of operation we are prepared to carry out on the matrix. We consider first the use of a Schur decomposition, which reduces the problem to that of computing the logarithm of a triangular matrix, and then consider working entirely with full matrices.

11.5.1. Schur Decomposition: Triangular Matrices

If an initial Schur decomposition is carried out then the problem becomes that of computing the logarithm of a triangular matrix $T \in \mathbb{C}^{n\times n}$. Our task is to decide how many square roots to take and what degree of Padé approximant to use, with the aim of finding a compromise between minimal cost and maximal accuracy. Two key facts are that a square root of a triangular matrix can be computed in $n^3/3$ flops (by Algorithm 6.3) and that $r_m(T)$ can be evaluated by the partial fraction expansion (11.18) in $mn^3/3$ flops (see Table C.1). Hence an extra square root is worth taking if it reduces the degree m of the required Padé approximant by more than 1. The required m will be determined from (11.19).

Table 11.1 reports, for a range of m from 1 to 64 and to three significant figures, the largest $\|X\|$, denoted by θ_m, such that the bound (11.19) guarantees $\|r_m(X) - \log(I+X)\| \le u = 2^{-53} \approx 1.1\times 10^{-16}$. By Problem 10.8, for $m \lesssim 12$ the absolute error bound (11.19) with $\|X\| \le \theta_m$ is also essentially a relative error bound. The table also reports the bound in (11.20) for $\kappa(q_m(X))$ and the bound in (11.21) for $\phi_m = \max_j \kappa\bigl(I + \beta_j^{(m)} X\bigr)$, both for $\|X\| = \theta_m$. Note first that the bounds for $\kappa(q_m(X))$ grow rapidly for $m \ge 10$, which is a concern if Horner's method is used

Table 11.1. *Maximal values θ_m of $\|X\|$ such that the bound* (11.19) *ensures $\|r_m(X) - \log(I + X)\|$ does not exceed $u = 2^{-53}$, along with upper bound* (11.20) *for $\kappa(q_m(X))$ and upper bound* (11.21) *for ϕ_m, both with $\|X\| = \theta_m$.*

m	1	2	3	4	5	6	7	8	9
θ_m	1.10e-5	1.82e-3	1.62e-2	5.39e-2	1.14e-1	1.87e-1	2.64e-1	3.40e-1	4.11e-1
$\kappa(q_m)$	1.00e0	1.00e0	1.05e0	1.24e0	1.77e0	3.09e0	6.53e0	1.63e1	4.62e1
ϕ_m	1.00e0	1.00e0	1.03e0	1.11e0	1.24e0	1.44e0	1.69e0	2.00e0	2.36e0

m	10	11	12	13	14	15	16	32	64
θ_m	4.75e-1	5.31e-1	5.81e-1	6.24e-1	6.62e-1	6.95e-1	7.24e-1	9.17e-1	9.78e-1
$\kappa(q_m)$	1.47e2	5.07e2	1.88e3	7.39e3	3.05e4	1.31e5	5.80e5	>1e16	>1e16
ϕ_m	2.76e0	3.21e0	3.71e0	4.25e0	4.84e0	5.47e0	6.14e0	2.27e1	8.85e1

to evaluate r_m. However, we will use the partial fraction expansion, which the ϕ_m values show will deliver a fully accurate evaluation for all the m of interest.

In interpreting the data in the table we need to know the effect of taking a square root of T on the required m. This can be determined irrespective of the triangularity of T. Since $\big(I - A^{1/2^{k+1}}\big)\big(I + A^{1/2^{k+1}}\big) = I - A^{1/2^k}$ and $A^{1/2^k} \to I$ as $k \to \infty$ it follows that for large k,

$$\|I - A^{1/2^{k+1}}\| \approx \frac{1}{2}\|I - A^{1/2^k}\|, \tag{11.23}$$

so that once $A^{1/2^k}$ has norm approximately 1 a further square root should approximately halve the distance to the identity matrix.[11]

Since $\theta_m/2 < \theta_{m-2}$ for $m > 7$, to minimize the cost we should keep taking square roots until $\|X\| \le \theta_7$, since each such square root costs only half the resultant saving (at least two multiple right-hand side solves) in the cost of evaluating the Padé approximant. The question is whether this choice needs to be moderated by stability considerations. We have already noted from the ϕ_m values in Table 11.1 that the evaluation of r_m for such X is as accurate as can be hoped. The only danger in scaling X to have norm at most θ_7, rather than some larger value, is that the greater number of square roots may increase the error in the matrix whose Padé approximant is computed. However, the later square roots are of matrices close to I, which are well conditioned in view of (6.2), so these extra square roots should be innocuous.

We are now ready to state an algorithm. Rather than tailor the algorithm to the particular precision, we express it in such a way that the same logic applies whatever the precision, with changes needed only to the integer constants and the θ_i.

Algorithm 11.9 (inverse scaling and squaring algorithm with Schur decomposition). Given $A \in \mathbb{C}^{n\times n}$ with no eigenvalues on $\mathbb{R}^-$ this algorithm computes $X = \log(A)$ via a Schur decomposition and the inverse scaling and squaring method. It uses the constants θ_i given in Table 11.1. The algorithm is intended for IEEE double precision arithmetic.

1 Compute a (complex) Schur decomposition $A = QTQ^*$.

[11] Briggs cleverly exploited (11.23) by computing differences that enabled him to reduce the number of square root evaluations needed for his logarithm tables.

2 $k = 0,\ p = 0$
3 while true
4 $\quad \tau = \|T - I\|_1$
5 $\quad$ if $\tau \le \theta_7$
6 $\qquad p = p + 1$
7 $\qquad j_1 = \min\{\, i : \tau \le \theta_i,\ i = 3\colon 7\,\}$
8 $\qquad j_2 = \min\{\, i : \tau/2 \le \theta_i,\ i = 3\colon 7\,\}$
9 $\qquad$ if $j_1 - j_2 \le 1$ or $p = 2$, $m = j_1$, goto line 14, end
10 $\quad$ end
11 $\quad T \leftarrow T^{1/2}$ using Algorithm 6.3.
12 $\quad k = k + 1$
13 end
14 Evaluate $U = r_m(T - I)$ using the partial fraction expansion (11.18).
15 $X = 2^k Q U Q^*$

Cost: $25n^3$ flops for the Schur decomposition plus $(k + m)n^3/3$ flops for U and $3n^3$ to form X: about $(30 + \frac{k}{3})n^3$ flops in total.

Assuming that (11.23) is a reasonable approximation and that $k > 0$, Algorithm 11.9 will take $m = 5$, 6, or 7 (see Problem 11.5). In the unlikely event that (11.23) is not a good approximation, the algorithm still uses a degree m appropriate to the desired accuracy but may take one more square root than necessary (the test "$p = 2$" on line 9 limits the number of unnecessary square roots).

At the start of this section we mentioned the danger of overscaling. Writing $T = \left[\begin{smallmatrix} T_{11} & T_{12} \\ 0 & T_{22} \end{smallmatrix}\right]$ in Algorithm 11.9, a large $\|T_{12}\|$ may cause more square roots to be taken than are necessary for accurate computation of the logarithms of the diagonal blocks. This phenomenon is mitigated in the Schur–Parlett algorithm of Section 11.6.2 below, which applies inverse scaling and squaring to diagonal blocks rather than the whole of T.

11.5.2. Full Matrices

Suppose that we are not willing to Schur factorize A. Square roots must therefore be computed using one of the Newton methods from Section 6.3. The economics of the method are now quite different, because the square roots are relatively expensive to compute and the cost depends on the number of iterations required.

For any of the Newton iterations for the square root, the number of iterations required will vary from typically up to 10 on the first few iterations to 4 or 5 for the last few. From Table 6.1 we see that the cost of one iteration is $4n^3$ flops for the DB and product DB iterations. Evaluating the partial fraction expansion (11.18) costs $8mn^3/3$ flops. So a later square root costs about $16n^3$ flops, while decreasing m by 1 decreases the cost of evaluating r_m by $8n^3/3$ flops. Therefore once we are in the regime where Padé approximation can be applied it is likely to be worth taking a further square root only if it allows m to be decreased by at least 7. From Table 11.1 and further θ_m values not shown here we find that such a decrease is achieved when $\|X\| \ge \theta_{16}$ (assuming that (11.23) is a good approximation). Our algorithm predicts whether another square root is worthwhile by assuming that the next Newton iteration would require the same number of iterations as the last one and thereby uses the values of θ_m and $\|X\|$ to compare the total cost with and without a further square root.

Algorithm 11.10 (inverse scaling and squaring algorithm). Given $A \in \mathbb{C}^{n\times n}$ with no eigenvalues on $\mathbb{R}^-$ this algorithm computes $X = \log(A)$ by the inverse scaling and squaring method. It uses the constants θ_i given in Table 11.1. The algorithm is intended for IEEE double precision arithmetic.

1 $k = 0$, $\mathrm{it}_0 = 5$, $p = 0$
2 while true
3 $\quad\tau = \|A - I\|_1$
4 $\quad$if $\tau < \theta_{16}$
5 $\quad\quad p = p + 1$
6 $\quad\quad j_1 = \min\{\, i : \tau \le \theta_i,\ i = 3\colon 16\,\}$
7 $\quad\quad j_2 = \min\{\, i : \tau/2 \le \theta_i,\ i = 3\colon 16\,\}$
8 $\quad\quad$if $2(j_1 - j_2)/3 \le \mathrm{it}_k$ or $p = 2$, $m = j_1$, goto line 13, end
9 $\quad$end
10 $\quad A \leftarrow A^{1/2}$ using the scaled product DB iteration (6.29); let it_{k+1} be the number of iterations required.
11 $\quad k = k + 1$
12 end
13 Evaluate $Y = r_m(A - I)$ using the partial fraction expansion (11.18).
14 $X = 2^k Y$

Cost: $\left(\sum_{i=1}^{k} \mathrm{it}_i\right) 4n^3 + 8mn^3/3$ flops.

11.6. Schur Algorithms

We now describe two other algorithms that make use of a Schur decomposition $A = QTQ^*$.

11.6.1. Schur–Fréchet Algorithm

Kenney and Laub [348, 1998] develop an algorithm for computing $\log(A)$ that computes the off-diagonal blocks of $\log(T)$ via Fréchet derivatives, using Algorithm 11.12 below. The algorithm is entirely analogous to the Schur–Fréchet algorithm for the exponential described in Section 10.4.2. For the details see [348, 1998].

11.6.2. Schur–Parlett Algorithm

Algorithm 9.6, the general Schur–Parlett algorithm, is not applicable to the matrix logarithm because the Taylor series of the logarithm has a finite (and small) radius of convergence. However, we can adapt it to the logarithm by providing a means to compute the logarithms of the diagonal blocks.

The only difficulty in computing the logarithm of a 2×2 upper triangular matrix, $F = \log(\left[\begin{smallmatrix} t_{11} & t_{12} \\ 0 & t_{22}\end{smallmatrix}\right])$, is in obtaining an accurate (1,2) element. The formula $f_{12} = t_{12}(\log\lambda_2 - \log\lambda_1)/(\lambda_2 - \lambda_1)$ from (4.16) can suffer damaging subtractive cancellation when $\lambda_1 \approx \lambda_2$. To avoid it we write

$$\begin{aligned}\log\lambda_2 - \log\lambda_1 &= \log\left(\frac{\lambda_2}{\lambda_1}\right) + 2\pi i\,\mathcal{U}(\log\lambda_2 - \log\lambda_1)\\ &= \log\left(\frac{1+z}{1-z}\right) + 2\pi i\,\mathcal{U}(\log\lambda_2 - \log\lambda_1),\end{aligned}$$

where U is the unwinding number (11.2) and $z = (\lambda_2 - \lambda_1)/(\lambda_2 + \lambda_1)$. Using the hyperbolic arc tangent, defined by

$$\operatorname{atanh}(z) := \frac{1}{2}\log\left(\frac{1+z}{1-z}\right), \tag{11.24}$$

we have

$$f_{12} = t_{12}\frac{2\operatorname{atanh}(z) + 2\pi i\mathcal{U}(\log\lambda_2 - \log\lambda_1)}{\lambda_2 - \lambda_1}. \tag{11.25}$$

It is important to note that atanh can be defined in a different way, as recommended in [81, 2002] and [327, 1987]:

$$\operatorname{atanh}(z) := \frac{1}{2}(\log(1+z) - \log(1-z)). \tag{11.26}$$

With this definition a slightly more complicated formula than (11.25) is needed:

$$f_{12} = t_{12}\frac{2\operatorname{atanh}(z) + 2\pi i\big[\mathcal{U}(\log\lambda_2 - \log\lambda_1) + \mathcal{U}(\log(1+z) - \log(1-z))\big]}{\lambda_2 - \lambda_1}. \tag{11.27}$$

MATLAB's `atanh` function is defined by (11.24). Of course, the use of (11.25) or (11.27) presupposes the availability of an accurate `atanh` implementation. Our overall formula is as follows, where we avoid the more expensive atanh formula when λ_1 and λ_2 are sufficiently far apart:

$$f_{12} = \begin{cases} \dfrac{t_{12}}{\lambda_1}, & \lambda_1 = \lambda_2, \\ t_{12}\dfrac{\log\lambda_2 - \log\lambda_1}{\lambda_2 - \lambda_1}, & |\lambda_1| < |\lambda_2|/2 \text{ or } |\lambda_2| < |\lambda_1|/2\ , \\ \text{(11.25) or (11.27)}, & \text{otherwise.} \end{cases} \tag{11.28}$$

Algorithm 11.11 (Schur–Parlett algorithm for matrix logarithm). Given $A \in \mathbb{C}^{n\times n}$ this algorithm computes $F = \log(A)$ via a Schur decomposition.

1 Execute Algorithm 9.6 with line 7 replaced by
"7 Evaluate $F_{ii} = \log(T_{ii})$ directly if T_{ii} is 1×1, using (11.28) if T_{ii} is 2×2, or else by Algorithm 11.9 (inverse scaling and squaring method)."

Algorithm 11.11 is our preferred alternative to Algorithm 11.9.

11.7. Numerical Experiment

We describe an experiment analogous to that for the matrix exponential in Section 10.5. The matrices are the exponentials of most of those used in the latter experiment, supplemented with matrices from the $\log(A)$ literature and some 2×2 triangular matrices; all are size 10×10 or less.

The methods are

1. MATLAB 7.5 (R2007b)'s `logm`, denoted `logm_old`, which implements a more basic version of Algorithm 11.11 that does not treat 2×2 diagonal blocks specially and uses a version of Algorithm 11.9 that employs a Padé approximant of fixed degree $m = 8$ that is used once $\|I - T^{1/2^k}\| \le 0.25$;

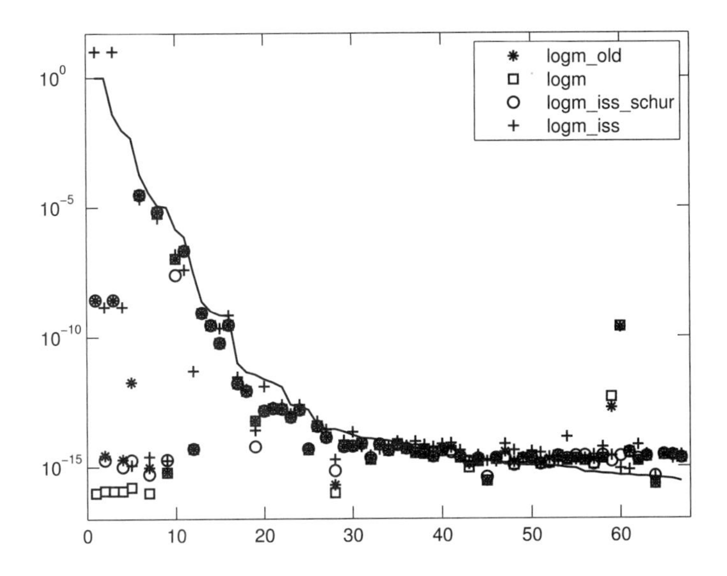

Figure 11.2. *Normwise relative errors for MATLAB's* `logm`, `logm_old`, `logm_iss_schur`, *and* `logm_iss`; *the solid line is* $\kappa_{\log}(A)u$.

2. the `logm` function in MATLAB 7.6 (R2008a) onwards, which implements Algorithm 11.11;

3. `logm_iss_schur`, which implements Algorithm 11.9;

4. `logm_iss`, which implements Algorithm 11.10. The convergence test for the scaled product DB iteration is based on $\|M_k - I\|$.

Figure 11.2 shows the normwise relative errors $\|\widehat{X} - \log(A)\|_F/\|\log(A)\|_F$, where $\widehat{X}$ is the computed logarithm and $\log(A)$ is computed at 100 digit precision, along with $\kappa_{\log}(A)u$, where $\kappa_{\log} = \mathrm{cond}_{\mathrm{rel}}(\log, A)$ is computed with the aid of Algorithm 11.12. Figure 11.3 presents the same data in the form of a performance profile.

Several observations can be made.

- `logm` is clearly superior to `logm_old`. Further investigation shows that both the special treatment of 2×2 diagonal blocks and the use of Algorithm 11.9 contribute to the improvement in accuracy.

- `logm_iss_schur` is even more markedly superior to `logm_iss`, showing the benefits for accuracy of applying the inverse scaling and squaring method to triangular matrices.

11.8. Evaluating the Fréchet Derivative

The relationship between the Fréchet derivatives of the exponential and the logarithm can be summarized in the equation

$$L = \int_0^1 e^{A(1-s)} M e^{As}\, ds,$$

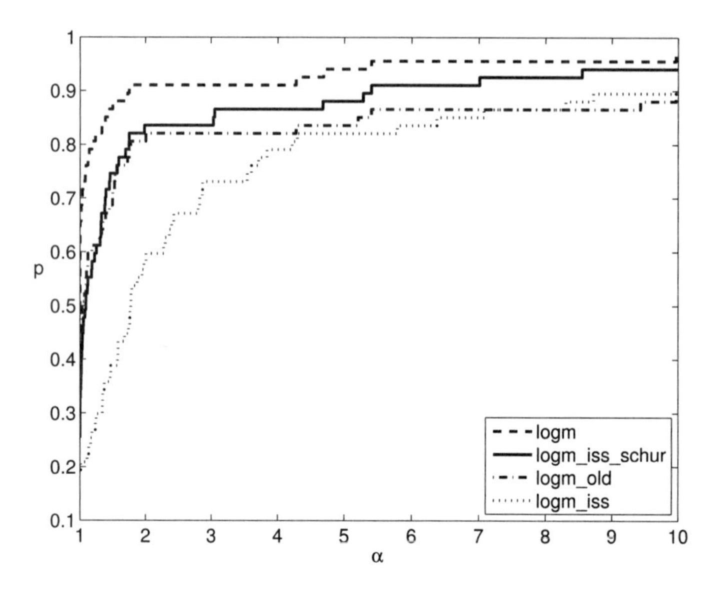

Figure 11.3. *Same data as in Figure* 11.2 *presented as a performance profile.*

where L is the Fréchet derivative of the exponential at A in the direction M, and M is the Fréchet derivative of the logarithm at e^A in the direction L (see (10.15) and (11.9)). The formula (10.49) therefore translates into

$$\operatorname{vec}(L_{\log}(A,E)) = 2\tau\big(\tfrac{1}{2}[X^T \oplus (-X)]\big)^{-1}(A^T \oplus A)^{-1}\operatorname{vec}(E), \quad X = \log(A), \tag{11.29}$$

where $\tau(x) = \tanh(x)/x$ and $\|X\| \le 1$ is assumed for some consistent matrix norm. That the inverses in (11.29) exist is verified in Problem 11.8.

In order to use this formula we first need to ensure that $X = \log(A)$ has norm at most 1. This is achieved by taking repeated square roots: if $\|I - A^{1/2^s}\| \le 1 - 1/e$ then $\|\log(A^{1/2^s})\| \le 1$ (see Problem 11.4), and so we can apply the formula to $A^{1/2^s}$. To take account of the square root phase we note that applying the chain rule (Theorem 3.4) to $\log(A) = 2\log(A^{1/2})$ gives $L_{\log}(A,E) = 2L_{\log}\big(A^{1/2}, L_{x^{1/2}}(A,E)\big)$. Hence, by recurring this relation, we have $L_{\log}(A,E) = L_{\log}(A^{1/2^s}, E_s)$, where

$$E_0 = 2^s E, \tag{11.30a}$$

$$A^{1/2^i}E_i + E_iA^{1/2^i} = E_{i-1}, \qquad i = 1\colon s. \tag{11.30b}$$

We obtain an algorithm for evaluating $L_{\log}(A,E)$ by using this Sylvester recurrence and reversing the steps in the derivation of Algorithm 10.27.

Algorithm 11.12 (Fréchet derivative of matrix logarithm). Given $A \in \mathbb{C}^{n\times n}$ this algorithm evaluates the Fréchet derivative $L = L(A,E)$ of the matrix logarithm via (11.29), using scaling and squaring and the [8/8] Padé approximant to $\tau(x) = \tanh(x)/x$. The parameters $\{\alpha_i,\beta_i\}_{i=1}^{8}$ are given in Table 10.5. The algorithm is intended for IEEE double precision arithmetic.

1 $B = A^{1/2^s}$ with $s \ge 0$ a minimal integer such that $\|I - A^{1/2^s}\|_1 \le 1 - 1/e$.

2 $E_0 = 2^s E$
3 for $i = 1\colon s$
4 Solve for E_i the Sylvester equation $A^{1/2^i} E_i + E_i A^{1/2^i} = E_{i-1}$.
5 end
6 Solve for G_9 the Sylvester equation $BG_9 + G_9 B = E_s$.
7 $X = \log(B)$
8 for $i = 8\colon -1\colon 1$
9 Solve for G_i the Sylvester equation $(I + X/\alpha_i)G_i + G_i(I - X/\alpha_i) = (I + X/\beta_i)G_{i+1} + G_{i+1}(I - X/\beta_i)$.
10 end
11 $L = 2G_0$

Cost: $16M$, s matrix square roots, 1 matrix logarithm, and the solution of $s + 9$ Sylvester equations.

In practice, it is advisable to combine the square root computations in line 1 with the loop in lines 3–5 in order to avoid having to store the intermediate square roots.

As with Algorithm 10.27 for the exponential, a Schur reduction of A to triangular form should be exploited in conjunction with Algorithm 11.12 to reduce the cost.

Algorithm 11.12 can combined with Algorithm 11.9, with the number of square roots determined by Algorithm 11.9, which has the more stringent requirement on $\|T - I\|$.

It is also possible to estimate the Fréchet derivative by quadrature. For one approach, see Dieci, Morini, and Papini [154, 1996].

An obvious analogue of Algorithm 10.28 can be formulated for computing and estimating the condition number of the matrix logarithm.

11.9. Notes and References

The integral representation in Theorem 11.1 was obtained by Richter [490, 1949] for A having distinct eigenvalues and by Wouk [617, 1965] without this restriction.

The unwinding number was introduced by Corless and Jeffrey [116, 1996] and is further investigated by Bradford, Corless, Davenport, Jeffrey, and Watt [81, 2002].

Theorem 11.4 is from Cheng, Higham, Kenney, and Laub [108, 2001].

Von Neumann [604, 1929, pp. 11–12] derives the sufficient conditions $\|B - I\| < 1$, $\|C - I\| < 1$, and $\|BC - I\| < 1$ for $\log(BC) = \log(B) + \log(C)$ when B and C commute. These are much stronger conditions than those in Theorems 11.3 and 11.4 and stem from the fact that von Neumann defined the matrix logarithm via the Mercator series (11.15).

The formula (11.10) for the Fréchet derivative is due to Dieci, Morini, and Papini [154, 1996].

Good references for history of the Mercator series (11.12), the Gregory series (11.13), and related series are Goldstine [222, 1977] and Hairer and Wanner [238, 1996].

The continued fraction expansion (11.17) can be found in Baker and Graves-Morris [39, 1996, p. 174].

Lemma 11.5, Theorem 11.6, and Lemma 11.7 are from Kenney and Laub [341, 1989].

Henry Briggs (1556–1630) was a contemporary of John Napier (1550–1617), the inventor of logarithms. More details of Briggs' calculations can be found in Goldstine [222, 1977] and Phillips [472, 2000].

The inverse scaling and squaring method was proposed by Kenney and Laub [340, 1989] for use with a Schur decomposition. It was further developed by Cheng, Higham, Kenney, and Laub [108, 2001] in a transformation-free form that takes as a parameter the desired accuracy. The two key ideas therein are prediction of the most efficient Padé degree, based on the bound (11.19), and the use of an incomplete form of the product DB iteration, with analysis to show how early the iteration can be terminated without compromising the overall accuracy; see Problem 11.7. The implementation in Kenney and Laub [340, 1989] requires $\|X\| = \|I - A^{1/2^s}\| \le 0.25$ and uses fixed Padé degree 8, while Dieci, Morini, and Papini [154, 1996] require $\|X\| \le 0.35$ and take Padé degree 9. Cardoso and Silva Leite [95, 2006] propose a version of the inverse scaling and squaring method based on Padé approximants to $\log\bigl((1+x)/(1-x)\bigr)$. It explicitly forms a matrix Cayley transform and requires a matrix squaring at each square root step, so its numerical stability is questionable and it appears to offer no general cost advantages over the use of $\log(1+x)$ (cf. Lemma 11.8).

The inverse scaling and squaring method is used by Arsigny, Commowick, Pennec, and Ayache [19, 2006] to compute the logarithm of a diffeomorphism, which they then use to compute the log-Euclidean mean (2.29) of three-dimensional geometrical transformations.

The formula (11.25) can be found in Kahan and Fateman [329, 1999].

Section 11.8 is based on Kenney and Laub [348, 1998].

In connection with overscaling in the inverse scaling and squaring method, Dieci and Papini [156, 2000] consider a block 2×2 upper triangular $T = (T_{ij}) \in \mathbb{C}^{n \times n}$ and obtain a bound for $\|r_m(T-I) - \log(T)\|$ of the form $\|T_{12}\|$ multiplied by a function of $\max_{i=1,2} \|T_{ii} - I\|$ (assumed less than 1) but having no direct dependence on $\|T - I\|$. This bound potentially allows a reduction in the number of square roots required by the inverse scaling and squaring method compared with the use of (11.19), since $\max_{i=1,2} \|T_{ii} - I\|$ may be less than 1 when $\|T - I\|$ greatly exceeds 1, due to a large T_{12}. Related results are obtained by Cardoso and Silva Leite [95, 2006] for the Padé approximants s_m defined in Lemma 11.8. A difficulty in applying these bounds is that it is not clear how best to choose the blocking. A more natural approach is Algorithm 11.11, which automatically blocks the Schur form and applies the inverse scaling and squaring method to the (usually small) diagonal blocks.

For Hermitian positive definite matrices, Lu [394, 1998] proposes an algorithm entirely analogous to those of his described in Sections 6.10 and 10.7.1 for the square root and the exponential, based on the linear partial fraction form of Padé approximants to $\log(1+x)$.

Problems

11.1. Show that for any λ not lying on $\mathbb{R}^-$ the (principal) logarithm $\log(J(\lambda))$ of a Jordan block $J(\lambda) \in \mathbb{C}^{m \times m}$ can be defined via the series (11.15). Thus explain how (11.15) can be used to define $\log(A)$ for any $A \in \mathbb{C}^{n \times n}$ having no eigenvalues on $\mathbb{R}^-$, despite the finite radius of convergence of the series.

11.2. Derive the formula (11.10) for the Fréchet derivative of the logarithm, using the definition (3.6) and the integral (11.1).

11.3. Show that $\|I - A\| < 1$ for some consistent matrix norm is sufficient for the convergence of the Gregory series (11.16).

11.4. Let $\|\cdot\|$ be a consistent norm. Show that if $\|I-M\| \le 1-1/e$ then $\|\log(M)\| \le 1$.

11.5. Show that if Algorithm 11.9 takes at least one square root and the approximation (11.23) is exact once $\|T-I\|_1 \approx \theta_7$ then the algorithm will take $m = 5$, 6, or 7.

11.6. Show that for $A \in \mathbb{C}^{n\times n}$ with no eigenvalues on $\mathbb{R}^-$, $\log(A) = \log(A/\|A\|) + \log(\|A\|I)$. This relation could be used prior to calling Algorithm 11.9 or Algorithm 11.10 on the matrix $A/\|A\|$ whose norm is 1. Is this worthwhile?

11.7. (Cheng, Higham, Kenney, and Laub [108, 2001]) Show that the iterates from the product DB iteration (6.17) satisfy $\log(A) = 2\log(X_k) - \log(M_k)$. Suppose we terminate the iteration after k iterations and set $X^{(1)} = X_k$, and now apply the product DB iteration to $X^{(1)}$, again for a finite number of iterations. Show that continuing this process leads after s steps to

$$\log(A) = 2^s \log(X^{(s)}) - \log(M^{(1)}) - 2\log(M^{(2)}) - \cdots - 2^{s-1}\log(M^{(s)}), \qquad (11.31)$$

where $X^{(i)}$ and $M^{(i)}$ are the final iterates from the product DB iteration applied to $X^{(i-1)}$. The (unknown) $\log(M^{(i)})$ terms on the right-hand side can be bounded using Problem 11.4.

11.8. Let $A \in \mathbb{C}^{n\times n}$ have no eigenvalues on $\mathbb{R}^-$. Let $X = \log(A)$ and assume $\|X\| \le 1$ for some consistent matrix norm.

(a) Show that $A^T \oplus A$ is nonsingular.
(b) Using the expansion $\tau(x) = \tanh(x)/x = \prod_{k=1}^{\infty}(\pi^2 + x^2/k^2)/(\pi^2 + 4x^2/(2k-1)^2)$ show that $\tau\big(\frac{1}{2}[X^T \oplus (-X)]\big)$ is nonsingular.

11.9. Let R be the upper triangular Cholesky factor of the $n\times n$ Pascal matrix (see Section 6.11). Show that $\log(R)$ is zero except on the first superdiagonal, which comprises $1, 2, \ldots, n-1$. Thus for $n = 4$,

$$\log\left(\begin{bmatrix} 1 & 1 & 1 & 1 & 1 \\ 0 & 1 & 2 & 3 & 4 \\ 0 & 0 & 1 & 3 & 6 \\ 0 & 0 & 0 & 1 & 4 \\ 0 & 0 & 0 & 0 & 1 \end{bmatrix}\right) = \begin{bmatrix} 0 & 1 & 0 & 0 & 0 \\ 0 & 0 & 2 & 0 & 0 \\ 0 & 0 & 0 & 3 & 0 \\ 0 & 0 & 0 & 0 & 4 \\ 0 & 0 & 0 & 0 & 0 \end{bmatrix}.$$

Explain why all logarithms of R are upper triangular and why this is the only real logarithm of R.

11.10. Let $\widehat{X}$ be an approximate logarithm of A. Derive a formula for the relative backward error of $\widehat{X}$.

11.11. (Kenney) Let $A \in \mathbb{C}^{n\times n}$ be nonsingular. Consider the iterations

$$X_{k+1} = X_k - I + e^{-X_k}A, \qquad (11.32)$$

$$X_{k+1} = X_k + \frac{1}{2}\big(e^{-X_k}A - A^{-1}e^{X_k}\big). \qquad (11.33)$$

Show that if $X_0A = AX_0$ then (11.32) is Newton's method for $e^X - A = 0$ and (11.33) is Newton's method for $(e^X - A)(e^X + A)^{-1} = 0$, and that $AX_k = X_kA$ and $X_kX_{k+1} = X_{k+1}X_k$ for all k in both cases. Show that if X_0 is a function of A then

(11.32) has local *quadratic* convergence to a primary logarithm of A while (11.33) has local *cubic* convergence to a primary logarithm of A. Which of the two iterations should be preferred in practice? Iteration (11.32) is used in multiple precision libraries to compute the logarithm of a scalar to high precision [36, 1993], [528, 1991].

In 1614 *John Napier* (1550–1617), *of Merchiston Castle, near* (*now* in) *Edinburgh, published his book* Mirifici Logarithmorum Canonis Descriptio, *in which he gives a table of logarithms and an account of how he computed them. If anyone is entitled to use the word "logarithm" it is Napier, since he coined the word, deriving it from the two Greek words* λόγος (*meaning "reckoning," in this context*) *and* ἀριθμός (*meaning "number"*).

— GEORGE M. PHILLIPS, *Two Millennia of Mathematics: From Archimedes to Gauss* (2000)

Another common felony is to exploit either of the transformations $\ln(uv) \rightleftharpoons \ln u + \ln v$ *even when both* u *and* v *could be negative. For example,*
$\ln((-1)(-1)) \to \ln 1 \to 0$, *whereas*
$\ln(-1) + \ln(-1) \to \pi i + \pi i \to 2\pi i$.

— DAVID R. STOUTEMYER, *Crimes and Misdemeanors in the Computer Algebra Trade* (1991)

Briggs must be viewed as one of the great figures in numerical analysis.

— HERMAN H. GOLDSTINE, *A History of Numerical Analysis from the 16th through the 19th Century* (1977)

For values of `x` *close to zero,* `'LNP1(x)'` *returns a more accurate result than does* `'LN(x+1)'`. *Using* `LNP1` *allows both the argument and the result to be near zero... The calculator can express numbers within* 10^{-449} *of zero, but within only* 10^{-11} *of* 1.

— *HP* 48 *Programmer's Reference Manual* (1990)

Pre-programmed functions let you perform log, trig and exponential calculations 10 *times faster than with a slide rule.*

— *The Hewlett-Packard HP-*35 *Scientific Pocket Calculator* (1973)

A real *Floquet factorization exists with* T*-periodic* $L_{F_Y}(t,0)$ *if and* only if $\Phi \equiv \Phi(T,0)$ *has a real logarithm.*

— PIERRE MONTAGNIER, CHRISTOPHER C. PAIGE, and RAYMOND J. SPITERI, *Real Floquet Factors of Linear Time-Periodic Systems* (2003)

Chapter 12
Matrix Cosine and Sine

We now turn our attention to the two most important trigonometric functions: the cosine and the sine. We saw in Section 2.1 that the matrix sine and cosine arise in the solution of the second order differential system (2.7). More general problems of this type, with a forcing term $f(t)$ on the right-hand side, arise from semidiscretization of the wave equation and from mechanical systems without damping, and their solutions can be expressed in terms of integrals involving the sine and cosine [514, 1979].

This chapter begins with a review of addition formulae and normwise upper and lower bounds. Expressions for the Fréchet derivatives are then derived. A detailed derivation is given of an algorithm for $\cos(A)$ that employs variable degree Padé approximation in conjunction with the double angle formula for the cosine. A numerical illustration is given of the performance of that algorithm and two alternatives. Then an algorithm is developed for computing $\cos(A)$ and $\sin(A)$ simultaneously that intertwines the cosine and sine double angle recurrences. Finally, preprocessing is briefly discussed.

12.1. Basic Properties

The matrix cosine and matrix sine can be defined for all $A \in \mathbb{C}^{n\times n}$ by

$$\cos(A) = I - \frac{A^2}{2!} + \frac{A^4}{4!} - \frac{A^6}{6!} + \cdots,$$
$$\sin(A) = A - \frac{A^3}{3!} + \frac{A^5}{5!} - \frac{A^7}{7!} + \cdots.$$

A fundamental relationship is the matrix analogue of Euler's formula,

$$e^{iA} = \cos(A) + i\sin(A), \tag{12.1}$$

which yields

$$\cos(A) = \frac{e^{iA} + e^{-iA}}{2}, \qquad \sin(A) = \frac{e^{iA} - e^{-iA}}{2i} \tag{12.2}$$

and

$$\cos^2(A) + \sin^2(A) = I. \tag{12.3}$$

For real A, we can write $\cos(A) = \operatorname{Re} e^{iA}$, $\sin(A) = \operatorname{Im} e^{iA}$.

A number of basic properties are analogous to those for the matrix exponential.

Theorem 12.1 (addition formulae). *For $A, B \in \mathbb{C}^{n\times n}$, the addition formulae*

$$\cos((A+B)t) = \cos(At)\cos(Bt) - \sin(At)\sin(Bt), \tag{12.4}$$
$$\sin((A+B)t) = \sin(At)\cos(Bt) + \cos(At)\sin(Bt) \tag{12.5}$$

hold for all t if and only if $AB = BA$.

Proof. The result follows from Theorem 10.2 on using (12.1). □

As a special case of the theorem we have the double angle formulae

$$\cos(2A) = 2\cos(A)^2 - I, \tag{12.6}$$

$$\sin(2A) = 2\sin(A)\cos(A). \tag{12.7}$$

The addition formulae hold for all A and B in the Kronecker sense.

Theorem 12.2 (Kronecker addition formulae). *Let $A \in \mathbb{C}^{n\times n}$ and $B \in \mathbb{C}^{m\times m}$. Then $f(A \otimes I) = f(A) \otimes I$ and $f(I \otimes B) = I \otimes f(B)$ for $f = \cos, \sin,$ and*

$$\cos(A \oplus B) = \cos(A) \otimes \cos(B) - \sin(A) \otimes \sin(B), \tag{12.8}$$

$$\sin(A \oplus B) = \sin(A) \otimes \cos(B) + \cos(A) \otimes \sin(B). \tag{12.9}$$

Proof. The result follows from Theorem 10.9 on using (12.2). □

The next result provides some easily evaluated upper and lower bounds on the norm of the sine and cosine.

Theorem 12.3 (norm bounds). *For $A \in \mathbb{C}^{n\times n}$ and any subordinate matrix norm we have*

$$\begin{aligned} 2 - \cosh(\|A\|) &\le 2 - \cosh(\|A^2\|^{1/2}) \\ &\le \|\cos(A)\| \le \cosh(\|A^2\|^{1/2}) \le \cosh(\|A\|), \end{aligned} \tag{12.10}$$

$$\|A\| - \frac{\|A\|^3}{6(1 - \|A\|^2/20)} \le \|\sin(A)\| \le \sinh(\|A\|). \tag{12.11}$$

Proof. We have the bound $\|\cos(A)\| \le \sum_{k=0}^{\infty} \|A^{2k}\|/(2k)! \le \sum_{k=0}^{\infty} \|A^2\|^k/(2k)! = \cosh(\|A^2\|^{1/2})$. Similarly,

$$\begin{aligned} \|\cos(A)\| &\ge 1 - \sum_{k=1}^{\infty} \|A^{2k}\|/(2k)! \ge 1 - \sum_{k=1}^{\infty} \|A^2\|^k/(2k)! \\ &= 1 - (\cosh(\|A^2\|^{1/2}) - 1) = 2 - \cosh(\|A^2\|^{1/2}). \end{aligned}$$

This gives (12.10), since $\|A^2\|^{1/2} \le \|A\|$ and hence $\cosh(\|A^2\|^{1/2}) \le \cosh(\|A\|)$. The upper bound for $\|\sin(A)\|$ is straightforward. For the lower bound we have

$$\begin{aligned} \|\sin(A)\| &\ge \|A\| - \sum_{k=1}^{\infty} \frac{\|A^{2k+1}\|}{(2k+1)!} \ge \|A\| - \frac{\|A^3\|}{6}\left(1 + \frac{\|A^2\|}{4\cdot 5} + \frac{\|A^4\|}{4\cdot 5\cdot 6\cdot 7} + \cdots\right) \\ &\ge \|A\| - \frac{\|A^3\|}{6}\left(1 + \frac{\|A^2\|}{4\cdot 5} + \left(\frac{\|A^2\|}{4\cdot 5}\right)^2 + \cdots\right) \ge \|A\| - \frac{\|A\|^3}{6(1 - \|A\|^2/20)}. \end{aligned}$$ □

Since $\|A^2\| \le \|A\|^2$ can be an arbitrarily weak inequality for nonnormal A (consider involutory matrices, for example), the inner bounds in (12.10) can be smaller than the outer ones by an arbitrary factor. Numerical methods are likely to form A^2 when evaluating $\cos(A)$, since cos is an even function, enabling the tighter bound to be evaluated at no extra cost (and in some cases, such as (2.8), A^2 is the given matrix).

12.2. Conditioning

By using (10.15) and (12.2) we can obtain expressions for the Fréchet derivatives of the cosine and sine functions:

$$L_{\cos}(A,E) = -\int_0^1 \big[\cos(A(1-s))E\sin(As) + \sin(A(1-s))E\cos(As)\big]\,ds, \quad (12.12)$$

$$L_{\sin}(A,E) = \int_0^1 \big[\cos(A(1-s))E\cos(As) - \sin(A(1-s))E\sin(As)\big]\,ds. \quad (12.13)$$

If A and E commute then these formulae reduce to $L_{\cos}(A,E) = -E\sin(A) = -\sin(A)E$ and $L_{\sin}(A,E) = E\cos(A) = \cos(A)E$. Setting $E = I$ we obtain (recalling (3.7)) the relations $\|L_{\cos}(A)\| \ge \|\sin(A)\|$ and $\|L_{\sin}(A)\| \ge \|\cos(A)\|$, and hence the relative condition numbers satisfy

$$\kappa_{\cos}(A) \ge \frac{\|\sin(A)\|\,\|A\|}{\|\cos(A)\|}, \qquad \kappa_{\sin}(A) \ge \frac{\|\cos(A)\|\,\|A\|}{\|\sin(A)\|}. \quad (12.14)$$

Just as for the exponential and the logarithm we can obtain explicit expressions for the Fréchet derivatives with the aid of vec and the Kronecker product.

Theorem 12.4 (Kronecker representation of Fréchet derivatives). *For $A, E \in \mathbb{C}^{n\times n}$ we have* $\mathrm{vec}(L_{\cos}(A,E)) = K_{\cos}(A)\,\mathrm{vec}(E)$ *and* $\mathrm{vec}(L_{\sin}(A,E)) = K_{\sin}(A)\,\mathrm{vec}(E)$, *where*

$$K_{\cos}(A) = -\big[\mathrm{sinc}(A^T \oplus (-A))\sin(I\otimes A) + f(A^T \oplus (-A))\cos(I\otimes A)\big], \quad (12.15)$$
$$K_{\sin}(A) = \mathrm{sinc}(A^T \oplus (-A))\cos(I\otimes A) - f(A^T \oplus (-A))\sin(I\otimes A), \quad (12.16)$$

where $f(x) = (1-\cos x)/x$.

Proof. From (12.12) and (12.9) we have

$$\begin{aligned} -\mathrm{vec}(L_{\cos}(A,E)) &= \int_0^1 \big[\sin(A^T s)\otimes\cos(A(1-s)) \\ &\qquad + \cos(A^T s)\otimes\sin(A(1-s))\big]\,ds\cdot\mathrm{vec}(E) \\ &= \int_0^1 \sin\big(A^T s \oplus A(1-s)\big)\,ds\cdot\mathrm{vec}(E) \\ &= \int_0^1 \sin\big([A^T\oplus(-A)]s + I\otimes A\big)\,ds\cdot\mathrm{vec}(E). \end{aligned}$$

Since $A^T \oplus (-A) = A^T\otimes I - I\otimes A$ and $I\otimes A$ commute we can use the addition formula (12.5) to obtain

$$K_{\cos}(A) = -\int_0^1 \big[\sin\big([A^T\oplus(-A)]s\big)\cos(I\otimes A) + \cos\big([A^T\oplus(-A)]s\big)\sin(I\otimes A)\big]\,ds.$$

The formula (12.15) now follows on using $\int_0^1 \sin(as)\,ds = (1-\cos(a))/a = f(a)$ and $\int_0^1 \cos(as)\,ds = \sin(a)/a \equiv \mathrm{sinc}(a)$. The proof of (12.16) is analogous. □

Table 12.1. *Number of matrix multiplications π_{2m} required to evaluate $p_{2m}(A)$ and $q_{2m}(A)$.*

$2m$	2	4	6	8	10	12	14	16	18	20	22	24	26	28	30
π_{2m}	1	2	3	4	5	5	6	6	7	7	8	8	9	9	9

12.3. Padé Approximation of Cosine

It is not known whether $[k/m]$ Padé approximants r_{km} of $\cos(x)$ exist for all m, though formulae of Magnus and Wynn [401, 1975] give the coefficients of the diagonal approximants r_m as ratios of determinants of matrices whose entries involve binomial coefficients. Since cos is an even function we need consider only even degrees $2m$. Padé approximants are readily computed symbolically, for example using MATLAB's Extended Symbolic Math Toolbox. The first two nontrivial Padé approximants are

$$r_2(x) = \frac{1 - \frac{5}{12}x^2}{1 + \frac{1}{12}x^2}, \qquad r_4(x) = \frac{1 - \frac{115}{252}x^2 + \frac{313}{15120}x^4}{1 + \frac{11}{252}x^2 + \frac{13}{15120}x^4}.$$

Thereafter the numerators and denominators of the rational coefficients grow rapidly in size; for example,

$$r_8(x) = \frac{1 - \frac{260735}{545628}x^2 + \frac{4375409}{141863280}x^4 - \frac{7696415}{13108167072}x^6 + \frac{80737373}{23594700729600}x^8}{1 + \frac{12079}{545628}x^2 + \frac{34709}{141863280}x^4 + \frac{109247}{65540835360}x^6 + \frac{11321}{1814976979200}x^8}.$$

There is no convenient continued fraction or partial fraction form for r_m.

To evaluate r_m we will explicitly evaluate $p_{2m} = \sum_{i=0}^{m} a_{2i}x^{2i}$ and $q_{2m} = \sum_{i=0}^{m} b_{2i}x^{2i}$ and then solve the multiple right-hand side system $q_{2m}r_{2m} = p_{2m}$. The most efficient evaluation scheme is to treat p_{2m} and q_{2m} as degree m polynomials in A^2 and apply the Paterson–Stockmeyer method, as described in Section 4.4.3. Of equal cost for $8 \le 2m \le 28$ are the schemes illustrated for $2m = 12$ by

$$A_2 = A^2, \quad A_4 = A_2^2, \quad A_6 = A_2A_4, \tag{12.17a}$$
$$p_{12} = a_0I + a_2A_2 + a_4A_4 + a_6A_6 + A_6(a_8A_2 + a_{10}A_4 + a_{12}A_6), \tag{12.17b}$$
$$q_{12} = b_0I + b_2A_2 + b_4A_4 + b_6A_6 + A_6(b_8A_2 + b_{10}A_4 + b_{12}A_6). \tag{12.17c}$$

Table 12.1 summarizes the cost of evaluating p_{2m} and q_{2m} for $2m = 2{:}\,2{:}\,30$.

12.4. Double Angle Algorithm for Cosine

When $\|A\| \lesssim 1$, $\cos(A)$ is readily approximated using a Taylor or Padé approximation, as we have already seen in Section 4.3 in the case of the Taylor series. For large $\|A\|$ we can reduce the norm of the matrix whose cosine is approximated by using an analogue of the scaling and squaring method for the matrix exponential. Let us define

$$C_i = \cos(2^{i-s}A).$$

The integer s is chosen so that $2^{-s}A$ has norm small enough that we can obtain a good approximation of $C_0 = \cos(2^{-s}A)$ at reasonable cost. By applying the cosine double angle formula (12.6) we can compute $C_s = \cos(A)$ from C_0 using the recurrence

$$C_{i+1} = 2C_i^2 - I, \quad i = 0\colon s-1. \tag{12.18}$$

To develop an algorithm we need to decide how to choose s and how to approximate C_0, taking into account computational cost and the effects of truncation and rounding errors.

We begin by examining the propagation of errors in the double angle recurrence. The following result describes the overall effect of the error in approximating C_0 and the rounding errors in the evaluation of the recurrence.

Theorem 12.5 (Higham and Smith). *Let $\widehat{C}_i = C_i + E_i$, where $C_0 = \cos(2^{-s}A)$, the C_i satisfy (12.18), and $\widehat{C}_{i+1} = fl(2\widehat{C}_i^2 - I)$. For the 1-, $\infty-$, and Frobenius norms, assuming that $\|E_i\| \le 0.05\|C_i\|$ for all i, we have*

$$\begin{aligned}\|E_i\| \le{}& (4.1)^i \|E_0\|\,\|C_0\|\,\|C_1\| \dots \|C_{i-1}\| \\ &+ \widetilde{\gamma}_{n+1} \sum_{j=0}^{i-1} 4.1^{i-j-1}(2.21\alpha_n \|C_j\|^2 + 1)\,\|C_{j+1}\| \dots \|C_{i-1}\|,\end{aligned} \tag{12.19}$$

where $\alpha_n = n$ for the 2-norm and $\alpha_n = 1$ otherwise.

Proof. We have

$$\widehat{C}_{i+1} = fl(2\widehat{C}_i^2 - I) = 2\widehat{C}_i^2 - I + R_i, \tag{12.20}$$

where

$$\|R_i\| \le \widetilde{\gamma}_{n+1}(2\alpha_n \|\widehat{C}_i\|^2 + 1). \tag{12.21}$$

We can rewrite (12.20) as

$$E_{i+1} = 2(E_i^2 + E_iC_i + C_iE_i) + R_i.$$

Taking norms, we obtain

$$\|E_{i+1}\| \le 2\|E_i\|(\|E_i\| + 2\|C_i\|) + \|R_i\|. \tag{12.22}$$

The assumption on $\|E_i\|$ then gives

$$\|E_{i+1}\| \le 4.1\|E_i\|\,\|C_i\| + \|R_i\|. \tag{12.23}$$

This recurrence is easily solved to give

$$\begin{aligned}\|E_{i+1}\| &\le 4.1^{i+1}\|E_0\|\,\|C_0\|\,\|C_1\| \dots \|C_i\| + \sum_{j=0}^{i} 4.1^{i-j}\|R_j\|\,\|C_{j+1}\| \dots \|C_i\| \\ &\le 4.1^{i+1}\|E_0\|\,\|C_0\|\,\|C_1\| \dots \|C_i\| \\ &\quad + \widetilde{\gamma}_{n+1} \sum_{j=0}^{i} 4.1^{i-j}(2.21\alpha_n \|C_j\|^2 + 1)\,\|C_{j+1}\| \dots \|C_i\|. \qquad \square\end{aligned}$$

Table 12.2. *Maximum value θ_{2m} of θ such that the absolute error bound (12.24) does not exceed $u = 2^{-53}$.*

$2m$	2	4	6	8	10	12	14	16	18	20	22	24	26	28	30
θ_{2m}	6.1e-3	0.11	0.43	0.98	1.7	2.6	3.6	4.7	5.9	7.1	8.3	9.6	10.9	12.2	13.6

Consider the special case where A is normal with real eigenvalues, λ_i. Then $\|C_i\|_2 = \max_{1\le j\le n} |\cos(2^{i-s}\lambda_j)| \le 1$ for all i and (12.19) yields

$$\|E_s\|_2 \le 4.1^s(\|E_0\|_2 + n\widetilde{\gamma}_{n+1}).$$

This bound reflects the fact that, because the double angle recurrence multiplies the square of the previous iterate by 2 at each stage, the errors could grow by a factor 4 at each stage, though this worst-case growth is clearly extremely unlikely.

It is natural to require the approximation $\widehat{C}_0$ to satisfy a relative error bound of the form $\|C_0 - \widehat{C}_0\|/\|C_0\| \le u$. The lower bound in (12.10) guarantees that the denominator is nonzero if $\|A\|$ or $\|A^2\|^{1/2}$ is less than $\cosh^{-1}(2) = 1.317\ldots$. Consider the term $\|E_0\|$ in the bound (12.19). With the relative error bound and the norm restriction just described, we have $\|E_0\| \le u\|C_0\| \le 2u$, which is essentially the same as if we imposed an absolute error bound $\|E_0\| \le u$. An absolute bound has the advantage of putting no restrictions on $\|A\|$ or $\|A^2\|^{1/2}$ and thus potentially allowing a smaller s, which means fewer double angle recurrence steps. The norms of the matrices $C_0, \ldots, C_{s-1}$ are different in the two cases, and we can expect an upper bound for $\|C_0\|$ to be larger in the absolute case. But if the absolute criterion permits fewer double angle steps then, as is clear from (12.19), significant gains in accuracy could accrue. In summary, the error analysis provides support for the use of an absolute error criterion if $\|C_0\|$ is not too large. Algorithms based on a relative error bound were developed by Higham and Smith [287, 2003] and Hargreaves and Higham [248, 2005], but experiments in the latter paper show that the absolute bound leads to a more efficient and more accurate algorithm. We now develop an algorithm based on an absolute error bound.

As for the exponential and the logarithm, Padé approximants to the cosine provide greater efficiency than Taylor series for the same quality of approximation. The truncation error for r_{2m} has the form

$$\cos(A) - r_{2m}(A) = \sum_{i=2m+1}^{\infty} c_{2i}A^{2i}.$$

Hence

$$\|\cos(A) - r_{2m}(A)\| \le \sum_{i=2m+1}^{\infty} |c_{2i}|\theta^{2i}, \tag{12.24}$$

where

$$\theta = \theta(A) = \|A^2\|^{1/2}.$$

Define θ_{2m} to be the largest value of θ such that the bound in (12.24) does not exceed u. Using the same mixture of symbolic and high precision calculation as was employed in Section 10.3 for the exponential, we found the values of θ_{2m} listed in Table 12.2.

Table 12.3. *Upper bound for $\kappa(q_{2m}(A))$ when $\theta \le \theta_{2m}$, based on* (12.26) *and* (12.27), *where the θ_{2m} are given in Table* 12.2. *The bound does not exist for* $2m \ge 26$.

$2m$	2	4	6	8	10	12	14	16	18	20	22	24
Bound	1.0	1.0	1.0	1.0	1.1	1.2	1.4	1.8	2.4	3.5	7.0	9.0e1

Table 12.4. *Upper bounds for $\|\widetilde{p}_{2m}\|_\infty$ and $\|\widetilde{q}_{2m}\|_\infty$ for $\theta \le \theta_{2m}$.*

$2m$	2	4	6	8	10	12	14	16	18	20	22	24
$\|\widetilde{p}_{2m}\|_\infty$	1.0	1.0	1.1	1.5	2.7	6.2	1.6e1	4.3e1	1.2e2	3.7e2	1.2e3	3.7e3
$\|\widetilde{q}_{2m}\|_\infty$	1.0	1.0	1.0	1.0	1.1	1.1	1.2	1.3	1.4	1.6	1.7	2.0

We now need to consider the effects of rounding errors on the evaluation of r_{2m}. Consider, first, the evaluation of p_{2m} and q_{2m}, and assume initially that A^2 is evaluated exactly. Let $g_{2m}(A^2)$ denote either of the even polynomials $p_{2m}(A)$ and $q_{2m}(A)$. It follows from Theorem 4.5 that for the 1- and ∞-norms,

$$\|g_{2m}(A^2) - fl(g_{2m}(A^2))\| \le \widetilde{\gamma}_{mn}\widetilde{g}_{2m}(\|A^2\|), \tag{12.25}$$

where $\widetilde{g}_{2m}$ denotes g_{2m} with its coefficients replaced by their absolute values. When we take into account the error in forming A^2 we find that the bound (12.25) is multiplied by a term that is approximately $\mu(A) = \||A|^2\|/\|A^2\| \ge 1$. The quantity μ can be arbitrarily large. However, $\mu_\infty(A) \le (\|A\|_\infty/\theta_\infty(A))^2$, so a large μ implies that basing the algorithm on $\theta(A)$ rather than $\|A\|$ produces a smaller s, which means that potentially increased rounding errors in the evaluation of p_{2m} and q_{2m} are balanced by potentially decreased error propagation in the double angle phase.

Since we obtain r_{2m} by solving a linear system with coefficient matrix $q_{2m}(A)$, we require $q_{2m}(A)$ to be well conditioned to be sure that the system is solved accurately. From $q_{2m}(A) = \sum_{i=0}^m b_{2i}A^{2i}$, we have

$$\|q_{2m}(A)\| \le \sum_{i=0}^m |b_{2i}|\theta^{2i}. \tag{12.26}$$

Using the inequality $\|(I+E)^{-1}\| \le (1-\|E\|)^{-1}$ for $\|E\| < 1$ gives

$$\|q_{2m}(A)^{-1}\| \le \frac{1}{|b_0| - \|\sum_{i=1}^m b_{2i}A^{2i}\|} \le \frac{1}{|b_0| - \sum_{i=1}^m |b_{2i}|\theta^{2i}}. \tag{12.27}$$

Table 12.3 tabulates the bound for $\kappa(q_{2m}(A))$ obtained from (12.26) and (12.27), the latter being finite only for $2m \le 24$. It shows that q_{2m} is well conditioned for all m in this range.

Now we consider the choice of m. In view of Table 12.3, we will restrict to $2m \le 24$. Table 12.4, which concerns the error bound (12.25) for the evaluation of p_{2m} and q_{2m}, suggests further restricting $2m \le 20$, say. From Table 12.1 it is then clear that we need consider only $2m = 2, 4, 6, 8, 12, 16, 20$. Dividing A (and hence θ) by 2 results in one extra matrix multiplication in the double angle phase, whereas for $\theta \le \theta_{2m}$ the cost of evaluating the Padé approximant increases by one matrix multiplication with each increase in m in our list of considered values. Since the numbers θ_{12}, θ_{16}, and θ_{20}

Table 12.5. *Logic for choice of scaling and Padé approximant degree $d \equiv 2m$. Assuming A has already been scaled, if necessary, so that $\theta \le \theta_{20} = 7.1$, further scaling should be done to bring θ within the range for the indicated value of d.*

Range of θ	d
$[0, \theta_{16}] = [0, 4.7]$	smallest $d \in \{2, 4, 6, 8, 12, 16\}$ such that $\theta \le \theta_d$
$(\theta_{16}, 2\theta_{12}] = (4.7, 5.2]$	12 (scale by 1/2)
$(2\theta_{12}, \theta_{20}] = (5.2, 7.1]$	20 (no scaling)

differ successively by less than a factor 2, the value of $d \equiv 2m$ that gives the minimal work depends on θ. For example, if $\theta = 7$ then $d = 20$ is best, because nothing would be gained by a further scaling by 1/2, but if $\theta = 5$ then scaling by 1/2 enables us to use $d = 12$, and the whole computation then requires one less matrix multiplication than if we immediately applied $d = 20$. Table 12.5 summarizes the relevant logic. The tactic, then, is to scale so that $\theta \le \theta_{20}$ and to scale further only if a reduction in work is achieved.

With this scaling strategy we have, by (12.10), $\|C_0\| \le \cosh(\theta_{20}) \le 606$. Since this bound is not too much larger than 1, the argument given at the beginning of this section provides justification for the following algorithm.

Algorithm 12.6 (double angle algorithm). Given $A \in \mathbb{C}^{n\times n}$ this algorithm approximates $C = \cos(A)$. It uses the constants θ_{2m} given in Table 12.2. The algorithm is intended for IEEE double precision arithmetic. At lines 5 and 12 the Padé approximants are to be evaluated via a scheme of form (12.17), making use of the matrix B.

1 $B = A^2$
2 $\theta = \|B\|_\infty^{1/2}$
3 for $d = [2\ 4\ 6\ 8\ 12\ 16]$
4   if $\theta \le \theta_d$
5         $C = r_d(A)$
6   quit
7   end
8 end
9 $s = \lceil \log_2(\theta/\theta_{20}) \rceil$ % Find minimal integer s such that $2^{-s}\theta \le \theta_{20}$.
10 Determine optimal d from Table 12.5 (with $\theta \leftarrow 2^{-s}\theta$); increase s as necessary.
11 $B \leftarrow 4^{-s}B$
12 $C = r_d(2^{-s}A)$
13 for $i = 1\colon s$
14     $C \leftarrow 2C^2 - I$
15 end

Cost: $\big(\pi_d + \lceil \log_2(\|A^2\|_\infty^{1/2}/\theta_d) \rceil\big)M + D$, where π_d and θ_d are tabulated in Tables 12.1 and 12.2, respectively.

In the next section we will compare Algorithm 12.6 with the following very simple competitor based on (12.2).

Algorithm 12.7 (cosine and sine via exponential). Given $A \in \mathbb{C}^{n\times n}$ this algorithm computes $C = \cos(A)$ and $S = \sin(A)$ via the matrix exponential.

1 $X = e^{iA}$
2 If A is real
3 $\quad C = \operatorname{Re} X$
4 $\quad S = \operatorname{Im} X$
5 else
6 $\quad C = (X + X^{-1})/2$
7 $\quad S = (X - X^{-1})/(2i)$
8 end

An obvious situation in which Algorithm 12.7 is likely to produce an inaccurate computed C is when $\|C\| \ll \|S\| \approx \|e^{iA}\|$, or conversely with the roles of C and S interchanged. However, (12.14) shows that in this situation $\cos(A)$ is ill conditioned, so an accurate result cannot be expected.

12.5. Numerical Experiment

We describe an experiment that compares the accuracy of three methods for computing $\cos(A)$.

1. Algorithm 12.6.
2. MATLAB's `funm`, called as `funm(A,@cos)`. This is Algorithm 9.6: the Schur–Parlett algorithm with the cosines of the diagonal blocks of the triangular matrix evaluated by Taylor series.
3. Algorithm 12.7, with the exponential evaluated by MATLAB's `expm` (which implements Algorithm 10.20).

We used a set of 55 test matrices including matrices from MATLAB, from the Matrix Computation Toolbox [264], and from [287, 2003]. The matrices are mostly 10×10 and their norms range from order 1 to 10^7, though more than half have ∞-norm 10 or less. We evaluated the relative error $\|\widehat{C} - C\|_F/\|C\|_F$, where $\widehat{C}$ is the computed approximation to C and $C = \cos(A)$ is computed at 100-digit precision. Figure 12.1 shows the results. The solid line is the unit roundoff multiplied by an estimate of $\kappa_{\cos}$ obtained using Algorithm 3.20 with finite differences. Figure 12.2 gives a performance profile representing the same data.

Some observations on the results:

- The three algorithms are all behaving in a numerically stable way apart from occasional exceptions for each algorithm.
- The average cost of Algorithm 12.6 in this experiment is about 9 matrix multiplications, or $18n^3$ flops, which compares favourably with the cost of at least $28n^3$ flops of `funm`.
- It is reasonable to conclude that on this test set Algorithm 12.7 is the most accurate solver overall, followed by Algorithm 12.6 and then `funm`.
- Algorithm 12.7 is expensive because it requires a matrix exponential evaluation in complex arithmetic: complex arithmetic costs significantly more than real arithmetic (see Section B.15) and a complex matrix requires twice the storage of a real matrix. The good performance of the algorithm is dependent on that of

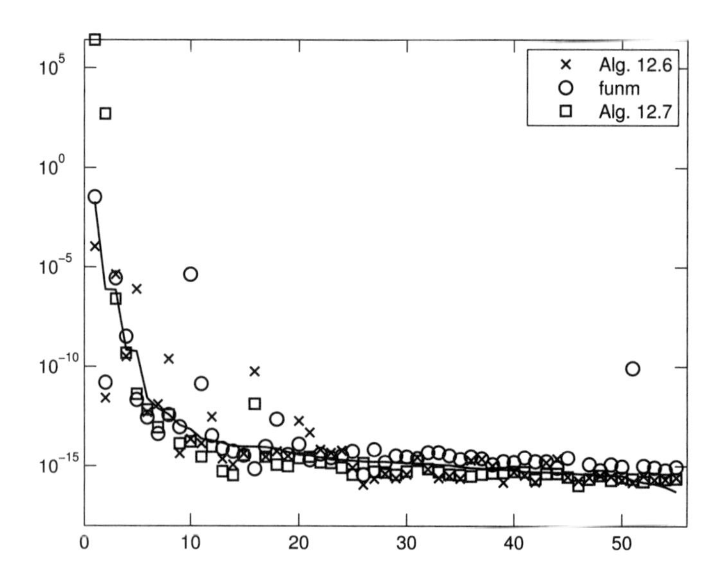

Figure 12.1. *Normwise relative errors for Algorithm* 12.6, *MATLAB's* `funm`, *and Algorithm* 12.7; *the solid line is an estimate of* $\kappa_{\cos}(A)u$.

`expm`: when MATLAB's `expmdemo1` (described in Section 10.5) is used in place of `expm` we find that Algorithm 12.7 jumps from having the best performance profile to the worst.

12.6. Double Angle Algorithm for Sine and Cosine

There is no natural analogue of Algorithm 12.6 for the sine, because the corresponding double angle recurrence $\sin(2A) = 2\sin(A)\cos(A)$ requires cosines. However, computing the sine reduces to computing the cosine through $\sin(A) = \cos(A - \frac{\pi}{2}I)$.

Suppose now that we wish to compute both $\sin(A)$ and $\cos(A)$. Algorithm 12.7 provides one way to do so, but requires complex arithmetic even when A is real. We will develop an analogue of Algorithm 12.6 that scales A by a power of 2, computes Padé approximants to both the sine and cosine of the scaled matrix, and then applies the double angle formulae $\cos(2A) = 2\cos^2(A) - I$ and $\sin(2A) = 2\sin(A)\cos(A)$. Computational savings are possible in the evaluation of the Padé approximants and in the double angle recurrences by reusing the cos terms.

Denote the $[m/m]$ Padé approximant to the sine function by $\widetilde{r}_m(x) = \widetilde{p}_m(x)/\widetilde{q}_m(x)$. Then $\widetilde{p}_m$ is odd and $\widetilde{q}_m$ even, their degrees being, respectively, m and $m-1$ if m is odd, otherwise $m-1$ and m [401, 1975], so the error in $\widetilde{r}_m$ has the form

$$\sin(A) - \widetilde{r}_m(A) = \sum_{i=m}^{\infty} c_{2i+1}A^{2i+1}.$$

Since this expansion contains only odd powers of A we bound the series in terms of $\|A\|$ instead of $\theta(A)$:

$$\|\sin(A) - \widetilde{r}_m(A)\| \le \sum_{i=m}^{\infty} |c_{2i+1}|\beta^{2i}, \qquad \beta = \|A\|. \tag{12.28}$$

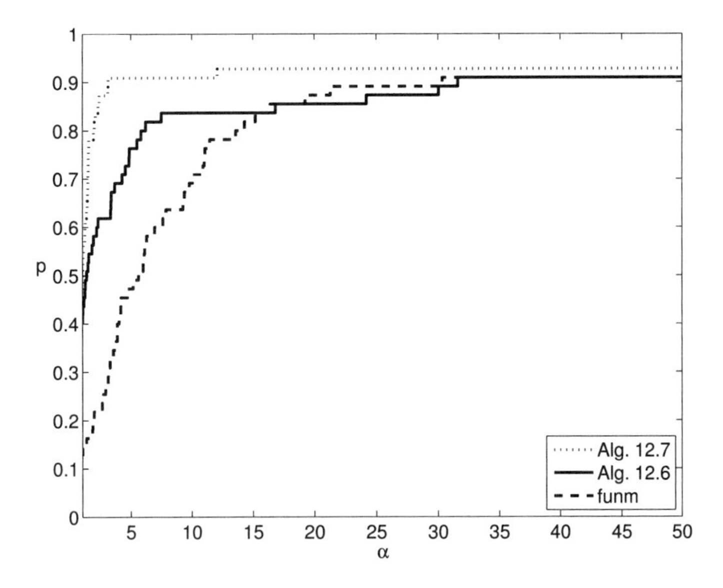

Figure 12.2. *Same data as in Figure* 12.1 *presented as a performance profile.*

Table 12.6. *Maximum value* β_m *of* $\|A\|$ *such that the absolute error bound* (12.28) *does not exceed* $u = 2^{-53}$.

m	2	3	4	5	6	7	8	9	10	11	12	
β_m	1.4e-3	1.8e-2	6.4e-2	1.7e-1	0.32	0.56	0.81	1.2	1.5	2.0	2.3	
m	13	14	15	16	17	18	19	20	21	22	23	24
β_m	2.9	3.3	3.9	4.4	5.0	5.5	6.2	6.7	7.4	7.9	8.7	9.2

Define β_m to be the largest value of β such that the bound (12.28) does not exceed u. Using the same technique as for the cosine, we computed the values shown in Table 12.6. These values of β_m can be compared with the values of θ_{2m} in Table 12.2.

On comparing Table 12.6 with Table 12.2 we see that for $4 \le 2m \le 22$ we have $\beta_{2m} < \theta_{2m} < \beta_{2m+1}$. We could therefore scale so that $\|2^{-s}A\| \le \beta_{2m}$ and then use the $[2m/2m]$ Padé approximants to the sine and cosine, or scale so that $\|2^{-s}A\| \le \theta_{2m}$ and use the $[2m/2m]$ Padé approximant to the cosine and the $[2m+1/2m+1]$ Padé approximant to the sine. Since we can write an odd polynomial in A ($\widetilde{p}_m(A)$) as A times an even polynomial of degree one less, it turns out to be as cheap to evaluate $\widetilde{r}_{2m+1}$ and r_{2m} as to evaluate $\widetilde{r}_{2m}$ and r_{2m}. Therefore we will scale so that $\|2^{-s}A\| \le \theta_{2m}$ and then evaluate r_{2m} for the cosine and $\widetilde{r}_{2m+1}$ for the sine. Evaluating p_{2m}, q_{2m}, $\widetilde{p}_{2m+1}$, and $\widetilde{q}_{2m+1}$ reduces to evaluating four even polynomials of degree $2m$. This can be done by forming the powers A^2, A^4, $\dots$, A^{2m}, at a total cost of $m+1$ multiplications. However, for $2m \ge 20$ it is more efficient to use the schemes of the form (12.17). We summarize the cost of evaluating p_{2m}, q_{2m}, $\widetilde{p}_{2m+1}$, and $\widetilde{q}_{2m+1}$ for $m = 2\colon 2\colon 24$ in Table 12.7.

Now we consider the choice of degree, $d \equiv 2m$. Bounds analogous to those in Table 12.3 show that $\widetilde{q}_{j+1}$ is well conditioned for $d \le 24$, and bounds for $\widetilde{p}_{j+1}$ and $\widetilde{q}_{j+1}$ analogous to those in Table 12.4 suggest restricting to $d \le 20$ (the same restriction

Table 12.7. *Number of matrix multiplications* $\widetilde{\pi}_{2m}$ *to evaluate* $p_{2m}(A)$, $q_{2m}(A)$, $\widetilde{p}_{2m+1}(A)$, *and* $\widetilde{q}_{2m+1}(A)$.

$2m$	2	4	6	8	10	12	14	16	18	20	22	24
$\widetilde{\pi}_{2m}$	2	3	4	5	6	7	8	9	10	10	11	11

that was made in Section 12.4 for the Padé approximants for the cosine). It is then clear from Table 12.7 that we need only consider $d = 2, 4, 6, 8, 10, 12, 14, 16, 20$. Noting that dividing A by 2 results in two extra multiplications in the double-angle phase and that increasing from one value of d to the next in our list of considered values increases the cost of evaluating the Padé approximants by one multiplication, we can determine the most efficient choice of d by a similar argument to that in the previous section. The result is that we should scale so that $\theta \le \theta_{20}$, and scale further according to exactly the same strategy as in Table 12.5 except that in the first line of the table "10" and "14" are added to the set of possible d values.

The algorithm can be summarized as follows.

Algorithm 12.8 (double angle algorithm for sine and cosine). Given $A \in \mathbb{C}^{n\times n}$ this algorithm approximates $C = \cos(A)$ and $S = \sin(A)$. It uses the constants θ_{2m} given in Table 12.2. The algorithm is intended for IEEE double precision arithmetic.

1 for $d = [2\ 4\ 6\ 8\ 10\ 12\ 14\ 16]$
2   if $\|A\|_\infty \le \theta_d$
3       $C = r_d(A)$, $S = \widetilde{r}_{d+1}(A)$
4   quit
5   end
6 end
7 $s = \lceil \log_2(\|A\|_\infty/\theta_{20}) \rceil$ % Find minimal integer s such that $2^{-s}\|A\|_\infty \le \theta_{20}$.
8 Determine optimal d from modified Table 12.5 (with $\theta = 2^{-s}\|A\|_\infty$)
   and increase s as necessary.
9 $C = r_d(2^{-s}A)$, $S = \widetilde{r}_{d+1}(2^{-s}A)$
10 for $i = 1\colon s$
11    $S \leftarrow 2CS$, $C \leftarrow 2C^2 - I$
12 end

Cost: $(\widetilde{\pi}_d + \lceil \log_2(\|A\|_\infty/\theta_d) \rceil)M + 2D$, where θ_d and $\widetilde{\pi}_d$ are tabulated in Tables 12.2 and 12.7, respectively.

How much work does Algorithm 12.8 save compared with separate computation of $\cos(A)$ and $\sin(A) = \cos(A - \frac{\pi}{2}I)$ by Algorithm 12.6? The answer is roughly $2\pi_d - \widetilde{\pi}_d$ matrix multiplies, which rises from 1 when $d = 4$ to 4 when $d = 20$; the overall saving is therefore up to about 29%.

We tested Algorithm 12.8 on the same set of test matrices as in Section 12.5. Figure 12.3 compares the relative errors for the computed sine and cosine with the corresponding errors from `funm`, invoked as `funm(A,@sin)` and `funm(A,@cos)`; with Algorithm 12.7; and with $\sin(A)$ computed as the shifted cosine $\sin(A) = \cos(A - \frac{\pi}{2}I)$ using Algorithm 12.6. Note that the cost of the two `funm` computations can be reduced by using the same Schur decomposition for $\sin(A)$ as for $\cos(A)$. Algorithm 12.8 provides similar or better accuracy to `funm` and the shifted cosine on this test set. Its cost varies from 9 matrix multiplies and solves to 55, with an average of 17.

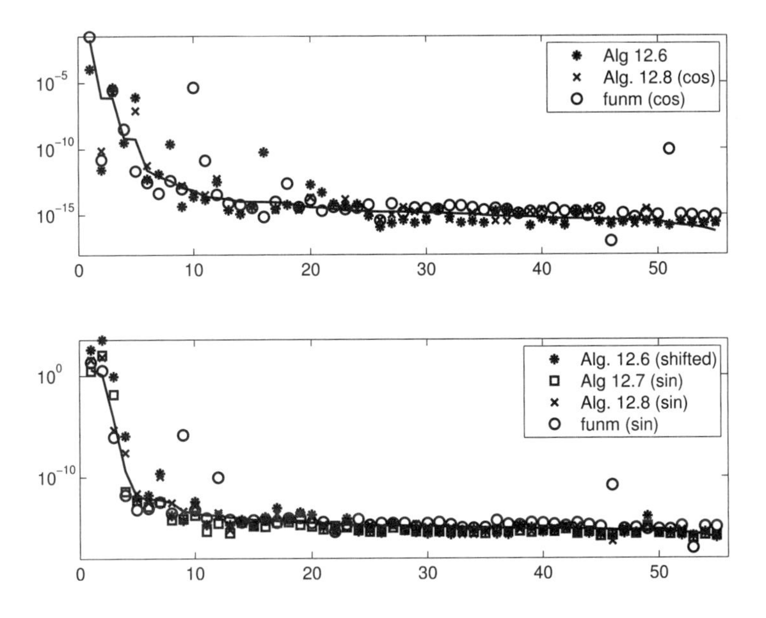

Figure 12.3. *Normwise relative errors for Algorithm* 12.6, *Algorithm* 12.7, *Algorithm* 12.8, `funm`, *and sine obtained as shifted cosine from Algorithm* 12.6. *The solid line is an estimate of* $\kappa_{\cos}(A)u$ *(top) and* $\kappa_{\sin}(A)u$ *(bottom).*

12.6.1. Preprocessing

Balancing and shifting can be used to reduce the norm prior to applying any of the algorithms in this chapter. We can exploit the periodicity relation (4.32) to shift A by πjI. Theorem 4.21 motivates taking j as whichever of the two integers nearest to $\mathrm{trace}(A)/(n\pi)$ gives the smaller value of $\|A-\pi jI\|_F$, or as 0 if neither choice reduces the Frobenius norm of A. Thus preprocessing consists of computing $\widetilde{A} = D^{-1}(A - \pi qI)D$, where D represents balancing, and recovering $\cos(A) = (-1)^q D\cos(\widetilde{A})D^{-1}$ and $\sin(A) = (-1)^q D\sin(\widetilde{A})D^{-1}$.

12.7. Notes and References

The bounds (12.10) are from Hargreaves and Higham [248, 2005]. The Fréchet derivative formulae in Section 12.2 appear to be new.

Serbin and Blalock [515, 1980] suggest computing $\cos(A)$ by a Taylor or Padé approximation in conjunction with the double angle recurrence, but do not propose a specific algorithm. Higham and Smith [287, 2003] develop an algorithm that chooses s so that $\|2^{-s}A\|_\infty \le 1$, approximates $\cos(A) \approx r_8(2^{-s}A)$, and then uses the double angle recurrence. Ad hoc analysis in [287, 2003] shows that the [8/8] Padé approximant provides full normwise relative accuracy in IEEE double precision. Hargreaves and Higham [248, 2005] make several improvements. They phrase truncation error bounds in terms of $\|A^2\|^{1/2}$ instead of $\|A\|$; choose the degree of the Padé approximant adaptively to minimize the computational cost subject to achieving a desired truncation error; and use an absolute, rather than relative, error criterion in the choice of Padé approximant. Theorem 12.5 is from Higham and Smith [287, 2003] and Algorithms 12.6 and 12.8 from Hargreaves and Higham [248, 2005].

The matrix cosine and sine are used in a matrix multiplication-dominated algorithm proposed by Yau and Lu [619, 1993] for solving the symmetric eigenvalue problem $Ax = \lambda x$. This algorithm requires $\cos(A)$ and $\sin(A)$, which are approximated using Chebyshev expansions in conjunction with the cosine double angle recurrence. For more details see [619, 1993] and Tisseur [568, 1997].

Problems

12.1. Show that if $A \in \mathbb{C}^{n\times n}$ is involutory ($A^2 = I$) then $\cos(k\pi A) = (-1)^k I$ for all integers k.

12.2. (Pólya and Szegö [475, 1998, Problem 210]) Does the equation $\sin(A) = \left[\begin{smallmatrix} 1 & 1 \\ 0 & 1 \end{smallmatrix}\right]$ have a solution?

12.3. Derive (12.1), (12.3), (12.4), and (12.5) using the results of Section 1.3.

12.4. Assuming that X_0 is nonsingular, show that X_1 obtained from one step of the Newton sign iteration (5.16) can be written $X_1 = \cos(i^{-1}\log(X_0))$. (Cf. line 6 of Algorithm 12.7.)

12.5. Suppose we know $\cos(A)$. Can we determine $\sin(A)$ from (12.3), i.e., from the relation $\cos^2(A) + \sin^2(A) = I$? Consider separately the cases where A is a general matrix and A is triangular.

The reader is expected not to be worried by expressions such as the sine of a matrix A.

— SIMON L. ALTMANN, *Rotations, Quaternions, and Double Groups* (1986)

Rational approximations for the cosine function play an important role in the linear stability analysis of numerical methods for initial-value problems of the special form $y'' = f(x, y),\ y(x_0) = y_0,\ y'(x_0) = z_0.$

— JOHN P. COLEMAN, *Rational Approximations for the Cosine Function; P-Acceptability and Order* (1992)

Chapter 13
Function of Matrix Times Vector: $f(A)b$

In some applications it is not $f(A)$ that is required but rather the action of $f(A)$ on a vector: $f(A)b$. Indeed, if A is sparse then $f(A)$ may be much denser than A and for large dimensions it may be too expensive to compute or store $f(A)$, while computing and storing $f(A)b$ may be feasible. In this chapter we consider various facets of the $f(A)b$ problem. Our treatment is relatively concise, for two reasons. First, some of the techniques used for $f(A)b$ are closely related to those for $f(A)$ that have been treated earlier in the book. Second, the $f(A)b$ problem overlaps with the theory and practice of Krylov subspace methods and of the numerical evaluation of contour integrals. A thorough treatment of these latter topics is outside the scope of this book. Moreover, the application of these techniques to the $f(A)b$ problem is a relatively new and active topic of research and the state of the art is advancing rapidly.

The development of a method for $f(A)b$ must be guided by the allowable operations with A. A minimal assumption is that matrix–vector products can be formed. Another possible assumption is that A is a large sparse matrix for which linear systems $(\alpha A + \beta I)x = c$ can be solved efficiently by sparse direct methods but the computation of the Hessenberg and Schur forms are impractical. Alternatively, we may assume that any necessary factorization can be computed.

We begin this chapter by obtaining a b-specific representation of $f(A)b$ in terms of Hermite interpolation. Then we consider computational approaches based on Krylov subspace methods, quadrature, and differential equations.

We emphasize that the methods described in this chapter are not intended for $f(A) = A^{-1}$, for which the special form of f can usually be exploited to derive more effective variants of these methods.

13.1. Representation via Polynomial Interpolation

Definition 1.4 specifies $f(A) = p(A)$, where p is a Hermite interpolating polynomial determined by the values of f on the spectrum of A. The degree of p is the degree of the minimal polynomial ψ_A of A and it may be greater than necessary in order to produce $f(A)b$. To see why, consider the *minimal polynomial of A with respect to b*, which is the unique monic polynomial $\psi_{A,b}$ of lowest degree such that $\psi_{A,b}(A)b = 0$. The results developed in Section 1.2.2 generalize from ψ_A to $\psi_{A,b}$. Specifically, $\psi_{A,b}$ divides any polynomial p for which $p(A)b = 0$; $\psi_{A,b}$ has the form

$$\psi_{A,b}(t) = \prod_{i=1}^{s}(t - \lambda_i)^{\ell_i}, \qquad 0 \le \ell_i \le n_i,$$

where ℓ_i depends on b as well as A and $\lambda_1, \dots, \lambda_s$ are the distinct eigenvalues of A, with n_i the dimension of the largest Jordan block in which λ_i appears; and we have the following generalization of Theorem 1.3.

Theorem 13.1. *For polynomials p and q and $A \in \mathbb{C}^{n\times n}$, $p(A)b = q(A)b$ if and only if $p^{(j)}(\lambda_i) = q^{(j)}(\lambda_i)$, $j = 0\colon \ell_i - 1$, $i = 1\colon s$.* □

With this background we can show that the required polynomial is of degree only $\deg \psi_{A,b} \le \deg \psi_A$.

Theorem 13.2. *Let f be defined on the spectrum of $A \in \mathbb{C}^{n\times n}$ and let $\psi_{A,b}$ be the minimal polynomial of A with respect to b. Then $f(A)b = q(A)b$, where q is the unique Hermite interpolating polynomial of degree less than $\sum_{i=1}^{s} \ell_i = \deg \psi_{A,b}$ that satisfies the interpolation conditions*

$$q^{(j)}(\lambda_i) = f^{(j)}(\lambda_i), \qquad j = 0\colon \ell_i - 1, \quad i = 1\colon s. \tag{13.1}$$

Proof. Let p be the polynomial given by Definition 1.4 such that $f(A) = p(A)$ and consider q defined by (13.1). By (1.7), (13.1), and Theorem 13.1, $q(A)b = p(A)b = f(A)b$. □

Of course, the polynomial q in Theorem 13.2 depends on both A and b, and $q(A)c = f(A)c$ is not guaranteed for $c \ne b$.

The kth *Krylov subspace* of $A \in \mathbb{C}^{n\times n}$ and a nonzero vector $b \in \mathbb{C}^n$ is defined by

$$\mathcal{K}_k(A, b) = \operatorname{span}\{b, Ab, \dots, A^{k-1}b\},$$

and it can also be written as

$$\mathcal{K}_k(A, b) = \operatorname{span}\{\, q(A)b : q \text{ is a polynomial of degree} \le k - 1 \,\}.$$

Theorem 13.2 says that $f(A)b \in \mathcal{K}_d(A, b)$, where $d = \deg \psi_{A,b}$. Thus the size of a Krylov subspace necessary to capture $f(A)b$ depends on both A and b. Note also that $\deg \psi_{A,b}$ can be characterized as the smallest k such that $\mathcal{K}_k(A, b) = \mathcal{K}_{k+1}(A, b)$.

We consider Krylov subspace methods in the next section.

13.2. Krylov Subspace Methods

13.2.1. The Arnoldi Process

The Arnoldi process for $A \in \mathbb{C}^{n\times n}$ attempts to compute the Hessenberg reduction $Q^*AQ = H$, where $Q \in \mathbb{C}^{n\times n}$ is unitary and $H \in \mathbb{C}^{n\times n}$ is upper Hessenberg. Writing $Q = [q_1, \dots, q_n]$ and equating kth columns in $AQ = QH$ gives

$$Aq_k = \sum_{i=1}^{k+1} h_{ik} q_i, \qquad k = 1\colon n - 1. \tag{13.2}$$

This may be rewritten as

$$h_{k+1,k} q_{k+1} = Aq_k - \sum_{i=1}^{k} h_{ik} q_i =: r_k, \tag{13.3}$$

where, using the fact that Q is unitary,

$$h_{ik} = q_i^* A q_k, \qquad i = 1\colon k. \tag{13.4}$$

Provided that $r_k \neq 0$, $q_{k+1} = r_k / h_{k+1,k}$ with $h_{k+1,k} = \|r_k\|_2$.

From (13.2) it follows by induction that

$$\operatorname{span}\{q_1, \dots, q_k\} = \operatorname{span}\{q_1, Aq_1, \dots, A^{k-1} q_1\},$$

that is, the Arnoldi vectors $\{q_i\}_{i=1}^k$ form an orthonormal basis for the Krylov subspace $\mathcal{K}_k(A, q_1)$. The Arnoldi process produces the factorization

$$AQ_k = Q_k H_k + h_{k+1,k} q_{k+1} e_k^T, \tag{13.5}$$

where $Q_k = [q_1, \dots, q_k]$ and $H_k = (h_{ij})$ is $k \times k$ upper Hessenberg. Columns 1 to $k-1$ of this factorization are just (13.2), while column k is (13.3). Note that (13.4) can be written

$$Q_k^* A Q_k = H_k, \tag{13.6}$$

which says that H_k is the orthogonal projection of A onto $\operatorname{span}\{q_1, \dots, q_k\} = \mathcal{K}_k(A, q_1)$.

We noted above that $\mathcal{K}_m(A, q_1) = \mathcal{K}_{m+1}(A, q_1)$ for $m = \deg \psi_{A,q_1}$, so after generating the orthonormal basis $q_1, \dots, q_m$ for $\mathcal{K}_m(A, q_1)$ the Arnoldi process must produce $r_m = 0$, i.e., the process terminates.

The Arnoldi process can be summarized as follows.

Algorithm 13.3 (Arnoldi process). Given $A \in \mathbb{C}^{n \times n}$ and $q_1 \in \mathbb{C}^n$ of unit 2-norm, this algorithm uses the Arnoldi process to compute the factorization $AQ = QH$, where $Q \in \mathbb{C}^{n \times m}$ ($m = \deg \psi_{A,q_1} \leq n$) has orthonormal columns and $H \in \mathbb{C}^{m \times m}$ is upper Hessenberg.

```
1  for k = 1: n
2      z = Aq_k
3      for i = 1: k
4          h_ik = q_i^* z
5          z = z − h_ik q_i    % modified Gram–Schmidt
6      end
7      h_{k+1,k} = ||z||_2
8      if h_{k+1,k} = 0, m = k, quit, end
9      q_{k+1} = z/h_{k+1,k}
10 end
```

A few comments on the Arnoldi process are in order.

Implicit A. The matrix A does not need to be known or stored explicitly, since it appears in the Arnoldi process only in the matrix–vector products Aq_k. Therefore A can be given in any implicit form that permits efficient evaluation of the action of A on a vector.

Orthogonality. Algorithm 13.3 uses the modified Gram–Schmidt process to orthogonalize Aq_k against $q_1, \dots, q_k$. Classical Gram–Schmidt can be used instead, in which case line 2 is replaced by the assignments $r_k = Aq_k$, $z = r_k$ and in line 5 onwards z is replaced by r_k. Loss of orthogonality of the Arnoldi vectors can occur in finite precision arithmetic for both Gram–Schmidt processes (the more so for classical Gram–Schmidt), and this can be cured by reorthogonalization. Loss of orthogonality can alternatively be avoided completely by using an implementation of the Arnoldi process based on Householder transformations.

Hermitian A. When A is Hermitian most of the inner products in the inner loop of the Arnoldi process are zero and so need not be computed. Indeed for Hermitian A the Arnoldi process reduces to the Lanczos process, which involves a three-term recurrence. The Lanczos process yields real, symmetric tridiagonal H_k.

13.2.2. Arnoldi Approximation of $f(A)b$

To approximate $f(A)b$ with the Arnoldi process we take $q_1 = b/\|b\|_2$ and, for some appropriate k,

$$\begin{aligned} f_k &:= \|b\|_2 Q_k f(H_k) e_1, \\ &= Q_k f(H_k) Q_k^* b. \end{aligned} \tag{13.7}$$

We are effectively evaluating f on the smaller Krylov subspace $\mathcal{K}_k(A, q_1)$ and then expanding the result back onto the original space $\mathbb{C}^n$.

When is (13.7) an exact approximation? Clearly, if $k = n$ is reached then $Q_k \in \mathbb{C}^{n\times n}$ is unitary and $f(A)b = f(Q_n H_n Q_n^*)b = Q_n f(H_n) Q_n^* b = Q_n f(H_n)\|b\|_2 e_1 = f_k$. More precisely, upon termination of the Arnoldi process on step $m = \deg \psi_{A,b}$, we have $f_m = f(A)b$ (see Problem 13.1).

Some more insight into the approximation (13.7) is provided by the next two results. The first says that the approximation is exact if f is a polynomial of sufficiently low degree.

Lemma 13.4 (Saad). *Let $A \in \mathbb{C}^{n\times n}$ and Q_k, H_k be the result of k steps of the Arnoldi process on A. Then for any polynomial p_j of degree $j \le k-1$ we have*

$$p_j(A)q_1 = Q_k p_j(H_k) e_1.$$

Proof. It suffices to prove the result for $p_j(t) = t^j$, which we do by induction. The result is trivially true for $j = 0$. Assume that it is true for j with $j \le k-2$. Then

$$\begin{aligned} A^{j+1} q_1 = A \cdot A^j q_1 &= A Q_k H_k^j e_1 \\ &= (Q_k H_k + h_{k+1,k}\, q_{k+1} e_k^T) H_k^j e_1 \\ &= Q_k H_k^{j+1} e_1 + h_{k+1,k}\, q_{k+1} e_k^T H_k^j e_1 \\ &= Q_k H_k^{j+1} e_1, \end{aligned}$$

since $e_k^T H_k^j e_1 = 0$ for $j \le k-2$. Thus the result is true for $j+1$, as required. □

The next result identifies the approximation (13.7) as being obtained from a polynomial that interpolates f on the spectrum of H_k rather than the spectrum of A.

Theorem 13.5 (Saad). *Let Q_k, H_k be the result of k steps of the Arnoldi process on $A \in \mathbb{C}^{n\times n}$ and $b \in \mathbb{C}^n$. Then*

$$\|b\|_2 Q_k f(H_k) e_1 = \widetilde{p}_{k-1}(A) b,$$

where $\widetilde{p}_{k-1}$ is the unique polynomial of degree at most $k-1$ that interpolates f on the spectrum of H_k.

Proof. Note first that the upper Hessenberg matrix H_k has nonzero subdiagonal elements $h_{j+1,j}$, which implies that it is nonderogatory (see Problem 13.3). Consequently, the minimal polynomial of H is the characteristic polynomial and so has degree k. By Definition 1.4, $f(H_k) = \widetilde{p}_{k-1}(H_k)$ for a polynomial $\widetilde{p}_{k-1}$ of degree at most $k-1$. Hence

$$\|b\|_2 Q_k f(H_k)e_1 = \|b\|_2 Q_k \widetilde{p}_{k-1}(H_k)e_1 = \|b\|_2 \widetilde{p}_{k-1}(A)q_1 = \widetilde{p}_{k-1}(A)b,$$

using Lemma 13.4 for the middle equality. $\square$

We briefly discuss some of the important issues associated with the approximation (13.7).

Restarted Arnoldi. As k increases so does the cost of storing the Arnoldi vectors $q_1, \dots, q_k$ and computing H_k. It is standard practice in the context of linear systems ($f(x) = 1/x$) and the eigenvalue problem to restart the Arnoldi process after a fixed number of steps with a judiciously chosen vector that incorporates information gleaned from the computations up to that step. For general f restarting is more difficult because of the lack of a natural residual. Eiermann and Ernst [174, 2006] show how restarting can be done for general f by a technique that amounts to changing at the restart both the starting vector and the function. This method requires the evaluation of f at a block Hessenberg matrix of size km, where m is the restart length. This expense is avoided in a different implementation by Afanasjew, Eiermann, Ernst, and Güttel [3, 2007], which employs a rational approximation to f and requires only the solution of k Hessenberg linear systems of size m. Further analysis of restarting is given by Afanasjew, Eiermann, Ernst, and Güttel [4, 2007].

Existence of $f(H_k)$. For $k < \deg \psi_{A,b}$ it is possible that $f(H_k)$ is not defined, even though $f(A)$ is. From (13.6) it follows that the field of values of H_k is contained in the field of values of A. Therefore a sufficient condition for $f(H_k)$ to be defined is that f and its first n derivatives are defined at all points within the field of values of A.

Computing $f(H_k)$. The Arnoldi approximation (13.7) requires the computation of $f(H_k)e_1$, which is the first column of $f(H_k)$. Since k is of moderate size we can compute the whole matrix $f(H_k)$ by any of the methods described in earlier chapters of this book, taking advantage of the Hessenberg structure if possible. Whether or not it is possible or advantageous to apply a method specific to this $f(H_k)b$ problem—such as those described in the following sections—depends on f, k, and the desired accuracy.

Convergence, error bounds, and stopping test. The quality of the approximations (13.7) depends on how well $\widetilde{p}_{k-1}(A)b$ approximates $f(A)b$, where $\widetilde{p}_{k-1}$ is defined in Theorem 13.5. It is well known that the eigenvalues in the outermost part of the spectrum tend to approximated first in the Arnoldi process, so we might expect the components of the approximations (13.7) in the eigenvectors corresponding to the outermost part of the spectrum to be the most accurate. We state a result that describes convergence for the exponential function and Hermitian matrices with spectrum in the left half-plane.

Theorem 13.6. *Let A be a Hermitian negative semidefinite matrix with eigenvalues on $[-4\rho, 0]$, where $\rho > 0$, and set $\beta = \|b\|_2$. Then the error in the approximation*

(13.7) *of* $e^A b$ *is bounded by*

$$\|e^A b - \beta Q_k e^{H_k} e_1\|_2 \le \begin{cases} 10\beta e^{-k^2/(5\rho)}, & 2\sqrt{\rho} \le k \le 2\rho, \\ \dfrac{10\beta}{\rho} e^{-\rho} \left(\dfrac{e\rho}{k}\right)^k, & k \ge 2\rho. \end{cases}$$

Proof. See Hochbruck and Lubich [291, 1997]. □

The bound of the theorem shows that rapid convergence is obtained for $k \ge \|A\|_2^{1/2}$. Similar convergence bounds for $f(x) = e^x$ are given by Hochbruck and Lubich in [291, 1997] for general A under the assumption that the field of values of A lies in a disk or a sector. However, convergence for general A and general f is poorly understood (see Problem 13.5).

We are not aware of a criterion for general f and general A for determining when the approximation (13.7) is good enough and hence when to terminate the Arnoldi process. For the exponential, an error estimate

$$\|e^A b - Q_k e^{H_k} Q_k^* b\|_2 \approx \|b\|_2 h_{k+1,k} |e_k^T e^{H_k} e_1|$$

is suggested by Saad [499, 1992, Sec. 5.2].

For Hermitian A and a given rational approximation r to f expressed in partial fraction form, Frommer and Simoncini [197, 2007] develop error bounds suitable for terminating the conjugate gradient method applied to the resulting shifted linear systems.

13.2.3. Lanczos Biorthogonalization

An alternative to the Arnoldi process for general A is Lanczos biorthogonalization. It constructs biorthogonal bases for $\mathcal{K}_m(A, v)$ and $\mathcal{K}_m(A^*, w)$ using two three-term recurrences involving matrix–vector products with A and A^*, producing two matrices $V_k, W_k \in \mathbb{C}^{n\times k}$ such that $W_k^* V_k = I_k$ and $W_k^* A V_k =: T_k$ is tridiagonal. An advantage over Arnoldi is the requirement to store fewer vectors. A major disadvantage is that the process can break down. Analogues of (13.5) and the approximation (13.7) apply to Lanczos biorthogonalization, and similar error bounds apply, though the results are weaker because the matrices V_k and W_k do not have orthonormal columns.

13.3. Quadrature

13.3.1. On the Real Line

Suppose we have a representation $f(A) = \int_0^a g(A, t)\, dt$ for some rational function g and $a \in \mathbb{R}$. Then the use of a quadrature formula

$$\int_0^a g(t)\, dt \approx \sum_{k=1}^m c_k g(t_k) \tag{13.8}$$

is natural, yielding

$$f(A)b \approx \sum_{k=1}^m g(A, t_k) b. \tag{13.9}$$

We have seen that the matrix sign function, matrix pth roots, the unitary polar factor, and the matrix logarithm all have the required integral representations with a rational function g (see (5.3), (7.1), (8.7), and (11.1)) and so are amenable to this approach. Because g is rational, (13.9) reduces to matrix–vector products and the solution of linear systems. If a Hessenberg reduction $A = QHQ^*$ can be computed then (13.9) transforms to

$$f(A)b \approx Q \sum_{k=1}^{m} g(H, t_k)d, \qquad d = Q^*b. \tag{13.10}$$

Since a Hessenberg system can be solved in $O(n^2)$ flops, as opposed to $O(n^3)$ flops for a dense system, (13.10) will be more efficient than (13.9) for sufficiently large m or sufficiently many different vectors b that need to be treated with the same A; indeed, ultimately the most efficient approach will be to employ a Schur decomposition instead of a Hessenberg decomposition.

The quadrature formula (13.8) might be a Gauss rule, a repeated rule, such as the repeated trapezium or repeated Simpson rule, or the result of applying adaptive quadrature with one of these rules. Gaussian quadrature has close connections with Padé approximation, and we noted in Section 11.4 that for the logarithm the use of Gauss–Legendre quadrature with the integral (11.1) leads to the diagonal Padé approximants to the logarithm, via (11.18). Focusing on $\log(I + X)$ for the moment, Theorem 11.6 shows that an error bound for the Padé approximant, and hence for Gaussian quadrature, is available for $\|X\| < 1$. For $\|X\| > 1$ the inverse scaling and squaring approach is not attractive in the context of $\log(A)b$, since it requires matrix square roots. However, adaptive quadrature can be used and automatically chooses the location and number of the quadrature points in order to achieve the desired accuracy. For more details, including numerical experiments, see Davies and Higham [136, 2005].

13.3.2. Contour Integration

For general f we can represent $y = f(A)b$ using the Cauchy integral formula (1.12):

$$y = \frac{1}{2\pi i} \int_\Gamma f(z)(zI - A)^{-1}b \, dz, \tag{13.11}$$

where f is analytic on and inside a closed contour Γ that encloses the spectrum of A. Suppose we take for the contour Γ a circle with centre α and radius β, $\{ z : z - \alpha = \beta e^{i\theta},\ 0 \le \theta \le 2\pi \}$, and then approximate the integral using the repeated trapezium rule. Using $dz = i\beta e^{i\theta} d\theta = id\theta(z(\theta) - \alpha)$ and writing the integrand in (13.11) as $g(z)$, we obtain

$$\int_\Gamma g(z)dz = i \int_0^{2\pi} (z(\theta) - \alpha)g(z(\theta))\, d\theta. \tag{13.12}$$

The integral in (13.12) is a periodic function of θ with period 2π. Applying the m-point repeated trapezium rule to (13.12) gives

$$\int_\Gamma g(z)\, dz \approx \frac{2\pi i}{m} \sum_{k=0}^{m-1} (z_k - \alpha) g(z_k),$$

where $z_k - \alpha = \beta e^{2\pi k i/m}$, that is, $z_0, \dots, z_m$ are equally spaced points on the contour Γ (note that since Γ is a circle we have $z_0 = z_m$). When A is real and we take α real it suffices to use just the z_k in the upper half-plane and then take the real part of the result. The attraction of the trapezium rule is that it is exponentially accurate when applied to an analytic integrand on a periodic domain [142, 1984, Sec. 4.6.5], [571, 2007].

Although this is a natural approach, it is in general very inefficient unless A is well conditioned, as shown by Davies and Higham [136, 2005] and illustrated below. The problem is that Γ is contained in a very narrow annulus of analyticity. However, by using a suitable change of variables in the complex plane, carefully constructed based on knowledge of the extreme points of the spectrum and any branch cuts or singularities of f, quadrature applied to the Cauchy integral can be very effective, as shown by Hale, Higham, and Trefethen [240, 2007].

For illustration consider the computation of $A^{1/2}$ for A the 5×5 Pascal matrix, which is symmetric positive definite with spectrum on the interval $[0.01, 92.3]$. First, we choose for Γ the circle with centre $(\lambda_{\min} + \lambda_{\max})/2$ and radius $\lambda_{\max}/2$, which is a compromise between enclosing the spectrum and avoiding the negative real axis. We find that about 32,000 and 262,000 points are needed to provide 2 and 13 decimal digits of accuracy, respectively. As an alternative we instead conformally map the region of analyticity of f and the resolvent to an annulus, with the interval containing the spectrum mapping to the inner boundary circle and the negative real axis mapping to the outer boundary circle. Then we apply the trapezium rule over a circle in the annulus. With the mapping constructed as described in [240, 2007], we need just 5 and 35 points to provide 2 and 13 decimal digits of accuracy, respectively—a massive improvement, due to the enlarged annulus of analyticity. By further exploiting the properties of the square root function these numbers of points can be reduced even more, with just 20 points yielding 14 digits. Theorems are available to explain this remarkably fast (geometric) convergence [240, 2007].

When a quadrature rule is applied to a (transformed) Cauchy integral it produces an answer that can be interpreted as the exact integral of a rational function whose poles are the nodes and whose residues are the weights. It is therefore natural to ask what rational approximations the trapezium rule is producing in the above example. It turns out the last of the approaches mentioned above reproduces what is essentially the best rational approximation to $x^{-1/2}$ in Theorem 5.15 identified by Zolotarev. However, conformal mapping and the trapezium rule can be applied to arbitrary functions and to matrices with spectra that are not all real, for which associated best rational approximations are usually not known.

Numerical evaluation of the ψ functions (see Section 10.7.4) via contour integrals, in the context of exponential integrators, is discussed by Kassam and Trefethen [336, 2005] and Schmelzer and Trefethen [503, 2006]. For the case of the exponential and matrices with negative real eigenvalues, see Trefethen, Weideman, and Schmelzer [574, 2006].

13.4. Differential Equations

Another approach to computing $y = f(A)b$ is to express y as the solution of an initial value differential equation problem and apply an ODE solver. As an example, consider $f(A) = A^\alpha$ for real α.

Theorem 13.7 (Davies and Higham). *For $A \in \mathbb{C}^{n\times n}$ with no eigenvalues on $\mathbb{R}^-$ and $\alpha \in \mathbb{R}$, the initial value ODE problem.*

$$\frac{dy}{dt} = \alpha(A-I)\big[t(A-I)+I\big]^{-1}y, \qquad y(0)=b, \qquad 0 \le t \le 1, \tag{13.13}$$

has the unique solution $y(t) = \big[t(A-I)+I\big]^{\alpha}b$, *and hence* $y(1) = A^{\alpha}b$.

Proof. The existence of a unique solution follows from the fact that the ODE satisfies a Lipschitz condition with Lipschitz constant $\sup_{0\le t\le 1}\|(A-I)\big[t(A-I)+I\big]^{-1}\| < \infty$. It is easy to check that $y(t)$ is this solution. $\square$

Thus $y(1) = A^{\alpha}b$ can be obtained by applying an ODE solver to (13.13). The initial value problem can potentially be stiff, depending on α, the matrix A, and the requested accuracy, so some care is needed in choosing a solver. Again, a Hessenberg or Schur reduction of A can be used to reduce the cost of evaluating the differential equation.

13.5. Other Methods

In this section we briefly outline some other approaches to the estimation or computation of $f(A)b$.

Another class of methods forms approximations $f(A)b \approx p_k(A)b$ where p_k is a polynomial obtained by truncating an expansion of f in terms of some suitable infinite sequence of polynomials satisfying a short (usually three-term) recurrence. This class of approximations is well developed in the case of linear systems. The use of Chebyshev series for Hermitian matrices was suggested by Druskin and Knizhnerman [167, 1989]. For non-Hermitian matrices the Faber series are more appropriate, and their use has been proposed by Moret and Novati [440, 2001]. Novati [450, 2003] chooses p_k as an interpolant to f at the Fejér points. These methods all need knowledge of the extremal eigenvalues of A.

13.6. Notes and References

The original reference for the Arnoldi process is Arnoldi [18, 1951].

Our treatment of Krylov subspace methods is necessarily very brief. We give here a few pointers to the literature, without trying to cite all the relevant papers.

Excellent general references on Krylov subspace methods include Bai, Demmel, Dongarra, Ruhe, and Van der Vorst [30, 2000], Golub and Van Loan [224, 1996, Chap. 9], Saad [500, 2003, Chaps. 6, 7], Watkins [608, 2007, Chap. 9].

The earliest uses in the numerical analysis literature of the approximation (13.7) appear to be by Knizhnerman [356, 1991] for general functions and by Gallopoulos and Saad [200, 1992], [498, 1992] for the exponential.

Lemma 13.4 is from Saad [498, 1992, Lem. 3.1]. Theorem 13.5 is a trivial generalization of Saad [498, 1992, Thm. 3.3], which applies to $f(x) = e^x$.

Error analysis of the modified Gram–Schmidt process is summarized in Higham [276, 2002, Sec. 19.8]. For classical Gram–Schmidt see Smoktunowicz, Barlow, and Langou [532, 2006]. For the Householder-based implementation of the Arnoldi process see Saad [500, 2003, Sec. 6.3.2]. Reorthogonalization is described in Stewart [538, 2001, Sec. 5.1] and in [30, 2000].

For more on Lanczos biorthogonalization see Saad [500, 2003, Chap. 7] or Watkins [608, 2007, Sec. 9.6].

An early treatment of the $f(A)b$ problem is that of Van Der Vorst [587, 1987]; see also [588, 2000] and [589, 2003, Chap. 11].

For a summary of a $\mathrm{sign}(A)b$ problem arising in lattice QCD see Section 2.7.

The effects of finite precision arithmetic on approximation of $f(A)b$ for Hermitian A by the Lanczos approach is analyzed by Druskin, Greenbaum, and Knizhnerman [166, 1998]. A survey of matrix functions with particular emphasis on Krylov methods is given by Frommer and Simoncini [198, 2008].

Theorem 13.7 is from Davies and Higham [136, 2005]. In the case $\alpha = 1/2$ and A symmetric positive definite, the theorem and its use with an ODE initial value solver are given by Allen, Baglama and Boyd [10, 2000], who describe an application in the numerical solution of stochastic differential equations. Another differential equations application in which $A^{\pm 1/2}b$ is needed for symmetric positive definite A is described by Hughes, Levit, and Winget [304, 1983].

Problems

13.1. Show that after $m = \deg \psi_{A,b}$ steps of the Arnoldi process with $q_1 = b/\|b\|_2$ we have $f_m = \|b\|_2 Q f(H_m) e_1 \equiv f(A)b$.

13.2. Show that $\dim(\mathcal{K}_m(A,b)) = \min(m, \deg \psi_{A,b})$.

13.3. Show that if $H \in \mathbb{C}^{n \times n}$ is upper Hessenberg with $h_{j+1,j} \neq 0$ for all j then H is nonderogatory, that is, no eigenvalue appears in more than one Jordan block.

13.4. (T. Lippert, communicated by A. Frommer) The following recursive algorithm computes $x = X_k b$, where X_k is the kth iterate from the Newton–Schulz iteration (5.22) for the matrix sign function, using only matrix–vector products.

1 function $x = NS(A, b, k)$
2 if $k = 1$
3 $\quad x = \frac{1}{2}A(3b - A(Ab))$
4 else
5 $\quad x = NS(A, b, k-1)$
6 $\quad x = NS(A, x, k-1)$
7 $\quad x = 3b - x$
8 $\quad x = \frac{1}{2}NS(A, x, k-1)$
9 end

Explain why the algorithm works and evaluate its cost, comparing it with the cost of computing X_k and then forming $X_k b$.

13.5. (RESEARCH PROBLEM) Obtain error bounds for the Arnoldi approximation (13.7) for general A and general f.

13.6. (RESEARCH PROBLEM) For $f(x) = e^x$ the Arnoldi approximation (13.7) requires the computation of e^H, where H is upper Hessenberg. Can Algorithm 10.20 (scaling and squaring) be usefully adapted to take advantage of the Hessenberg structure?

The basic idea of the Krylov subspace techniques considered in this paper is to approximately project the exponential of the large matrix onto a small Krylov subspace. The only matrix exponential operation performed is therefore with a much smaller matrix.

— Y. SAAD, *Analysis of Some Krylov Subspace Approximations to the Matrix Exponential Operator* (1992)

We can consider each Krylov subspace method *as theoretically made from two main parts, a Krylov subspace* process, *and a* subproblem solution.

— C. C. PAIGE, *Krylov Subspace Processes, Krylov Subspace Methods, and Iteration Polynomials* (1994)

The Gaussian rules were out of favor in the days of paper-and-pencil scientific computation, as the numbers involved were helter-skelter irrational decimals, impossible to remember and difficult to enter on a keyboard without error.

— PHILIP J. DAVIS and AVIEZRI S. FRAENKEL, *Remembering Philip Rabinowitz* (2007)

Chapter 14
Miscellany

This final chapter treats a few miscellaneous topics that do not fit elsewhere in the book.

14.1. Structured Matrices

In many applications the matrices that arise are structured and when structure is present it is natural to try to exploit it. The benefits to be gained include faster and more accurate algorithms and reduced storage, as well as more physically meaningful solutions in the presence of rounding and truncation errors. The interaction of structure with matrix functions is a deep and fascinating subject. A great deal is known, but many open questions remain. A thorough treatment of structure is outside the scope of this book, and indeed no single book could adequately cover all its aspects. In this section we mention just a few examples of structured $f(A)$ problems and give pointers to the literature. The following list is not meant to be complete, but is intended to give a flavor of this very active area of research.

14.1.1. Algebras and Groups

If $A \in \mathbb{C}^{n\times n}$ is unitary then $A^{-1} = A^*$. Assuming that A has no eigenvalues on $\mathbb{R}^-$, we can take the (principal) square root in this relation to obtain

$$A^{-1/2} = (A^{-1})^{1/2} = (A^*)^{1/2} = (A^{1/2})^*,$$

which shows that $A^{1/2}$ is also unitary. Therefore the square root function preserves the property of being unitary. However, underlying this result is a much more general one. Suppose that $M^{-1}A^*M = A^{-1}$ for some nonsingular $M \in \mathbb{C}^{n\times n}$. Then we have

$$A^{-1/2} = \left(M^{-1}A^*M\right)^{1/2} = M^{-1}\left(A^*\right)^{1/2}M = M^{-1}\left(A^{1/2}\right)^*M, \tag{14.1}$$

which shows that $A^{1/2}$ satisfies a relation of the same form. Thus the same proof has shown preservation of structure for a much wider class of matrices. This line of thinking suggests that there are gains to be made by carrying out analysis in a suitably general setting. A setting that has proved very fruitful is a scalar product space and its associated structures.

Let $\mathbb{K} = \mathbb{R}$ or $\mathbb{C}$ and consider a scalar product on $\mathbb{K}^n$, that is, a bilinear or sesquilinear form $\langle\cdot,\cdot\rangle_{\mathrm{M}}$ defined by any nonsingular matrix M: for $x, y \in \mathbb{K}^n$,

$$\langle x,y\rangle_{\mathrm{M}} = \begin{cases} x^TMy & \text{for real or complex bilinear forms,} \\ x^*My & \text{for sesquilinear forms.} \end{cases}$$

Table 14.1. *Structured matrices associated with some scalar products.*

$$R = \begin{bmatrix} & & 1 \\ & \cdot^{\cdot^{\cdot}} & \\ 1 & & \end{bmatrix}, \quad J = \begin{bmatrix} 0 & I_n \\ -I_n & 0 \end{bmatrix}, \quad \Sigma_{p,q} = \begin{bmatrix} I_p & 0 \\ 0 & -I_q \end{bmatrix} \text{ with } p+q=n.$$

Space	M	Automorphism Group $\mathbb{G} = \{G : G^\star = G^{-1}\}$	Jordan Algebra $\mathbb{J} = \{S : S^\star = S\}$	Lie Algebra $\mathbb{L} = \{K : K^\star = -K\}$
		Bilinear forms		
$\mathbb{R}^n$	I	Real orthogonals	Symmetrics	Skew-symmetrics
$\mathbb{C}^n$	I	Complex orthogonals	Complex symmetrics	Cplx skew-symmetrics
$\mathbb{R}^n$	$\Sigma_{p,q}$	Pseudo-orthogonals	Pseudosymmetrics	Pseudoskew-symmetrics
$\mathbb{C}^n$	$\Sigma_{p,q}$	Cplx pseudo-orthogonals	Cplx pseudo-symm.	Cplx pseudo-skew-symm.
$\mathbb{R}^n$	R	Real perplectics	Persymmetrics	Perskew-symmetrics
$\mathbb{R}^{2n}$	J	Real symplectics	Skew-Hamiltonians	Hamiltonians
$\mathbb{C}^{2n}$	J	Complex symplectics	Cplx J-skew-symm.	Complex J-symmetrics
		Sesquilinear forms		
$\mathbb{C}^n$	I	Unitaries	Hermitian	Skew-Hermitian
$\mathbb{C}^n$	$\Sigma_{p,q}$	Pseudo-unitaries	Pseudo-Hermitian	Pseudoskew-Hermitian
$\mathbb{C}^{2n}$	J	Conjugate symplectics	J-skew-Hermitian	J-Hermitian

The *adjoint* of A with respect to the scalar product $\langle\cdot,\cdot\rangle_{\mathrm{M}}$, denoted by $A^\star$, is uniquely defined by the property $\langle Ax, y\rangle_{\mathrm{M}} = \langle x, A^\star y\rangle_{\mathrm{M}}$ for all x, $y \in \mathbb{K}^n$. It can be shown that the adjoint is given explicitly by

$$A^\star = \begin{cases} M^{-1}A^TM & \text{for bilinear forms,} \\ M^{-1}A^*M & \text{for sesquilinear forms.} \end{cases} \tag{14.2}$$

Associated with $\langle\cdot,\cdot\rangle_{\mathrm{M}}$ is an *automorphism group* $\mathbb{G}$, a *Lie algebra* $\mathbb{L}$, and a *Jordan algebra* $\mathbb{J}$, which are the subsets of $\mathbb{K}^{n\times n}$ defined by

$$\mathbb{G} := \{G : \langle Gx, Gy\rangle_{\mathrm{M}} = \langle x, y\rangle_{\mathrm{M}} \ \forall x, y \in \mathbb{K}^n\} = \{G : G^\star = G^{-1}\}, \tag{14.3}$$

$$\mathbb{L} := \{L : \langle Lx, y\rangle_{\mathrm{M}} = -\langle x, Ly\rangle_{\mathrm{M}} \ \forall x, y \in \mathbb{K}^n\} = \{L : L^\star = -L\}, \tag{14.4}$$

$$\mathbb{J} := \{S : \langle Sx, y\rangle_{\mathrm{M}} = \langle x, Sy\rangle_{\mathrm{M}} \ \forall x, y \in \mathbb{K}^n\} = \{S : S^\star = S\}. \tag{14.5}$$

$\mathbb{G}$ is a multiplicative group, while $\mathbb{L}$ and $\mathbb{J}$ are linear subspaces. Table 14.1 shows a sample of well-known structured matrices in $\mathbb{G}$, $\mathbb{L}$, or $\mathbb{J}$ associated with some scalar products.

In this language, (14.1) and its analog with "$*$" replaced by "T", shows that $A \in \mathbb{G}$ implies $A^{1/2} \in \mathbb{G}$, that is, the square root function preserves matrix automorphism groups. One can ask which other functions preserve $\mathbb{G}$. A thorough investigation of this question is given by Higham, Mackey, Mackey, and Tisseur [283, 2005].

Mappings between $\mathbb{G}$, $\mathbb{L}$, and $\mathbb{J}$ provide examples of interplay between these structures. For example, if $A \in \mathbb{L}$ then $X = e^A \in \mathbb{G}$, because $X^\star = (e^A)^\star = e^{A^\star} = e^{-A} = X^{-1}$. This generalizes the familiar property that the exponential of a skew-Hermitian matrix is unitary. A more general setting in which similar questions can be considered is a Lie group and its associated Lie algebra (not necessarily arising from a scalar product), and the observation just made is a special case of the well-known one that the exponential map takes the Lie algebra into the corresponding Lie group. This mapping is important in the numerical solution of ODEs on Lie groups by geometric integration methods. For details, see Hairer, Lubich, and Wanner [239, 2002], Iserles, Munthe-Kaas, Nørsett, and Zanna [313, 2000], and Iserles and Zanna [314, 2005].

Two particularly important classes in Table 14.1 are the Hamiltonian matrices and the symplectic matrices. Results on logarithms of such matrices can be found in Dieci [152, 1996] and [153, 1998]. The result that every real skew-Hamiltonian matrix has a real Hamiltonian square root is proved by Faßbender, Mackey, Mackey, and Xu [182, 1999], while Ikramov proves the corresponding result for complex matrices [308, 2001].

Some of the techniques developed in these contexts can be applied even more generally. For example, $A \in \mathbb{C}^{n\times n}$ is centrosymmetric if $JAJ = A$ (where J is defined in Table 14.1) and by the same argument as above, $JA^{1/2}J = A^{1/2}$, so the square root function preserves centrosymmetry.

Perturbation analysis of matrix functions can be done in such way as to restrict perturbations to those that maintain the structure of the matrices being perturbed, yielding structured condition numbers. Such analysis has been carried out by Davies for the Jordan and Lie algebras $\mathbb{J}$ and $\mathbb{L}$ [134, 2004].

The polar decomposition $A = UH$ interacts with the above structures in an interesting way. First, various results can be proved about how U and H inherit structure from A. Second, the polar decomposition can be generalized to a decomposition $A = UH$ in which $U \in \mathbb{G}$ and $H \in \mathbb{J}$, subject to suitable conditions on A. For details and further references see Higham [277, 2003], Higham, Mackey, Mackey, and Tisseur [282, 2004], [283, 2005], and Mackey, Mackey, and Tisseur [400, 2006] (see, in particular, the discussion in Section 6), and for similar considerations in a Lie group setting see Iserles and Zanna [314, 2005] and Munthe-Kaas, Quispel, and Zanna [443, 2001].

14.1.2. Monotone Functions

A function $f : \mathbb{C}^{n\times n} \to \mathbb{C}^{n\times n}$ is *nonnegative* if $f(A) \ge 0$ whenever $A \ge 0$ and *monotone* if $f(A) \ge f(B)$ whenever $A \ge B$, where the ordering is the positive semidefinite ordering on Hermitian matrices (see Section B.12). Much is known about functions that are monotone or nonnegative, and this area overlaps substantially with the topic of matrix inequalities. A sample result is that the function $f(A) = A^r$ is monotone for $r \in [0, 1]$. We cite just the books by Donoghue [163, 1974] and Bhatia [64, 1997], [65, 2007], which contain many further references to the literature.

14.1.3. Other Structures

A *Stieltjes matrix* is a symmetric positive definite matrix $A \in \mathbb{R}^{n\times n}$ such that $a_{ij} \le 0$ for $i \ne j$. Micchelli and Willoughby [427, 1979] characterize functions f that preserve the class of Stieltjes matrices.

$A \in \mathbb{C}^{n\times n}$ is *coninvolutory* if $A\overline{A} = I$. Horn and Johnson [296, 1991, Cor. 6.4.22] shows that every coninvolutory matrix has a coninvolutory square root.

Toeplitz matrix structure is not in general preserved by matrix functions. But a primary matrix function of a triangular Toeplitz matrix is again triangular and Toeplitz, which follows from the facts that $f(A)$ is a polynomial in A and that the sum and product of two triangular Toeplitz matrices are triangular and Toeplitz.

More generally, a primary matrix function of a circulant matrix is a circulant. This follows from the fact that circulant matrices $C \in \mathbb{C}^{n\times n}$ are precisely the matrices diagonalized by the discrete Fourier transform (DFT) matrix (1.17):

$$F_n C F_n^{-1} = D = \mathrm{diag}(d_i).$$

Hence $f(C) = F_n^{-1} f(D) F_n$ is a circulant.

14.1.4. Data Sparse Representations

Some other types of structure—very different to those considered in the previous sections—can be put under the heading of "data sparse", which describes a matrix that is not necessarily sparse in the usual sense but can be represented by a relatively small number of parameters. One such class of matrices is the *semiseparable matrices*, which can be defined as the matrices whose upper and lower triangles agree with the corresponding triangles of two rank r matrices, where r is the semiseparability rank [597, 2005]. An example of such a matrix is the inverse of a tridiagonal matrix having no zeros on the subdiagonal or superdiagonal, which is semiseparable with $r = 1$ (see, e.g., Meurant [425, 1992]). There are other ways in which low rank blocks can render a matrix data sparse, and it may be sufficient that low rank approximations can be found. The hierarchical, $\mathcal{H}$-matrix format is a recursive representation that employs low rank approximations, and matrices arising in certain applications are well suited to this representation; for an overview see Börm, Grasedyck, and Hackbusch [76, 2003].

14.1.5. Computing Structured $f(A)$ for Structured A

Assuming we know that A and $f(A)$ are both structured (possibly with different, but related structures), what methods are available that exploit the structure? This is a very general question about which much is known. We give a small selection of references.

For algorithms for computing the square root or logarithm of orthogonal and unitary matrices, or more generally matrices in a group $\mathbb{G}$, see Cardoso, Kenney, and Silva Leite [93, 2003], Cheng, Higham, Kenney, and Laub [107, 2000], and Higham, Mackey, Mackey, and Tisseur [283, 2005].

If f preserves a certain structure, do the Padé approximants to f also preserve the structure? It is well known that diagonal Padé approximants r_m to the exponential have the property that $r(A)$ is unitary if A is skew-Hermitian. Dieci [152, 1996, Thm. 2.2] shows a kind of converse: that diagonal Padé approximants to the logarithm yield skew-Hermitian matrices when evaluated at unitary matrices.

Computing functions of matrices stored in a data sparse format in a way that exploits the structure is relatively little explored to date, other than of course for the matrix inverse. For some work on functions of $\mathcal{H}$-matrices see Baur and Benner [47, 2006], Gavrilyuk, Hackbusch, and Khoromskij [209, 2002], and Grasedyck, Hackbusch, and Khoromskij [228, 2003].

14.2. Exponential Decay of Functions of Banded Matrices

Let A be the 6×6 symmetric tridiagonal matrix

$$A = \begin{bmatrix} 4 & 1 & 0 & 0 & 0 & 0 \\ 1 & 4 & 1 & 0 & 0 & 0 \\ 0 & 1 & 4 & 1 & 0 & 0 \\ 0 & 0 & 1 & 4 & 1 & 0 \\ 0 & 0 & 0 & 1 & 4 & 1 \\ 0 & 0 & 0 & 0 & 1 & 4 \end{bmatrix}.$$

The magnitudes of three functions of A are as follows, where the elements are shown to one significant figure:

$$e^A = \begin{bmatrix} \text{9e+1} & \text{8e+1} & \text{3e+1} & \text{1e+1} & \text{3e+0} & \text{5e-1} \\ \text{8e+1} & \text{1e+2} & \text{9e+1} & \text{4e+1} & \text{1e+1} & \text{3e+0} \\ \text{3e+1} & \text{9e+1} & \text{1e+2} & \text{9e+1} & \text{4e+1} & \text{1e+1} \\ \text{1e+1} & \text{4e+1} & \text{9e+1} & \text{1e+2} & \text{9e+1} & \text{3e+1} \\ \text{3e+0} & \text{1e+1} & \text{4e+1} & \text{9e+1} & \text{1e+2} & \text{8e+1} \\ \text{5e-1} & \text{3e+0} & \text{1e+1} & \text{3e+1} & \text{8e+1} & \text{9e+1} \end{bmatrix},$$

$$\log(A) = \begin{bmatrix} \text{1e+0} & \text{3e-1} & \text{3e-2} & \text{6e-3} & \text{1e-3} & \text{2e-4} \\ \text{3e-1} & \text{1e+0} & \text{3e-1} & \text{4e-2} & \text{6e-3} & \text{1e-3} \\ \text{3e-2} & \text{3e-1} & \text{1e+0} & \text{3e-1} & \text{4e-2} & \text{6e-3} \\ \text{6e-3} & \text{4e-2} & \text{3e-1} & \text{1e+0} & \text{3e-1} & \text{3e-2} \\ \text{1e-3} & \text{6e-3} & \text{4e-2} & \text{3e-1} & \text{1e+0} & \text{3e-1} \\ \text{2e-4} & \text{1e-3} & \text{6e-3} & \text{3e-2} & \text{3e-1} & \text{1e+0} \end{bmatrix},$$

$$A^{1/2} = \begin{bmatrix} \text{2e+0} & \text{3e-1} & \text{2e-2} & \text{2e-3} & \text{4e-4} & \text{7e-5} \\ \text{3e-1} & \text{2e+0} & \text{3e-1} & \text{2e-2} & \text{2e-3} & \text{4e-4} \\ \text{2e-2} & \text{3e-1} & \text{2e+0} & \text{3e-1} & \text{2e-2} & \text{2e-3} \\ \text{2e-3} & \text{2e-2} & \text{3e-1} & \text{2e+0} & \text{3e-1} & \text{2e-2} \\ \text{4e-4} & \text{2e-3} & \text{2e-2} & \text{3e-1} & \text{2e+0} & \text{3e-1} \\ \text{7e-5} & \text{4e-4} & \text{2e-3} & \text{2e-2} & \text{3e-1} & \text{2e+0} \end{bmatrix}.$$

In all three cases there is decay of the elements away from the main diagonal, the rate of decay depending on the function. This is a general phenomenon for symmetric band matrices and is not limited to matrices that are diagonally dominant.

Theorem 14.1. *Let f be analytic in an ellipse containing $\Lambda(A)$ and let $A \in \mathbb{C}^{n\times n}$ be Hermitian and of bandwidth m ($a_{ij} = 0$ for $|i-j| > m$). Then $f(A) = (f_{ij})$ satisfies $|f_{ij}| \le C\rho^{|i-j|}$, where C is a constant and $\rho = q^{1/m}$, where $q \in (0,1)$ depends only on f.*

Proof. See Benzi and Golub [57, 1999]. □

The theorem shows that the elements of $f(A)$ are bounded in an exponentially decaying manner away from the diagonal, with the bound decreasing as the bandwidth decreases. Note that this does not necessarily mean that "decay to zero" is observed in practice, as the following example illustrates:

$$A = \begin{bmatrix} 0 & 0.5 & 0 & 0 & 0 & 0 \\ 0.5 & 1 & 3 & 0 & 0 & 0 \\ 0 & 3 & 1 & 1 & 0 & 0 \\ 0 & 0 & 1 & 1 & 10 & 0 \\ 0 & 0 & 0 & 10 & 1 & 10 \\ 0 & 0 & 0 & 0 & 10 & 2 \end{bmatrix}, \quad e^A = \begin{bmatrix} \text{2e+0} & \text{1e+1} & \text{5e+1} & \text{6e+2} & \text{8e+2} & \text{6e+2} \\ \text{1e+1} & \text{3e+2} & \text{1e+3} & \text{2e+4} & \text{3e+4} & \text{2e+4} \\ \text{5e+1} & \text{1e+3} & \text{6e+3} & \text{9e+4} & \text{1e+5} & \text{9e+4} \\ \text{6e+2} & \text{2e+4} & \text{9e+4} & \text{1e+6} & \text{2e+6} & \text{1e+6} \\ \text{8e+2} & \text{3e+4} & \text{1e+5} & \text{2e+6} & \text{2e+6} & \text{2e+6} \\ \text{6e+2} & \text{2e+4} & \text{9e+4} & \text{1e+6} & \text{2e+6} & \text{1e+6} \end{bmatrix}$$

Further decay results applying to general A and with a certain "graph theoretic distance" replacing $|i-j|$ in the exponent are provided by Benzi and Razouk [58, 2007].

Bounds of the same form as in Theorem 14.1 specialized to the exponential and for general A are obtained by Iserles [312, 2000].

14.3. Approximating Entries of Matrix Functions

A variation on the $f(A)$ problem is to approximate not $f(A)$ or $f(A)b$ but rather individual entries of $f(A)$, or perhaps a function of those entries such as a bilinear form $g(u,v) = u^T f(A)v$. For $u = e_i$ and $v = e_j$, $g(u,v) = f(A)_{ij}$.

An approach to this problem developed by Golub and his collaborators for Hermitian matrices expresses $g(u,v)$ as a Riemann–Stieltjes integral with respect to a suitable measure and approximates the integral by Gauss-type quadrature rules. The Lanczos process provides an efficient means to construct these rules using only matrix–vector products with A. See Bai, Fahey, and Golub [32, 1996], Bai, Fahey, Golub, Menon, and Richter [33, 1998], Bai and Golub [34, 1997], [35, 1999], and Golub and Meurant [223, 1994].

Extensions to non-Hermitian matrices using the Arnoldi process have been developed by Calvetti, Kim, and Reichel [91, 2005] and Guo and Renaut [235, 2004].

Appendix A
Notation

Page numbers denote the pages on which a definition can be found.

$A \geq B$	componentwise inequality, p. 329; also positive (semi)definite ordering (Löwner (partial) ordering), p. 330
$A^{1/2}$	the principal square root, p. 20
$\sqrt{A}$	(arbitrary) square root, p. 133
$A^{1/p}$	principal pth root, p. 174
A^{+}	pseudoinverse, p. 325
$\lceil x \rceil$	ceiling function, p. 321
$\mathbb{C}$	the complex numbers
$\mathrm{cond}_{\mathrm{abs}}$	absolute condition number, p. 56
$\mathrm{cond}_{\mathrm{rel}}$	relative condition number, p. 55
conv	convex hull, p. 102
$\det(A)$	determinant
e (vector)	$[1, 1, \ldots, 1]^T$
$\{\lambda_i(A)\}$	the set of eigenvalues of A
$f[\ldots]$	divided difference, p. 332
$\#$	geometric mean, p. 46
$\circ$	Hadamard (Schur) product, p. 62; also composition of functions
$\lfloor x \rfloor$	floor function, p. 321
$\kappa(A)$	matrix condition number, p. 328
$\otimes$	Kronecker product, p. 331
$\oplus$	Kronecker sum, p. 331
L_f	Fréchet derivative of f, p. 56
$\Lambda(A)$	spectrum (set of eigenvalues), p. 322
$\Lambda_\epsilon(A)$	ϵ-pseudospectrum, p. 47
$\log(A)$	principal logarithm, p. 20
n_i	the index of an eigenvalue λ_i, pp. 3, 322
$\mathrm{null}(A)$	null space of matrix, p. 321
$\mathbb{R}$	the real numbers
$\mathbb{R}^-$	closed negative real axis
$\mathrm{range}(A)$	range of matrix, p. 321
$\rho(A)$	spectral radius, p. 322
$\mathrm{sign}(A)$	matrix sign function, pp. 39, 107
$\mathrm{trace}(A)$	trace of matrix, p. 321
vec	vec operator, p. 331
$\mathbb{Z}$	the integers

Appendix B
Background: Definitions and Useful Facts

This appendix collects together a variety of basic definitions, properties, and results from matrix analysis and numerical analysis that are needed in the main text. Terms being defined are in boldface. Where specific references are not given, general sources of further information are Horn and Johnson [295, 1985] and Lancaster and Tismenetsky [371, 1985].

B.1. Basic Notation

$\mathbb{R}$ denotes the **real numbers** and $\mathbb{R}^-$ the **closed negative real axis.** $\mathbb{C}$ denotes the **complex numbers.** $\mathbb{Z}$ denotes the **integers.**

The **real** and **imaginary parts** of a complex number z are denoted by $\operatorname{Re} z$ and $\operatorname{Im} z$, respectively.

The **identity matrix** is $I_{m,n} = (\delta_{ij}) \in \mathbb{R}^{m\times n}$ and $I_n \equiv I_{n,n}$. The ith column of I_n is denoted by e_i. Similarly, the $m \times n$ zero matrix is $0_{m,n}$.

For $A = (a_{ij}) \in \mathbb{C}^{m\times n}$, A^T denotes the **transpose** (a_{ji}) and A^* denotes the **conjugate transpose** $\overline{A}^T = (\overline{a}_{ji})$.

The **trace** of $A \in \mathbb{C}^{n\times n}$ is $\operatorname{trace}(A) = \sum_{i=1}^n a_{ii}$. Two key properties are that $\operatorname{trace}(A)$ is the sum of the eigenvalues of A and that $\operatorname{trace}(AB) = \operatorname{trace}(BA)$ for all $A \in \mathbb{C}^{m\times n}$ and $B \in \mathbb{C}^{n\times m}$.

The **determinant** of $A \in \mathbb{C}^{n\times n}$ is denoted by $\det(A)$. Key properties are $\det(AB) = \det(A)\det(B)$, $\det(\alpha A) = \alpha^n \det(A)$ $(\alpha \in \mathbb{C})$, and that $\det(A)$ is the product of the eigenvalues of A.

The **floor** and **ceiling functions** are defined by $\lfloor x \rfloor = \max\{\, n \in \mathbb{Z} : n \le x \,\}$ and $\lceil x \rceil = \min\{\, n \in \mathbb{Z} : x \le n \,\}$, respectively.

The **empty set** is denoted by $\emptyset$.

The **range** of $A \in \mathbb{C}^{m\times n}$ is $\operatorname{range}(A) = \{\, Ax : x \in \mathbb{C}^n \,\}$ and the **null space** of A is $\operatorname{null}(A) = \{\, x \in \mathbb{C}^n : Ax = 0 \,\}$.

The **dimension** $\dim(V)$ of a vector space V is the maximum number of linearly independent vectors in V.

The notation $1\colon n$ denotes the sequence $1, 2, 3, \dots, n$. $A(i\colon j, r\colon s)$ denotes the submatrix of A comprising the intersection of rows i to j and columns r to s.

The notation $X = O(\|E\|)$ denotes that $\|X\| \le c\|E\|$ for some constant c for all sufficiently small $\|E\|$, while $X = o(\|E\|)$ means that $\|X\|/\|E\| \to 0$ as $E \to 0$.

B.2. Eigenvalues and Jordan Canonical Form

Let $A \in \mathbb{C}^{n\times n}$. The scalar $\lambda \in \mathbb{C}$ is an **eigenvalue** with corresponding eigenvector $x \ne 0$ if $Ax = \lambda x$. The eigenvalues are the zeros of the **characteristic polynomial**

$q(t) = \det(tI - A)$, which has degree n. The **minimal polynomial** of A is the unique monic polynomial ψ of lowest degree such that $\psi(A) = 0$.

The set of eigenvalues of A, called the **spectrum**, is denoted by $\Lambda(A) = \{\lambda_1, \ldots, \lambda_n\}$. We sometimes denote by $\lambda_i(A)$ the ith eigenvalue of A in some (usually arbitrary) ordering. The **eigenspace** of A corresponding to an eigenvalue λ is the vector space $\{ x \in \mathbb{C}^n : Ax = \lambda x \}$.

The **algebraic multiplicity** of λ is its multiplicity as a zero of the characteristic polynomial $q(t)$. The **geometric multiplicity** of λ is $\dim(\mathrm{null}(A - \lambda I))$, which is the dimension of the eigenspace of A corresponding to λ, or equivalently the number of linearly independent eigenvectors associated with λ.

Any matrix $A \in \mathbb{C}^{n\times n}$ can be expressed in the **Jordan canonical form**

$$Z^{-1}AZ = J = \mathrm{diag}(J_1, J_2, \ldots, J_p), \tag{B.1a}$$

$$J_k = J_k(\lambda_k) = \begin{bmatrix} \lambda_k & 1 & & \\ & \lambda_k & \ddots & \\ & & \ddots & 1 \\ & & & \lambda_k \end{bmatrix} \in \mathbb{C}^{m_k \times m_k}, \tag{B.1b}$$

where Z is nonsingular and $m_1 + m_2 + \cdots + m_p = n$. The λ_k are the eigenvalues of A. The matrices J_k are called **Jordan blocks**. The Jordan matrix J is unique up to the ordering of the blocks J_k, but the transforming matrix Z is not unique.

In terms of the Jordan canonical form (B.1), the algebraic multiplicity of λ is the sum of the dimensions of the Jordan blocks in which λ appears, while the geometric multiplicity is the number of Jordan blocks in which λ appears.

An eigenvalue is called **semisimple** if its algebraic and geometric multiplicities are the same or, equivalently, if it occurs only in 1×1 Jordan blocks. An eigenvalue is **defective** if is not semisimple, that is, if it appears in a Jordan block of size greater than 1, or, equivalently, if its algebraic multiplicity exceeds its geometric multiplicity. A matrix is **defective** if it has a defective eigenvalue, or, equivalently, if it does not have a complete set of linearly independent eigenvectors.

A matrix is **derogatory** if in the Jordan canonical form an eigenvalue appears in more than one Jordan block. Equivalently, a matrix is derogatory if its minimal polynomial (see page 4) has degree less than that of the characteristic polynomial. The properties of being defective and derogatory are independent. A matrix that is not derogatory is called **nonderogatory**.

A matrix is **diagonalizable** if in the Jordan canonical form the Jordan matrix J is diagonal, that is, $m_k \equiv 1$ and $p = n$ in (B.1).

The **index of an eigenvalue** is the dimension of the largest Jordan block in which it appears. The **index of a matrix** is the index of its zero eigenvalue, which can be characterized as the smallest nonnegative integer k such that $\mathrm{rank}(A^k) = \mathrm{rank}(A^{k+1})$.

The **spectral radius** $\rho(A) = \max\{ |\lambda| : \det(A - \lambda I) = 0 \}$.

The **field of values** of $A \in \mathbb{C}^{n\times n}$ is the set of all Rayleigh quotients:

$$F(A) = \left\{ \frac{z^*Az}{z^*z} : 0 \neq z \in \mathbb{C}^n \right\}.$$

The set $F(A)$ is convex, and when A is normal it is the convex hull of the eigenvalues. For a Hermitian matrix $F(A)$ is a segment of the real axis and for a skew-Hermitian matrix it is a segment of the imaginary axis.

The following eigenvalue localization theorem is often useful.

Theorem B.1 (Gershgorin, 1931). *The eigenvalues of $A \in \mathbb{C}^{n\times n}$ lie in the union of the n disks in the complex plane*

$$D_i = \Big\{ z \in \mathbb{C} : |z - a_{ii}| \le \sum_{\substack{j=1\\ j\ne i}}^{n} |a_{ij}| \Big\}, \qquad i = 1\colon n. \qquad \square$$

B.3. Invariant Subspaces

A subspace $\mathcal{X}$ of $\mathbb{C}^n$ is an **invariant subspace** for A if $A\mathcal{X} \subseteq \mathcal{X}$, that is, $x \in \mathcal{X}$ implies $Ax \in \mathcal{X}$.

Let the columns of $X \in \mathbb{C}^{n\times p}$ form a basis for $\mathcal{X}$. Then $\mathcal{X}$ is an invariant subspace for A if and only if $AX = XB$ for some $B \in \mathbb{C}^{p\times p}$. When the latter equation holds, the spectrum of B is contained within that of A. More specifically (see Problem B.1), if (λ, x) is an eigenpair of A with $x \in \mathcal{X}$ then (λ, X^+x) is an eigenpair of B (where X^+ is the pseudoinverse of X, defined in Section B.6.1), while if (λ, z) is an eigenpair of B then (λ, Xz) is an eigenpair of A. References: Stewart [538, 2001, Sec. 4.1] and Watkins [607, 2002, Sec. 6.1], [608, 2007, Sec. 2.1].

B.4. Special Classes of Matrices

$T \in \mathbb{C}^{n\times n}$ is **upper (lower) triangular** if $t_{ij} = 0$ for $i > j$ ($i < j$). A triangular matrix T is **strictly triangular** if $t_{ii} = 0$ for all i. T is **upper trapezoidal** if $t_{ij} = 0$ for $i > j$. $T = (T_{ij})$ is **upper quasi-triangular** if it is block upper triangular with the diagonal blocks T_{ii} either 1×1 or 2×2.

$A \in \mathbb{C}^{2n\times 2n}$ is **Hamiltonian** if JA is Hermitian, where $J = \left[\begin{smallmatrix} 0 & I_n \\ -I_n & 0 \end{smallmatrix}\right]$, and **symplectic** if $A^TJA = J$.

Let $A \in \mathbb{C}^{n\times n}$ for the rest of this section.

A is **diagonalizable** if there exists a nonsingular X such that $X^{-1}AX = D$ is diagonal.

A is **symmetric** if $A^T = A$. A is **Hermitian** if $A^* = A$. A Hermitian matrix A is **positive definite** if $x^*Ax > 0$ for all nonzero $x \in \mathbb{C}^n$ and **positive semidefinite** if $x^*Ax \ge 0$ for all nonzero $x \in \mathbb{C}^n$. Equivalent conditions for a Hermitian A to be positive (semi)definite are that $\lambda_i(A) > 0$ for all i ($\lambda_i(A) \ge 0$ for all i).

The product of two Hermitian positive definite matrices has positive eigenvalues (proof: $AB = A^{1/2}\cdot A^{1/2}BA^{1/2}\cdot A^{-1/2}$, so AB is similar to a Hermitian positive definite matrix).

$A \in \mathbb{R}^{n\times n}$ is **orthogonal** if $A^TA = I$, that is, $A^T = A^{-1}$. $A \in \mathbb{C}^{n\times n}$ is **unitary** if $A^*A = I$, that is, $A^* = A^{-1}$.

A **permutation matrix** is a matrix obtained by reordering the columns of the identity matrix. A permutation matrix is orthogonal.

A is **normal** if $AA^* = A^*A$, or, equivalently, if A is diagonalizable by a unitary matrix. There are many other equivalent conditions for a matrix to be normal [231, 1987], [295, 1985, Sec. 2.5].

A is **nilpotent** if $A^k = 0$ for some $k \ge 0$. Such a matrix has only zero eigenvalues and its index of nilpotency is the index of the zero eigenvalue, which is the smallest k such that $A^k = 0$.

A is **involutory** if $A^2 = I$ and **idempotent** if $A^2 = A$.

We write block diagonal matrices

$$\mathrm{diag}(D_1,\dots,D_m) = \mathrm{diag}(D_i) = \begin{bmatrix} D_1 & & \\ & \ddots & \\ & & D_m \end{bmatrix},$$

which illustrates the convention that blank entries in a matrix denote zeros.

A **Toeplitz matrix** is one for which the entries along each diagonal are constant: $a_{ij} = r_{i-j}$ for some parameters r_k. A **circulant matrix** is a special type of Toeplitz matrix in which each row is a cyclic permutation one element to the right of the row above. Examples:

$$T = \begin{bmatrix} a & b & c \\ d & a & b \\ e & d & a \end{bmatrix}, \qquad C = \begin{bmatrix} a & b & c \\ c & a & b \\ b & c & a \end{bmatrix}.$$

$A \in \mathbb{C}^{n\times n}$ is **strictly diagonally dominant (by rows)** if

$$\sum_{j\neq i} |a_{ij}| < |a_{ii}|, \quad i = 1\colon n. \tag{B.2}$$

A **Vandermonde matrix** has the form

$$V = V(\alpha_1,\alpha_2,\dots,\alpha_n) = \begin{bmatrix} 1 & 1 & \dots & 1 \\ \alpha_1 & \alpha_2 & \dots & \alpha_n \\ \vdots & \vdots & \dots & \vdots \\ \alpha_1^{n-1} & \alpha_2^{n-1} & \dots & \alpha_n^{n-1} \end{bmatrix} \in \mathbb{C}^{n\times n}.$$

A linear system $V^T x = b$ is called a dual Vandermonde system. A **confluent Vandermonde matrix** V_c is obtained when some successive α_i repeat and each column corresponding to a repeated α_i is obtained by "differentiating" the previous one. For example,

$$V_c(\alpha_1,\alpha_1,\alpha_1,\alpha_2) = \begin{bmatrix} 1 & 0 & 0 & 1 \\ \alpha_1 & 1 & 0 & \alpha_2 \\ \alpha_1^2 & 2\alpha_1 & 2 & \alpha_2^2 \\ \alpha_1^3 & 3\alpha_1^2 & 6\alpha_1 & \alpha_2^3 \end{bmatrix}.$$

For more details see [276, 2002, Chap. 22].

B.5. Matrix Factorizations and Decompositions

An **LU factorization** of $A \in \mathbb{C}^{n\times n}$ is a factorization $A = LU$, where L is unit lower triangular (lower triangular with 1s on the diagonal) and U is upper triangular. In practice, unless A has special properties that guarantee the existence and numerical stability of the factorization, partial pivoting is used, which produces the factorization $PA = LU$, where P is a permutation matrix.

Every Hermitian positive definite $A \in \mathbb{C}^{n\times n}$ has a unique **Cholesky factorization** $A = R^*R$, where $R \in \mathbb{C}^{n\times n}$ is upper triangular with positive diagonal elements.

Any $A \in \mathbb{C}^{m\times n}$ with $m \geq n$ has a **QR factorization** $A = QR$, where $Q \in \mathbb{C}^{m\times m}$ is unitary and R is upper trapezoidal, that is, $R = [R_1^T\ 0]^T$, where $R_1 \in \mathbb{C}^{n\times n}$ is upper triangular.

Any $A \in \mathbb{C}^{n\times n}$ has a **Schur decomposition** $Q^*AQ = T$, where Q is unitary and T upper triangular. The eigenvalues of A appear on the diagonal of T. For each k, the leading k columns of Q span an invariant subspace of A.

Any $A \in \mathbb{R}^{n\times n}$ has a **real Schur decomposition** $Q^TAQ = R$, where Q is real orthogonal and R is real upper quasi-triangular with any 2×2 diagonal blocks having complex conjugate eigenvalues. If A is normal then the 2×2 blocks R_{ii} have the form [295, 1985, Thm. 2.5.8]

$$R_{ii} := \begin{bmatrix} a & b \\ -b & a \end{bmatrix}, \qquad b \neq 0.$$

Any Hermitian $A \in \mathbb{C}^{n\times n}$ has a **spectral decomposition** $A = Q\Lambda Q^*$, where Q is unitary and $\Lambda = \mathrm{diag}(\lambda_i)$, with λ_i the eigenvalues of A, which are real.

Any $A \in \mathbb{C}^{m\times n}$ has a **singular value decomposition (SVD)**

$$A = U\Sigma V^*, \quad \Sigma = \mathrm{diag}(\sigma_1, \sigma_2, \dots, \sigma_p) \in \mathbb{R}^{m\times n}, \quad p = \min(m, n),$$

where $U \in \mathbb{C}^{m\times m}$ and $V \in \mathbb{C}^{n\times n}$ are unitary and $\sigma_1 \geq \sigma_2 \geq \cdots \geq \sigma_p \geq 0$. The σ_i are the **singular values** of A, and they are the nonnegative square roots of the p largest eigenvalues of A^*A. The columns of U and V are the **left and right singular vectors** of A, respectively. The **rank** of A is equal to the number of nonzero singular values. If A is real, U and V can be taken to be real. The essential SVD information is contained in the compact SVD $A = U\Sigma V^*$, where $U \in \mathbb{C}^{m\times r}$, $\Sigma = \mathrm{diag}(\sigma_1, \dots, \sigma_r)$, $V \in \mathbb{C}^{n\times r}$, and $r = \mathrm{rank}(A)$. The extent to which the SVD is unique is described in Problem B.11.

For more details on the above factorizations and decompositions see Demmel [145, 1997], Higham [276, 2002], Stewart [538, 2001], Trefethen and Bau [572, 1997], and Watkins [607, 2002].

B.6. Pseudoinverse and Orthogonality

B.6.1. Pseudoinverse

The **pseudoinverse** $X \in \mathbb{C}^{n\times m}$ of $A \in \mathbb{C}^{m\times n}$ is the unique matrix satisfying the four **Moore–Penrose conditions**

$$\begin{array}{ll} \text{(i)} \ \ AXA = A, & \text{(ii)} \ \ XAX = X, \\ \text{(iii)} \ \ AX = (AX)^*, & \text{(iv)} \ \ XA = (XA)^*. \end{array} \tag{B.3}$$

The pseudoinverse is denoted by A^+. If $A = U\Sigma V^*$ is an SVD then $A^+ = V\Sigma^+U^*$, where $\Sigma^+ = \mathrm{diag}(\sigma_1^{-1}, \dots, \sigma_r^{-1}, 0, \dots, 0) \in \mathbb{R}^{n\times m}$ and $r = \mathrm{rank}(A)$. When A has rank n, $A^+ = (A^*A)^{-1}A^*$.

In general it is *not* the case that $(AB)^+ = B^+A^+$ for $A \in \mathbb{C}^{m\times n}$, $B \in \mathbb{C}^{n\times p}$. A sufficient condition for this equality to hold is that $\mathrm{rank}(A) = \mathrm{rank}(B) = n$; see, e.g., Campbell and Meyer [92, 1979, Sec. 1.4].

For any $A \in \mathbb{C}^{m\times n}$ we have (see Problem B.3)

$$\mathrm{range}(A^*A) = \mathrm{range}(A^*). \tag{B.4}$$

For more on the pseudoinverse see any textbook on numerical linear algebra or Ben-Israel and Greville [52, 2003].

B.6.2. Projector and Orthogonal Projector

Let S be a subspace of $\mathbb{C}^m$ and let $P_S \in \mathbb{C}^{m\times m}$.

- P_S is the **projector** onto S if $\mathrm{range}(P_S) = S$ and $P_S^2 = P_S$. The projector is not unique.

- P_S is the **orthogonal projector** onto S if $\mathrm{range}(P_S) = S$, $P_S^2 = P_S$, and $P_S^* = P_S$. The orthogonal projector is unique (see Problem B.9). In terms of the pseudoinverse, $P_{\mathrm{range}(A)} = AA^+$ and $P_{\mathrm{range}(A^*)} = A^+A$ (see Problem B.10).

An excellent treatment of projectors can be found in Meyer [426, 2000, p. 386].

B.6.3. Partial Isometry

$U \in \mathbb{C}^{m\times n}$ is a **partial isometry** (or subunitary matrix) if $\|Ux\|_2 = \|x\|_2$ for all $x \in \mathrm{range}(U^*)$; in words: U is norm-preserving on the orthogonal complement of its null space. Several equivalent conditions for a matrix to be a partial isometry are collected in the following result.

Lemma B.2. *For $U \in \mathbb{C}^{m\times n}$ each of the following conditions is equivalent to U being a partial isometry:*

(a) $U^+ = U^*$,

(b) $UU^*U = U$,

(c) *the singular values of U are all 0 or 1.*

Proof. The equivalences are straightforward and can be obtained, for example, using Problem B.4. □

For more about partial isometries see Erdelyi [177, 1966] and Campbell and Meyer [92, 1979, Chap. 4].

B.7. Norms

A **matrix norm** on $\mathbb{C}^{m\times n}$ is a function $\|\cdot\| : \mathbb{C}^{m\times n} \to \mathbb{R}$ satisfying the following conditions:

1. $\|A\| \ge 0$ with equality if and only if $A = 0$.

2. $\|\alpha A\| = |\alpha|\,\|A\|$ for all $\alpha \in \mathbb{C}$, $A \in \mathbb{C}^{m\times n}$.

3. $\|A + B\| \le \|A\| + \|B\|$ for all $A, B \in \mathbb{C}^{m\times n}$ (the triangle inequality).

Given a vector norm on $\mathbb{C}^n$, the corresponding **subordinate matrix norm** is defined by

$$\|A\| = \max_{x\neq 0} \frac{\|Ax\|}{\|x\|}.$$

The most important case is where the vector norm is the **p-norm**, defined by $\|x\|_p = \left(\sum_{i=1}^n |x_i|^p\right)^{1/p}$ $(1 \le p < \infty)$ and $\|x\|_\infty = \max_{i=1:n} |x_i|$, which gives the matrix p-norm $\|A\|_p = \max_{x\neq 0} \|Ax\|_p/\|x\|_p$. The three most useful subordinate matrix norms

Table B.1. *Constants α_{pq} such that $\|A\|_p \le \alpha_{pq}\|A\|_q$, $A \in \mathbb{C}^{m\times n}$.*

p \ q	1	2	∞	F
1	1	$\sqrt{m}$	m	$\sqrt{m}$
2	$\sqrt{n}$	1	$\sqrt{m}$	1
∞	n	$\sqrt{n}$	1	$\sqrt{n}$
F	$\sqrt{n}$	$\sqrt{\mathrm{rank}(A)}$	$\sqrt{m}$	1

are

$$\|A\|_1 = \max_{1\le j\le n} \textstyle\sum_{i=1}^m |a_{ij}|, \qquad \text{"max column sum"},$$
$$\|A\|_2 = \big(\rho(A^*A)\big)^{1/2} = \sigma_{\max}(A), \qquad \text{spectral norm},$$
$$\|A\|_\infty = \max_{1\le i\le m} \textstyle\sum_{j=1}^n |a_{ij}| = \|A^*\|_1, \qquad \text{"max row sum"},$$

These norms are all within a constant factor (depending only on the dimensions of A) of each other, as summarized in Table B.1.

The **Frobenius norm** for $A \in \mathbb{C}^{m\times n}$ is

$$\|A\|_F = \left(\sum_{i=1}^m\sum_{j=1}^n |a_{ij}|^2\right)^{1/2} = \big(\mathrm{trace}(A^*A)\big)^{1/2}. \tag{B.5}$$

For any subordinate matrix norm $\|I_n\| = 1$, whereas$\|I_n\|_F = \sqrt{n}$. When results in this book assume the norm to be subordinate it is usually because they use $\|I\| = 1$.

If $\|AB\| \le \|A\|\,\|B\|$ for all $A \in \mathbb{C}^{m\times n}$ and $B \in \mathbb{C}^{n\times p}$ then the norm is called **consistent**. (Note that there can be three different norms in this relation, so we should say the *norms* are consistent.) All subordinate matrix norms are consistent, as is the Frobenius norm.

A vector norm on $\mathbb{C}^n$ is **absolute** if $\|\,|x|\,\| = \|x\|$ for all $x \in \mathbb{C}^n$. A matrix norm subordinate to an absolute vector norm satisfies [295, 1985, Thm. 5.6.37]

$$\|\,\mathrm{diag}(d_i)\| = \max_i |d_i|. \tag{B.6}$$

A norm on $\mathbb{C}^{m\times n}$ for which $\|UAV\| = \|A\|$ for all unitary $U \in \mathbb{C}^{m\times m}$ and $V \in \mathbb{C}^{n\times n}$ and all $A \in \mathbb{C}^{m\times n}$ is called a **unitarily invariant norm**. Such a norm is a function only of the singular values of A, and hence it satisfies $\|A\| = \|A^*\|$. To be more precise, for a unitarily invariant norm $\|A\|$ is a **symmetric gauge function** of the singular values. A symmetric gauge function is an absolute, permutation-invariant norm on $\mathbb{R}^n$, that is, a vector norm g such that $g(|x|) = g(x)$, where $|x| = (|x_i|)$, and $g(x) = g(Px)$ for any permutation matrix P. A vector norm is absolute if and only if it is **monotone**, where a monotone norm is one for which $|x| \le |y| \Rightarrow \|x\| \le \|y\|$. For more details, see Horn and Johnson [295, 1985, Sec. 7.4] or Stewart and Sun [539, 1990].

For any unitarily invariant norm [296, 1991, Cor. 3.5.10],

$$\|ABC\| \le \|A\|_2\,\|B\|\,\|C\|_2, \quad A \in \mathbb{C}^{r\times m},\ B \in \mathbb{C}^{m\times n},\ C \in \mathbb{C}^{n\times s} \tag{B.7}$$

(in fact, any two of the norms on the right-hand side can be 2-norms).

The following result is used in Chapter 8.

Theorem B.3 (Mirsky). *Let $A, B \in \mathbb{C}^{m\times n}$ have SVDs with diagonal matrices $\Sigma_A, \Sigma_B \in \mathbb{R}^{m\times n}$, where the diagonal elements are arranged in nonincreasing order. Then $\|A - B\| \ge \|\Sigma_A - \Sigma_B\|$ for every unitarily invariant norm.*

Proof. See Mirsky [432, 1960, Thm. 5] or Horn and Johnson [295, 1985, Thm. 7.4.51]. □

The **condition number** (with respect to inversion) of $A \in \mathbb{C}^{n\times n}$ is $\kappa(A) = \|A\|\,\|A^{-1}\|$.

For any consistent norm and $A \in \mathbb{C}^{n\times n}$

$$\rho(A) \le \|A\| \tag{B.8}$$

(see Problem B.5).

For any $A \in \mathbb{C}^{n\times n}$ and $\epsilon > 0$ there is a consistent matrix norm (depending on A) such that $\|A\| \le \rho(A) + \epsilon$ [295, 1985, Lem. 5.6.10]. In particular, if $\rho(A) < 1$ there is a consistent matrix norm such that $\|A\| < 1$. This fact provides an easy way to prove the result that

$$\rho(A) < 1 \quad \Rightarrow \quad \lim_{k\to\infty} A^k = 0. \tag{B.9}$$

Lemma B.4. *For $A, B \in \mathbb{C}^{n\times n}$ and any consistent matrix norm we have*

$$\|A^m - B^m\| \le \|A - B\|\bigl(\|A\|^{m-1} + \|A\|^{m-2}\|B\| + \cdots + \|B\|^{m-1}\bigr).$$

Proof. The bound follows from the identity $A^m - B^m = \sum_{i=0}^{m-1} A^i(A-B)B^{m-1-i}$, which can be proved by induction. The inductive step uses $A^m - B^m = (A^{m-1} - B^{m-1})B + A^{m-1}(A - B)$. □

B.8. Matrix Sequences and Series

Let $A_k = (a_{ij}^{(k)}) \in \mathbb{C}^{m\times n}$. The sequence of matrices $\{A_k\}_{k=1}^{\infty}$ is said to converge to $A = \lim_{k\to\infty} A_k$ if $a_{ij}^{(k)} \to a_{ij}$ as $k \to \infty$ for all i and j. Equivalently, $\lim_{k\to\infty} \|A - A_k\| = 0$ for some matrix norm.

The notation $\sum_{k=1}^{\infty} A_k$ is an abbreviation for $\lim_{k\to\infty} \sum_{i=1}^{k} A_i$. A sufficient condition for convergence of a matrix series is given in the following result [433, 1961, Thm. 11.2.1].

Lemma B.5. *The series $\sum_{k=1}^{\infty} A_k$ converges if the series $\sum_{k=1}^{\infty} \|A_k\|$ converges for some matrix norm.*

B.9. Perturbation Expansions for Matrix Inverse

Let $\|\cdot\|$ be any subordinate matrix norm. If $\|E\| < 1$ then $I + E$ is nonsingular and $\|(I + E)^{-1}\| \le 1/(1 - \|E\|)$. If A is nonsingular and $\|A^{-1}E\| < 1$ then $A + E$ is nonsingular and

$$(A + E)^{-1} = A^{-1} - A^{-1}EA^{-1} + O(\|E\|^2). \tag{B.10}$$

A useful identity related to (B.10) is

$$A^{-1} - B^{-1} = A^{-1}(B - A)B^{-1}.$$

B.10. Sherman–Morrison–Woodbury Formula

If $A \in \mathbb{C}^{n\times n}$ is nonsingular and $v^*A^{-1}u \neq -1$ then $A + uv^*$ is nonsingular and

$$(A + uv^*)^{-1} = A^{-1} - \frac{A^{-1}uv^*A^{-1}}{1 + v^*A^{-1}u}. \tag{B.11}$$

This is known as the **Sherman–Morrison formula**.

More generally, If $U, V \in \mathbb{C}^{n\times p}$ and $I + V^*A^{-1}U$ is nonsingular then $A + UV^*$ is nonsingular and

$$(A + UV^*)^{-1} = A^{-1} - A^{-1}U(I + V^*A^{-1}U)^{-1}V^*A^{-1}, \tag{B.12}$$

which is the **Sherman–Morrison–Woodbury formula**.

For historical background on these formulae see Zhang [624, 2005, Chaps 1 and 6].

B.11. Nonnegative Matrices

For $A \in \mathbb{R}^{m\times n}$, $|A|$ denotes the matrix with (i,j) element $|a_{ij}|$. For $A \in \mathbb{R}^{m\times n}$, $A \geq 0$ ($A > 0$) denotes that $a_{ij} \geq 0$ ($a_{ij} > 0$) for all i and j. We say A is **nonnegative** if $A \geq 0$ and **positive** if $A > 0$.

For $A, B \in \mathbb{R}^{n\times n}$,

$$|A| \leq B \quad \Rightarrow \quad \rho(A) \leq \rho(|A|) \leq \rho(B) \tag{B.13}$$

[295, 1985, Thm. 8.1.18]. A corollary of this result is that $\rho(A) \geq \max_i a_{ii}$ [295, 1985, Cor. 8.1.20].

$A \in \mathbb{R}^{n\times n}$ is a **stochastic matrix** if $A \geq 0$ and the row sums are all equal to 1 (i.e., $Ae = e$, where $e = [1, 1, \ldots, 1]^T$).

$A \in \mathbb{C}^{n\times n}$ with $n \geq 2$ is **reducible** if there is a permutation matrix P such that

$$P^TAP = \begin{bmatrix} A_{11} & A_{12} \\ 0 & A_{22} \end{bmatrix},$$

where A_{11} and A_{22} are square, nonempty submatrices. A is **irreducible** if it is not reducible.

The following two results summarize some of the key spectral properties of nonnegative matrices proved by Perron and Frobenius.

Theorem B.6 (Perron–Frobenius). *If $A \in \mathbb{R}^{n\times n}$ is nonnegative and irreducible then*

(a) $\rho(A) > 0$,

(b) *$\rho(A)$ is an eigenvalue of A,*

(c) *there is an $x > 0$ such that $Ax = \rho(A)x$,*

(d) *$\rho(A)$ is an eigenvalue of algebraic multiplicity* 1. □

Theorem B.7 (Perron–Frobenius). *If $A \in \mathbb{R}^{n\times n}$ is nonnegative then*

(a) *$\rho(A)$ is an eigenvalue of A,*

(b) *there is a nonnegative eigenvector of A corresponding to $\rho(A)$.* □

For more details of Perron–Frobenius theory see Berman and Plemmons [60, 1994, Chap. 2], Horn and Johnson [295, 1985, Chap. 8], or Lancaster and Tismenetsky [371, 1985, Chap. 15].

B.12. Positive (Semi)definite Ordering

For Hermitian matrices X and Y, $X \geq Y$ denotes that $X - Y$ is positive semidefinite while $X > Y$ denotes that $X - Y$ is positive definite. (Note that this usage of inequality signs differs from that in the previous section. Which usage is in effect will usually be clear from the context.) $X \leq Y$ ($X < Y$) is equivalent to $Y \geq X$ ($Y > X$). The ordering $\geq$ on positive semidefinite matrices is sometimes called the **Löwner (partial) ordering**. Some of the properties of inequalities on real numbers carry over to Hermitian matrices. For this book we need the results below.

Lemma B.8. *Let $A, B, C \in \mathbb{C}^{n\times n}$ be Hermitian. Then*

(a)

$$A \geq 0, \quad B \geq 0, \quad AB = BA \quad \Rightarrow \quad AB \geq 0; \tag{B.14}$$

(b)

$$A \geq 0, \quad B \leq C, \quad AB = BA, \quad AC = CA \quad \Rightarrow \quad AB \leq AC; \tag{B.15}$$

(c) *if $\{X_k\}$ is a sequence of Hermitian matrices satisfying $B \leq X_{k+1} \leq X_k$ for all k then $X_k \to X_* \geq B$ as $k \to \infty$.*

Proof. (a) AB is Hermitian, since $(AB)^* = B^*A^* = BA = AB$. Let X and Y be the Hermitian positive semidefinite square roots of A and B. X and Y are functions of A and B, so $X = p(A)$, $Y = q(B)$ for some polynomials p and q, and therefore X and Y commute. Then $AB = X^2Y^2 = (XY)^2$, which implies that the eigenvalues of AB are nonnegative, as required.

(b) The inequality can be written $A(C - B) \geq 0$, which follows from (a).

(c) The following proof is due to Roy Mathias. For arbitrary $x \in \mathbb{C}^n$ consider the scalar sequence $\alpha_k(x) = x^*X_kx$. Since $x^*Bx \leq \alpha_{k+1}(x) \leq \alpha_k(x)$, we have $\alpha_k(x) \to \alpha_*(x) \geq x^*Bx$. Suppose, first, that the X_k are all real. Consider the polarization identity

$$\operatorname{Re} x^*Ay = \frac{1}{4}\big((x+y)^*A(x+y) - (x-y)^*A(x-y)\big)$$

and set $A = X_k$. For any given $x, y \in \mathbb{R}^n$ the right-hand side has a limit, $\alpha_*(x,y)$, as $k \to \infty$, from the result just shown. Hence $x^*X_ky \to \alpha_*(x,y)$. Now put $x = e_i$, $y = e_j$, to deduce that $(X_k)_{ij} \to \alpha_*(e_i, e_j) =: (X_*)_{ij}$. Clearly, $X_* \geq B$. If the X_k are not all real then an analogous proof holds using the complex polarization identity

$$\begin{aligned} x^*Ay = \frac{1}{4}\big[&(x+y)^*A(x+y) - (x-y)^*A(x-y) \\ &- i\big[(x+iy)^*A(x+iy) - (x-iy)^*A(x-iy)\big]\big]. \end{aligned}$$

□

Theorem B.9. *If $A \geq B \geq 0$ then $A^p \geq B^p$ for $p \in [0,1]$.*

Proof. See, e.g., Bhatia [64, 1997, Thm. V.1.9], [65, 2007, Thm. 4.2.1] or Zhan [623, 2002, Thm. 1.1].

B.13. Kronecker Product and Sum

The **Kronecker product** of $A \in \mathbb{C}^{m\times n}$ and $B \in \mathbb{C}^{p\times q}$ is $A \otimes B = (a_{ij}B) \in \mathbb{C}^{mp\times nq}$. A key property of the Kronecker product is $(A \otimes B)(C \otimes D) = AC \otimes BD$.

The **vec operator** stacks the columns of a matrix into one long vector: if $A = [a_1, a_2, \dots, a_m]$ then $\mathrm{vec}(A) = [a_1^T a_2^T \dots a_m^T]^T$.

The **vec-permutation matrix** Π is the permutation matrix defined by the property that $\mathrm{vec}(A^T) = \Pi\,\mathrm{vec}(A)$. It satisfies $(A \otimes B)\Pi = \Pi(B \otimes A)$.

We have

$$\mathrm{vec}(AXB) = (B^T \otimes A)\,\mathrm{vec}(X). \tag{B.16}$$

If $A \in \mathbb{C}^{n\times n}$ has eigenvalues λ_r and $B \in \mathbb{C}^{m\times m}$ has eigenvalues μ_s then [371, 1985, Thm. 12.2.1]

$$\Lambda\Big(\sum_{i,j=0}^{k} c_{ij}\,A^i \otimes B^j\Big) = \sum_{i,j=0}^{k} c_{ij}\,\lambda_r^i \mu_s^j, \qquad r = 1\colon n, \quad s = 1\colon m. \tag{B.17}$$

The **Kronecker sum** of $A \in \mathbb{C}^{m\times m}$ and $B \in \mathbb{C}^{n\times n}$ is defined by $A \oplus B = A \otimes I_n + I_m \otimes B$. By (B.17), the eigenvalues of $A \oplus B$ are $\lambda_{ij} = \lambda_i(A) + \lambda_j(B)$, $i = 1\colon m$, $j = 1\colon n$.

For any p-norm we have $\|A \otimes B\|_p = \|A\|_p \|B\|_p$ and $\|A \oplus B\|_p \le \|A\|_p + \|B\|_p$.

For more details of Kronecker products and sums, see Horn and Johnson [296, 1991, Chap. 4] or Lancaster and Tismenetsky [371, 1985, Chap. 12]. Note that some authors define $A \oplus B = I_n \otimes A + B \otimes I_m$, which is $B \oplus A$ in our notation; the elegant statement of Theorem 10.9 is a reason for preferring our definition.

B.14. Sylvester Equation

The linear matrix equation

$$AX - XB = C, \tag{B.18}$$

where $A \in \mathbb{R}^{m\times m}$, $B \in \mathbb{R}^{n\times n}$, and $C \in \mathbb{R}^{m\times n}$ are given and $X \in \mathbb{R}^{m\times n}$ is to be determined, is called the **Sylvester equation**. By applying the vec operator, it can be rewritten in the form

$$(I_n \otimes A - B^T \otimes I_m)\,\mathrm{vec}(X) = \mathrm{vec}(C). \tag{B.19}$$

The mn eigenvalues of the coefficient matrix are given by

$$\lambda_{ij}(I_n \otimes A - B^T \otimes I_m) = \lambda_i(A) - \lambda_j(B), \quad i = 1\colon m, \quad j = 1\colon n, \tag{B.20}$$

and hence the Sylvester equation is nonsingular precisely when A and B have no eigenvalue in common.

B.15. Floating Point Arithmetic

All rounding error analysis in the book is based on the standard model of floating point arithmetic,

$$fl(x \mathbin{\mathrm{op}} y) = (x \mathbin{\mathrm{op}} y)(1+\delta), \qquad |\delta| \le u, \quad \mathrm{op} = +, -, *, /, \tag{B.21}$$

where fl denotes the computed value of an expression and u is the **unit roundoff**. In addition to (B.21), we can also use the variant

$$fl(x \operatorname{op} y) = \frac{x \operatorname{op} y}{1+\delta}, \qquad |\delta| \le u. \tag{B.22}$$

In IEEE double precision arithmetic, $u = 2^{-53} \approx 1.11 \times 10^{-16}$, while for IEEE single precision arithmetic, $u = 2^{-24} \approx 5.96 \times 10^{-8}$.

This model applies to real x and y. For complex data the model must be adjusted by increasing the bound for $|\delta|$ slightly [276, 2002, Lem. 3.5].

Error analysis results are often stated in terms of the constants

$$\gamma_n = \frac{nu}{1-nu}, \qquad \widetilde{\gamma}_n = \frac{cnu}{1-cnu},$$

where c is a small integer constant whose precise value is unimportant. For all the algorithms in this book we can state error bounds that are valid for both real and complex arithmetic by using the $\widetilde{\gamma}$ notation.

The following lemma [276, 2002, Lem. 3.1] is fundamental for carrying out rounding error analysis.

Lemma B.10. *If $|\delta_i| \le u$ and $\rho_i = \pm 1$ for $i = 1\colon n$, and $nu < 1$, then*

$$\prod_{i=1}^{n} (1+\delta_i)^{\rho_i} = 1 + \theta_n,$$

where

$$|\theta_n| \le \frac{nu}{1-nu} = \gamma_n.$$

Complex arithmetic costs significantly more than real arithmetic: in particular, for scalars, a complex addition requires 3 real additions and a complex multiplication requires 4 real multiplications and 2 real additions, though the latter can be reduced at the cost of some stability [272, 1992]. The ratio of execution time of a numerical algorithm implemented in real versus complex arithmetic depends on both the algorithm and the computer, but a ratio of 4 is reasonably typical.

For the computed product $\widehat{C} = fl(AB)$ of $A, B \in \mathbb{C}^{n\times n}$ we have $\|C - \widehat{C}\|_p \le \widetilde{\gamma}_n \|A\|_p \|B\|_p$, $p = 1, \infty, F$.

For more on floating point arithmetic and rounding error analysis see Higham [276, 2002].

B.16. Divided Differences

Let $x_0, x_1, \dots, x_n \in \mathbb{C}$ be ordered so that equal points are contiguous, that is,

$$x_i = x_j \quad (i < j) \quad \Rightarrow \quad x_i = x_{i+1} = \cdots = x_j. \tag{B.23}$$

Divided differences of a function f at the points x_k are defined recursively by

$$f[x_k] = f(x_k),$$

$$f[x_k, x_{k+1}] = \begin{cases} \dfrac{f(x_{k+1}) - f(x_k)}{x_{k+1} - x_k}, & x_k \ne x_{k+1}, \\ f'(x_{k+1}), & x_k = x_{k+1}, \end{cases}$$

$$f[x_0, x_1, \dots, x_{k+1}] = \begin{cases} \dfrac{f[x_1, x_2, \dots, x_{k+1}] - f[x_0, x_1, \dots, x_k]}{x_{k+1} - x_0}, & x_0 \neq x_{k+1}, \\ \dfrac{f^{(k+1)}(x_{k+1})}{(k+1)!}, & x_0 = x_{k+1}. \end{cases} \tag{B.24}$$

One of their main uses is in constructing the Newton divided difference form of the interpolating polynomial, details of which can be found in numerical analysis textbooks, such as Neumaier [447, 2001, Sec. 3.1] and Stoer and Bulirsch [542, 2002, Sec. 2.1.5].

Another important representation of a divided difference is as a multiple integral. Assume f is k times continuously differentiable. It is easy to check that $f[x_0, x_1] = \int_0^1 f'(x_0 + (x_1 - x_0)t)\, dt$. This formula generalizes to the Genocchi–Hermite formula

$$f[x_0, x_1, \dots, x_k] = \int_0^1 \int_0^{t_1} \int_0^{t_2} \cdots \int_0^{t_{k-1}} f^{(k)}\Big(x_0 + \sum_{j=1}^{k} t_j (x_j - x_{j-1})\Big)\, dt_k \dots dt_2\, dt_1, \tag{B.25}$$

which expresses $f[x_0, x_1, \dots, x_k]$ as an average of $f^{(k)}$ over the simplex with vertices $x_0, x_1, \dots, x_k$. Note that this formula, like (B.29) below, does not require the x_i to be ordered as in (B.23), so it gives a meaning to divided differences for an arbitrary ordering.

It follows from (B.25) that if f is k-times continuously differentiable then the divided difference $f[x_0, x_1, \dots, x_k]$ is a continuous function of its arguments. Moreover, for *real* points x_i,

$$f[x_0, x_1, \dots, x_k] = \frac{f^{(k)}(\xi)}{k!} \quad \text{for some } \xi \in [\min_i x_i, \max_i x_i]. \tag{B.26}$$

Hence for confluent arguments we recover the confluent case in (B.24):

$$f[\underbrace{x, x, \dots, x}_{k+1 \text{ times}}] = \frac{f^{(k)}(x)}{k!}. \tag{B.27}$$

No result of the form (B.26) holds for complex x_i, even for $k = 1$ [178, 1992]. It nevertheless follows from (B.25) that if Ω is a closed convex set containing $x_0, x_1, \dots, x_k \in \mathbb{C}$ then

$$|f[x_0, x_1, \dots, x_k]| \le \frac{\max_{z \in \Omega} |f^{(k)}(z)|}{k!}. \tag{B.28}$$

Yet another definition of divided differences can be given in terms of contour integration. Let f be analytic inside and on a closed contour Γ that encloses $x_0, x_1, \dots, x_k$. Then

$$f[x_0, x_1, \dots, x_k] = \frac{1}{2\pi i} \int_\Gamma \frac{f(z)}{(z - x_0)(z - x_1)\dots(z - x_k)}\, dz. \tag{B.29}$$

The properties (B.24)–(B.27) are readily deduced from this definition.

For more details of divided difference see, for example, Conte and de Boor [113, 1980, Thm. 2.5], Isaacson and Keller [311, 1966, Sec. 6.1], and Neumaier [447, 2001, Thm. 3.1.8]. The formulae (B.25) and (B.29) are less commonly treated, hence we give a list of references in which they can be found: de Boor [143, 2005], Gel'fond [210, 1971, Sec. 1.4.3], Horn and Johnson [296, 1991, Sec. 6.1], Isaacson and Keller [311, 1966, Sec. 6.1], Kahan [328, 2006], Kahan and Fateman [329, 1999], Milne-Thompson [428, 1933, Chap. 1], and Ostrowski [455, 1973].

Problems

B.1. Prove the eigenpair relations between A and B stated in Section B.3.

B.2. Use the Moore–Penrose conditions (B.3) to show that if A is Hermitian then so is its pseudoinverse. Deduce that if A is Hermitian then A commutes with A^+.

B.3. Show that for $A \in \mathbb{C}^{m\times n}$, $\mathrm{range}(A^*A) = \mathrm{range}(A^*)$.

B.4. Show that every partial isometry $A \in \mathbb{C}^{m\times n}$ has the form $A = P\left[\begin{smallmatrix} I_r & 0 \\ 0 & 0_{m-r,n-r} \end{smallmatrix}\right]Q^*$, for some unitary $P \in \mathbb{C}^{m\times m}$ and $Q \in \mathbb{C}^{n\times n}$, where $r = \mathrm{rank}(A)$.

B.5. Show that for any $A \in \mathbb{C}^{n\times n}$ and any consistent matrix norm, $\rho(A) \le \|A\|$, where ρ is the spectral radius.

B.6. Let $A \in \mathbb{C}^{m\times n}$. Show that $\left\|\left[\begin{smallmatrix} A \\ 0 \end{smallmatrix}\right]\right\| = \|A\|$ for any unitarily invariant norm.

B.7. Let $A \in \mathbb{C}^{n\times n}$ and $U \in \mathbb{C}^{m\times n}$, where $U^*U = I_n$. Show that $\|A\| = \|UA\|$ for any unitarily invariant norm. (This result is immediate from the definition of unitarily invariant norm if $m = n$.)

B.8. Show that for $A \in \mathbb{C}^{m\times n}$ and $B \in \mathbb{C}^{n\times p}$, $\|AB\|_F \le \|A\|_2\|B\|_F$ and $\|AB\|_F \le \|A\|_F\|B\|_2$ (special cases of (B.7)).

B.9. Show that the orthogonal projector onto a subspace S is unique.

B.10. Show that AA^+ and A^+A are the orthogonal projectors onto $\mathrm{range}(A)$ and $\mathrm{range}(A^*)$, respectively.

B.11. Let $A \in \mathbb{C}^{n\times n}$ have the SVD $A = P\Sigma Q^*$ with $n\times n$ factors, and let $r = \mathrm{rank}(A)$. Show that if $A = \widetilde{P}\Sigma\widetilde{Q}^*$ is another SVD then $\widetilde{P} = P\,\mathrm{diag}(D, W_1)$ and $\widetilde{Q}^* = \mathrm{diag}(D^*, W_2)Q^*$, where $D \in \mathbb{C}^{r\times r}$ is unitary and block diagonal with $d_{ij} = 0$ if $\sigma_i \ne \sigma_j$, and $W_1, W_2 \in \mathbb{C}^{(n-r)\times(n-r)}$ are arbitrary unitary matrices.

Appendix C
Operation Counts

In this appendix we summarize the cost of some standard matrix computations. The unit of measure is the flop, which denotes any of the four elementary scalar operations $+$, $-$, $*$, $/$. The operation counts assume the use of substitution for solving triangular systems and LU factorization, and appropriate symmetric variants thereof in the symmetric case, for solving linear systems. When complex arithmetic is involved, the flops should be interpreted as counting operations on complex numbers.

For details of the relevant algorithms see, for example, Golub and Van Loan [224, 1996], Higham [276, 2002], Trefethen and Bau [572, 1997], or Watkins [607, 2002].

It is important to stress that the run time of an algorithm in a particular computing environment may or may not be well predicted by the flop count. Different matrix operations may run at different speeds, depending on the machine architecture. In particular, matrix multiplication exploits hierarchical memory architectures better than matrix inversion, so it is generally desirable to make algorithms matrix multiplication-rich. See Dongarra, Duff, Sorensen and Van der Vorst [162, 1998] and Golub and Van Loan [224, 1996] for more details.

Table C.1. *Cost of some matrix computations for real, $n \times n$ matrices. Here A and B are general, nonsymmetric; T is triangular; H and M are symmetric. X denotes a matrix to be determined; Y denotes a symmetric matrix to be determined. x, b are n-vectors.*

Operation	Number of flops
AB	$2n^3$
A^{-1}	$2n^3$
H^{-1}	n^3
T^{-1}	$n^3/3$
AT	n^3
HM	$n^3/3$
T_1T_2	$\begin{cases} 2n^3/3 & \text{if } T_1T_2 \text{ is full} \\ n^3/3 & \text{if } T_1T_2 \text{ is triangular} \end{cases}$
LU	$2n^3/3$
$Tx = b$	n^2
$TX = B$	$\begin{cases} n^3 & \text{for general } B \\ n^3/3 & \text{if } B \text{ has same triangular structure as } T \end{cases}$
$TY = B$	n^3
$AX = B$	$8n^3/3$ [a]
$HX = B$	$7n^3/3$ [b]
$HY = B$	n^3 [b]
$AX + XB = C$	$60n^3$ [c]
$T_1X + XT_2 = C$	$2n^3$

[a] Assuming an LU factorization of A is computed.
[b] Assuming an LDL^T or Cholesky factorization of H is computed.
[c] Assuming Schur decompositions of A and B are computed.

Table C.2. *Cost of some matrix factorizations and decompositions. A is $n \times n$, except for QR and SVD, where it is $m \times n$ ($m \geq n$).*

Factorization/decomposition	Number of flops
LU factorization with partial pivoting ($PA = LU$)	$2n^3/3$
LU factorization with partial pivoting of upper Hessenberg matrix ($PA = LU$)	n^2
Cholesky factorization ($A = R^*R$)	$n^3/3$
Householder QR factorization ($A = QR$)	$2n^2(m - n/3)$ for R; $4(m^2n - mn^2 + n^3/3)$ for $m \times m$ Q; $2n^2(m - n/3)$ for $m \times n$ Q; $2np(2m - n)$ for QB with $m \times p$ B and Q held in factored form.
SVD[a] ($A = P\Sigma Q^*$)	$14mn^2 + 8n^3$ ($P(:, 1\colon n)$, Σ, and Q)[b] $6mn^2 + 20n^3$ ($P(:, 1\colon n)$, Σ, and Q)[c]
Hessenberg decomposition ($A = QHQ^*$)	$14n^3/3$ (Q and H), $10n^3/3$ (H only)
Schur decomposition[a] ($A = QTQ^*$)	$25n^3$ (Q and T), $10n^3$ (T only)
For Hermitian A:	
Tridiagonal reduction ($A = QTQ^*$)	$8n^3/3$ (Q and T), $4n^3/3$ (T only)
Spectral decomposition ($A = QDQ^*$)	$9n^3$ (Q and D), $4n^3/3$ (D only)

[a]These costs are estimates taken from Golub and Van Loan [224, 1996].
[b]Golub–Reinsch SVD.
[c]Golub–Reinsch SVD with preliminary QR factorization.

Appendix D
Matrix Function Toolbox

The Matrix Function Toolbox contains MATLAB implementations of many of the algorithms described in this book. The toolbox is intended to facilitate understanding of the algorithms through MATLAB experiments, to be useful for research in the subject, and to provide a basis for the development of more sophisticated implementations. The codes are "plain vanilla" versions; they contain the core algorithmic aspects with a minimum of inessential code. In particular, the following features should be noted.

- The codes have little error checking of input arguments.
- The codes do not print intermediate results or the progress of an iteration.
- For the iterative algorithms a convergence tolerance is hard-coded (in function `mft_tolerance`). For greater flexibility this tolerance could be made an input argument.
- The codes are designed for simplicity and readability rather than maximum efficiency.
- Algorithmic options such as preprocessing are omitted.
- The codes are intended for double precision matrices. Those algorithms in which the parameters can be adapted to the precision have not been written to take advantage of single precision inputs.

The contents of the toolbox are listed in Table D.1, along with a reference to the corresponding algorithms, theorems, or sections in the book. We have not provided codes for algorithms that are already provided as part of MATLAB. Such matrix-function M-files are listed in Table D.2, along with a reference to the corresponding algorithm in this book.

The Matrix Function Toolbox is available from

`http://www.ma.man.ac.uk/~higham/mftoolbox`

To test that the toolbox is working correctly, run the function `mft_test` one or more times.

Table D.1. *Contents of Matrix Function Toolbox and corresponding parts of this book.*

`arnoldi`	Arnoldi iteration	Algorithm 13.3
`ascent_seq`	Ascent sequence for square (singular) matrix.	Theorem 1.22
`cosm`	Matrix cosine by double angle algorithm.	Algorithm 12.6
`cosm_pade`	Evaluate Padé approximation to the matrix cosine.	Section 12.3
`cosmsinm`	Matrix cosine and sine by double angle algorithm.	Algorithm 12.8
`cosmsinm_pade`	Evaluate Padé approximations to matrix cosine and sine.	Section 12.6
`expm_cond`	Relative condition number of matrix exponential.	Algorithms 3.17 and 10.27
`expm_frechet_pade`	Fréchet derivative of matrix exponential via Padé approximation.	Algorithm 10.27
`expm_frechet_quad`	Fréchet derivative of matrix exponential via quadrature.	Algorithm 10.26
`fab_arnoldi`	$f(A)b$ approximated by Arnoldi method.	Section 13.2.2
`funm_condest1`	Estimate of 1-norm condition number of matrix function.	Algorithm 3.22
`funm_condest_fro`	Estimate of Frobenius norm condition number of matrix function.	Algorithm 3.20
`funm_ev`	Evaluate general matrix function via eigensystem.	Section 4.5
`funm_simple`	Simplified Schur–Parlett method for function of a matrix.	Section 4.6
`logm_cond`	Relative condition number of matrix logarithm.	Section 11.8
`logm_frechet_pade`	Fréchet derivative of matrix logarithm via Padé approximation.	Algorithm 11.12
`logm_iss`	Matrix logarithm by inverse scaling and squaring method.	Algorithm 11.10
`logm_pade_pf`	Evaluate Padé approximation to matrix logarithm by partial fraction form.	Section 11.4
`mft_test`	Test the Matrix Function Toolbox.	
`mft_tolerance`	Convergence tolerance for matrix iterations.	
`polar_newton`	Polar decomposition by scaled Newton iteration.	Algorithm 8.20
`polar_svd`	Canonical polar decomposition via singular value decomposition.	Theorem 8.1
`polyvalm_ps`	Evaluate polynomial at matrix argument by Paterson–Stockmeyer algorithm.	Section 4.2
`power_binary`	Power of matrix by binary powering (repeated squaring).	Algorithm 4.1
`quasitriang_struct`	Block structure of upper quasitriangular matrix.	
`riccati_xaxb`	Solve Riccati equation $XAX = B$ in positive definite matrices.	Algorithm 6.22
`rootpm_newton`	Coupled Newton iteration for matrix pth root.	(7.19)
`rootpm_real`	Pth root of real matrix via real Schur form.	Algorithm 7.6
`rootpm_schur_newton`	Matrix pth root by Schur–Newton method.	Algorithm 7.15
`rootpm_sign`	Matrix pth root via matrix sign function.	Algorithm 7.17

Table D.1. (*continued*)

`signm`	Matrix sign decomposition.	Algorithm 5.5
`signm_newton`	Matrix sign function by Newton iteration.	Algorithm 5.14
`sqrtm_db`	Matrix square root by Denman–Beavers iteration.	(6.15), (6.28)
`sqrtm_dbp`	Matrix square root by product form of Denman–Beavers iteration.	(6.17), (6.29)
`sqrtm_newton`	Matrix square root by Newton iteration (unstable).	(6.12), (6.27)
`sqrtm_newton_full`	Matrix square root by full Newton method.	(6.11)
`sqrtm_pd`	Square root of positive definite matrix via polar decomposition.	Algorithm 6.21
`sqrtm_pulay`	Matrix square root by Pulay iteration.	(6.48)
`sqrtm_real`	Square root of real matrix by real Schur method.	Algorithm 6.7
`sqrtm_triang_min_norm`	Estimated minimum norm square root of triangular matrix.	Algorithm 6.23
`sylvsol`	Solve Sylvester equation.	

Table D.2. *Matrix-function-related M-files in MATLAB and corresponding algorithms in this book.*

`expm`	Matrix exponential.	Algorithm 10.20
`funm`	Evaluate general matrix function.	Algorithm 9.6
`logm`	Matrix logarithm.	Algorithm 11.11
`polyvalm`	Evaluate polynomial with matrix argument.	Algorithm 4.2
`sqrtm`	Matrix square root.	Algorithm 6.3

Appendix E
Solutions to Problems

Remember that very little is gained by reading the solution to a problem before seriously attempting to solve it.
— WILLIAM BRIGGS, *Ants, Bikes, and Clocks: Problem Solving for Undergraduates* (2005)

If repeated efforts have been unsuccessful, the reader can afterwards analyze the solution ... with more incisive attention, bring out the actual principle which is the salient point, assimilate it, and commit it to his memory.
— GEORGE PÓLYA AND GABOR SZEGÖ, *Problems and Theorems in Analysis I* (1998)

1.1. Consider two Jordan canonical forms

$$A = ZJZ^{-1} = WJW^{-1} \tag{E.1}$$

(by incorporating a permutation matrix in W we can assume without loss of generality that J is the same matrix in both cases). The definition gives $f_1(A) = Zf(J)Z^{-1}$, $f_2(A) = Wf(J)W^{-1}$ and we need to show that $f_1(A) = f_2(A)$, that is, $W^{-1}Zf(J)Z^{-1}W = f(J)$, or $X^{-1}f(J)X = f(J)$ where $X = Z^{-1}W$. Now by (E.1) we have $X^{-1}JX = J$, which implies $f(J) = f(X^{-1}JX) = X^{-1}f(J)X$, the last equality following from Definition 1.2. Hence $f_1(A) = f_2(A)$, as required.

Further insight can be gained by noting that the general form of X is given by Theorem 1.25 and that this form commutes with $f(J)$ given by (1.4).

1.2. We have $-J_k = DJ_k(-\lambda_k)D$, where $D = \operatorname{diag}(1, -1, 1, \dots, (-1)^{m_k-1})$. Hence $f(-J_k) = Df(J_k(-\lambda_k))D$, from which the result follows. Alternatively, we can write $-J_k = -\lambda_k I + N_k$, where N_k is zero except for a superdiagonal of -1s, and expand $f(-\lambda_k I + N_k)$ in a Taylor series (cf. (1.5)).

1.3. By Theorem 1.38, it suffices to show that $AX = XA$ implies $f(A)X = Xf(A)$. If X is nonsingular then $Xf(A)X^{-1} = f(XAX^{-1}) = f(AXX^{-1}) = f(A)$, so $Xf(A) = f(A)X$. If X is singular we can choose $\epsilon > 0$ such that $X + \epsilon I$ is nonsingular. Clearly, $X + \epsilon I$ commutes with A, and so by the previous part $(X + \epsilon I)f(A) = f(A)(X + \epsilon I)$, which implies $Xf(A) = f(A)X$, as required.

1.4. We will use properties from Theorem 1.13 without comment.

(a) The Jordan canonical form of A can be ordered so that $A = ZJZ^{-1}$ with $j_{11} = \lambda$ and $Z(:,1) = x$. Now $f(A) = Zf(J)Z^{-1}$ and so $f(A)Z = Zf(J)$. The first column of this equation gives $f(A)x = f(\lambda)x$, as required. For an alternative proof, let p interpolate to the values of f on the spectrum of A. Then $p(\lambda) = f(\lambda)$. Since $A^k x = \lambda^k x$ for all k, $f(A)x = p(A)x = p(\lambda)x = f(\lambda)x$, as required.

(b) Setting $(\lambda, x) \equiv (\alpha, e)$ in (a) gives the row sum result. If A has column sums α then $A^T e = \alpha e$ and applying the row sum result to A^T gives $f(\alpha)e = f(A^T)e = f(A)^T e$. So $f(A)$ has column sums $f(\alpha)$.

1.5. There is clearly a lower bound on the degrees of polynomials p such that $p(A) = 0$, and by the Cayley–Hamilton theorem the characteristic polynomial is a candidate for having minimal degree, so the minimum is at most n. Hence ψ exists, as we can always normalize to obtain a monic polynomial. Let p_1 and p_2 be two monic polynomials of lowest degree. Then $p_3 = p_2 - p_1$ is a polynomial of degree less than p_1 and p_2 and $p_3(A) = 0$; hence $p_3 = 0$, i.e., $p_1 = p_2$. So ψ is unique.

1.6. Let $B = I - \frac{1}{3}A$. From $\psi(A) = 0$ we obtain $9B^2 - 15B + 5I = 0$. Premultiplying by B^{-1} and rearranging gives $B^{-1} = (3A + 6I)/5$, as required.

1.7. It is easiest to use the polynomial interpolation definition (1.4), which says that $\cos(\pi A) = p(A)$, where $p(1) = \cos\pi = -1$, $p'(1) = -\pi\sin\pi = 0$, $p(2) = \cos 2\pi = 1$. Writing $p(t) = a + bt + ct^2$ we have

$$\begin{bmatrix} 1 & 1 & 1 \\ 0 & 1 & 2 \\ 1 & 2 & 4 \end{bmatrix} \begin{bmatrix} a \\ b \\ c \end{bmatrix} = \begin{bmatrix} -1 \\ 0 \\ 1 \end{bmatrix},$$

which can be solved to give $p(t) = 1 - 4t + 2t^2$. Hence

$$\cos(\pi A) = p(A) = I - 4A + 2A^2 = \begin{bmatrix} -3 & 0 & 0 & 4 \\ 0 & -1 & 0 & 0 \\ 0 & 0 & -1 & 0 \\ -2 & 0 & 0 & 3 \end{bmatrix}.$$

Evaluating $\cos(\pi A)$ from its power series would be much more complicated.

1.8. From the spectral properties of uv^* identified in Section 1.2.5 we deduce that the characteristic polynomial is $p(t) = t^{n-1}(t - v^*u)$ and the minimal polynomial is $\psi(t) = t(t - v^*u)$. As a check, we have $\psi(uv^*) = uv^*(uv^* - (v^*u)I) = v^*u(uv^* - uv^*) = 0$.

1.9. Let the eigenvalues of A be λ_1 and λ_2. We have the explicit formula

$$f(A) = f(\lambda_1)I + f[\lambda_1, \lambda_2](A - \lambda_1 I). \tag{E.2}$$

To understand the formula it is useful to examine the Jordan form of A.

The matrix $A = \left[\begin{smallmatrix} a & b \\ c & d \end{smallmatrix}\right]$ has eigenvalues $\lambda = (a + d \pm \sqrt{(a-d)^2 + 4bc})/2$. Hence A has distinct eigenvalues unless $(a-d)^2 + 4bc = 0$. Suppose A has a double eigenvalue, λ. If A is diagonalizable then $A = X\operatorname{diag}(\lambda, \lambda)X^{-1} = \lambda I$, so A must be diagonal. This means that we can characterize the Jordan structure in terms of the elements of A as follows:

```
1  if (a − d)^2 + 4bc ≠ 0
2     A has distinct eigenvalues
3  else if b = c = 0
4     A has a double eigenvalue in two 1 × 1 Jordan blocks and A = λI (λ = a = d)
5  else
6     A has a Jordan block of size 2
7  end
```

In the second of the three cases, the derivative $f'(\lambda)$ appears to enter in the formula (E.2), even though the theory says it shouldn't. The appearance is illusory, because $A - \lambda_1 I = 0$.

1.10. The formula is trivial for $\beta = 0$, so assume $\beta \neq 0$. The matrix $A = \alpha I + \beta J$ is symmetric, with $n-1$ eigenvalues α and one eigenvalue $\alpha+n\beta$. The interpolation conditions (1.7) are $p(\alpha) = f(\alpha)$, $p(\alpha + n\beta) = f(\alpha + n\beta)$. Hence

$$p(t) = \frac{t-(\alpha+n\beta)}{-n\beta} f(\alpha) + \frac{t-\alpha}{n\beta} f(\alpha+n\beta).$$

Thus

$$f(A) = \frac{\beta J - n\beta I}{-n\beta} f(\alpha) + \frac{\beta J}{n\beta} f(\alpha+n\beta) = f(\alpha)I + n^{-1}(f(\alpha+n\beta) - f(\alpha))J.$$

Of course, the formula is a special case of (1.16).

1.11. With $p(t) = t$ we have, in the notation of (1.7), $p(\lambda_i) = \lambda_i$, $i = 1\colon s$; $p'(\lambda_i) = 1$ if $n_i \geq 1$, $i = 1\colon s$; $p^{(j)}(\lambda_i) = 0$ for $2 \leq j \leq n_i - 1$, $i = 1\colon s$.

1.12. The interpolating polynomial in Definition 1.4 is

$$p(t) = f(\lambda_1)\frac{t-\lambda_2}{\lambda_1-\lambda_2} + f(\lambda_2)\frac{t-\lambda_1}{\lambda_2-\lambda_1}.$$

Hence

$$f(A) = p(A) = \frac{f(\lambda_1)}{\lambda_1-\lambda_2}(A-\lambda_2 I) + \frac{f(\lambda_2)}{\lambda_2-\lambda_1}(A-\lambda_1 I).$$

1.13. For the Jordan canonical form definition (1.2), use Theorem 1.25 to write down the form of a general matrix B that commutes with A. The verification of $f(A)B = Bf(A)$ then reduces to the fact that upper triangular Toeplitz matrices commute.

For the interpolation definition (1.4) the result is immediate: $f(A)$ is a polynomial in A and so commutes with B since A does.

For the Cauchy integral definition (1.11) we have, using $(zI - A)B = B(zI - A)$,

$$\begin{aligned} f(A)B &= \frac{1}{2\pi i}\int_\Gamma f(z)(zI-A)^{-1}B\,dz \\ &= \frac{1}{2\pi i}B\int_\Gamma f(z)(zI-A)^{-1}\,dz = Bf(A). \end{aligned}$$

1.14. The Jordan structure of A cannot be reliably computed in floating point arithmetic, but the eigenvalues can (in the sense of obtaining backward stable results). Therefore we can find the Hermite interpolating polynomial p satisfying $p^{(j)}(\lambda_i) = f^{(j)}(\lambda_i)$, $j = 0\colon j_i - 1$, $i = 1\colon s$, where $\lambda_1, \ldots, \lambda_s$ are the distinct eigenvalues and λ_i has *algebraic* multiplicity j_i. This polynomial p satisfies more interpolation conditions than in (1.7), but nevertheless $p(A) = f(A)$ (see Remark 1.5). Perhaps the most elegant way to obtain p is in the divided difference form (1.10), which needs no attempt to identify repeated eigenvalues.

1.15. The Jordan canonical form can be written as

$$A = uv^* = [\,u \quad v/(v^*v) \quad X\,]\begin{bmatrix} 0 & 1 & 0 \\ 0 & 0 & 0 \\ 0 & 0 & 0 \end{bmatrix}\begin{bmatrix} u^*/(u^*u) \\ v^* \\ Y \end{bmatrix},$$

where X and Y are chosen so that $AX = 0$, $[\,u \quad v/(v^*v) \quad X\,]$ is nonsingular, and

$$[\,u \quad v/(v^*v) \quad X\,]\begin{bmatrix} u^*/(u^*u) \\ v^* \\ Y \end{bmatrix} = I. \tag{E.3}$$

Hence

$$f(A) = [u \quad v/(v^*v) \quad X] \begin{bmatrix} f(0) & f'(0) & 0 \\ 0 & f(0) & 0 \\ 0 & 0 & f(0) \end{bmatrix} \begin{bmatrix} u^*/(u^*u) \\ v^* \\ Y \end{bmatrix}$$
$$= f(0)\frac{uu^*}{u^*u} + f'(0)uv^* + f(0)\frac{vv^*}{v^*v} + f(0)XY.$$

But $XY = I - uu^*/(u^*u) - vv^*/(v^*v)$ from (E.3), and hence $f(A) = f(0)I + f'(0)uv^*$.

1.16. The formula can be obtained from the polynomial interpolation definition by noting that:

(a) For $v^*u \neq 0$, $A = \alpha I + uv^*$ has a semisimple eigenvalue α of multiplicity $n-1$ and an eigenvalue $\alpha + v^*u$.

(b) For $v^*u = 0$, A has n eigenvalues α; $n-2$ of these are in 1×1 Jordan blocks and there is one 2×2 Jordan block.

The Sherman–Morrison formula is obtained by writing $(A+uv^*)^{-1} = \big(A(I+A^{-1}u\cdot v^*)\big)^{-1} = (I + A^{-1}u\cdot v^*)^{-1}A^{-1}$ and applying (1.16) with $f(x) = x^{-1}$.

1.17. Write $A = \lambda I_n + \left[\begin{smallmatrix} c \\ 0 \end{smallmatrix}\right] e_n^T$. Then, by (1.16),

$$f(A) = f(\lambda)I_n + f[\lambda,\lambda]\begin{bmatrix} c \\ 0 \end{bmatrix} e_n^T = \begin{bmatrix} f(\lambda)I_{n-1} & f'(\lambda)c \\ 0 & f(\lambda) \end{bmatrix}.$$

This of course agrees with the result of applying Theorem 1.21.

1.18. It is easy to see that for scalars x and y, $p(x)-p(y) = q(x,y)(x-y)$ for some polynomial q. We can substitute tI for x and A for y to obtain

$$p(t)I - p(A) = q(tI, A)(tI - A). \tag{E.4}$$

If $p(A) = 0$ then we have $p(t)(tI-A)^{-1} = q(tI,A)$, so that $p(t)(tI-A)^{-1}$ is a polynomial in t. Conversely, if $p(t)(tI-A)^{-1}$ is a polynomial in t then from (E.4) it follows that $p(A)(tI-A)^{-1} = p(t)(tI-A)^{-1} - q(tI,A)$ is a polynomial. Since $p(A)$ is a constant this implies that $p(A) = 0$.

To obtain the Cayley–Hamilton theorem set $p(t) = \det(tI - A)$. From the formula $B^{-1} = \mathrm{adj}(B)/\det(B)$, where the adjugate adj is the transpose of the matrix of cofactors, it follows that $p(t)(tI-A)^{-1} = \mathrm{adj}(tI - A)$ is a polynomial in t, so $p(A) = 0$ by the first part.

1.19. By Theorem 1.40 there is a unitary U such that $U^*AU = R$ and $U^*BU = S$ with R and S upper triangular. Thus $f(x,y) = \det(U(xR-yS)U^*) = \det(xR-yS) = \prod_{i=1}^n (xr_{ii} - ys_{ii})$. Hence

$$f(B,A) = \prod_{i=1}^n (Br_{ii} - As_{ii}) = \prod_{i=1}^n (U(Sr_{ii} - Rs_{ii})U^*) = U\cdot \prod_{i=1}^n (Sr_{ii} - Rs_{ii})\cdot U^*.$$

Now $f(B,A)$ is of the form $U\cdot\prod_{i=1}^n T_i\cdot U^*$, where T_i is upper triangular with zero (i,i) element, and it follows from [295, 1985, Lem. 2.4.1] that $\prod_{i=1}^n T_i = 0$. If A and B do not commute we can modify the proof to use the generalized Schur decomposition $Q^*AZ = T$, $Q^*BZ = S$, with Q and Z unitary and T and S upper triangular, but the proof then fails at the second equality in the displayed equation.

1.20. Let $Z^{-1}AZ = J = \mathrm{diag}(J_k)$ be a Jordan canonical form. Then $Z^{-1}(-A)Z = -J$ and so $Z^{-1}f(-A)Z = f(-J) = \mathrm{diag}(f(-J_k))$. Now $f(-z) = \pm f(z)$ implies $f^{(j)}(-z) = \pm(-1)^j f^{(j)}(z)$, and from (1.34) we see that $f(-J_k) = \pm f(J_k)$. Hence $\mathrm{diag}(f(-J_k)) = \mathrm{diag}(\pm f(J_k)) = \pm f(J)$, and the result follows.

If f is analytic, a shorter proof is obtained from the Cauchy integral definition. Let Γ be a closed contour enclosing $\Lambda(A)$ and Γ_1 its reflection in the imaginary axis, which encloses $\Lambda(-A)$. From (1.12),

$$\begin{aligned} f(-A) &= \frac{1}{2\pi i}\int_{\Gamma_1} f(z)(zI+A)^{-1}\,dz \\ &= \frac{1}{2\pi i}\int_{\Gamma} f(-w)(wI-A)^{-1}\,dw \qquad (w=-z) \\ &= \pm\frac{1}{2\pi i}\int_{\Gamma} f(w)(wI-A)^{-1}\,dw = \pm f(A). \end{aligned}$$

1.21. Assume $P \neq 0$ and $P \neq I$, since otherwise the result is immediate. Then the minimal polynomial is $t^2 - t = 0$, which implies that $Z^{-1}PZ = \left[\begin{smallmatrix} I & 0 \\ 0 & 0 \end{smallmatrix}\right]$ for some nonsingular Z. Hence $aI + bP = Z\left[\begin{smallmatrix} (a+b)I & 0 \\ 0 & aI \end{smallmatrix}\right]Z^{-1}$, so

$$\begin{aligned} f(aI+bP) &= Z\begin{bmatrix} f(a+b)I & 0 \\ 0 & f(a)I \end{bmatrix} Z^{-1} = Z\left(f(a)I + \begin{bmatrix} [f(a+b)-f(a)]I & 0 \\ 0 & 0 \end{bmatrix}\right) Z^{-1} \\ &= f(a)I + (f(a+b)-f(a))P. \end{aligned}$$

1.22. No. $f(A)$ is a polynomial in A and so cannot equal A^* in general (consider triangular A). However, if A is normal then $A = QDQ^*$ for some unitary Q and $D = \mathrm{diag}(\lambda_i)$, and for $f(\lambda) = \overline{\lambda}$ we have $f(A) = Qf(D)Q^* = Q\overline{D}Q^* = A^*$. As a check, for unitary matrices we have $f(A) = A^* = A^{-1}$ and for skew-Hermitian matrices $f(A) = A^* = -A$, so f is clearly a matrix function in both cases.

1.23. We have

$$(zI-A)^{-1} = \begin{bmatrix} z^{-1} & z^{-2} & \dots & z^{-n} \\ & z^{-1} & \dots & z^{-n+1} \\ & & \ddots & \vdots \\ & & & z^{-1} \end{bmatrix}.$$

Hence

$$f(A) = \frac{1}{2\pi i}\int_{\Gamma} z^j (zI-A)^{-1}\,dz = \frac{1}{2\pi i}\int_{\Gamma} \begin{bmatrix} z^{j-1} & z^{j-2} & \dots & z^{j-n} \\ & z^{j-1} & \dots & z^{j-n+1} \\ & & \ddots & \vdots \\ & & & z^{j-1} \end{bmatrix} dz$$

$$= \begin{bmatrix} 0 & \dots & \overset{j+1}{1} & \dots & 0 \\ & 0 & & \ddots & \vdots \\ & & \ddots & & 1 \\ & & & \ddots & \vdots \\ & & & & 0 \end{bmatrix} = A^j,$$

using the Cauchy residue theorem.

1.24. The formula (1.4) provides two upper triangular square roots, X_1 and X_2, corresponding to the two different branches of f at λ_k. We have to show that these are the only upper triangular square roots. Let X be an upper triangular square root of J_k. Then, equating (i,i) and $(i,i+1)$ elements in $X^2 = J_k$ gives $x_{ii}^2 = \lambda_k$ and $(x_{ii} + x_{i+1,i+1})x_{i,i+1} = 1$, $i = 1\colon m_k - 1$. The second equation implies $x_{ii} + x_{i+1,i+1} \neq 0$, so from the first, $x_{11} = x_{22} = \dots = x_{m_k,m_k} = \pm\lambda_k^{1/2}$. Since $x_{ii} + x_{jj} \neq 0$ for all i and j, X is uniquely determined by its diagonal elements (see Algorithm 4.13 or Algorithm 6.3); these are the same as those of X_1 or X_2, so $X = X_1$ or $X = X_2$.

1.25. $A^n = 0$ implies that all the eigenvalues of A are zero, and $A^{n-1} \neq 0$ then implies that the Jordan form of A comprises just a single $n \times n$ Jordan block. Hence A has no square root by Theorem 1.22. Or, in a more elementary fashion, if $A = X^2$ then $X^{2n} = A^n = 0$, which implies $X^n = 0$ since X is $n \times n$. But then $A^{n-1} = X^{2(n-1)} = X^n X^{n-2} = 0$ since $n \geq 2$, which is a contradiction. For the last part, we have $(A+cA^{n-1})^2 = A^2+2cA^n+c^2A^{2n-2} = A^2$ for any c.

1.26. Yes, if A is nonderogatory, but in general no: Theorem 1.25 gives the general form, from which it is clear that $Z^{-1}XZ$ is necessarily block diagonal if and only if A is nonderogatory.

1.27. We can write the Jordan canonical form of A as $A = Z \operatorname{diag}(J_1, 0)Z^{-1}$, where J_1 contains the Jordan blocks corresponding to the nonzero eigenvalues. With f denoting the square root function, any primary square root of A has the form

$$f(A) = Zf(\operatorname{diag}(J_1, 0))Z^{-1} = Z \operatorname{diag}(f(J_1), f(0))Z^{-1} = Z \operatorname{diag}(f(J_1), 0)Z^{-1}.$$

By Theorem 1.29, $f(J_1) = J_1^{1/2}$ is the unique square root of J_1 all of whose eigenvalues lie in the open right half-plane, and it is a primary matrix function of A. Hence $Z \operatorname{diag}(J_1^{1/2}, 0)Z^{-1}$ is the unique square root of the required form (being a primary square root it is independent of the choice of Z). That this square root is real when A is real can be seen with the aid of the real Schur decomposition or from Theorem 1.18 or Remark 1.9.

1.28. By direct calculation we find that all upper triangular square roots of A are of the form

$$X(\theta) = \pm \begin{bmatrix} 0 & 1 & \theta \\ & 1 & 1 \\ & & 0 \end{bmatrix},$$

where $\theta \in \mathbb{C}$ is arbitrary. Now A is involutory ($A = A^2$) so any polynomial in A has the form $p(A) = \alpha I + \beta A$, which has equal $(1,2)$ and $(1,3)$ elements. It follows that $\pm X(1)$ are the only primary square roots of A.

Since $\dim(\operatorname{null}(A)) = 2$, 0 is a semisimple eigenvalue of A and hence A is diagonalizable. Indeed

$$V^{-1}AV = \operatorname{diag}(1, 0, 0), \qquad V = \begin{bmatrix} 1 & -1 & 0 \\ 1 & -1 & -1 \\ 0 & 1 & 1 \end{bmatrix}.$$

A family of nonprimary square roots is obtained as

$$Y = V \operatorname{diag}(1, \left[\begin{smallmatrix} 0 & \theta \\ 0 & 0 \end{smallmatrix}\right])V^{-1} = \begin{bmatrix} -\theta & 1+\theta & 1 \\ -\theta & 1+\theta & 1 \\ \theta & -\theta & 0 \end{bmatrix}.$$

Note that for $\theta \neq 0$, Y has a different Jordan structure than A—a phenomenon that for matrix square roots can happen only when A is singular.

1.29. Note that $A = \operatorname{diag}(J_2(0), 0)$. Let $X^2 = A$ and consider the possible Jordan block structures of X. Applying Theorem 1.36 with $f(x) = x^2$ to any Jordan block of X we find that $\ell = 2$ and case b(ii) must pertain, with $r = 3$ and $p = q = 1$. Hence $X = ZJ_3(0)Z^{-1}$ for some nonsingular Z. To determine (as far as possible) Z we write $X^2 = A$ as $ZJ_3(0)^2 = AZ$ and examine the resulting equations, which force Z to have the form

$$Z = \begin{bmatrix} a & b & c \\ 0 & 0 & a \\ 0 & d & e \end{bmatrix},$$

where $\det(Z) = -a^2 d$ must be nonzero. Evaluating $X = ZJ_3(0)Z^{-1}$ we find that some of the remaining parameters in Z are redundant and the general form of X is

$$X = \begin{bmatrix} 0 & y & x \\ 0 & 0 & 0 \\ 0 & 1/x & 0 \end{bmatrix}, \qquad x \neq 0.$$

1.30. Let X be a square root of A and let X have Jordan matrix

$$\operatorname{diag}(J_1(0), \dots, J_k(0), 0, \dots, 0, J_{r+1}(\lambda_{r+1}), \dots, J_p(\lambda_p)),$$

where $J_i(0)$ is of size at least 2 for $i = 1{:}\,k$ and $\lambda_i \neq 0$ for $i \geq r+1$. By Theorem 1.36, $J_i(0)$ splits into smaller Jordan blocks when squared for $i = 1{:}\,k$, since $f'(0) = 0$ for $f(x) = x^2$. Therefore $A = X^2$ has more Jordan blocks than X. But any polynomial in A has no more Jordan blocks than A. Therefore X cannot be a polynomial in A.

1.31. The form of X is rather surprising. Since any primary square root of A is a polynomial in A, a first reaction might be to think that X is a nonprimary square root. However, X and A are both symmetric and structural considerations do not rule out X being a polynomial in A. In fact, A has distinct eigenvalues (known to be $0.25\sec(i\pi/(2n+1))^2$, $i = 1{:}\,n$ [189, 1997]), so all its square roots are primary. X is clearly not $A^{1/2}$, since X has zero elements on the diagonal and so is not positive definite. In fact, X has $\lceil n/2 \rceil$ positive eigenvalues and $\lfloor n/2 \rfloor$ negative eigenvalues (which follows from the inertia properties of a 2×2 block symmetric matrix—see, for example, Higham and Cheng [279, 1998, Thm. 2.1]). X is an indefinite square root that "just happens" to have a very special structure.

1.32. For any square root X we have $XA = X \cdot X^2 = X^2 \cdot X = AX$. Likewise, a logarithm X of A is a solution of $e^X = A$, so $XA = Xe^X = e^X X = AX$. If A is nonderogatory then by Theorem 1.37, X is a polynomial in A.

1.33. $A = e^B$ where B is the Jordan block $J_n(0)$, as is easily seen from (1.4). $B + 2k\pi i$ is also a logarithm for any $k \in \mathbb{Z}$, and these are all the logarithms, as can be seen from Theorem 1.28 on noting that A has just one block in its Jordan canonical form since $\operatorname{rank}(A - I) = n - 1$.

1.34. Let $X = \log A$ and $Y = e^{X/2}$. Then $Y^2 = e^{X/2}e^{X/2} = e^X = A$, using Theorem 10.2. So Y is some square root of A. The eigenvalues of Y are of the form $e^{\lambda_i/2}$, where λ_i is an eigenvalue of X and has $\operatorname{Im}\lambda_i \in (-\pi, \pi)$, and so $-\pi/2 < \arg\lambda_i/2 < \pi/2$. Thus the spectrum of Y lies in the open right half-plane, which means that $Y = A^{1/2}$.

1.35. $A^{1/2}$ is a polynomial in A and $B^{1/2}$ is a polynomial in B, so $A^{1/2}$ commutes with $B^{1/2}$. Therefore $(A^{1/2}B^{1/2})^2 = A^{1/2}B^{1/2}A^{1/2}B^{1/2} = A^{1/2}A^{1/2}B^{1/2}B^{1/2} = AB$. Thus $A^{1/2}B^{1/2}$ is some square root of AB. By Corollary 1.41 the eigenvalues of $A^{1/2}B^{1/2}$ are of the form $\lambda_i(A^{1/2})\lambda_i(B^{1/2})$ and so lie in the open right half-plane if the eigenvalues of A and B lie in the open right half-plane. The latter condition is needed to ensure the desired equality, as is clear by taking $A = B$.

1.36. Let $C = e^A$. The condition on $\Lambda(A)$ is equivalent to A being a primary logarithm of C, since the nonprimary logarithms are characterized by two copies of a repeated eigenvalue being mapped to different logarithms, which must differ by a nonzero integer multiple of $2\pi i$. So $A = \log(C) = p(C) = p(e^B) = \widetilde{p}(B)$ for some polynomials p and $\widetilde{p}$, and A therefore commutes with B.

An example showing the necessity of the condition on $\Lambda(A)$ is

$$A = \begin{bmatrix} 0 & 2\pi - 1 & 1 \\ -2\pi & 0 & 0 \\ -2\pi & 0 & 0 \end{bmatrix}, \qquad B = \begin{bmatrix} 0 & 2\pi & 1 \\ -2\pi & 0 & 0 \\ 0 & 0 & 0 \end{bmatrix},$$

for which $e^A = e^B = I$, both matrices have spectrum $\{0, 2\pi i, -2\pi i\}$, and $AB \neq BA$. Such examples are easily constructed using the real Jordan form analogue of Theorem 1.27.

Schmoeger [505, 2003] investigates what can be concluded from $e^A = e^B$ when the condition on $\Lambda(A)$ is not satisfied but A is normal.

1.37. Let $\lambda_1, \dots, \lambda_s$ be the distinct eigenvalues of $C = e^A$ and denote by q_i the algebraic multiplicity of λ_i. The q_k copies of λ_k are mapped to $\log \lambda_k + 2\pi r_j i$ in A and $\log \lambda_k + 2\pi s_j i$ in B, for some integers r_j and s_j, $j = 1\colon q_k$. The given condition implies that $r_j = s_j \equiv t_k$ for all j, so that all copies of λ_k are mapped to the same logarithm. This is true for all k, so A and B are primary logarithms with the same spectrum, which means that they are the same matrix.

1.38. Suppose, first, that f is even. From Theorem 1.26 we know that for any square root X of A, $f(X)$ has the form

$$f(X) = ZU \operatorname{diag}(f(L_k^{(j_k)}))U^{-1}Z^{-1},$$

where $A = ZJZ^{-1}$ is a Jordan canonical form with $J = \operatorname{diag}(J_k)$, $L_k^{(j_k)}$ is a square root of J_k, $j_k = 1, 2$, and U commutes with J. But $L_k^{(1)} = -L_k^{(2)}$ implies

$$f(L_k^{(1)}) = f(L_k^{(2)}) \tag{E.5}$$

by Problem 1.20. Hence $U \operatorname{diag}(f(L_k^{(j)}))U^{-1} = \operatorname{diag}(f(L_k^{(j)}))$ (cf. the proof of Theorem 1.26). Thus $f(X) = Z \operatorname{diag}(f(L_k^{(j_k)}))Z^{-1}$, which is the same for all choices of the j_k by (E.5).

The proof for f odd is very similar, reducing to the observation that $(L_k^{(j_k)})^{\pm 1} f(L_k^{(j_k)})$ is the same for $j_k = 1$ as for $j_k = 2$.

1.39. If $A = \log(e^A)$ then $\max\{\,|\operatorname{Im}(\lambda_i)| : \lambda_i \in \Lambda(A)\,\} < \pi$ is immediate from the definition of the principal logarithm. Suppose the latter eigenvalue condition holds and let $X = e^A$. X has no eigenvalues on $\mathbb{R}^-$ and so $\log(X)$ is defined. A is clearly some logarithm of X and its eigenvalues satisfy $\operatorname{Im} \lambda \in (-\pi, \pi)$. By Theorem 1.28, every logarithm other than $\log(X)$ has at least one eigenvalue with $|\operatorname{Im} \lambda| \geq \pi$. Therefore A must be the principal logarithm, $\log(X)$.

1.40. By applying g to the equation $f(A)f(B)f(A)^{-1} = f(B)$ and using Theorem 1.13 (c) we obtain $f(A)Bf(A)^{-1} = B$, which can be rewritten as $B^{-1}f(A)B = f(A)$. Applying g to this equation gives $B^{-1}AB = A$, or $AB = BA$, as required.

1.41. Let A have the spectral decomposition $A = Q\Lambda Q^*$, where, without loss of generality, we can suppose that the eigenvalues are ordered so that $\Lambda = \operatorname{diag}(\lambda_1 I_1, \dots, \lambda_m I_m)$, with $\lambda_1, \dots, \lambda_m$ distinct. Suppose X is a Hermitian positive definite square root of A, so that $A = X^2$. Then $\Lambda = Q^*AQ = Q^*X^2Q = (Q^*XQ)^2 =: Y^2$, where Y is Hermitian positive definite. Now Y clearly commutes with $\Lambda = Y^2$, so $Y = \operatorname{diag}(Y_k)$, where the blocking is conformable with that of Λ. It remains to determine the Y_k, which satisfy $Y_k^2 = \lambda_k I$. Now the eigenvalues of Y_k must be $\pm\lambda_k^{1/2}$, and since Y_k is positive definite, the eigenvalues must all be $\lambda_k^{1/2}$. But the only Hermitian matrix with all its eigenvalues equal to $\lambda_k^{1/2}$ is $Y_k = \lambda_k^{1/2} I_k$. Hence $X = QYQ^* = Q \operatorname{diag}(\lambda_k^{1/2} I_k)Q^* = Q\Lambda^{1/2}Q^*$ is the unique Hermitian positive definite square root of A.

1.42. Let Y be another square root of A with eigenvalues in the open right half-plane. Since $YA = Y^3 = AY$, Y commutes with A, and hence with any polynomial in A; in particular Y commutes with X. Therefore

$$(X + Y)(X - Y) = X^2 - Y^2 - XY + YX = X^2 - Y^2 = A - A = 0.$$

Since X and Y commute, the eigenvalues of $X + Y$ are of the form $\lambda_i(X) + \lambda_i(Y)$, by Corollary 1.41, and these are all nonzero since the spectra of X and Y are in the open right half-plane. Hence $X + Y$ is nonsingular and thus $X - Y = 0$, as required.

1.43. Using (1.29), we have

$$A(\alpha I - BA)^k = (\alpha I - AB)^k A.$$

Let $\alpha \neq 0$, $x \neq 0$, and $(\alpha I - BA)^k x = 0$. Then, on premultiplying the latter equation by A, we obtain $(\alpha I - AB)^k Ax = 0$. Now $Ax \neq 0$, since otherwise $\alpha x = 0$, which would be a contradiction. Hence $0 \neq x \in \operatorname{null}((\alpha I - BA)^k)$ implies $0 \neq Ax \in \operatorname{null}((\alpha I - AB)^k)$. This relation obviously holds with A and B interchanged. Hence $\operatorname{null}((\alpha I - BA)^k)$ is isomorphic to $\operatorname{null}((\alpha I - AB)^k)$ for all k when $\alpha \neq 0$. Taking α to be the nonzero eigenvalues gives the first part of the result.

For the last part, suppose first that $m > n$. Since $Bx = 0$ implies $ABx = 0$, AB has at least $m - n$ linearly independent null vectors, namely those of B, and hence AB has a zero eigenvalue with geometric multiplicity at least $m - n$. For $m < n$ the analogous argument applies.

1.44. Since the matrix $\left[\begin{smallmatrix} I_m & A \\ 0 & I_n \end{smallmatrix}\right]$ is nonsingular, the two outer matrices in (1.35) are similar, and so by applying f to this identity and using Theorem 1.13 (c) and (f) we obtain

$$\begin{aligned}\begin{bmatrix} f(AB) & 0 \\ X & f(0) \end{bmatrix}\begin{bmatrix} I_m & A \\ 0 & I_n \end{bmatrix} &= f\left(\begin{bmatrix} AB & 0 \\ B & 0 \end{bmatrix}\right)\begin{bmatrix} I_m & A \\ 0 & I_n \end{bmatrix} = \begin{bmatrix} I_m & A \\ 0 & I_n \end{bmatrix} f\left(\begin{bmatrix} 0 & 0 \\ B & BA \end{bmatrix}\right) \\ &= \begin{bmatrix} I_m & A \\ 0 & I_n \end{bmatrix}\begin{bmatrix} f(0) & 0 \\ Y & f(BA) \end{bmatrix}\end{aligned}$$

for some X and Y. Equating (1,2) blocks gives $f(AB)A = Af(BA)$.

This proof obviously requires that $f(0)$ be defined, which need not be the case in Corollary 1.34 when $m = n$. The proof requires, moreover, that f be defined on the spectrum of $\left[\begin{smallmatrix} AB & 0 \\ B & 0 \end{smallmatrix}\right]$. Now from

$$\begin{bmatrix} AB & 0 \\ B & 0 \end{bmatrix} = \begin{bmatrix} A & 0 \\ I & 0 \end{bmatrix}\begin{bmatrix} B & 0 \\ 0 & 0 \end{bmatrix} =: CD, \quad DC = \begin{bmatrix} B & 0 \\ 0 & 0 \end{bmatrix}\begin{bmatrix} A & 0 \\ I & 0 \end{bmatrix} = \begin{bmatrix} BA & 0 \\ 0 & 0 \end{bmatrix}$$

together with Theorem 1.32, it follows that for f to be defined on the spectrum of $\left[\begin{smallmatrix} AB & 0 \\ B & 0 \end{smallmatrix}\right]$ for all A and B we need that $f^{(k)}(0)$ be defined, where k is one more than the dimension of the largest Jordan block corresponding to zero in BA. Consider, for example, the scalar case $A = 0$, $B = 1$. The same requirement applies to AB. To summarize, this proof requires the existence of one extra derivative at 0, and so yields a slightly weaker result.

1.45. In a way similar to the proof of Problem 1.44, we find

$$\begin{bmatrix} f(AB + \alpha I_m) & 0 \\ X_1 & f(\alpha) I_n \end{bmatrix}\begin{bmatrix} I_m & A \\ 0 & I_n \end{bmatrix} = \begin{bmatrix} I_m & A \\ 0 & I_n \end{bmatrix}\begin{bmatrix} f(\alpha) I_m & 0 \\ X_2 & f(BA + \alpha I_n) \end{bmatrix} \tag{E.6}$$

for some X_1 and X_2. Equating (1,1) blocks gives $f(AB + \alpha I_m) = f(\alpha) I_m + AX_2$. To determine X_2 we use Theorem 1.13 (a) to obtain

$$\begin{aligned}\begin{bmatrix} \alpha I_m & 0 \\ B & BA + \alpha I_n \end{bmatrix}\begin{bmatrix} f(\alpha) I_m & 0 \\ X_2 & f(BA + \alpha I_n) \end{bmatrix} &= \begin{bmatrix} f(\alpha) I_m & 0 \\ X_2 & f(BA + \alpha I_n) \end{bmatrix} \\ &\quad \times \begin{bmatrix} \alpha I_m & 0 \\ B & BA + \alpha I_n \end{bmatrix}.\end{aligned}$$

Equating $(2,1)$ blocks gives $f(\alpha)B + (BA + \alpha I_n)X_2 = \alpha X_2 + f(BA + \alpha I_n)B$, or $X_2 = (BA)^{-1}(f(BA + \alpha I_n) - f(\alpha) I_n)B$. The result follows. As in Problem 1.44 this proof requires an extra assumption: in this case the existence of an extra derivative of f at α.

1.46. Setting $\alpha = 0$ in (1.31) gives

$$f(AB) = f(0)I_m + A(BA)^{-1}\bigl(f(BA) - f(0)I_n\bigr)B.$$

Postmultiplying by A gives

$$\begin{aligned} f(AB)A &= f(0)A + A(BA)^{-1}\bigl(f(BA) - f(0)I_n\bigr)BA \\ &= f(0)A + A(BA)^{-1}f(BA)BA - f(0)A \\ &= Af(BA), \end{aligned}$$

using Theorem 1.13 (a). Note that this proof requires the existence of $f(0)$, which is not needed for Corollary 1.34.

1.47. The formula (1.31) does not generalize in this way because $h(X) = f(D + X)$ is not a function of X according to our definition—more precisely, it does not correspond to a scalar "stem function" evaluated at X, because of the presence of D. In the "$\alpha I \to D$" generalization, the right-hand side of (1.31) would contain $f(D_n + BA)$ where D_n is an $n \times n$ diagonal matrix obtained from D, yet there is no reasonable way to define D_n.

1.48. Yes for $n \le 2$; no for $n > 2$. AB must have a Jordan form with eigenvalues all zero, these eigenvalues appearing in 1×1 or 2×2 blocks. BA has the same eigenvalues as AB, so the question is whether BA has Jordan blocks of dimension only 1 or 2. This can be answered using Theorem 1.32. But working instead from first principles, note that the dimensions of the Jordan blocks cannot exceed 3, because $BABABA = B(ABAB)A = 0$. There are obviously no counterexamples with $n = 2$, but for $n = 3$ we find in MATLAB

```
A =
     0     0     1
     0     0     0
     0     1     0
B =
     0     0     1
     1     0     0
     0     0     0
>> [A*B  A*B*A*B]
ans =
     0     0     0     0     0     0
     0     0     0     0     0     0
     1     0     0     0     0     0
>> [B*A B*A*B*A]
ans =
     0     1     0     0     0     1
     0     0     1     0     0     0
     0     0     0     0     0     0
```

1.49. Theorem 1.32 shows that AB and BA have the same nonzero eigenvalues, so $I_m + AB$ and $I_n + BA$ have the same nonunit eigenvalues. The result follows, since the determinant is the product of the eigenvalues. Alternatively, take determinants in (1.36).

1.50. The answer is no. The problem can be answered from first principles, using just the power series definition of $\sin A$. Assume the equation has a solution A. The eigenvalues of $\sin A$ are 1 and 1; hence the eigenvalues of A must be of the form $d_1 = (2p + \frac{1}{2})\pi$, $d_2 = (2q + \frac{1}{2})\pi$ for some integers p and q. If $p \ne q$ then A has distinct eigenvalues and so is diagonalizable: $A = XDX^{-1}$ for some X, with $D = \mathrm{diag}(d_i)$. But then $\sin A = X\sin(D)X^{-1} = XIX^{-1} = I$, which is a contradiction. Hence $p = q$, and by the same argument the equation cannot be satisfied if A is diagonalizable. Therefore $A = X\left[\begin{smallmatrix}\theta & 1\\ 0 & \theta\end{smallmatrix}\right]X^{-1}$ for some X, with $\theta = (2p + \frac{1}{2})\pi$. Then, either from the power series or from (4.17), $\sin A = X\left[\begin{smallmatrix}1 & \cos\theta\\ 0 & 1\end{smallmatrix}\right]X^{-1} = X\left[\begin{smallmatrix}1 & 0\\ 0 & 1\end{smallmatrix}\right]X^{-1} = I$, which is again a contradiction.

1.51. Denote the given matrix by B, and note that B has just one $n \times n$ Jordan block in its Jordan form, since $\operatorname{rank}(B - I) = n - 1$. Let $f(x) = \cosh(x)$. The eigenvalues of A must be $f^{-1}(1) = 0$. We have $f'(0) = \sinh(0) = 0$, $f''(0) = \cosh(0) = 1$. Theorem 1.36(b) (with $\ell = 2$) says that $f(A)$ must have more than one Jordan block. This contradicts the Jordan structure of B and hence the equation has no solution.

1.52. (a) $f(z) = 1/z$ for $z \neq 0$ and $f^{(j)}(0) = 0$ for all j. Hence f is discontinuous at zero. Nevertheless, $f(A)$ is defined.

(b) This formula is readily verified by direct computation using (1.37). Of course, Definition 1.4 directly yields A^D as a polynomial in A, and this polynomial may be of much smaller degree than $x^k p(x)^{k+1}$. We note also that if $B^{-k-1} = q(B)$ then $A^D = A^k q(A)$, which provides an alternative formula.

(c) As explained in Section 1.2.5, the index of uv^* is 1 if $v^*u \neq 0$ and 2 otherwise. If the index is 1 then $(uv^*)^D = 0$. Otherwise, $(uv^*)^D = uv^*/(v^*u)^2$, as can be obtained from (1.14), for example.

For more on the Drazin inverse, see Campbell and Meyer [92, 1979].

1.53. If $A \in \mathbb{C}^{m\times n}$ has the (compact) SVD $A = P\Sigma Q^*$ with $P \in \mathbb{C}^{m\times r}$, $\Sigma = \operatorname{diag}(\sigma_i) \in \mathbb{R}^{r\times r}$, $Q \in \mathbb{C}^{r\times n}$, where $r = \operatorname{rank}(A)$, we can define $f(A) = Pf(\Sigma)Q^*$, where $f(\Sigma) = \operatorname{diag}(f(\sigma_i))$. An alternative representation is $f(A) = Uf(H)$, where $A = UH$ is a polar decomposition (see Chapter 8) and $f(H)$ is the usual function of a matrix. This definition, which is given and investigated by Ben-Israel and Greville [52, 2003, Sec. 6.7], does not reduce to the usual definition when $m = n$. The definition does not appear to lead to any useful new insights or have any significant applications.

2.1. We have

$$\frac{d}{dt}\big(e^{-At}y(t)\big) = -Ae^{-At}y + e^{-At}y' = e^{-At}(y' - Ay) = e^{-At}f(t,y).$$

Hence $e^{-As}y(s) - y(0) = \int_0^s e^{-At}f(t,y)\,dt$. Multiplying through by e^{As} and interchanging the roles of s and t gives the result.

2.2. The function $g(x) = \cos(xt) + x^{-1}\sin(xt)$ is an even function of x. Hence by Problem 1.38, $g(\sqrt{A})$ is the same for all square roots $\sqrt{A}$ of A. Even more directly, expanding g as a power series shows that $g(\sqrt{A})$ is a power series in A and hence is independent of the choice of $\sqrt{A}$.

2.3. Let A have the Jordan canonical form

$$A = ZJZ^{-1} = Z\begin{bmatrix} J_1 & 0 \\ 0 & J_2 \end{bmatrix} Z^{-1},$$

where the eigenvalues of $J_1 \in \mathbb{C}^{p\times p}$ lie in the left half-plane and those of J_2 lie in the right half-plane. Then $\operatorname{sign}(A) = Z\operatorname{diag}(-I_p, I_{n-p})Z^{-1}$ and, as we noted in Section 2.5, range(W) is the invariant subspace corresponding to the eigenvalues in the right half-plane. Write $Q = [\,Q_1 \quad Q_2\,]$, where $Q_1 \in \mathbb{C}^{n\times q}$. Then

$$W\Pi = [\,Q_1 \quad Q_2\,]\begin{bmatrix} R_{11} & R_{12} \\ 0 & 0 \end{bmatrix} = Q_1\,[\,R_{11} \quad R_{12}\,].$$

Hence Q_1 is an orthogonal basis for range(W), which means that range(Q_1), too, is an invariant subspace corresponding to the eigenvalues in the right half-plane, and hence $q = n - p$. Thus $AQ_1 = Q_1 Y$, for some Y whose eigenvalues lie in the right half-plane. Hence

$$Q^T AQ = \begin{bmatrix} Q_1^T AQ_1 & Q_1^T AQ_2 \\ Q_2^T AQ_1 & Q_2^T AQ_2 \end{bmatrix} = \begin{bmatrix} Y & Q_1^T AQ_2 \\ (Q_2^T Q_1)Y & Q_2^T AQ_2 \end{bmatrix} = \begin{bmatrix} Y & Q_1^T AQ_2 \\ 0 & Q_2^T AQ_2 \end{bmatrix},$$

since $Q_2^T Q_1 = 0$. The result follows.

2.4. Since A and B commute, $A^{1/2}$ and $B^{1/2}$ commute. Hence $X = A^{1/2}B^{1/2}$ satisfies $XA^{-1}X = A^{1/2}B^{1/2}A^{-1}A^{1/2}B^{1/2} = B^{1/2}A^{1/2}A^{-1}A^{1/2}B^{1/2} = B$. Moreover, X is clearly Hermitian, and it has positive eigenvalues because it is the product of positive definite matrices.

For the log-Euclidean geometric mean E we have, since $\log(A)$ and $\log(B)$ commute,

$$E(A,B) = e^{\frac{1}{2}\log(A)+\frac{1}{2}\log(B)} = e^{\frac{1}{2}\log(A)}e^{\frac{1}{2}\log(B)} = e^{\log(A^{1/2})}e^{\log(B^{1/2})} = A^{1/2}B^{1/2},$$

where we have used Theorems 10.2 and 11.2 (or Problem 1.34).

2.5. $A \# A = A$ follows from any of the formulae. Next, using (2.26),

$$(A \# B)^{-1} = (B(B^{-1}A)^{1/2})^{-1} = (B^{-1}A)^{-1/2}B^{-1} = (A^{-1}B)^{1/2}B^{-1} = A^{-1} \# B^{-1}.$$

Inverting both sides of $XA^{-1}X = B$ gives $X^{-1}AX^{-1} = B^{-1}$ and thence $XB^{-1}X = A$, which shows that $A \# B = B \# A$.

Finally, using the first formula in (2.26), the given inequality is

$$B^{1/2}(B^{-1/2}AB^{-1/2})^{1/2}B^{1/2} \le \frac{1}{2}(A + B).$$

Premultiplying and postmultiplying by $B^{-1/2}$ gives the equivalent inequality $C^{1/2} \le \frac{1}{2}(C + I)$, where $C = (B^{-1/2}AB^{-1/2})^{1/2}$ is Hermitian positive definite, and this may be rewritten $(C^{1/2} - I)^2 \ge 0$, which is trivially true.

2.6. Since $A \# B = (\frac{A}{\alpha} \# \frac{B}{\beta})(\alpha\beta)^{1/2}$ we can assume without loss of generality that $\alpha = \beta = 1$. From $A \# B = B(B^{-1}A)^{1/2}$ and Problem 6.2 we have

$$A \# B = \frac{A + \sqrt{\det(B^{-1}A)}\,B}{\sqrt{\operatorname{trace}(B^{-1}A) + 2\sqrt{\det(B^{-1}A)}}}.$$

Let $\Lambda(B^{-1}A) = \{\lambda_1, \lambda_2\}$. Since $\det(B^{-1}A) = \det(A)/\det(B) = \alpha^2/\beta^2 = 1$, we have $\lambda_1 = 1/\lambda_2 \equiv \lambda$. Hence $\operatorname{trace}(B^{-1}A) + 2\sqrt{\det(B^{-1}A)} = \lambda + \lambda^{-1} + 2$. But $\det(A + B) = \det(B)\det(B^{-1}A + I) = (\lambda + 1)(\lambda^{-1} + 1) = \lambda + \lambda^{-1} + 2$, which yields the result.

2.7. We have $XR^*RX = B$ and hence $(RXR^*)^2 = RBR^*$, so $RXR^* = (RBR^*)^{1/2}$. Hence $X = R^{-1}(RBR^*)^{1/2}R^{-*}$, which is clearly Hermitian positive definite. Any of the methods from Chapter 6 can be used to evaluate $(RBR^*)^{1/2}$. This given formula is more efficient to evaluate than the formulae given in Section 2.9, but Algorithm 6.22 is even better.

3.1. By determining the linear part of $f(X + E) - f(X)$, we find that $L(X, E) = 0$, E, and $-\sin(X)E$, respectively, using (12.4) in the latter case. If the second of these expressions seems counterintuitive, note that $L(X, E) = E$ says that $L(X)$ is the identity *operator*.

3.2. We have

$$\begin{aligned} L(X,E) &= \lim_{t\to 0} \frac{f(X + tE) - f(X)}{t} \\ &= Q\left(\lim_{t\to 0} \frac{f(T + t\widetilde{E}) - f(T)}{t}\right)Q^* \\ &= QL(T, \widetilde{E})Q^*, \end{aligned}$$

where $\widetilde{E} = Q^*EQ$, as required.

3.3. If L and M are both Fréchet derivatives of f then

$$f(X+E)-f(X)=L(X,E)+o(\|E\|)=M(X,E)+o(\|E\|),$$

and so $L(X,E)-M(X,E)=o(\|E\|)$. Let $E=t\widetilde{E}$ with $\widetilde{E}$ fixed. Then, by the linearity of the Fréchet derivative, $L(X,E)-M(X,E)=t(L(X,\widetilde{E})-M(X,\widetilde{E}))=o(t\|\widetilde{E}\|)$, that is, $L(X,\widetilde{E})-M(X,\widetilde{E})=t^{-1}o(t\|\widetilde{E}\|)$, Taking the limit as $t\to 0$ gives $L(X,\widetilde{E})=M(X,\widetilde{E})$. Since $\widetilde{E}$ is arbitrary, $L(X)=M(X)$.

3.4. The Fréchet derivative is defined by

$$f(X+E)-f(X)-L(X,E)=o(\|E\|).$$

Setting $E=tF$, for fixed F and varying t, we have, since $L(X)$ is a linear operator,

$$f(X+tF)-f(X)-tL(X,F)=f(X+tF)-f(X)-L(X,tF)=o(\|tF\|),$$

which implies

$$\lim_{t\to 0}\Big(\frac{f(X+tF)-f(X)}{t}-L(X,F)\Big)=0,$$

showing that $L(X,F)$ is the derivative in the direction F.

3.5. The maximum Jordan block size is n. The maximum is attained when A_{11} and A_{22} are Jordan blocks corresponding to the same eigenvalue λ and $A_{12}=e_{n_1}e_{n_2}^T$. For example, with $n_1=2$, $n_2=3$,

$$A=\left[\begin{array}{cc|ccc} \lambda & 1 & 0 & 0 & 0\\ 0 & \lambda & 1 & 0 & 0\\ \hline & & \lambda & 1 & 0\\ & & & \lambda & 1\\ & & & & \lambda \end{array}\right]$$

and $\operatorname{rank}(A-\lambda I)=4$, so A has a single 5×5 Jordan block corresponding to λ. For a more precise result of this form see Mathias [412, 1996, Lem. 3.1].

3.6. It suffices to check that

$$f(X+E)=f(X)+\sum_{i=1}^{\infty}a_i\sum_{j=1}^{i}X^{j-1}EX^{i-j}+O(\|E\|^2),$$

since $L(X,E)$ is the linear term in this expansion. The matrix power series has the same radius of convergence as the given scalar series (see Theorem 4.7), so if $\|X\|<r$ we can scale $E\to\theta E$ so that $\|X+\theta E\|<r$ and the expansion is valid. But $L(X,\theta E)=\theta L(X,E)$, so the scaled expansion yields $L(X,E)$. $K(X)$ is obtained by using (B.16).

3.7. Straightforward from (3.24).

3.8. If f is a polynomial, $p(t)=\sum_{i=0}^m a_it^i$, then as a special case of (3.24) we see that $L(X,E)=E\sum_{i=1}^m ia_iX^{i-1}=Ef'(X)=f'(X)E$. For general f, Theorem 3.7 implies that the Fréchet derivative $L(X,E)$ of f is the same as that of the polynomial $p_{X\oplus X}$ and so the result follows from the polynomial case.

3.9. From the Cauchy integral definition (1.11) we have, for small enough $\|E\|$,

$$f(X+E)=\frac{1}{2\pi i}\int_{\Gamma}f(z)(zI-X-E)^{-1}\,dz.$$

By expanding the inverse we find that

$$\begin{aligned} f(X+E)-f(X) &= \frac{1}{2\pi i}\int_\Gamma f(z)\big[(zI-X-E)^{-1}-(zI-X)^{-1}\big]\,dz \\ &= \frac{1}{2\pi i}\int_\Gamma f(z)(zI-X)^{-1}E(zI-X)^{-1}\,dz + O(\|E\|^2). \end{aligned}$$

The expression for $L(X,E)$ follows. If X commutes with E then all the terms in the integral commute and by using Cauchy's integral formula for derivatives we find that this expression is equal to $f'(X)E$ and $Ef'(X)$.

3.10. Let X have the Jordan canonical form $X = ZJZ^{-1}$, where $J = \operatorname{diag}(J_k)$. Without loss of generality we can assume that $u = Ze_i$ and $v^T = e_j^T Z^{-1}$ for some i and j. Then $Je_i = \lambda e_i$ and $e_j^T J = \mu e_j^T$, which means that the (i,i) entry of J marks the start of a Jordan block and the (j,j) entry marks the end of a Jordan block. There are now two cases.

(a) $i=j$. In this case $\lambda=\mu$ and J has a 1×1 Jordan block λ in the (i,i) position. With $E = uv^T$,

$$f(X+tE) = f(Z(J+te_ie_i^T)Z^{-1}) = Zf(J+te_ie_i^T)Z^{-1}.$$

Now $f(J+te_ie_i^T)$ agrees with $f(J)$ except in the (i,i) position, where it equals $f(\lambda+t) = f(\lambda)+tf'(\lambda)+O(t^2)$. Hence

$$\begin{aligned} f(X+tE) &= Zf(J)Z^{-1} + [tf'(\lambda)+O(t^2)]Ze_ie_i^TZ^{-1} \\ &= f(X) + [tf'(\lambda)+O(t^2)]uv^T \\ &= f(X) + f'(\lambda)tE + O(t^2), \end{aligned}$$

which shows that $L(X,E) = f'(\lambda)E$.

(b) $i\ne j$. With $E = uv^T$, we have $f(X+tE) = Zf(J+te_ie_j^T)Z^{-1}$. It is easy to see from Theorem 4.11 that $f(J+te_ie_j^T)$ agrees with $f(J)$ except in the (i,j) entry, which equals $f[\lambda,\mu]$. Hence

$$f(X+tE) = f(X) + f[\lambda,\mu]tZe_ie_j^TZ^{-1} = f(X) + f[\lambda,\mu]tE,$$

which means that $L(X,E) = f[\lambda,\mu]E$.

4.1. The formula for $f(C)$ is obtained by forming the powers of C, its validity following from Theorem 4.7. We can express e^C concisely as

$$e^C = \begin{bmatrix} \cosh\sqrt{AB} & A(\sqrt{BA})^{-1}\sinh\sqrt{BA} \\ B(\sqrt{AB})^{-1}\sinh\sqrt{AB} & \cosh\sqrt{BA} \end{bmatrix}, \tag{E.7}$$

where $\sqrt{X}$ denotes any square root of X.

By introducing $z = y'$, we can rewrite (2.7) as

$$\begin{bmatrix} z \\ y \end{bmatrix}' = \begin{bmatrix} 0 & -A \\ I & 0 \end{bmatrix}\begin{bmatrix} z \\ y \end{bmatrix}, \qquad \begin{bmatrix} z(0) \\ y(0) \end{bmatrix} = \begin{bmatrix} y_0' \\ y_0 \end{bmatrix}.$$

Hence

$$\begin{bmatrix} z \\ y \end{bmatrix} = \exp\left(\begin{bmatrix} 0 & -A \\ I & 0 \end{bmatrix}\right)\begin{bmatrix} y_0' \\ y_0 \end{bmatrix},$$

and (2.8) follows on using (E.7) and $\cosh(iA) = \cos(A)$, $\sinh(iA) = i\sin(A)$.

4.2. Let p_{km}/q_{km} and $\widetilde{p}_{km}/\widetilde{q}_{km}$ be two $[k/m]$ Padé approximants to $f(x)$. Then from the definition (4.12) it follows that

$$\frac{p_{km}\widetilde{q}_{km} - \widetilde{p}_{km}q_{km}}{q_{km}\widetilde{q}_{km}} = \frac{p_{km}}{q_{km}} - \frac{\widetilde{p}_{km}}{\widetilde{q}_{km}} = O(x^{k+m+1}). \tag{E.8}$$

But $q_{km}\widetilde{q}_{km} = O(1)$, since $q_{km}(0) = \widetilde{q}_{km}(0) = 1$, and $p_{km}\widetilde{q}_{km} - \widetilde{p}_{km}q_{km}$ is a polynomial of degree at most $k+m$. Hence (E.8) can hold only if $p_{km}\widetilde{q}_{km} - \widetilde{p}_{km}q_{km} = 0$, that is, $p_{km}/q_{km} = \widetilde{p}_{km}/\widetilde{q}_{km}$.

4.3. Obtaining (4.46) is straightforward. To show the equivalence of (4.44) and (4.46) we just have to show that F_{12} given by (4.46) satisfies (4.44), since we know that (4.44) has a unique solution. Using $f(T_{ii})T_{ii} = T_{ii}f(T_{ii})$, we have

$$\begin{aligned} T_{11}F_{12} - F_{12}T_{22} &= T_{11}f(T_{11})X - T_{11}Xf(T_{22}) - f(T_{11})XT_{22} + Xf(T_{22})T_{22} \\ &= f(T_{11})(T_{11}X - XT_{22}) - (T_{11}X - XT_{22})f(T_{22}) \\ &= f(T_{11})T_{12} - T_{12}f(T_{22}) \end{aligned}$$

by (4.45), so that F_{12} satisfies (4.44).

4.4. We have $Y = f(TT^{-1}ST^{-1}) = Tf(T^{-1}S)T^{-1} = TXT^{-1}$. By Theorem 1.13 (a), $T^{-1}SX = XT^{-1}S$, i.e., $SX = TXT^{-1}S = YS$. Similarly, $ST^{-1}Y = YST^{-1}$, i.e., $T^{-1}YT = S^{-1}YS = X$, or $YT = TX$.

Now $x_{ii} = y_{ii} = f(s_{ii}/t_{ii})$, $i = 1\colon n$. The off-diagonal elements are computed using an analogue of the Parlett recurrence obtained from (4.47):

$$\begin{bmatrix} s_{ii} & -s_{jj} \\ t_{ii} & -t_{jj} \end{bmatrix} \begin{bmatrix} x_{ij} \\ y_{ij} \end{bmatrix} = \begin{bmatrix} s_{ij}(y_{ii} - x_{jj}) + \sum_{k=i+1}^{j-1}(y_{ik}s_{kj} - s_{ik}x_{kj}) \\ t_{ij}(f_{ii} - f_{jj}) + \sum_{k=i+1}^{j-1}(f_{ik}t_{kj} - t_{ik}f_{kj}) \end{bmatrix}.$$

This system can be solved provided the coefficient matrix is nonsingular, that is, $s_{ii}t_{jj} - s_{jj}t_{ii} \neq 0$.

4.5. Let $J(\lambda) = \left[\begin{smallmatrix} 1 & 1 \\ 0 & 1 \end{smallmatrix}\right]$, $\lambda = 1$, $\phi_0(\lambda) = \lambda$. For $g(x) = x^2$ we have $x_* \equiv x_k = 1$ yet $X_k = \left[\begin{smallmatrix} 1 & 2^k \\ 0 & 1 \end{smallmatrix}\right]$ diverges. On the other hand, for $g(x) = (x^2+1)/2$ we have $x_* \equiv x_k = 1$ yet $X_k \equiv J(\lambda) = X_*$.

4.6. Let $\delta^{-1}A = XJX^{-1} = X(\delta^{-1}D+M)X^{-1}$ be the Jordan canonical form of $\delta^{-1}A$, where $\delta > 0$, with $D = \mathrm{diag}(\lambda_i)$ containing the eigenvalues of A and M the off-diagonal part of the Jordan form. Then $A = X(D + \delta M)X^{-1}$, so $A^k = X(D + \delta M)^k X^{-1}$ and hence

$$\|A^k\|_p \leq \kappa_p(X)(\rho(A) + \delta)^k \tag{E.9}$$

for any p-norm. If $\rho(A) < 1$ we can choose $\delta > 0$ such that $\rho(A) + \delta < 1$ and so $\|A^k\|_p$ is bounded for all k. By the equivalence of matrix norms (see, e.g., [537, 1998, Thm. 4.6]) the result holds for any matrix norm. The bound (E.9) is a special case of a result of Ostrowski [455, 1973, Thm. 20.1]. Notice that this argument actually shows that $A^k \to 0$ if $\rho(A) < 1$.

More precisely, A is power bounded if and only if $\rho(A) \leq 1$ and for any eigenvalue λ such that $|\lambda| = 1$, λ is semisimple (i.e., λ appears only in 1×1 Jordan blocks). However, for our purposes of obtaining sufficient conditions for stability, the sufficient condition $\rho(A) < 1$ is all we need.

4.7. Any limit y_* must satisfy $y_* = cy_* + d$, so that $y_* = d/(1 - c)$, and

$$y_{k+1} - y_* = c_k(y_k - y_*) + \underbrace{y_*(c_k - c) + d_k - d}_{\to 0},$$

so it suffices to take $d = 0$ and show that $y_k \to 0$. There exist k and $q \in [0, 1)$ such that $|c_i| \leq q < 1$ for $i \geq k$; let $D = \max_{i \geq k} |d_i|$. For $E \geq D/(1 - q)$, if $|y_i| \leq E$ and $i \geq k$ then $|y_{i+1}| \leq qE + D \leq E$. Hence with $M = \max\{E, \max_{i \leq k} |y_i|\}$ we have $|y_i| \leq M$ for all i.

Given $\epsilon > 0$, there exists $n(\epsilon)$ such that for all $i \geq n(\epsilon)$, $|c_i| \leq q < 1$ and $|d_i| \leq \epsilon$. Then, for $i \geq n(\epsilon)$, $|y_{i+1}| \leq q|y_i| + \epsilon$, and so $|y_{n(\epsilon)+j}| \leq q^j M + \epsilon/(1 - q) \leq 2\epsilon/(1 - q)$ for large enough j. It follows that $y_i \to 0$, as required.

4.8. This result is immediate from the Newton identities (4.20).

5.1. From (5.2) we have $\mathrm{sign}(A) = A(A^2)^{-1/2} = A \cdot I^{-1/2} = A$.

5.2. $\operatorname{sign}(A) = \operatorname{sign}(A^{-1})$. The easiest way to see this is from the Newton iteration (5.16), because both $X_0 = A$ and $X_0 = A^{-1}$ lead to $X_1 = \frac{1}{2}(A+A^{-1})$ and hence the same sequence $\{X_k\}_{k\geq 1}$.

5.3. Applying the Cauchy integral formula with $f(z) = z^{-1/2}$ and using a Hankel contour that goes from $-\infty - 0i$ to 0 then around the origin and back to $-\infty + 0i$, with $t = iz^{1/2}$, so that $dt = \frac{1}{2}iz^{-1/2}dz$, we have

$$A(A^2)^{-1/2} = 2 \cdot \frac{1}{2\pi i} A \int_0^\infty z^{-1/2}(zI - A^2)^{-1}\, dz = \frac{1}{\pi} A \int_0^\infty (t^2 I + A^2)^{-1}\, dt.$$

5.4. Using the matrix analogue of the formula $\int (x^2 + a^2)^{-1}\, dx = a^{-1} \arctan(x/a)$, from (5.3) we have

$$\operatorname{sign}(A) = \frac{2}{\pi} \arctan(tA^{-1})\Big|_0^\infty = \lim_{t\to\infty} \frac{2}{\pi} \arctan(tA^{-1}).$$

The result follows, since $\operatorname{sign}(A) = \operatorname{sign}(A^{-1})$.

5.5. No: A^2 differs from I in the (1,3) and (2,3) entries. A quick way to arrive at the answer without computing A^2 is to note that if A is the sign of some matrix then since $a_{22} = a_{33} = 1$ we must have $A(2\colon 3, 2\colon 3) = I$ (see the discussion following Algorithm 5.5), which is a contradiction.

5.6. The result follows from Theorem 5.2, since $C = B(A^{-1}B)^{-1/2} = B(B^{-1}A)^{1/2}$.

5.8. Let Y be an approximate solution to $X^2 = I$ and write $Y + E = S$, where $S = \operatorname{sign}(A)$. Then $I = S^2 = (Y + E)^2 = Y^2 + YE + EY + E^2$. Dropping the second order term in E gives Newton's method, which defines E as the solution to the Sylvester equation $YE + EY = I - Y^2$. If Y is nonsingular then a solution to this equation (the unique solution if $\lambda_i(Y) + \lambda_j(Y) \neq 0$ for all i and j) is $E = \frac{1}{2}(Y^{-1} - Y)$, whence $Y + E = \frac{1}{2}(Y + Y^{-1})$, which leads to the iteration (5.16) if we take $X_0 = A$.

5.9. Let $F = I - (S + E)^2 = -(SE + ES + E^2)$. Then, for sufficiently small E,

$$\begin{aligned}
\operatorname{sign}(S + E) &= (S + E)(I - F)^{-1/2} \\
&= (S + E)(I + \tfrac{1}{2}F + O(\|F\|^2)) \\
&= S + E - \frac{1}{2}(E + SES) + O(\|E\|^2) \\
&= S + \frac{1}{2}(E - SES) + O(\|E\|^2),
\end{aligned}$$

and the result follows.

5.10. Let $x_k = \coth\theta_k$. Then

$$\begin{aligned}
\coth\theta_{k+1} &= \frac{1}{2}(\coth\theta_k + \tanh\theta_k) = \frac{\cosh^2\theta_k + \sinh^2\theta_k}{2\cosh\theta_k \sinh\theta_k} \\
&= \frac{\cosh 2\theta_k}{\sinh 2\theta_k} = \coth 2\theta_k.
\end{aligned}$$

Hence $\theta_{k+1} = 2\theta_k$, and so $\theta_k = 2^k\theta_0$ follows by induction. Now $\coth x = (e^x + e^{-x})/(e^x - e^{-x})$, so $\coth 2^k\theta_0 \to 1$ or -1 as $k \to \infty$ according as $\operatorname{Re}\theta_0 > 0$ or $\operatorname{Re}\theta_0 < 0$, or equivalently, $\operatorname{Re} x_0 > 0$ or $\operatorname{Re} x_0 < 0$.

5.11. Let $x_0 = ir_0$ with r_0 real. It is easy to show by induction that $x_k = ir_k$, where r_k is real and

$$r_{k+1} = \frac{1}{2}\left(r_k - \frac{1}{r_k}\right).$$

The x_k cannot converge because they are pure imaginary and the only possible limits are ± 1. Setting $r_k = -\cot(\pi\theta_k)$, we have

$$\begin{aligned}
-\cot(\pi\theta_{k+1}) = r_{k+1} &= \frac{1}{2}\left(-\cot(\pi\theta_k) + \frac{1}{\cot(\pi\theta_k)}\right) \\
&= \frac{1}{2}\frac{-\cos(\pi\theta_k)^2 + \sin(\pi\theta_k)^2}{\cos(\pi\theta_k)\sin(\pi\theta_k)} \\
&= -\frac{1}{2}\frac{\cos(2\pi\theta_k)}{\frac{1}{2}\sin(2\pi\theta_k)} = -\cot(2\pi\theta_k).
\end{aligned}$$

So $\theta_{k+1} = 2\theta_k$, or equivalently, given the periodicity of cot,

$$\theta_{k+1} = 2\theta_k \bmod 1. \tag{E.10}$$

The behaviour of the r_k is completely described by this simple iteration. If θ_0 has a periodic binary expansion then periodic orbits are produced. If θ_0 has a terminating binary expansion then eventually $\theta_k = 0$, that is, $r_k = \infty$. Irrational θ_0 lead to sequences r_k in which the same value never occurs twice. The mapping (E.10) is known as the Bernoulli shift [165, 1992, Ex. 3.8].

5.13. For $\mu > 0$ we have

$$g(\mu) := d(\mu X) = \sum_{i=1}^{n}(\log\mu + \log|\lambda_i|)^2,$$

and hence

$$g'(\mu) = \frac{2}{\mu}\sum_{i=1}^{n}(\log\mu + \log|\lambda_i|).$$

Solving $g'(\mu) = 0$ gives $\log(|\lambda_1|\dots|\lambda_n|) = -\log\mu^n$, or $\mu = (|\lambda_1|\dots|\lambda_n|)^{-1/n} = |\det(X)|^{-1/n}$. The last part is trivial.

5.14. (a) If a is real, $x_1 = \operatorname{sign}(a)$. Otherwise, $a = re^{i\theta}$ and $\gamma_0 x_0 = e^{i\theta}$ lies on the unit circle, $x_1 = \cos\theta$ is real, and hence $x_2 = \operatorname{sign}(a)$.

(b) In view of Theorem 5.11, it suffices to consider the case where $A \in \mathbb{R}^{2\times 2}$ has a complex conjugate pair of eigenvalues, $\lambda = re^{\pm i\theta}$, $\theta \in (0,\pi)$. Then $\mu_0 X_0$ has eigenvalues $e^{\pm i\theta}$ and X_1 has equal eigenvalues $\cos\theta$. The next iterate, X_2, has eigenvalues ± 1 and the iteration has converged, since the Jordan form of A is necessarily diagonal.

5.15. We make use of the observation that if $|x| < 1$ then $(1+x)^{1/2}$ has a convergent Maclaurin series $1 + \sum_{k=1}^{\infty} a_k x^k$ such that $\sum_{k=1}^{\infty}|a_k||x|^k = 1 - \sqrt{1-x}$. Since $\operatorname{sign}(A) = I$ we have $A = (A^2)^{1/2}$ and hence $A = (I+E)^{1/2} = I + \sum_{k=1}^{\infty} a_k E^k$, since $\|E\| < 1$. Then

$$\|A - I\| = \left\|\sum_{k=1}^{\infty} a_k E^k\right\| \le \sum_{k=1}^{\infty}|a_k|\|E\|^k = 1 - \sqrt{1-\|E\|} = \frac{\|E\|}{1+\sqrt{1-\|E\|}} < \|E\|.$$

The upper bound in (5.40), when specialized to $\operatorname{sign}(A) = I$, is $\|A - I\| \le \|E\|$, which is up to a factor 2 weaker. The conditions under which the two bounds hold, $\|I - A\| < 1$ and $\|I - A^2\| < 1$, are not easily compared, but the latter can hold when the former is far from holding.

5.16. The test (5.46), based on the fact that $X_k^2 \to I$, is a weaker version of (5.45). While $\operatorname{trace}(X_k^2)$ can be computed on $O(n^2)$ flops, $\|X_k^2 - I\|$ can be estimated in $O(n^2)$ flops by Algorithm 3.21, so there is no real advantage to (5.46) for iterations that do not form X_k^2.

The test (5.47), which exploits the fact that $\operatorname{trace}(\operatorname{sign}(S))$ is an integer, is at first sight attractive, as it is of negligible cost and may appear to be immune to the effects of roundoff. The major drawback of the test is that since it is based on the (sum of the) eigenvalues of X_k, it ignores nonnormality in X_k, and so may stop too early (see Theorem 5.10). In addition, the test could be satisfied by chance, well before convergence.

5.17. Consider the sign iteration (5.16) with $X_0 = W$. It is easy to check that the X_k are all Hamiltonian. Write the iteration as

$$X_{k+1} = \frac{1}{2}(X_k + (JX_k)^{-1}J), \qquad X_0 = W,$$

or

$$Y_{k+1} = \frac{1}{2}(Y_k + JY_k^{-1}J), \qquad Y_0 = JW,$$

where the matrix $Y_k = JX_k$ is Hermitian and is just X_k with its blocks rearranged and their signs changed.

6.1. Straightforward on evaluating the (1,2) block of (5.3) with $A \leftarrow \begin{bmatrix} 0 & A \\ I & 0 \end{bmatrix}$.

6.2. Let $A = \begin{bmatrix} a & b \\ c & d \end{bmatrix}$ and $\delta = \det(A)^{1/2} = (ad - bc)^{1/2}$. Then $\operatorname{trace}(X) = \pm(a + d \pm 2\delta)^{1/2}$. Hence

$$X = \pm(a + d \pm 2\delta)^{-1/2} \begin{bmatrix} a \pm \delta & b \\ c & d \pm \delta \end{bmatrix}.$$

If $a + d \pm 2\delta \neq 0$ then A has distinct eigenvalues and all four square roots are given by this formula. Otherwise, A has repeated eigenvalues and the formula breaks down for at least one of the choices of sign in the term $\pm 2\delta$. In this case there may be no square roots; there may be just two (when the Jordan form of A has one 2×2 Jordan block), which the formula gives; or there may be infinitely many square roots (in this case $A = aI$) and the formula gives just $\pm a^{1/2} I$.

6.3. Let $A(\epsilon) = \begin{bmatrix} \epsilon & 1 \\ 0 & \epsilon \end{bmatrix}$. Then

$$A(\epsilon)^{1/2} = \begin{bmatrix} \epsilon^{1/2} & \dfrac{1}{2\epsilon^{1/2}} \\ 0 & \epsilon^{1/2} \end{bmatrix}.$$

So $\|A(\epsilon)^{1/2}\|^2 / \|A(\epsilon)\| \approx \epsilon^{-1}/4 \to \infty$ as $\epsilon \to 0$.

6.4. Let X be a primary square root of A. If $z \in \operatorname{range}(A)$ then, for some y, $z = Ay = X^2 y = X(Xy) \in \operatorname{range}(X)$, so $\operatorname{range}(A) \subseteq \operatorname{range}(X)$. The converse inclusion (and indeed the desired equality) is trivial if A, and hence X, is nonsingular, so we can assume A is singular. If $z \in \operatorname{range}(X)$ then, for some y, $z = Xy = p(A)y$, for some polynomial p, by Definition 1.4. Now since 0 is an eigenvalue of A, one of the interpolation conditions (1.7) is $p(0) = f(0) = 0$, so p has zero constant term and thus has the form $p(t) = tq(t)$ for a polynomial q. Hence $z = Aq(A)y \in \operatorname{range}(A)$, giving $\operatorname{range}(X) \subseteq \operatorname{range}(A)$, as required.

If A is singular and the square root is nonprimary we can have $\operatorname{range}(A) \subset \operatorname{range}(X)$, as is the case for $A = 0$ and $X = \begin{bmatrix} 0 & 1 \\ 0 & 0 \end{bmatrix}$, for example.

6.5. If there is just one zero eigenvalue then it is easy to see that Algorithm 6.3 runs to completion and computes a primary square root. Otherwise, $u_{ii} = u_{jj} = 0$ for some $i \neq j$ and the algorithm breaks down with division by zero at the stage where it is solving (6.5). There are now two possibilities. First, (6.5) may be inconsistent. Second, (6.5) may be automatically satisfied because the right-hand side is zero; if so, what value to assign to u_{ij} in order to obtain a primary square root, or indeed any square root, may depend on information from the later steps of the algorithm.

For example, consider

$$T = \begin{bmatrix} 0 & 1 & 1 \\ 0 & 1 & 1 \\ 0 & 0 & 0 \end{bmatrix} = \begin{bmatrix} 0 & 1 & x \\ 0 & 1 & 1 \\ 0 & 0 & 0 \end{bmatrix} \begin{bmatrix} 0 & 1 & x \\ 0 & 1 & 1 \\ 0 & 0 & 0 \end{bmatrix} = U^2.$$

T has a semisimple zero eigenvalue. Algorithm 6.3 computes the first superdiagonal of U and then for $i = 1$, $j = 3$, (6.5) has the form $0 \cdot x = 1 - 1 = 0$. We can assign x any value, but only $x = 1$ produces a primary square root: if $x \neq 1$ then U has rank 2 and hence its zero eigenvalue appears in a Jordan block of size 2.

For the matrix

$$T = \begin{bmatrix} 0 & 1 & 0 & 1 & 0 \\ & 1 & 0 & 1 & 0 \\ & & 0 & 1 & 0 \\ & & & 1 & 0 \\ & & & & 0 \end{bmatrix},$$

which has semisimple zero eigenvalue of multiplicity 3, any upper triangular square root has the form

$$U = \begin{bmatrix} 0 & 1 & a & \frac{1}{2} - a & b \\ & 1 & 0 & \frac{1}{2} & 0 \\ & & 0 & 1 & c \\ & & & 1 & 0 \\ & & & & 0 \end{bmatrix},$$

where a, b, and c are arbitrary subject to $ac = 0$. But the constraint on a and c is not discovered until the last step of Algorithm 6.3, and for U to be a primary square root we need $\operatorname{rank}(U) = 2$ and hence $b = c = 0$ and $a = 0$.

The conclusion is that for singular matrices it is best to employ the reordered Schur form, as described in Section 6.2.

6.6. Consider Algorithm 6.3 and suppose, first, that $T \in \mathbb{R}^{n \times n}$. The diagonal elements satisfy

$$\widehat{u}_{ii} = \sqrt{t_{ii}}(1 + \delta_i), \qquad |\delta_i| \le u.$$

For the off-diagonal elements, using the analysis of inner products in [276, 2002, Sec. 3.1] we find that, whatever ordering is used in the summation,

$$(\widehat{u}_{ii} + \widehat{u}_{jj})\widehat{u}_{ij}(1 + \theta_3) = t_{ij} - \sum_{k=i+1}^{j-1} \widehat{u}_{ik}\widehat{u}_{kj}(1 + \theta_{n-2}),$$

where $|\theta_k| \le \gamma_k$. Hence $\widehat{U}^2 = T + \Delta T$, $|\Delta T| \le \gamma_{n-2}|\widehat{U}|^2$. The same analysis holds for complex data but the constants must be increased slightly. This can be accounted for by replacing γ_{n-2} by $\widetilde{\gamma}_{n-2}$, or $\widetilde{\gamma}_n$ for simplicity. For Algorithm 6.7 the errors in forming U_{ii} in (6.9) mean that only a normwise bound can be obtained.

6.7. We find $X_1 = \frac{1}{2}(A+I) = \frac{1}{2}\operatorname{diag}(a+1, b+1) = \frac{1}{2}\operatorname{diag}(a+1, -(a+1))$. Hence for $k = 1$, the linear system in (6.11) is singular (in view of (B.18)–(B.20)) and the method breaks down. However, the system is consistent and (6.12) generates one particular solution—the diagonal one.

6.8. Since X_0 does not commute with A we cannot invoke Theorem 6.9. Applying the iteration we find that

$$X_1 = \begin{bmatrix} 1 & \xi\theta \\ 0 & \mu \end{bmatrix}, \qquad \xi = \frac{1 - \mu}{2},$$

and hence

$$X_k = \begin{bmatrix} 1 & \xi^k\theta \\ 0 & \mu \end{bmatrix}.$$

Thus we have linear convergence to $A^{1/2}$ if $|\xi| < 1$ (except when $\xi = 0$, i.e., $\mu = 1$, which gives convergence in one step) and divergence if $|\xi| \ge 1$. For real μ, these two situations correspond to $-1 < \mu < 3$ and $\mu < -1$ or $\mu \ge 3$, respectively. Hence quadratic convergence, and even convergence itself, can be lost when X_0 does not commute with A.

6.9. (a) The result in the hint is proved by using the spectral decomposition $C = QDQ^*$ (Q unitary, $D = \mathrm{diag}(d_i) > 0$) to rewrite the equation as $\widetilde{X}D + D\widetilde{X} = \widetilde{H}$, where $\widetilde{X} = Q^*XQ$, $\widetilde{H} = Q^*HQ$. Then $\widetilde{X} = \widetilde{H} \circ D_1$, where $D_1 = ((d_i + d_j)^{-1})$ and $\circ$ is the Hadamard (entrywise) product. The matrix D_1 is a Cauchy matrix with positive parameters and hence is positive definite (as follows from its upper triangular LU factor having positive diagonal [276, 2002, Sec. 28.1]). The Hadamard product of a Hermitian positive definite matrix with a Hermitian positive (semi)definite matrix is Hermitian positive (semi)definite [296, 1991, Thm. 5.2.1], so $\widetilde{X}$ and hence X are Hermitian positive (semi)definite.

From (6.11) follow the three relations

$$X_k(X_{k+1} - X_k) + (X_{k+1} - X_k)X_k = A - X_k^2, \tag{E.11}$$

$$X_kX_{k+1} + X_{k+1}X_k = A + X_k^2, \tag{E.12}$$

$$X_{k+1}^2 - A = (X_{k+1} - X_k)^2. \tag{E.13}$$

From $X_0 > 0$, (E.12), and the hint, it follows that $X_1 > 0$, and (E.13) gives $X_1^2 \geq A$. Assume $X_k^2 \geq A$ and $X_k > 0$. Then

(i) $X_{k+1} \leq X_k$ by (E.11) and the hint,

(ii) $X_{k+1} > 0$ by (E.12) and the hint,

(iii) $X_{k+1}^2 \geq A$ by (E.13).

Hence all but one of the required inequalities follow by induction. The remaining inequality, $X_k \geq A^{1/2}$, follows from (iii) on invoking Theorem B.9.

The sequence X_k is nonincreasing in the positive semidefinite ordering and bounded below by $A^{1/2}$, so it converges to a limit, $X_* > 0$, by Lemma B.8 (c). From (E.13) it follows that $X_*^2 = A$. But $A^{1/2}$ is the only positive definite square root, so $X_* = A^{1/2}$.

(b) If X_0 commutes with A then by Lemma 6.8 the full and simplified Newton iterations generate exactly the same sequence, so monotonic convergence holds for (6.12). However, for arbitrary $X_0 > 0$ the simplified iteration (6.12) does not, in general, converge.

(c) Since $X^2 = A$ if and only if $(Z^{-1}XZ)^2 = Z^{-1}AZ$, part (a) can be applied to $\widetilde{A} = Z^{-1}AZ$ and $\widetilde{X}_0 = Z^{-1}X_0Z$, for which $\widetilde{X}_k = Z^{-1}X_kZ$. Monotonic convergence holds for the $\widetilde{X}_k > 0$.

6.10. If the iteration is to converge then it must converge on the spectrum of A, that is, the iterations

$$x_{k+1} = \frac{1}{2}\left(x_k + \frac{\lambda}{x_k}\right), \quad x_0 = \lambda$$

must converge for each eigenvalue λ of A. If $\lambda \in \mathbb{R}^-$ then $x_k \in \mathbb{R}$ for all k and so x_k cannot converge to either of the square roots of λ, both of which are pure imaginary. Hence the Newton iteration does not converge. In view of the relation with the Newton sign iteration given in Theorem 6.9 (which is valid for scalar a even when $a \in \mathbb{R}^-$), the behaviour of the iteration is described by Problem 5.11.

6.11. The symmetrization step is legal because $X_kA = AX_k \Rightarrow X_kA^{1/2} = A^{1/2}X_k$. The variables X_k and Y_k are easily seen to satisfy (6.15).

6.12. The modified iteration is only linearly convergent, as is easily verified numerically. The reasoning used to derive the modified iteration is dubious for an iteration that is already quadratically convergent.

6.13. The iterates satisfy $Y_k = A^{-1}X_k$, so to analyze the errors in the (exact) iteration we must set $F = A^{-1}E$, which gives

$$G(A^{1/2} + E, A^{-1/2} + F) = \frac{1}{2}\begin{bmatrix} E - A^{-1/2}EA^{1/2} \\ A^{-1}E - A^{-1/2}EA^{-1/2} \end{bmatrix} + O\left(\left\|\begin{bmatrix} E \\ F \end{bmatrix}\right\|^2\right).$$

Now all iterates X_k are functions of A, as is $A^{1/2}$, so E, which represents $X_k - A^{1/2}$, is a function of A and so commutes with A and $A^{1/2}$. Thus the first order term is zero and

$$G(A^{1/2} + E, A^{-1/2} + F) = O\left(\left\|\begin{bmatrix} E \\ F \end{bmatrix}\right\|^2\right),$$

which implies the quadratic convergence of the DB iteration (near the solution—this analysis does not prove global convergence).

6.14. The uncoupled recurrence is numerically unstable [274, 1997], so this "simplification" is not recommended. The iteration (6.63) has exactly the same stability properties as the Newton iteration (6.12).

6.15. $C \geq 0$ implies that $\rho(C)$ is an eigenvalue of C (see Theorem B.7), and since the cardioid extends only to 1 on the positive real axis, if $\Lambda(C)$ lies in the cardioid then $\rho(C) < 1$, so the spectrum of C must lie inside the unit circle. So the requirement on $\rho(C)$ cannot be relaxed.

6.16. We need show that X_k defined by (6.46) with $X_0 = D^{1/2}$ is related to B_k from (6.48) by $X_k = D^{1/2} + B_k$. This is clearly true for $k = 0$. Suppose it is true for k. Then the modified Newton iteration (6.46) is

$$D^{1/2}E_k + E_k D^{1/2} = A - X_k^2 = A - D - B_k^2 - D^{1/2}B_k - B_k D^{1/2},$$

that is,

$$D^{1/2}(E_k + B_k) + (E_k + B_k)D^{1/2} = A - D - B_k^2.$$

Comparing with (6.48), we see that $B_{k+1} = E_k + B_k = X_{k+1} - X_k + B_k = X_{k+1} - D^{1/2}$, as required.

6.17. The eigenvalues μ_i of C are $\mu_i = 1 - \lambda_i/s$. At a minimum we must have $1 - \lambda_1/s = \min_i \mu_i = -\max_i \mu_i = -(1 - \lambda_n/s)$, which yields the claimed values of s and $\rho(C)$.

6.18. We have $C = I - A/s \equiv I - \mu A$, where $\mu = 1/s$. Now $\|I - \mu A\|_F = \|\operatorname{vec}(I) - \operatorname{vec}(A)\mu\|_2$, so the problem is essentially a 1-variable linear least squares problem. The normal equations are $\operatorname{vec}(A)^* \operatorname{vec}(A)\mu = \operatorname{vec}(A)^* \operatorname{vec}(I)$, or $\operatorname{trace}(A^*A)\mu = \operatorname{trace}(A^*)$, as required.

6.19. For A with a semisimple zero eigenvalue linear convergence to $A^{1/2}$ holds if the nonzero eigenvalues of A satisfy the conditions of Theorem 6.16, with $A^{1/2}$ now denoting the matrix in Problem 1.27.

By using the Jordan form and Theorem 4.15 the convergence problem as a whole is reduced to showing the linear convergence to $\lambda^{1/2}$ of $x_{k+1} = x_k + \alpha(\lambda - x_k^2)$, $x_0 = (2\alpha)^{-1}$, for every eigenvalue λ of A. Only the convergence for a semisimple $\lambda = 0$ is in question. We just have to show that the iteration $x_{k+1} = x_k - \alpha x_k^2$, $x_0 = (2\alpha)^{-1}$ converges to 0 when $0 < \alpha \leq \rho(A)^{-1/2}$; in fact, irrespective of the value of $\alpha > 0$ we have $0 < x_{k+1} < x_k < \cdots < x_0$ and so x_k converges to the unique fixed point 0.

6.20. We are given that $A = sI - B$ with $B \geq 0$ and $s > \rho(B)$. Since $\operatorname{diag}(B) = \operatorname{diag}(s - a_{ii}) \geq 0$, $s \geq \max_i a_{ii}$ is necessary. Let $\alpha = \max_i a_{ii}$. Then $A = \alpha I - C$, where $C = B + (\alpha - s)I$. Now $C \geq 0$, since $c_{ij} = b_{ij} \geq 0$ for $i \neq j$ and $c_{ii} = \alpha - a_{ii} \geq 0$. Since $B \geq 0$ and $C \geq 0$, $\rho(B)$ is an eigenvalue of B and $\rho(C)$ an eigenvalue of C, by Theorem B.7. Hence $\rho(C) = \rho(B) + \alpha - s < \alpha$. Hence we can take $s = \alpha$ in (6.52).

6.21. One way to derive the algorithm is via the matrix sign function. We noted in Section 2.9 that X is the (1,2) block of $\operatorname{sign}(\left[\begin{smallmatrix} 0 & B \\ A & 0 \end{smallmatrix}\right])$. Given Cholesky factorizations $A = R^*R$ and $B = S^*S$ we have

$$\begin{bmatrix} 0 & B \\ A & 0 \end{bmatrix} = \begin{bmatrix} 0 & S^*S \\ R^*R & 0 \end{bmatrix} = \begin{bmatrix} R^{-1} & 0 \\ 0 & S^{-1} \end{bmatrix} \begin{bmatrix} 0 & RS^* \\ SR^* & 0 \end{bmatrix} \begin{bmatrix} R & 0 \\ 0 & S \end{bmatrix}.$$

Hence

$$\operatorname{sign}\left(\begin{bmatrix} 0 & B \\ A & 0 \end{bmatrix}\right) = \begin{bmatrix} R^{-1} & 0 \\ 0 & S^{-1} \end{bmatrix} \begin{bmatrix} 0 & U \\ U^{-1} & 0 \end{bmatrix} \begin{bmatrix} R & 0 \\ 0 & S \end{bmatrix},$$

where, in view of (8.6), $RS^* = UH$ is a polar decomposition. From the (1,2) block of this equation we obtain $X = R^{-1}US$. That X is Hermitian positive definite can be seen from $X = R^{-1}U \cdot S = S^*H^{-1} \cdot S$. In fact, we have obtained a variant of Algorithm 6.22. For any nonsingular A, the unitary polar factor of A is the conjugate transpose of the unitary polar factor of A^*. Hence it is equivalent to find the unitary polar factor $V = U^*$ of SR^* and then set $X = R^{-1}V^*S \equiv R^{-1}US$.

6.22. The principal square root is

$$\begin{bmatrix} 1 & -(2+\epsilon)^{-1} & \dfrac{2(1+\epsilon)}{\epsilon^2\,(2+\epsilon)\,(2+3\,\epsilon)} \\ 0 & 1+\epsilon & (2+3\,\epsilon)^{-1} \\ 0 & 0 & 1+2\,\epsilon \end{bmatrix},$$

which has norm $O(1/\epsilon^2)$. However, the non-principal square root

$$\begin{bmatrix} 1 & \epsilon^{-1} & 0 \\ 0 & -(1+\epsilon) & \epsilon^{-1} \\ 0 & 0 & 1+2\,\epsilon \end{bmatrix}$$

has norm only $O(1/\epsilon)$.

7.1. We use Theorem 7.3. As noted in Section 1.5, the ascent sequence is $d_1 = 1$, $d_2 = 1$, $\dots$, $d_m = 1$, $0 = d_{m+1} = d_{m+2} = \cdots$. Hence for $\nu = 0$ and $m > 1$ there is more than one element of the sequence in the interval $(p\nu, p(\nu+1)) = (0, p)$. Thus for $m > 1$ there is no pth root.

7.2. Assuming A has no eigenvalues on $\mathbb{R}^-$, we can define $A^\alpha = \exp(\alpha \log A)$, where the log is the principal logarithm. For $\alpha = 1/p$, with p an integer, this definition yields $A^{1/p}$, as defined in Theorem 7.2. To see this, note that, by commutativity, $(A^\alpha)^p = \exp(\alpha \log A)^p = \exp(p\alpha \log A) = \exp(\log A) = A$, so that A^α is *some* pth root of A. To determine which root it is we need to find its spectrum. The eigenvalues of A^α are of the form $e^{\frac{1}{p}\log\lambda}$, where λ is an eigenvalue of A. Now $\log \lambda = x + iy$ with $y \in (-\pi, \pi)$, and so $e^{\frac{1}{p}\log\lambda} = e^{x/p}e^{iy/p}$ lies in the segment $\{\, z : -\pi/p < \arg(z) < \pi/p \,\}$. The spectrum of A^α is therefore precisely that of $A^{1/p}$, so these two matrices are one and the same.

For $\alpha \in (0,1)$ we can also define A^α by (7.1) with $p = 1/\alpha$.

7.3. Let X have Jordan canonical form $X = ZJZ^{-1}$. Then $I_n = X^p = ZJ^pZ^{-1}$, that is, $J^p = I_n$. This implies that the eigenvalues λ (the diagonal elements of J) satisfy $\lambda^p = 1$, i.e., they are pth roots of unity. But then if the Jordan form is nontrivial we see from (1.4) that J^p has nonzero elements in the upper triangle. This contradicts $J^p = I_n$, so J must be diagonal.

7.4. From Theorem 3.5 with $f(x) = x^p$ and $f^{-1}(x) = x^{1/p}$ we have $L_{x^p}\left(X, L_{x^{1/p}}(X^p, E)\right) = E$. Now $L_{x^p}(X, E) = \sum_{j=1}^{p} X^{j-1}EX^{p-j}$ (cf. (3.24)) and so with $X^p = A$, $L = L_{x^{1/p}}(A, E)$ satisfies $\sum_{j=1}^{p} X^{j-1}LX^{p-j} = E$, or

$$\sum_{j=1}^{p} (X^{p-j})^T \otimes X^{j-1} \cdot \operatorname{vec}(L) = \operatorname{vec}(E). \tag{E.14}$$

The formula for $\operatorname{cond}(X)$ follows on solving for $\operatorname{vec}(L)$ and taking 2-norms. The eigenvalues of the coefficient matrix in (E.14) are, by (B.20), $\sum_{j=1}^{p} \mu_r^{p-j}\mu_s^{j-1} = (\mu_r^p - \mu_s^p)/(\mu_r - \mu_s)$

for $\mu_r \neq \mu_s$, where the μ_i are the eigenvalues of X. It follows that the coefficient matrix is nonsingular, and the condition number finite, precisely when every copy of an eigenvalue of A is mapped to the same pth root, which is equivalent to saying X that is a primary pth root, by Theorem 7.1.

7.5. The proof is analogous to that of Lemma 6.8.

7.7. Since $A^{1/p} = \|A\|_2^{1/p} C^{1/p}$ where $C = A/\|A\|_2 \leq I$, we can assume $A \leq I$ without loss of generality. Given $0 \leq Z \leq Y \leq I$ and $YZ = ZY$ we have, using (B.15),

$$Y^p - Z^p = (Y-Z)(Y^{p-1} + Y^{p-2}Z + \cdots + Z^{p-1}) \leq (Y-Z)(I+I+\cdots+I) = p(Y-Z). \quad \text{(E.15)}$$

Suppose $I \geq X_k \geq X_{k-1}$. Then

$$\begin{aligned} X_{k+1} - X_k &= X_k + \frac{1}{p}(A - X_k^p) - X_{k-1} - \frac{1}{p}(A - X_{k-1}^p) \\ &= X_k - X_{k-1} - \frac{1}{p}(X_k^p - X_{k-1}^p) \\ &\geq 0 \end{aligned}$$

by (E.15). Moreover, $X_0 \leq I$ and if $X_k \leq I$ then, using (E.15) again,

$$X_{k+1} = X_k + \frac{1}{p}(A - X_k^p) \leq X_k + \frac{1}{p}(I - X_k^p) \leq X_k + I - X_k = I.$$

Since $0 = X_0 < X_1 = A/p \leq I$, it follows that $0 \leq X_k \leq X_{k+1} \leq I$ for all k, by induction. The sequence X_k, being monotonically nondecreasing and bounded above, has a limit, $X > 0$ (see Lemma B.8 (c)). But X must satisfy $X = X + \frac{1}{p}(A - X^p)$, and so $X^p = A$, as required.

7.8. For $R_k = I - X_k A$ we have $R_{k+1} = I - (2I - X_k A)X_k A = I - 2X_k A + (X_k A)^2 = (I - X_k A)^2 = R_k^2$, and likewise for $R_k = I - AX_k$.

7.9. Since X_0 does not commute with A we cannot invoke Theorem 7.10 or Theorem 7.12. Applying the iteration we find that

$$X_1 = \begin{bmatrix} 1 & \xi\theta \\ 0 & \mu \end{bmatrix}, \qquad \xi = \frac{2\mu^2 - \mu - 1}{2\mu^2},$$

and hence

$$X_k = \begin{bmatrix} 1 & \xi^k\theta \\ 0 & \mu \end{bmatrix}.$$

Thus we have linear convergence to $A^{-1/2}$ if $|\xi| < 1$ (except when $\xi = 0$, which gives convergence in one step) and divergence if $|\xi| \geq 1$. For real μ, these two situations correspond to $\mu > -1$ and $\mu \leq -1$, respectively. This example shows that (quadratic) convergence can be lost when X_0 does not commute with A.

7.10. We find from the proof of Theorem 7.10 that $a_2 = (1/2)(1 + 1/p)$, which decreases from 3/4 for $p = 2$ to 1/2 as $p \to \infty$. So once the error is sufficiently small we can expect slightly faster convergence for larger values of p.

7.11. The formula $A^{1/p} = AX^{p-1}$ can be used.

7.13. Since X_0, and hence all the iterates and B, commute with A, we can use the given factorization with $x \leftarrow X_k$ and $b \leftarrow B$ to obtain

$$\begin{aligned} X_{k+1} - B &= \frac{X_k^{1-p}}{p}((p-1)X_k^p - pX_k^{p-1}B + A) \\ &= (X_k - B)^2 \frac{X_k^{1-p}}{p} \sum_{i=0}^{p-2} (i+1) X_k^i B^{p-2-i}. \end{aligned}$$

We conclude that

$$\|X_{k+1} - B\| \le \|X_k - B\|^2 \frac{\|X_k^{1-p}\|}{p} \sum_{i=0}^{p-2} (i+1)\|X_k\|^i \|B\|^{p-2-i}.$$

Hence we can take $c = (p-1)\|B^{1-p}\|\,\|B\|^{p-2} \le (p-1)\kappa(B)^{p-1}/\|B\|$.

7.14. The relation (7.11) holds for this iteration, too, if in (7.11) X/p is replaced by $-X/p$.

7.15. Gershgorin's theorem (Theorem B.1) shows that every eigenvalue lies in one of the disks $|z - a_{ii}| \le 1 - a_{ii}$, and by diagonal dominance we have $a_{ii} > 0.5$, so the spectrum lies in $E(1,p)$ in (7.17). Hence the iteration with $X_0 = I$ converges to $A^{-1/p}$ by Theorem 7.12.

7.16. We have

$$\mathrm{vec}(A - \widetilde{X}^p) = -\left(\sum_{i=0}^{p-1} (X^{p-1-i})^T \otimes X^i\right) \mathrm{vec}(E) + O(\|E\|^2).$$

Hence, taking the 2-norm,

$$\|A - \widetilde{X}^p\|_F \le \|E\|_F \left\|\sum_{i=0}^{p-1} (X^{p-1-i})^T \otimes X^i\right\|_2 + O(\|E\|_F^2)$$

is a sharp bound, to first order in E. In particular, if $\widetilde{X}$ is a correctly rounded pth root then we would expect $\rho_A(\widetilde{X})$ to be of order u. The alternative quantity $\|A - \widetilde{X}^p\|/\|\widetilde{X}^p\|$ is smaller and results from writing $\widetilde{X}^p = A + \sum_{i=0}^{p-1} X^i E X^{p-1-i} + O(\|E\|^2)$ and bounding $\|X^i E X^{p-1-i}\| \le \epsilon \|X\|^p$, which is not a sharp bound.

8.1. Direct computation shows that $H = (A^*A)^{1/2} = \mathrm{diag}(0,1,\dots,1)$. Then $A = UH$ implies that $U(:,2\colon m) = I(:,1\colon m-1)$. The first column of U is determined by the requirement $U^*U = I$ and must be of the form θe_m, where $|\theta| = 1$. If we require a real decomposition then $\theta = \pm 1$.

8.3. In both cases $H = 0$. For the polar decomposition U can be any matrix with orthonormal columns. For the canonical polar decomposition, $U = AH^+ = 0$.

8.4. Note that $V = Q^{1/2}$ is unitary with spectrum in the open right half-plane, so $V + V^* = V + V^{-1}$ is Hermitian and positive definite. Thus the polar decomposition is $I + Q = Q^{1/2} \cdot (Q^{1/2} + Q^{-1/2}) \equiv UH$. Hence $Q^{1/2}$ can be computed by applying (8.17) with $A = I + Q$. This result is a special case of a more general result applying to matrices Q in the automorphism group of a scalar product [283, 2005, Thm. 4.7].

8.5. Let $Q_1 + Q_2 = UH$. It is straightforward to verify that $H = (Q_1^*Q_2)^{1/2} + (Q_1^*Q_2)^{-1/2}$ satisfies $(Q_1 + Q_2)^*(Q_1 + Q_2) = 2I + Q_1^*Q_2 + Q_2^*Q_1 = H^2$. The given eigenvalue condition ensures that H is positive definite. Hence $U = (Q_1 + Q_2)H^{-1} = (Q_1 + Q_2) \cdot (Q_1 + Q_2)^{-1}Q_1(Q_2^*Q_2)^{1/2}$. This of course generalizes Problem 8.4.

8.6. The result is immediate from $\|Hx\|_2^2 = x^*H^2x = x^*A^*Ax = \|Ax\|_2^2$.

8.7. Let A have the SVD $A = P[\Sigma \ 0]Q^*$, where $\Sigma \in \mathbb{R}^{m\times m}$ is possibly singular. Then $A = P[I_m \ 0]Q^* \cdot Q\left[\begin{smallmatrix} \Sigma & 0 \\ 0 & G \end{smallmatrix}\right]Q^* \equiv UH$ is a polar decomposition for any Hermitian positive semidefinite $G \in \mathbb{C}^{(n-m)\times(n-m)}$. The decomposition always exists, but H is never unique. The nonuniqueness is clear in the extreme case $m = 1$, with $A = a^*$, $U = u^*$. Then

$$a^* = u^*H =: \frac{a^*\widetilde{H}^{-1}}{\|a^*\widetilde{H}^{-1}\|_2} \cdot \|a^*\widetilde{H}^{-1}\|_2 \widetilde{H}$$

is a polar decomposition for any Hermitian positive definite $\widetilde{H}$ (e.g., $\widetilde{H} = I$). Another polar decomposition is (8.4): $a^* = \|a\|_2^{-1}a^* \cdot \|a\|_2^{-1}aa^*$.

8.8. We are given that $A = UH$ with $U^* = U^+$ and $H = H^*$. The given condition range(U^*) = range(H) is equivalent to $U^+U = HH^+$, by Theorem 8.3. We know from the theorem that $A^+ = H^+U^+$. Thus $AA^+ = UHH^+U^+ = UU^+UU^+ = UU^+$, so that range$(A)$ = range(U).

Conversely, we are given range(U) = range(A), i.e., $UU^+ = AA^+$, and need to show that range(U^*) = range(H), i.e., (since $U^* = U^+$) $U^+U = HH^+$. Now, recalling that H and H^+ commute, since $U = AH^+$ we have

$$U^+U = U^*U = H^+A^*AH^+ = H^+H^2H^+ = HH^+HH^+ = HH^+,$$

as required.

8.9. Using the spectral decomposition of A, we have $A = Q\Lambda Q^* = Q\,\mathrm{Sign}(\Lambda)Q^* \cdot Q|\Lambda|Q^* \equiv UH$. Thus $U = f(A)$ with $f(z) = \mathrm{Sign}(z) = z/|z|$ (not the same sign function as in Chapter 5) and $H = g(A)$ with $g(z) = |z|$.

8.10. We use the same notation as in the proof given for the Frobenius norm. Note that $A^*E+E^*A = HU^*E+E^*UH = H(U^*Q-I_n)+(Q^*U-I_n)H = HU^*Q+Q^*UH-2H$. Let t be an eigenvalue of H that maximizes $(t-1)^2$ and let w with $\|w\|_2 = 1$ be a corresponding eigenvector. Then, from (8.9),

$$\|A - U\|_2^2 = (t-1)^2. \tag{E.16}$$

Since $\|B\|_2^2 = \max_{\|z\|_2=1} z^*B^*Bz$, (8.11) implies

$$\begin{aligned}\|A - Q\|_2^2 &\ge w^*(H - I_n)^2w - w^*(HU^*Q + Q^*UH - 2H)w\\ &= (t-1)^2 - (tw^*U^*Qw + tw^*Q^*Uw - 2t)\\ &= (t-1)^2 - 2t(\mathrm{Re}\, w^*U^*Qw - 1)\\ &\ge (t-1)^2 \end{aligned} \tag{E.17}$$

on using $\mathrm{Re}\, w^*U^*Qw \le |w^*U^*Qw| \le \|Uw\|_2\|Qw\|_2 = 1$. The result follows from (E.16) and (E.17).

8.11. The result follows from (8.14) and

$$\|B^*A - W\|_F^2 = \mathrm{trace}(A^*BB^*A + I) - \mathrm{trace}(A^*BW + W^*B^*A).$$

Note that while the minimizers are the same, the minimal values of the two objective functions are different.

8.12. Let $Q \in \mathbb{C}^{m\times n}$ be any partial isometry and let A have the SVD $A = P\left[\begin{smallmatrix}\Sigma_r & 0\\ 0 & 0_{m-r,n-r}\end{smallmatrix}\right]Q^*$, where $r = \mathrm{rank}(A)$. Then, by Theorem B.3,

$$\|A - Q\| \ge \|\,\mathrm{diag}(\sigma_1 - \mu_1, \dots, \sigma_r - \mu_r, -\mu_{r+1}, \dots, -\mu_{\min(m,n)})\|,$$

where the μ_i are the singular values of Q. Since $\mu_i \in \{0, 1\}$ for all i (see Lemma B.2) we have

$$\|A - Q\| \ge \|\,\mathrm{diag}(f(\sigma_1), \dots, f(\sigma_r), 0, \dots, 0)\|,$$

where $f(x) = \min\{x, |1-x|\}$. The lower bound is attained for $Q = P\,\mathrm{diag}(\mu_i)V^*$ with $\mu_i = 1$ if $\sigma_i \ge 1/2$, $\mu_i = 0$ if $\sigma_i \le 1/2$, $i = 1\colon r$, and $\mu_i = 0$ for $i > r$.

8.13. Substituting the SVD $A = P\Sigma Q^*$ we have to maximize

$$\begin{aligned} f(W) &= \mathrm{Re\,trace}(W^*A) = \mathrm{Re\,trace}(W^*P\Sigma Q^*) = \mathrm{Re\,trace}(Q^*W^*P\Sigma)\\ &=: \mathrm{Re\,trace}(V\Sigma),\end{aligned}$$

where $V = Q^*W^*P$ is unitary with $\det(V) = \det(PQ^*)\det(W^*)$. Let $r = \mathrm{rank}(A)$.

(a) Since V is unitary, $|v_{ii}| \le 1$, and so $f(W) \le \sum_{i=1}^{r} \sigma_r$. Equality is attained iff $V = \mathrm{diag}(I_r, Z)$, for some unitary Z, and since $W = PV^*Q^*$, (8.1) shows that this condition holds iff W is a unitary polar factor of A. If A is nonsingular then the unitary polar factor is unique and so W is the unique maximizer.

(b) If $\det(PQ^*) = 1$ then (a) shows that all solutions are given by unitary polar factors (8.1) of A with $\det(W) = \det(Z) = 1$. If $\sigma_{n-1} \ne 0$ then Z either is empty (if $r = n$) or else has just one degree of freedom ($v_{nn} = \pm 1$), which is used up by the condition $\det(W) = 1$, so the solution is unique.
If $\det(PQ^*) = -1$ then it is easy to see that $\det(W) = 1$ implies $f(W) = \mathrm{Re}\,\mathrm{trace}(V\Sigma) \le \sigma_1 + \cdots + \sigma_{n-1} - \sigma_n$, with equality when $V = \mathrm{diag}(1, \ldots, 1, -1)$, and this V is clearly unique if $\sigma_{n-1} > \sigma_n$. (For a detailed proof, see Hanson and Norris [246, 1981].) It remains to check that the expression for the maximizer is independent of the choice of SVD (i.e., of P and Q); this is easily seen using Problem B.11.

8.14. Define $Z = M^{1/2}Q$. Then the problem is

$$\min\{\, \|M^{1/2}A - Z\|_F : Z^*Z = I_n \,\},$$

which is just the problem in Theorem 8.4. So the solution is $Q = M^{-1/2}U$, where U is the unitary polar factor of $M^{1/2}A$. The solution can also be expressed as $Q = R^{-1}U$, where U is the unitary polar factor of RA and and $M = R^*R$ is the Cholesky factorization.

8.15. By using the spectral decomposition $A = Q\,\mathrm{diag}(\lambda_i)Q^*$ we can reduce (i) to the case where $A = \mathrm{diag}(\lambda_i)$. Then

$$\|f(A)X - Xf(A)\|_F^2 = \sum_{i,j} |(f(\lambda_i) - f(\lambda_j))x_{ij}|^2 \le c^2 \sum_{i,j} |\lambda_i - \lambda_j|^2 |x_{ij}|^2 = c^2 \|AX - XA\|_F^2.$$

Parts (ii) and (iv) are straightforward. Part (iii) is obtained by taking $X = I$ and $f(z) = |z|$ in part (ii).

8.16. We have

$$r_1 - r_2 e^{i(\theta_2 - \theta_1)} = e^{-i\theta_1}(z_1 - z_2). \tag{E.18}$$

Swapping the roles of z_1 and z_2 and taking the complex conjugate gives

$$r_2 - r_1 e^{i(\theta_2 - \theta_1)} = e^{i\theta_2}(\overline{z_2 - z_1}). \tag{E.19}$$

Adding these two equations yields

$$r_1(1 - e^{i(\theta_2 - \theta_1)}) + r_2(1 - e^{i(\theta_2 - \theta_1)}) = e^{-i\theta_1}(z_1 - z_2) + e^{i\theta_2}(\overline{z_2 - z_1}).$$

Hence

$$1 - e^{i(\theta_2 - \theta_1)} = \frac{e^{-i\theta_1}(z_1 - z_2) + e^{i\theta_2}(\overline{z_2 - z_1})}{r_1 + r_2},$$

which yields the result on multiplying through by $e^{i\theta_1}$ and taking absolute values. This proof is a specialization of Li's proof of the matrix version of the bound (Theorem 8.10).

8.17. We have

$$U = \begin{bmatrix} 1 & 0 \\ 0 & 1 \\ 0 & 0 \end{bmatrix}, \qquad \widetilde{U} = \begin{bmatrix} 1 & 0 \\ 0 & \epsilon/\sqrt{\epsilon^2 + \delta^2} \\ 0 & \delta/\sqrt{\epsilon^2 + \delta^2} \end{bmatrix},$$

so the (3,2) element of $U - \widetilde{U}$ is of order δ/ϵ, as required.

8.18. Let Y be an approximate solution to $X^*X = I$ and write $Y + E = U$, where U is the unitary polar factor of A. Then $I = U^*U = (Y+E)^*(Y+E) = Y^*Y + Y^*E + E^*Y + E^*E$. Dropping the second order term in E gives Newton's method, which defines E as the solution to the Sylvester equation $Y^*E + E^*Y = I - Y^*Y$. This equation is of the form $Z + Z^* = I - Y^*Y$ and so determines the Hermitian part of $Z = Y^*E$ as $\frac{1}{2}(I - Y^*Y)$. Setting the skew-Hermitian part to zero gives $E = \frac{1}{2}(Y^{-*} - Y)$. Then $Y + E = \frac{1}{2}(Y + Y^{-*})$, which leads to the iteration (8.17) if we take $Y_0 = A$ and recur this argument.

8.19. We have

$$X_1 = \frac{1}{2}(A + A^{-*}) = \frac{A^{-*}}{2}(A^*A + I) = Y_1^{-*}.$$

If $Y_k = X_k^{-*}$ then

$$X_{k+1} = \frac{1}{2}(X_k + X_k^{-*}) = \frac{1}{2}(Y_k^{-*} + Y_k) = \frac{1}{2}Y_k^{-*}(I + Y_k^*Y_k) = Y_{k+1}^{-*}.$$

The result follows by induction.

8.20. We will show that for the scalar Newton–Schulz iteration $x_{k+1} = g(x_k) = \frac{1}{2}x_k(3 - x_k^2)$ with $x_0 \in (0, \sqrt{3})$, $x_k \to 1$ quadratically as $k \to \infty$. This yields the quadratic convergence of the matrix iteration for $0 < \sigma_i(A) < \sqrt{3}$ by using the SVD of A to diagonalize the iteration. Now g agrees with f in (7.16) for $p = 2$, and the argument in the proof of Theorem 7.11 shows that $x_k \to 1$ as $k \to \infty$, with quadratic convergence described by $x_{k+1} - 1 = -\frac{1}{2}(x_k - 1)^2(x_k + 2)$. The latter equation leads to (8.36). The residual recurrence can be shown directly or deduced from Theorem 7.10 with $p = 2$.

8.21. The error bound (8.18) shows a multiplier in the quadratic convergence condition of $\|X_k^{-1}\|_2/2$, which converges to $1/2$ as $X_k \to U$. For the Newton–Schulz iteration the multiplier is $\|X_k + 2U\|_2/2$, which converges to $3/2$ in the limit. We conclude that the Newton iteration has an asymptotic error constant three times smaller than that for the Newton–Schulz iteration, and so can be expected to converge a little more quickly in general.

8.22. In view of (5.38), X_1 has singular values $1 \le \sigma_n^{(1)} \le \cdots \le \sigma_1^{(1)}$. For $k \ge 1$, X_k therefore has singular values $1 \le \sigma_n^{(k)} \le \cdots \le \sigma_1^{(k)} = f^{(k-1)}(\sigma_1^{(1)})$. The result follows.

8.23. (a) From (8.24), and using the trace characterization (B.5) of the Frobenius norm, we have

$$\|X_{k+1}\|_F^2 = \frac{1}{4}\left(\mu_k^2\|X_k\|_F^2 + 2\,\mathrm{Re}\,\mathrm{trace}(X_kX_k^{-*}) + \mu_k^{-2}\|X_k^{-1}\|_F^2\right).$$

Differentiating with respect to μ_k shows that the minimum is obtained at μ_k^F.

(b) Write $X \equiv X_k$. Using $\|X_{k+1}\| \le \frac{1}{2}\left(\mu_k\|X\| + \mu_k^{-1}\|X_k^{-*}\|\right)$ for the 1- and ∞-norms, and $\|A\|_1 = \|A^*\|_\infty$, we obtain

$$\|X_{k+1}\|_1\|X_{k+1}\|_\infty \le \frac{1}{4}\left(\mu_k^2\|X\|_1\|X\|_\infty + \|X\|_1\|X^{-1}\|_1 + \|X\|_\infty\|X^{-1}\|_\infty + \mu_k^{-2}\|X^{-1}\|_1\|X^{-1}\|_\infty\right).$$

Differentiating reveals the minimum at $\mu_k = \mu_k^{1,\infty}$.

8.24. Using Table C.2 we see that for $m > n$, the SVD approach requires about $\min(14mn^2 + 8n^3, 6mn^2 + 20n^3) + 2mn^2 + n^3$ flops, while Algorithm 8.20 requires $6mn^2 + (2k - 3\frac{1}{3})n^3 \le 6mn^2 + 17n^3$ flops. The SVD approach is clearly the more expensive. For $m = n$ the operation counts are $25n^3$ flops for the SVD versus at most $22n^3$ flops for Algorithm 8.20. Note the comments in Appendix C concerning the relevance of flop counts.

8.25. (a) For the first part, writing $c = \cos\theta$, $s = \sin\theta$, we find that

$$A' = G^T \begin{bmatrix} a_{ii} & a_{ij} \\ a_{ji} & a_{jj} \end{bmatrix} = \begin{bmatrix} ca_{ii} - sa_{ji} & ca_{ij} - sa_{jj} \\ sa_{ii} + ca_{ji} & sa_{ij} + ca_{jj} \end{bmatrix}.$$

The trace of A' is $f(\theta) = c(a_{ii} + a_{jj}) + s(a_{ij} - a_{ji})$. Since $f'(\theta) = -s(a_{ii} + a_{jj}) + c(a_{ij} - a_{ji})$ we see that $f'(\theta) = 0$ is precisely the condition for symmetry. It is easily checked that for a suitable choice of the signs of c and s, $\operatorname{trace}(A') > \operatorname{trace}(A)$ and hence the trace is maximized (rather than minimized).

(b) $\operatorname{trace}(GA) = \operatorname{trace}(A) - \operatorname{trace}(2vv^T/(v^Tv) \cdot A) = \operatorname{trace}(A) - 2v^TAv/(v^Tv)$. This quantity is maximized when the Rayleigh quotient $v^TAv/(v^Tv)$ is minimized, which occurs when v is an eigenvector corresponding to $\lambda_{\min}(A)$, which is negative by assumption. Hence $\max_v \operatorname{trace}(GA) = \operatorname{trace}(A) + 2|\lambda_{\min}(A)| > \operatorname{trace}(A)$.

(c) Details can be found in Smith [529, 2002, Chap. 3]. This idea was originally suggested by Faddeev and Faddeeva [180, 1964]; Kublanovskaya [364, 1963] had earlier investigated a symmetrization process based on just Givens rotations. The algorithm is only linearly convergent (with slow linear convergence) and a proof of global convergence appears difficult. Unfortunately, the idea does not live up to its promise.

8.26. (a) The blocks of Q satisfy

$$Q_1 = (I + A^*A)^{-1/2}M, \quad Q_2 = A(I + A^*A)^{-1/2}M,$$

giving

$$Q_2Q_1^* = A(I + A^*A)^{-1}.$$

Using this formula with $A \equiv X_k$ gives (8.38).

(b) The flop counts per iteration are $mn^2 + 7n^3/3$ for (8.37) and $6mn^2 + 8n^3/3$ for (8.38). If advantage is taken of the leading identity block in (8.38) the flop count can be reduced, but not enough to approach the operation count of (8.37).

(c) As discussed at the end of Section 8.6, we can scale (8.37) by setting $X_k \leftarrow \mu_k X_k$, with one of the scalings (8.25)–(8.27). The problem is how to compute μ_k, since X_k^{-1} is not available (and does not exist if $m \neq n$). Concentrating on (8.25), from (8.38a) we have $R_k^*R_k = I + X_k^*X_k$, and since R_k is triangular we can apply the power and inverse power methods to estimate the extremal singular values of R_k and hence those of X_k. Unfortunately, we obtain μ_k only after X_{k+1} has been (partly) computed, so we can only use μ_k to scale the next iterate: $X_{k+1} \leftarrow \mu_k X_{k+1}$.

9.1. ($\Leftarrow$) For any strictly triangular N let $D + N = Z \operatorname{diag}(J_1, \dots, J_q)Z^{-1}$ ($q \geq p$) be a Jordan canonical form of $D + N$ with Jordan blocks

$$J_i = \begin{bmatrix} \lambda_i & 1 & & & \\ & \lambda_i & 1 & & \\ & & \ddots & \ddots & \\ & & & \ddots & 1 \\ & & & & \lambda_i \end{bmatrix} \in \mathbb{C}^{m_i \times m_i},$$

where, necessarily, m_i does not exceed the k_j corresponding to λ_i. Then

$$f(D + N) = Z \operatorname{diag}(f(J_1), \dots, f(J_q))Z^{-1},$$

where, from (1.4),

$$f(J_i) = \begin{bmatrix} f(\lambda_i) & f'(\lambda_i) & \dots & \dots & \frac{f^{(m_i-1)}(\lambda_i)}{(m_i-1)!} \\ & f(\lambda_i) & f'(\lambda_i) & \dots & \vdots \\ & & \ddots & \ddots & \vdots \\ & & & \ddots & f'(\lambda_i) \\ & & & & f(\lambda_i) \end{bmatrix}. \tag{E.20}$$

Since the derivatives of f are zero on any repeated eigenvalue and $f(\lambda_i) = f(\lambda_1)$ for all i, $f(D+N) = Zf(D)Z^{-1} = Zf(\lambda_1)IZ^{-1} = f(\lambda_1)I = f(D)$.

($\Rightarrow$) Let $F = f(D+N)$, and note that by assumption $F = f(D)$ and hence F is diagonal. The equation $F(D+N) = (D+N)F$ reduces to $FN = NF$, and equating (i,j) elements for $j > i$ gives $(f_{ii} - f_{jj})n_{ij} = 0$. Since this equation holds for all strictly triangular N, it follows that $f_{ii} = f_{jj}$ for all i and j and hence that $F = f(\lambda_1)I$.

If at least one of the λ_i is repeated, then we can find a permutation matrix P and a strictly upper bidiagonal matrix B such that $PDP^T + B = P(D + P^TBP)P^T$ is nonderogatory and is in Jordan canonical form, and $N = P^TBP$ is strictly upper triangular. We have $\Lambda(D) = \Lambda(D+N)$ and so the requirement $f(D+N) = f(D)$ implies that $f(PDP^T + B) = Pf(D)P^T = f(\lambda_1)I$, and hence, in view of (E.20), (9.5) holds.

9.2. The proof is entirely analogous to that of Lemma 6.4. For the last part, a 2×2 real normal matrix with distinct eigenvalues $a \pm ib$ has the form $A = \left[\begin{smallmatrix} a & b \\ -b & a \end{smallmatrix}\right]$ and if $f(a+ib) = c+id$ then $f(A) = \left[\begin{smallmatrix} c & d \\ -d & c \end{smallmatrix}\right]$.

10.1. $X(t) = e^{(A+E)t}$ satisfies the differential equation $X'(t) = AX(t) + EX(t)$, $X(0) = I$. Hence from the matrix analogue of (2.3), we have

$$X(t) = e^{At} + \int_0^t e^{A(t-s)} EX(s)\, ds,$$

which is (10.13) on substituting for X.

10.2. Let $AB = BA$ and $f(t) = e^{(A+B)t} - e^{At}e^{Bt}$. Then $f(0) = 0$ and, because e^{tA} is a polynomial in tA (as is any function of tA) and hence commutes with B,

$$\begin{aligned} f'(t) &= (A+B)e^{t(A+B)} - Ae^{tA}e^{tB} - e^{tA}Be^{Bt} \\ &= (A+B)e^{t(A+B)} - (A+B)e^{tA}e^{tB} \\ &= (A+B)(e^{t(A+B)} - e^{tA}e^{tB}) \\ &= (A+B)f(t). \end{aligned}$$

The unique solution to this initial value problem is $f(t) \equiv 0$, as required.

10.3. From (10.13) with $t = 1$ and $B \equiv A + E$, we have

$$e^B = e^A + \int_0^1 e^{A(1-s)}(B-A)e^{Bs}\, ds.$$

Hence

$$\begin{aligned} \|e^A - e^B\| &\le \|A - B\| \int_0^1 e^{\|A\|(1-s)} e^{\|B\|s}\, ds \\ &\le \|A - B\| \int_0^1 e^{\max(\|A\|,\|B\|)}\, ds = \|A - B\|\, e^{\max(\|A\|,\|B\|)}. \end{aligned}$$

10.4. $\det(e^{A+B}) = \prod_{i=1}^n \lambda_i(e^{A+B}) = \prod_{i=1}^n e^{\lambda_i(A+B)} = e^{\mathrm{trace}(A+B)} = e^{\mathrm{trace}(A)+\mathrm{trace}(B)} = e^{\mathrm{trace}(A)}e^{\mathrm{trace}(B)} = \det(e^A)\det(e^B) = \det(e^Ae^B)$.

10.5. The bound is immediate from Lemma 10.15 and Theorem 10.10.

10.6. From (B.28) we have $|f[\lambda,\mu]| \le \max_{z\in\Omega} |f'(z)| = \max_{z\in\Omega} |e^z| = \max(\mathrm{Re}\,\lambda, \mathrm{Re}\,\mu) \le \alpha(A)$, where α is defined in (10.10) and Ω is the line joining λ and μ, and there is equality for $\lambda = \mu$. Hence $\max_{\lambda,\mu\in\Lambda(A)} |f[\lambda,\mu]| = \alpha(A) = \|e^A\|_2$. Thus the two different formulae are in fact equivalent.

10.7. By Corollary 3.10, since $e^{\lambda_i} \neq 0$ and $e^{\lambda_i} = e^{\lambda_j}$ if and only if $\lambda_i - \lambda_j = 2m\pi i$ for some integer m, $L(A)$ is nonsingular precisely when no two eigenvalues of A differ by $2m\pi i$ for some nonzero integer m.

10.8. Since $\log(I+G) = \sum_{j=1}^{\infty}(-1)^{j+1}G^j/j$ (see (1.1)),

$$\|\log(I+G)\| \le \sum_{j=1}^{\infty} \frac{\|G\|^j}{j} = -\log(1-\|G\|) \le \|G\| \sum_{j=0}^{\infty} \|G\|^j = \frac{\|G\|}{1-\|G\|}.$$

Similarly,

$$\begin{aligned}\|\log(I+G)\| &\ge \|G\| - \frac{\|G^2\|}{2} - \frac{\|G^3\|}{3} - \cdots \\ &\ge \|G\| - \frac{\|G^2\|}{2}(1+\|G\|+\|G\|^2+\cdots) = \|G\| - \frac{\|G\|^2}{2(1-\|G\|)}.\end{aligned}$$

10.9. From (10.23),

$$q_m(A) = I + \sum_{j=1}^{m} \frac{(2m-j)!\,m!}{(2m)!\,(m-j)!} \frac{(-A)^j}{j!} =: I + F.$$

Now

$$\|F\| \le \sum_{j=1}^{m} \frac{(2m-j)!\,m!}{(2m)!\,(m-j)!} \frac{\|A\|^j}{j!} = q_m(-\|A\|) - 1,$$

and so $\|q_m(A)^{-1}\| = \|(I+F)^{-1}\| \le 1/(1-\|F\|) \le 1/(2-q_m(-\|A\|))$.
Since, for $1 \le j \le m$,

$$\frac{(2m-j)!\,m!}{(2m)!\,(m-j)!} \le \left(\frac{m}{2m}\right)^j = 2^{-j},$$

we have

$$\|F\| \le \sum_{j=1}^{m} (\|A\|/2)^j \frac{1}{j!} = e^{\|A\|/2} - 1.$$

Hence $\|q_m(A)^{-1}\| \le 1/(1-\|F\|) \le 1/(2-e^{\|A\|/2})$ provided $\|A\| < \log 4$. The last part is straightforward to check.

10.10. We can rewrite (10.25) as

$$G = e^{-A} r_{km}(A) - I = \frac{(-1)^{m+1}}{(k+m)!} A^{k+m+1} q_{km}(A)^{-1} \int_0^1 e^{(t-1)A}(1-t)^k t^m \, dt.$$

Taking norms gives

$$\begin{aligned}\|G\| &\le \frac{\|A\|^{k+m+1}}{(k+m)!} \|q_{km}(A)^{-1}\| \int_0^1 e^{(1-t)\|A\|}(1-t)^k t^m \, dt \\ &= \frac{\|A\|^{k+m+1}}{(k+m)!} \|q_{km}(A)^{-1}\| \int_0^1 e^{t\|A\|} t^k (1-t)^m \, dt.\end{aligned}$$

Using (10.25) with A replaced by $\|A\|$, this bound can be rewritten as

$$\|G\| \le q_{mk}(\|A\|)\, \|q_{km}(A)^{-1}\| \left| e^{\|A\|} - r_{mk}(\|A\|) \right|.$$

Specializing to $k = m$ and using Problem 10.9 to rewrite the bound entirely in terms of $\|A\|$ gives (10.59).

10.11. The proof is by induction. We have

$$\|A^2 - fl(A^2)\| \le \gamma_n \|A\|^2 = nu\|A\|^2 + O(u^2),$$

so the result is true for $k = 1$. Assume the result is true for $k-1$ and write $\widehat{X}_k = fl(A^{2^k}) =: A^{2^k} + E_k$. We have

$$\widehat{X}_k = fl(\widehat{X}_{k-1}^2) = \widehat{X}_{k-1}^2 + F_k, \qquad \|F_k\| \le \gamma_n \|\widehat{X}_{k-1}\|^2.$$

Hence

$$\begin{aligned}\widehat{X}_k &= \left(A^{2^{k-1}} + E_{k-1}\right)^2 + F_k \\ &= A^{2^k} + A^{2^{k-1}} E_{k-1} + E_{k-1} A^{2^{k-1}} + E_{k-1}^2 + F_k.\end{aligned}$$

Hence

$$\begin{aligned}\|E_k\| &\le 2\|A^{2^{k-1}}\| \, \|E_{k-1}\| + \|E_{k-1}\|^2 + \|F_k\| \\ &\le 2(2^{k-1}-1)nu\|A^{2^{k-1}}\| \, \|A\|^2 \, \|A^2\| \, \|A^4\| \dots \|A^{2^{k-2}}\| + \gamma_n \|\widehat{X}_{k-1}\|^2 + O(u^2) \\ &= 2(2^{k-1}-1)nu\|A\|^2 \, \|A^2\| \, \|A^4\| \dots \|A^{2^{k-1}}\| + \gamma_n \|A^{2^{k-1}}\|^2 + O(u^2) \\ &\le (2^k - 1)nu\|A\|^2 \, \|A^2\| \, \|A^4\| \dots \|A^{2^{k-1}}\| + O(u^2).\end{aligned}$$

10.12. The form of the diagonal blocks is immediate. From Theorem 4.12 and (10.15) we see that the (1,2) block of e^A is the (1,2) block of

$$\int_0^1 \begin{bmatrix} e^{A_{11}(1-s)} & 0 \\ 0 & e^{A_{22}(1-s)} \end{bmatrix} \begin{bmatrix} 0 & A_{12} \\ 0 & 0 \end{bmatrix} \begin{bmatrix} e^{A_{11}s} & 0 \\ 0 & e^{A_{22}s} \end{bmatrix} ds,$$

which gives the result.

10.13. The formula can be obtained from the polynomial interpolation definition of e^A, but the following approach is more elegant. Since A is skew-symmetric we know that $Q^*AQ = \mathrm{diag}(0, i\theta, -i\theta) =: i\theta K$ for some $\theta \in \mathbb{R}$. Then $Q^* e^A Q = \mathrm{diag}(1, e^{i\theta}, e^{-i\theta}) = \mathrm{diag}(1, c+is, c-is) = M + cL + isK$, where $c = \cos\theta$, $s = \sin\theta$, $M = \mathrm{diag}(1,0,0)$, and $L = \mathrm{diag}(0,1,1)$. Hence $e^A = Q(M+cL+isK)Q^* = M + cQLQ^* + (s/\theta)A$. But $A^2 = Q(i\theta K)^2 Q^* = -\theta^2 QLQ^*$, so $cQLQ^* = -cA^2/\theta^2$ and $QMQ^* = Q(I-L)Q^* = I - QLQ^* = I + A^2/\theta^2$. It remains to note that $\|A\|_F^2 = 2\theta^2$.

Note that the Cayley–Hamilton theorem shows that $e^A \in \mathbb{R}^{3\times 3}$ must be a cubic polynomial in A; the simplification in (10.60) results from the fact that $A^3 = -\theta^2 A$, a relation specific to $n = 3$.

10.14. $A = t\left[\begin{smallmatrix} 0 & -R \\ R & 0 \end{smallmatrix}\right]$, where R is the 4×4 "reverse identity matrix" with $r_{ij} = 1$ for $i+j = 5$ and $r_{ij} = 0$ otherwise.

11.1. We have

$$J(\lambda) = \begin{bmatrix} \lambda & 1 & & \\ & \lambda & \ddots & \\ & & \ddots & 1 \\ & & & \lambda \end{bmatrix} = \lambda \begin{bmatrix} 1 & \lambda^{-1} & & \\ & 1 & \ddots & \\ & & \ddots & \lambda^{-1} \\ & & & 1 \end{bmatrix} = \lambda(I + N).$$

Hence $\log(J(\lambda)) = (\log\lambda)I + \log(I+N)$, since the conditions of Theorems 11.3 and 11.4 are trivially satisfied for $B = \lambda I$, $C = I + N$. Since N is nilpotent, $\log(I+N) = \sum_{i=1}^{m-1}(-1)^{i+1}N^i/i$. Hence

$$\log(J(\lambda)) = \begin{bmatrix} \log\lambda & \frac{1}{\lambda} & \dots & \frac{(-1)^m}{(m-1)\lambda^{m-1}} \\ & \log\lambda & \ddots & \vdots \\ & & \ddots & \frac{1}{\lambda} \\ & & & \log\lambda \end{bmatrix},$$

which of course agrees with (1.4). Via the Jordan canonical form, $\log(A)$ can then be defined for A with no eigenvalues on $\mathbb{R}^-$.

11.2. The proof is straightforward, by using the integral and (B.10) to obtain the first order part of the expansion of $\log(A+E) - \log(A)$.

11.3. We have $\max_i |1-\lambda_i(A)| = \rho(I-A) \le \|I-A\| < 1$, which implies that every eigenvalue of A lies in the open right half-plane, as required. The convergence also follows from

$$\|(I-A)(I+A)^{-1}\| = \|I-A\|\,\|(2I+A-I)^{-1}\| \le \frac{\|I-A\|}{2-\|I-A\|} < 1.$$

11.4. Let $X = M - I$. From (10.58), $\|\log(M)\| \le -\log(1-\|X\|)$. Thus if $\|I-M\| \le 1-1/e$ then $\|\log(M)\| \le |\log(1-(1-1/e)| = 1$.

11.5. The property follows from consideration of Table 11.1 and, in particular, these numbers:

θ_3	θ_4	θ_5	$\theta_7/2$	θ_6	$2\theta_5$	θ_7	$2\theta_6$	θ_8
0.0162	0.0539	0.114	0.132	0.187	0.228	0.264	0.374	0.340

11.6. The equality follows from Theorem 11.4. This preprocessing is rather pointless: we need $A \approx I$, but making $\|A\| = 1$ in general makes no useful progress towards this goal.

11.7. The individual eigenvalues of X_k follow the scalar iteration $x_{k+1} = (x_k + a x_k^{-1})/2$, $x_0 = a$, where a is an eigenvalue of A. With the transformation $y_k = a^{-1/2} x_k$ this becomes $y_{k+1} = (y_k + y_k^{-1})/2$, $y_0 = a^{1/2}$, and it follows that $-\pi/2 < \arg y_k < \pi/2$ for all k, or equivalently that x_k is in the open half plane of $a^{1/2}$ for all k.

From the derivation of the product DB iteration we know that $M_k = X_k A^{-1} X_k$ and that X_k and M_k are functions of A. By commutativity and Theorem 1.40 we know that for each eigenvalue λ_j of A there are corresponding eigenvalues μ_j of M_k and ξ_j of X_k, with $\mu_j = \xi_j^2/\lambda_j$. By Theorem 1.20 it suffices to show that $\log \mu_j = 2\log \xi_j - \log \lambda_j$ for all j. Let $z = \xi_j/\lambda_j^{1/2}$. Then $\arg z = \arg \xi_j - \arg \lambda_j^{1/2} + 2\pi k$ for some $k \in \mathbb{Z}$, and since $|\arg z| < \pi/2$ (as we just saw in the previous paragraph) and $|\arg \lambda_j^{1/2}| < \pi/2$, we have $k = 0$. Hence $|\arg \xi_j - \arg \lambda_j^{1/2}| < \pi$, which implies, by Theorem 11.3, $\log z = \log \xi_j - \log \lambda_j^{1/2} = \log \xi_j - \frac{1}{2}\log \lambda_j$. Thus $\log \mu_j = 2\log \xi_j - \log \lambda_j$, as required. Equation (11.31) follows by induction.

11.8. (a) Since $X = \log(A)$ we have $A = e^X$, so $\lambda \in \Lambda(A)$ implies $\lambda = e^\mu$ for some μ. The eigenvalues of $B = A^T \oplus A = A^T \otimes I + I \otimes A$ are of the form $\lambda_j + \lambda_k = e^{\mu_j} + e^{\mu_k}$. If B is singular then $e^{\mu_j} + e^{\mu_k} = 0$ for some j and k, that is, $e^{\mu_j - \mu_k} = -1$, or $\mu_j - \mu_k = (2\ell - 1)\pi i$ for some $\ell \in \mathbb{Z}$. But then $\|X\| \ge \rho(X) \ge \max_{j,k}(|\mu_j|, |\mu_k|) > 1$, which contradicts $\|X\| \le 1$. Hence B is nonsingular.

(b) The zeros of τ are $x = \pm ik\pi$ and satisfy $|x| \ge \pi$. Since $\rho(X) \le \|X\| \le 1$, no eigenvalue of X is a zero of τ and so $\tau(X)$ is nonsingular.

11.9. For the first part see Edelman and Strang [173, 2004]. Since $\mathrm{rank}(R-I) = n-1$, R is similar to a single Jordan block $J(0)$. Hence by Theorem 1.28, in which $p = 1$, all logarithms are primary matrix functions and have the form $\log(R) + 2j\pi i I$ for $j \in \mathbb{Z}$, which is always upper triangular and is real only for $j = 0$.

11.10. The relative backward error of $\widehat{X}$ is the smallest value of $\|E\|/\|A\|$ such that $\widehat{X} = \log(A+E)$. There is only one E satisfying the latter equation, namely $E = e^{\widehat{X}} - A$, so the relative backward error is $\|e^{\widehat{X}} - A\|/\|A\|$.

11.11. Consider first (11.32). Newton's method for $f(X) = 0$ solves for E in the first order perturbation expansion of $f(X+E) = 0$ and sets $X \leftarrow X+E$. For $f(X) = e^X - A$ we have, if $XE = EX$, $f(X+E) = e^X e^E - A = e^X(I+E) - A$ to first order, or $E = e^{-X}A - I$, which gives $X + E = X - I + e^{-X}A$. But $XA = AX$ clearly implies $XE = EX$, so under the assumption $XA = AX$, E is the Newton update (which we are implicitly assuming is unique). Since $X+E$ clearly commutes with A, this argument can be repeated and we have derived (11.32). The commutativity relations involving X_k are easily proved by induction.

To analyze the asymptotic convergence, let $A = e^X$ with X a primary logarithm of A, and write $X_{k+1} - X = X_k - X - I + e^{-X_k}A = X_k - X - I + e^{X-X_k}$, since X and X_k are functions of A and hence commute. Theorem 4.8 gives $e^{X-X_k} = I + (X - X_k) + \frac{1}{2}\max_{0\le t\le 1}\|(X-X_k)^2 e^{t(X-X_k)}\|$, which gives $\|X - X_{k+1}\| \le \frac{1}{2}\|X - X_k\|^2 e^{\|X-X_k\|}$ for any subordinate matrix norm.

The analysis for (11.33) is very similar, if a little more tedious. The convergence bound is $\|X - X_{k+1}\| \le \frac{1}{6}\|X - X_k\|^3 e^{\|X-X_k\|}$.

The extra cost of (11.33) over (11.32) is one matrix inversion per iteration, which is negligible compared with the cost of the matrix exponential. Therefore, all other things being equal, the cubically convergent iteration (11.33) should be preferred.

12.1. With $B = k\pi A$, $B^{2i} = (k\pi)^{2i}I$, so $\cos(B) = \sum_{i=0}^{\infty}\frac{(-1)^i}{(2i)!}B^{2i} = \bigl(\sum_{i=0}^{\infty}\frac{(-1)^i}{(2i)!}(k\pi)^{2i}\bigr)I = \cos(k\pi)I = (-1)^k I$.

12.2. From Theorem 1.36 the only possible Jordan form of A is $J_2(\lambda) = \left[\begin{smallmatrix}1&1\\0&1\end{smallmatrix}\right]$ with $\sin\lambda = 1$. But then $\cos(\lambda) = 0$ and so by Theorem 1.36 the Jordan block splits into two 1×1 blocks in $\sin(J_2(\lambda))$. Hence $\sin(A)$ cannot be equal to (or similar to) $\left[\begin{smallmatrix}1&1\\0&1\end{smallmatrix}\right]$.

12.3. All four identities follow from Theorems 1.16 and 1.20.

12.4. Let $B = i^{-1}\log(X_0)$. Then $e^{iB} = X_0$ and so we have $X_1 = \frac{1}{2}(X_0 + X_0^{-1}) = \frac{1}{2}(e^{iB} + e^{-iB}) = \cos(B)$, as required.

12.5. $\sin(A) = \sqrt{I - \cos^2(A)}$ for some square root, but without knowing the eigenvalues of A we cannot determine which is the required square root. But if A is triangular then the eigenvalues of A are known and we can proceed. Consider, first, the scalar case and note that

$$\begin{aligned}\cos(x+iy) &= \cos(x)\cos(iy) - \sin(x)\sin(iy) = \cos(x)\cosh(y) - i\sin(x)\sinh(y),\\ \sin(x+iy) &= \sin(x)\cos(iy) + \cos(x)\sin(iy) = \sin(x)\cosh(y) + i\cos(x)\sinh(y).\end{aligned}$$

Since $\cosh(y) \ge 0$, from x we can determine the sign of $\operatorname{Re}\sin(x+iy)$ and hence which square root to take in $\sin(z) = \sqrt{1 - \cos^2(z)}$. Returning to the matrix case, we can determine each diagonal element of $\sin(A) = \sqrt{I - \cos^2(A)}$ by the procedure just outlined. If the eigenvalues of A are distinct then we can use the Parlett recurrence to solve for the off-diagonal part of $\sin(A)$. However, if A has a repeated eigenvalue then the recurrence breaks down. Indeed, the equation $\cos^2(A)+\sin^2(A) = I$ may not contain enough information to determine $\sin(A)$. For example, from

$$A = \begin{bmatrix}0 & x\\ 0 & 0\end{bmatrix}, \quad \cos(A) = \begin{bmatrix}1 & 0\\ 0 & 1\end{bmatrix}, \quad \sin(A) = \begin{bmatrix}0 & x\\ 0 & 0\end{bmatrix},$$

it is clear that $\sin(A)$ cannot be recovered from $I - \cos^2(A) = 0$ without using knowledge of A.

13.1. By (13.5), $AQ_m = Q_m H_m$, where $Q_m \in \mathbb{C}^{n\times m}$, $H_m \in \mathbb{C}^{m\times m}$. Let U be such that $[Q_m\ U]$ is unitary. Then $[Q_m\ U]^* A [Q_m\ U] = \left[\begin{smallmatrix}H_m & 0\\ 0 & 0\end{smallmatrix}\right]$. Hence

$$\begin{aligned}f(A)b &= [Q_m\ U] f\left(\begin{bmatrix}H_m & 0\\ 0 & U^*AU\end{bmatrix}\right)\begin{bmatrix}Q_m^*\\ U^*\end{bmatrix} b\\ &= [Q_m\ U]\begin{bmatrix}f(H_m) & 0\\ 0 & f(U^*AU)\end{bmatrix}\|b\|_2 e_1 = \|b\|_2 Q_m f(H_m) e_1 = f_m.\end{aligned}$$

13.2. Let $d = \deg \psi_{A,b}$. If $m < d$ then $\dim(\mathcal{K}_m(A,b)) = m$. Indeed $\dim(\mathcal{K}_m(A,b)) \le m$ trivially, and if $\dim(\mathcal{K}_m(A,b)) < m$ then $\sum_{i=0}^m c_i A^i b = 0$ for some c_i not all zero, which contradicts the definition of $\psi_{A,b}$. If $m \ge d$ then $\dim(\mathcal{K}_m(A,b)) = d$ because $A^m b$ and higher powers can be expressed as linear combinations of $b, Ab, \dots, A^{d-1}b$ using the equation $\psi_{A,b}(A)b = 0$.

13.3. For any $\mu \in \mathbb{C}$, $H - \mu I$ is upper Hessenberg with nonzero subdiagonal elements and so $\operatorname{rank}(H - \mu I) \ge n - 1$. For μ an eigenvalue we therefore have $\operatorname{rank}(H - \mu I) = n - 1$, which implies that μ appears in only one Jordan block.

13.4. The correctness of the algorithm is easily proved by induction. For $k = 1$ the correctness is immediate. For general k it is easy to see that $x = \frac{1}{2}X_{k-1}(3I - X_{k-1}^2)b$.

Since the function makes 3 recursive calls, with 3 matrix–vector multiplications at the lowest level ($k = 1$), the cost is 3^k matrix–vector multiplications. This is to be compared with $2k$ matrix multiplications and 1 matrix–vector multiplication if we compute X_k and then $X_k b$. The recursive computation requires fewer flops if $3^k n^2 \lesssim 2kn^3$. If $k = 8$, this requires $n \gtrsim 410$. The temporary storage required by the recursive algorithm is just k vectors (one per level of recursion).

Note that this algorithm does not lend itself to dynamic determination of k, unlike for the usual Newton–Schulz iteration: if we know $X_{k-1}b$ then computing $X_k b$ costs 2/3 as much as computing $X_k b$ from scratch.

B.1. Since the columns of X form a basis they are independent, and so $X^+X = I$. Let (λ, x) be an eigenpair of A with $x \in \mathcal{X}$. Then $x = Xz$ for a unique $z \ne 0$, and $z = X^+x$. Hence

$$\lambda x = Ax = AXz = XBz.$$

Multiplying on the left by X^+ gives $\lambda z = \lambda X^+ x = Bz$, so (λ, z) is an eigenpair for B.

Let (λ, z) be an eigenpair for B. Then $AXz = XBz = \lambda Xz$, and $Xz \ne 0$ since the columns of X are independent, so (λ, Xz) is an eigenpair for A.

B.2. Let A be Hermitian and write $X = A^+$. Taking the conjugate transposes of (B.3) (i) and (B.3) (ii) yields $AX^*A = A$ and $X^*AX^* = X^*$. Using (B.3) (iii) we have $(X^*A)^* = AX = (AX)^* = X^*A$, and (B.3) (iv) yields $(AX^*)^* = XA = (XA)^* = AX^*$. So X^* satisfies the same four Moore–Penrose conditions as X, which means that $X = X^*$ by the uniqueness of the pseudoinverse. Finally, by (B.3) (iii) we have $AX = (AX)^* = XA$, so A and X commute.

B.3. If $x \in \operatorname{range}(A^*A)$ then, for some y, $x = A^*Ay = A^*(Ay) \in \operatorname{range}(A^*)$, so $\operatorname{range}(A^*A) \subseteq \operatorname{range}(A^*)$. We indicate three ways to obtain the reverse inclusion. (a) If $x \in \operatorname{range}(A^*)$ then, for some z and with $X = A^+$, $x = A^*z = A^*X^*A^*z = A^*(AX)^*z = A^*(AX)z = A^*A(Xz) \in \operatorname{range}(A^*A)$, so $\operatorname{range}(A^*) \subseteq \operatorname{range}(A^*A)$. (b) $A^*Az = 0$ implies $\|Az\|_2^2 = z^*A^*Az = 0$, so that $\operatorname{null}(A^*A) \subseteq \operatorname{null}(A)$, which implies that $\operatorname{rank}(A^*A) \ge \operatorname{rank}(A) = \operatorname{rank}(A^*)$ and hence $\operatorname{range}(A^*) \subseteq \operatorname{range}(A^*A)$. (c) Since $\operatorname{range}(A) \cup \operatorname{null}(A^*) = \mathbb{C}^m$, any $z \in \mathbb{C}^m$ can be written $z = Ax + w$, $w \in \operatorname{null}(A^*)$. Hence $A^*z = A^*Ax$, which implies $\operatorname{range}(A^*) \subseteq \operatorname{range}(A^*A)$.

B.4. Let A have the SVD $A = P\left[\begin{smallmatrix} \Sigma_r & 0 \\ 0 & 0_{m-r,n-r} \end{smallmatrix}\right]Q^*$ and partition $P = [P_1\ P_2]$ and $Q = [Q_1\ Q_2]$, where $P_1 \in \mathbb{C}^{m\times r}$ and $Q_1 \in \mathbb{C}^{n\times r}$ satisfies $Q_1^*Q_1 = I_r$. Then $A^* = Q_1\Sigma_r P_1^*$, so $\operatorname{range}(A^*) = \operatorname{range}(Q_1)$. So $x \in \operatorname{range}(A^*)$ iff $x = Q_1 y$ for some y. For such an x we have

$$\|Ax\|_2 = \|AQ_1y\|_2 = \left\| \begin{bmatrix} \Sigma_r & 0 \\ 0 & 0_{m-r,n-r} \end{bmatrix} \begin{bmatrix} I \\ 0 \end{bmatrix} y \right\|_2 = \|\Sigma_r y\|_2.$$

Since $\|x\|_2 = \|y\|_2$, A is a partial isometry iff $\|\Sigma_r y\|_2 = \|y\|_2$ for all y, or equivalently, $\Sigma_r = I_r$.

B.5. Let λ be an eigenvalue of A and x the corresponding eigenvector, and form the matrix $X = [x, x, \dots, x] \in \mathbb{C}^{n\times n}$. Then $AX = \lambda X$, so $|\lambda|\|X\| = \|AX\| \le \|A\|\,\|X\|$, showing that $|\lambda| \le \|A\|$. For a subordinate norm it suffices to take norms in the equation $Ax = \lambda x$.

B.6. Let $B = \left[\begin{smallmatrix} A \\ 0 \end{smallmatrix}\right]$. Since $B^*B = A^*A$, B and A have the same singular values, so $\|B\| = \|A\|$.

B.7. For any unitarily invariant norm, $\|B\|$ depends only on the singular values of B. Now $(UA)^*(UA) = A^*U^*UA = A^*A$, so UA and A have the same singular values. Hence $\|A\| = \|UA\|$. Alternatively, choose $V \in \mathbb{C}^{m\times(m-n)}$ so that $[U, V]$ is unitary and use Problem B.6 to deduce that

$$\|A\| = \left\|\begin{bmatrix} A \\ 0 \end{bmatrix}\right\| = \left\|[U \quad V]\begin{bmatrix} A \\ 0 \end{bmatrix}\right\| = \|UA\|.$$

B.8. If $A = PDQ^*$ is an SVD then

$$\begin{aligned}\|AB\|_F &= \|PDQ^*B\|_F = \|D(Q^*B)\|_F \\ &\le \|\operatorname{diag}(\max_i \sigma_i)(Q^*B)\|_F = (\max_i \sigma_i)\|Q^*B\|_F = \|A\|_2\|B\|_F.\end{aligned}$$

B.9. Let P_1 and P_2 be orthogonal projectors onto S. Since $\operatorname{range}(P_1) = \operatorname{range}(P_2)$, $P_2 = P_1X$ for some X. Then $P_1P_2 = P_1^2X = P_1X = P_2$. Likewise, $P_2P_1 = P_1$. Hence, for any z,

$$\begin{aligned}\|(P_1 - P_2)z\|_2^2 &= z^*(P_1 - P_2)(P_1 - P_2)z \\ &= z^*(P_1^2 + P_2^2 - P_1P_2 - P_2P_1)z \\ &= z^*(P_1 + P_2 - P_2 - P_1)z = 0.\end{aligned}$$

Therefore $P_1 - P_2 = 0$.

B.10. AA^+ is Hermitian by (B.3) (iii), and $(AA^+)^2 = AA^+AA^+ = AA^+$ by (B.3) (i). It remains to show that $\operatorname{range}(AA^+) = \operatorname{range}(A)$.

Let $x \in \operatorname{range}(A)$, so that $x = Ay$ for some y. Then, by (B.3) (i), $x = AA^+Ay = AA^+x$, so $x \in \operatorname{range}(AA^+)$. Conversely, $x \in \operatorname{range}(AA^+)$ implies $x = AA^+y$ for some y and then $x = A(A^+y) \in \operatorname{range}(A)$. Thus AA^+ is the orthogonal projector onto $\operatorname{range}(A)$.

By the first part, the orthogonal projector onto $\operatorname{range}(A^*)$ is $A^*(A^*)^+ = A^*(A^+)^* = (A^+A)^* = A^+A$.

B.11. We have $AA^* = P\Sigma^2P^* = \widetilde{P}\Sigma^2\widetilde{P}^*$, so $(\widetilde{P}^*P)\Sigma^2 = \Sigma^2(\widetilde{P}^*P)$. By equating elements on both sides (or invoking Theorem 1.25) we find that $\widetilde{P}^*P = D_1$, where D_1 is unitary and block diagonal with $(D_1)_{ij} = 0$ if $\sigma_i \ne \sigma_j$. Similarly, $A^*A = Q\Sigma^2Q^* = \widetilde{Q}\Sigma^2\widetilde{Q}^*$ implies $\widetilde{Q}^*Q = D_2$, where D_2 is unitary and block diagonal with $(D_2)_{ij} = 0$ if $\sigma_i \ne \sigma_j$. Then $P\Sigma Q^* = A = \widetilde{P}\Sigma\widetilde{Q}^* = PD_1^*\Sigma D_2Q^*$, which implies $\Sigma = D_1^*\Sigma D_2 = \Sigma D_1^*D_2$, since D_1 commutes with Σ^2 and hence with $(\Sigma^2)^{1/2}$. Write $\Sigma = \operatorname{diag}(\Sigma_1, 0_{n-r})$, where $\Sigma_1 \in \mathbb{R}^{r\times r}$ is nonsingular. Then it is easy to see that $D_1^*D_2 = \operatorname{diag}(I_r, W)$, where W is an arbitrary unitary matrix. The result follows.

Bibliography

You will find it a very good practice always to verify your references, sir.
— MARTIN JOSEPH ROUTH (1878)

The distribution of the year of publication of the references in this bibliography is shown in the following graph.

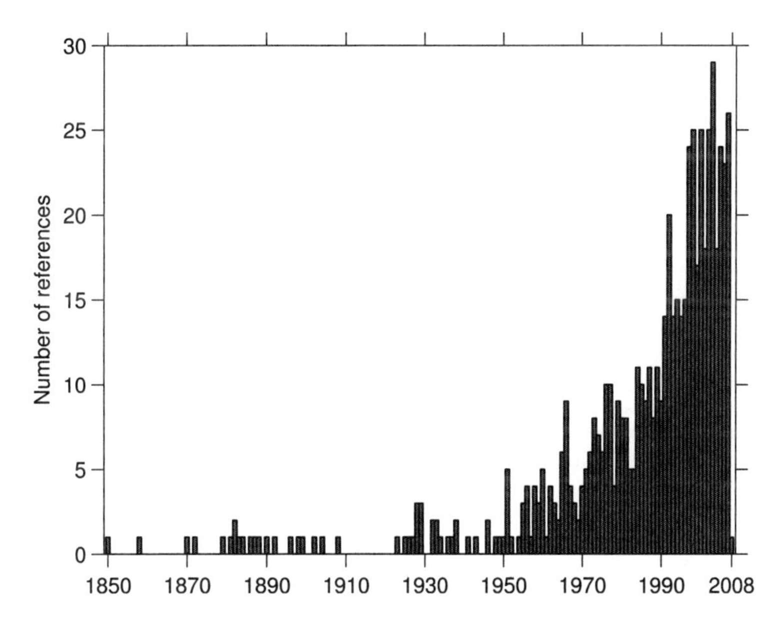

[1] N. I. Achieser. *Theory of Approximation.* Frederick Ungar Publishing Co., New York, 1956. x+307 pp. (Cited on p. 130.)

[2] Eva Achilles and Richard Sinkhorn. Doubly stochastic matrices whose squares are idempotent. *Linear and Multilinear Algebra*, 38:343–349, 1995. (Cited on p. 191.)

[3] Martin Afanasjew, Michael Eiermann, Oliver G. Ernst, and Stefan Güttel. Implementation of a restarted Krylov subspace method for the evaluation of matrix functions. Manuscript, July 2007. 27 pp. (Cited on p. 305.)

[4] Martin Afanasjew, Michael Eiermann, Oliver G. Ernst, and Stefan Güttel. On the steepest descent method for matrix functions. Manuscript, 2007. 19 pp. (Cited on p. 305.)

[5] S. N. Afriat. Analytic functions of finite dimensional linear transformations. *Proc. Cambridge Philos. Soc.*, 55(1):51–61, 1959. (Cited on p. 27.)

[6] Donald J. Albers and G. L. Alexanderson, editors. *Mathematical People: Profiles and Interviews.* Birkhäuser, Boston, MA, USA, 1985. xvi+372 pp. ISBN 0-8176-3191-7. (Cited on p. 171.)

[7] J. Albrecht. Bemerkungen zu Iterationsverfahren zur Berechnung von $A^{1/2}$ und A^{-1}. *Z. Angew. Math. Mech.*, 57:T262–T263, 1977. (Cited on p. 167.)

[8] G. Alefeld and N. Schneider. On square roots of M-matrices. *Linear Algebra Appl.*, 42:119–132, 1982. (Cited on p. 167.)

[9] Marc Alexa. Linear combination of transformations. *ACM Trans. Graphics*, 21(3): 380–387, 2002. (Cited on pp. 50, 53.)

[10] E. J. Allen, J. Baglama, and S. K. Boyd. Numerical approximation of the product of the square root of a matrix with a vector. *Linear Algebra Appl.*, 310:167–181, 2000. (Cited on p. 310.)

[11] Simon L. Altmann. *Rotations, Quaternions, and Double Groups.* Oxford University Press, 1986. xiv+317 pp. ISBN 0-19-855372-2. (Cited on p. 300.)

[12] E. Anderson, Z. Bai, C. H. Bischof, S. Blackford, J. W. Demmel, J. J. Dongarra, J. J. Du Croz, A. Greenbaum, S. J. Hammarling, A. McKenney, and D. C. Sorensen. *LAPACK Users' Guide.* Third edition, Society for Industrial and Applied Mathematics, Philadelphia, PA, USA, 1999. xxvi+407 pp. ISBN 0-89871-447-8. (Cited on pp. 100, 229.)

[13] T. Ando. Concavity of certain maps on positive definite matrices and applications to Hadamard products. *Linear Algebra Appl.*, 26:203–241, 1979. (Cited on p. 52.)

[14] T. Ando. *Operator-Theoretic Methods for Matrix Inequalities.* Hokusei Gakuen University, March 1998. 77 pp. (Cited on p. 52.)

[15] T. Ando, Chi-Kwong Li, and Roy Mathias. Geometric means. *Linear Algebra Appl.*, 385:305–334, 2004. (Cited on p. 52.)

[16] Huzihiro Araki and Shigeru Yamagami. An inequality for Hilbert–Schmidt norm. *Commun. Math. Phys.*, 81:89–96, 1981. (Cited on p. 215.)

[17] M. Arioli, B. Codenotti, and C. Fassino. The Padé method for computing the matrix exponential. *Linear Algebra Appl.*, 240:111–130, 1996. (Cited on pp. 104, 248.)

[18] W. E. Arnoldi. The principle of minimized iterations in the solution of the matrix eigenvalue problem. *Quart. Appl. Math.*, 9:17–29, 1951. (Cited on p. 309.)

[19] Vincent Arsigny, Oliver Commowick, Xavier Pennec, and Nicholas Ayache. A log–Euclidean framework for statistics on diffeomorphisms. In *Medical Image Computing and Computer-Assisted Intervention—MICCAI 2006*, Rasmus Larsen, Mads Nielsen, and Jon Sporring, editors, number 4190 in *Lecture Notes in Computer Science*, Springer-Verlag, Berlin, 2006, pages 924–931. (Cited on p. 284.)

[20] Vincent Arsigny, Pierre Fillard, Xavier Pennec, and Nicholas Ayache. Geometric means in a novel vector space structure on symmetric positive-definite matrices. *SIAM J. Matrix Anal. Appl.*, 29(1):328–347, 2007. (Cited on p. 47.)

[21] Ashkan Ashrafi and Peter M. Gibson. An involutory Pascal matrix. *Linear Algebra Appl.*, 387:277–286, 2004. (Cited on p. 165.)

[22] Kendall E. Atkinson and Weimin Han. *Theoretical Numerical Analysis: A Functional Analysis Framework.* Second edition, Springer-Verlag, New York, 2005. xviii+576 pp. ISBN 0-387-25887-6. (Cited on p. 69.)

[23] Jean-Pierre Aubin and Ivar Ekeland. *Applied Nonlinear Analysis.* Wiley, New York, 1984. xi+518 pp. ISBN 0-471-05998-6. (Cited on p. 69.)

[24] L. Autonne. Sur les groupes linéaires, réels et orthogonaux. *Bulletin de la Société Mathématique de France*, 30:121–134, 1902. (Cited on p. 213.)

[25] Ivo Babuška, Milan Práger, and Emil Vitásek. *Numerical Processes in Differential Equations.* Wiley, London, 1966. (Cited on p. 167.)

[26] Zhaojun Bai, Wenbin Chen, Richard Scalettar, and Ichitaro Yamazaki. Lecture notes on advances of numerical methods for Hubbard quantum Monte Carlo simulation. Manuscript, 2007. (Cited on pp. 44, 263.)

[27] Zhaojun Bai and James W. Demmel. Design of a parallel nonsymmetric eigenroutine toolbox, Part I. In *Proceedings of the Sixth SIAM Conference on Parallel Processing for Scientific Computing, Volume I*, Richard F. Sincovec, David E. Keyes, Michael R. Leuze, Linda R. Petzold, and Daniel A. Reed, editors, Society for Industrial and Applied Mathematics, Philadelphia, PA, USA, 1993, pages 391–398. Also available as Research Report 92-09, Department of Mathematics, University of Kentucky, Lexington, KY, USA, December 1992, 30 pp. (Cited on pp. 41, 51.)

[28] Zhaojun Bai and James W. Demmel. On swapping diagonal blocks in real Schur form. *Linear Algebra Appl.*, 186:73–95, 1993. (Cited on p. 229.)

[29] Zhaojun Bai and James W. Demmel. Using the matrix sign function to compute invariant subspaces. *SIAM J. Matrix Anal. Appl.*, 19(1):205–225, 1998. (Cited on pp. 41, 124, 125.)

[30] Zhaojun Bai, James W. Demmel, Jack J. Dongarra, Axel Ruhe, and Henk A. Van der Vorst, editors. *Templates for the Solution of Algebraic Eigenvalue Problems: A Practical Guide.* Society for Industrial and Applied Mathematics, Philadelphia, PA, USA, 2000. xxix+410 pp. ISBN 0-89871-471-0. (Cited on p. 309.)

[31] Zhaojun Bai, James W. Demmel, and Ming Gu. Inverse free parallel spectral divide and conquer algorithms for nonsymmetric eigenproblems. *Numer. Math.*, 76:279–308, 1997. (Cited on p. 130.)

[32] Zhaojun Bai, Mark Fahey, and Gene H. Golub. Some large-scale matrix computation problems. *J. Comput. Appl. Math.*, 74:71–89, 1996. (Cited on pp. 44, 318.)

[33] Zhaojun Bai, Mark Fahey, Gene H. Golub, M. Menon, and E. Richter. Computing partial eigenvalue sum in electronic structure calculations. Technical Report SCCM-98-03, SCCM, Stanford University, January 1998. 19 pp. (Cited on p. 318.)

[34] Zhaojun Bai and Gene H. Golub. Bounds for the trace of the inverse and the determinant of symmetric positive definite matrices. *Ann. Numer. Math.*, 4:29–38, 1997. (Cited on p. 318.)

[35] Zhaojun Bai and Gene H. Golub. Some unusual eigenvalue problems. In *Vector and Parallel Processing—VECPAR'98*, J. Palma, J. Dongarra, and V. Hernández, editors, volume 1573 of *Lecture Notes in Computer Science*, Springer-Verlag, Berlin, 1999, pages 4–19. (Cited on p. 318.)

[36] David H. Bailey. Algorithm 719: Multiprecision translation and execution of FORTRAN programs. *ACM Trans. Math. Software*, 19(3):288–319, 1993. (Cited on pp. 188, 286.)

[37] David H. Bailey, Yozo Hida, Xiaoye S. Li, and Brandon Thompson. ARPREC: An arbitrary precision computation package. Technical Report LBNL-53651, Lawrence Berkeley National Laboratory, Berkeley, California, March 2002. 8 pp. (Cited on p. 188.)

[38] George A. Baker, Jr. *Essentials of Padé Approximants*. Academic Press, New York, 1975. xi+306 pp. (Cited on pp. 80, 104, 264, 274.)

[39] George A. Baker, Jr. and Peter Graves-Morris. *Padé Approximants*, volume 59 of *Encyclopedia of Mathematics and Its Applications*. Second edition, Cambridge University Press, 1996. xiv+746 pp. (Cited on pp. 80, 104, 274, 283.)

[40] H. F. Baker. The reciprocation of one quadric into another. *Proc. Cambridge Philos. Soc.*, 23:22–27, 1925. (Cited on pp. 28, 52.)

[41] A. V. Balakrishnan. Fractional powers of closed operators and the semigroups generated by them. *Pacific J. Math.*, 10(2):419–437, 1960. (Cited on p. 187.)

[42] R. P. Bambah and S. Chowla. On integer cube roots of the unit matrix. *Science and Culture*, 12:105, 1946. (Cited on p. 189.)

[43] Itzhack Y. Bar-Itzhack, J. Meyer, and P. A. Fuhrmann. Strapdown matrix orthogonalization: The dual iterative algorithm. *IEEE Trans. Aerospace and Electronic Systems*, AES-12(1):32–37, 1976. (Cited on p. 215.)

[44] A. Y. Barraud. Investigations autour de la fonction signe d'une matrice application a l'équation de Riccati. *R.A.I.R.O. Automatique/Systems Analysis and Control*, 13(4): 335–368, 1979. (Cited on p. 130.)

[45] Anders Barrlund. Perturbation bounds on the polar decomposition. *BIT*, 30:101–113, 1990. (Cited on p. 215.)

[46] Friedrich L. Bauer. *Decrypted Secrets: Methods and Maxims of Cryptology*. Third edition, Springer-Verlag, Berlin, 2002. xii+474 pp. ISBN 3-540-42674-4. (Cited on pp. 165, 167.)

[47] U. Baur and Peter Benner. Factorized solution of Lyapunov equations based on hierarchical matrix arithmetic. *Computing*, 78:211–234, 2006. (Cited on pp. 51, 316.)

[48] Connice A. Bavely and G. W. Stewart. An algorithm for computing reducing subspaces by block diagonalization. *SIAM J. Numer. Anal.*, 16(2):359–367, 1979. (Cited on p. 89.)

[49] C. A. Beattie and S. W. Smith. Optimal matrix approximants in structural identification. *J. Optimization Theory and Applications*, 74(1):23–56, 1992. (Cited on p. 217.)

[50] Alfredo Bellen and Marino Zennaro. *Numerical Methods for Delay Differential Equations*. Oxford University Press, 2003. xiv+395 pp. ISBN 0-19-850654-6. (Cited on p. 51.)

[51] Richard Bellman. *Introduction to Matrix Analysis*. Second edition, McGraw-Hill, New York, 1970. xxiii+403 pp. Reprinted by Society for Industrial and Applied Mathematics, Philadelphia, PA, USA, 1997. ISBN 0-89871-399-4. (Cited on pp. 51, 105, 165, 167, 264, 267.)

[52] Adi Ben-Israel and Thomas N. E. Greville. *Generalized Inverses: Theory and Applications*. Second edition, Springer-Verlag, New York, 2003. xv+420 pp. ISBN 0-387-00293-6. (Cited on pp. 214, 325, 353.)

[53] Peter Benner and Ralph Byers. Disk functions and their relationship to the matrix sign function. In *Proceedings of the European Control Conference ECC97, Paper 936, BELWARE Information Technology, Waterloo, Belgium*, 1997. CD ROM. (Cited on p. 49.)

[54] Peter Benner, Ralph Byers, Volker Mehrmann, and Hongguo Xu. A unified deflating subspace approach for classes of polynomial and rational matrix equations. Preprint SFB393/00-05, Zentrum für Technomathematik, Universität Bremen, Bremen, Germany, January 2000. 29 pp. (Cited on pp. 49, 187.)

[55] Peter Benner and Enrique S. Quintana-Ortí. Solving stable generalized Lyapunov equations with the matrix sign function. *Numer. Algorithms*, 20(1):75–100, 1999. (Cited on pp. 51, 123.)

[56] Peter Benner, Enrique S. Quintana-Ortí, and Gregorio Quintana-Ortí. Solving stable Sylvester equations via rational iterative schemes. *J. Sci. Comput.*, 28:51–83, 2006. (Cited on p. 51.)

[57] Michele Benzi and Gene H. Golub. Bounds for the entries of matrix functions with applications to preconditioning. *BIT*, 39(3):417–438, 1999. (Cited on p. 317.)

[58] Michele Benzi and Nader Razouk. Decay bounds and $O(n)$ algorithms for approximating functions of sparse matrices. *Electron. Trans. Numer. Anal.*, 28:16–39, 2007. (Cited on p. 318.)

[59] Håvard Berland, Bård Skaflestad, and Will Wright. EXPINT—A MATLAB package for exponential integrators. *ACM Trans. Math. Software*, 33(1):Article 4, 2007. (Cited on p. 37.)

[60] Abraham Berman and Robert J. Plemmons. *Nonnegative Matrices in the Mathematical Sciences*. Society for Industrial and Applied Mathematics, Philadelphia, PA, USA, 1994. xx+340 pp. Corrected republication, with supplement, of work first published in 1979 by Academic Press. ISBN 0-89871-321-8. (Cited on pp. 159, 260, 329.)

[61] David S. Bernstein and Charles F. Van Loan. Rational matrix functions and rank-1 updates. *SIAM J. Matrix Anal. Appl.*, 22(1):145–154, 2000. (Cited on p. 28.)

[62] Rajendra Bhatia. Some inequalities for norm ideals. *Commun. Math. Phys.*, 111: 33–39, 1987. (Cited on p. 135.)

[63] Rajendra Bhatia. Matrix factorizations and their perturbations. *Linear Algebra Appl.*, 197/198:245–276, 1994. (Cited on pp. 215, 217.)

[64] Rajendra Bhatia. *Matrix Analysis*. Springer-Verlag, New York, 1997. xi+347 pp. ISBN 0-387-94846-5. (Cited on pp. 69, 187, 215, 217, 237, 315, 330.)

[65] Rajendra Bhatia. *Positive Definite Matrices*. Princeton University Press, Princeton, NJ, USA, 2007. ix+254 pp. ISBN 0-691-12918-5. (Cited on pp. 52, 315, 330.)

[66] Rajendra Bhatia and Fuad Kittaneh. Approximation by positive operators. *Linear Algebra Appl.*, 161:1–9, 1992. (Cited on p. 214.)

[67] Rajendra Bhatia and Kalyan K. Mukherjea. On weighted Löwdin orthogonalization. *Int. J. Quantum Chemistry*, 29:1775–1778, 1986. (Cited on p. 42.)

[68] M. D. Bingham. A new method for obtaining the inverse matrix. *J. Amer. Statist. Assoc.*, 36(216):530–534, 1941. (Cited on p. 90.)

[69] Dario A. Bini, Nicholas J. Higham, and Beatrice Meini. Algorithms for the matrix pth root. *Numer. Algorithms*, 39(4):349–378, 2005. (Cited on pp. 187, 188, 191.)

[70] Dario A. Bini, Guy Latouche, and Beatrice Meini. *Numerical Methods for Structured Markov Chains*. Oxford University Press, 2005. xi+327 pp. ISBN 0-19-852768-3. (Cited on p. 45.)

[71] R. E. D. Bishop. Arthur Roderick Collar. 22 February 1908–12 February 1986. *Biographical Memoirs of Fellows of the Royal Society*, 33:164–185, 1987. (Cited on p. 27.)

[72] Åke Björck and C. Bowie. An iterative algorithm for computing the best estimate of an orthogonal matrix. *SIAM J. Numer. Anal.*, 8(2):358–364, 1971. (Cited on p. 215.)

[73] Åke Björck and Sven Hammarling. A Schur method for the square root of a matrix. *Linear Algebra Appl.*, 52/53:127–140, 1983. (Cited on pp. 163, 166, 167, 171, 187.)

[74] Sergio Blanes and Fernando Casas. On the convergence and optimization of the Baker–Campbell–Hausdorff formula. *Linear Algebra Appl.*, 378:135–158, 2004. (Cited on p. 263.)

[75] Artan Boriçi. QCDLAB project. `http://phys.fshn.edu.al/qcdlab.html`. (Cited on p. 43.)

[76] Steffen Börm, Lars Grasedyck, and Wolfgang Hackbusch. Hierarchical matrices. Lecture Note No. 21, Max-Planck-Institute for Mathematics in the Sciences, Leipzig, Germany, 2003. 71 pp. Revised June 2006. (Cited on p. 316.)

[77] Jonathan M. Borwein, David Bailey, and Roland Girgensohn. *Experimentation in Mathematics: Computational Paths to Discovery.* A K Peters, Natick, Massachusetts, 2004. x+357 pp. ISBN 1-56881-136-5. (Cited on p. 33.)

[78] C. Bouby, D. Fortuné, W. Pietraszkiewicz, and C. Vallée. Direct determination of the rotation in the polar decomposition of the deformation gradient by maximizing a Rayleigh quotient. *Z. Angew. Math. Mech.*, 85(3):155–162, 2005. (Cited on p. 43.)

[79] David W. Boyd. The power method for ℓ^p norms. *Linear Algebra Appl.*, 9:95–101, 1974. (Cited on p. 69.)

[80] Geoff Boyd, Charles A. Micchelli, Gilbert Strang, and Ding-Xuan Zhou. Binomial matrices. *Adv. in Comput. Math.*, 14:379–391, 2001. (Cited on p. 166.)

[81] Russell J. Bradford, Robert M. Corless, James H. Davenport, David J. Jeffrey, and Stephen M. Watt. Reasoning about the elementary functions of complex analysis. *Annals of Mathematics and Artificial Intelligence*, 36:303–318, 2002. (Cited on pp. 280, 283.)

[82] T. J. Bridges and P. J. Morris. Differential eigenvalue problems in which the parameter appears nonlinearly. *J. Comput. Phys.*, 55:437–460, 1984. (Cited on p. 46.)

[83] William Briggs. *Ants, Bikes, and Clocks: Problem Solving for Undergraduates.* Society for Industrial and Applied Mathematics, Philadelphia, PA, USA, 2005. 168 pp. ISBN 0-89871-574-1. (Cited on p. 343.)

[84] A. Buchheim. On the theory of matrices. *Proc. London Math. Soc.*, 16:63–82, 1884. (Cited on p. 26.)

[85] A. Buchheim. An extension of a theorem of Professor Sylvester's relating to matrices. *Phil. Mag.*, 22(135):173–174, 1886. Fifth series. (Cited on p. 26.)

[86] G. J. Butler, Charles R. Johnson, and H. Wolkowicz. Nonnegative solutions of a quadratic matrix equation arising from comparison theorems in ordinary differential equations. *SIAM J. Alg. Discrete Methods*, 6(1):47–53, 1985. (Cited on p. 167.)

[87] Ralph Byers. A LINPACK-style condition estimator for the equation $AX - XB^T = C$. *IEEE Trans. Automat. Control*, AC-29(10):926–928, 1984. (Cited on p. 226.)

[88] Ralph Byers. Solving the algebraic Riccati equation with the matrix sign function. *Linear Algebra Appl.*, 85:267–279, 1987. (Cited on pp. 51, 125, 130, 132.)

[89] Ralph Byers, Chunyang He, and Volker Mehrmann. The matrix sign function method and the computation of invariant subspaces. *SIAM J. Matrix Anal. Appl.*, 18(3):615–632, 1997. (Cited on pp. 122, 124.)

[90] D. Calvetti, E. Gallopoulos, and L. Reichel. Incomplete partial fractions for parallel evaluation of rational matrix functions. *J. Comput. Appl. Math.*, 59:349–380, 1995. (Cited on p. 104.)

[91] Daniela Calvetti, Sun-Mi Kim, and Lothar Reichel. Quadrature rules based on the Arnoldi process. *SIAM J. Matrix Anal. Appl.*, 26(3):765–781, 2005. (Cited on p. 318.)

[92] S. L. Campbell and C. D. Meyer, Jr. *Generalized Inverses of Linear Transformations.* Pitman, London, 1979. xi+272 pp. Reprinted by Dover, New York, 1991. ISBN 0-486-66693-X. (Cited on pp. 325, 326, 353.)

[93] João R. Cardoso, Charles S. Kenney, and F. Silva Leite. Computing the square root and logarithm of a real P-orthogonal matrix. *Appl. Numer. Math.*, 46:173–196, 2003. (Cited on p. 316.)

[94] João R. Cardoso and F. Silva Leite. The Moser–Veselov equation. *Linear Algebra Appl.*, 360:237–248, 2003. (Cited on p. 46.)

[95] João R. Cardoso and F. Silva Leite. Padé and Gregory error estimates for the logarithm of block triangular matrices. *Appl. Numer. Math.*, 56:253–267, 2006. (Cited on p. 284.)

[96] Gianfranco Cariolaro, Tomaso Erseghe, and Peter Kraniauskas. The fractional discrete cosine transform. *IEEE Trans. Signal Processing*, 50(4):902–911, 2002. (Cited on p. 28.)

[97] Lennart Carleson and Theodore W. Gamelin. *Complex Dynamics.* Springer-Verlag, New York, 1993. ix+175 pp. ISBN 0-387-97942-5. (Cited on p. 156.)

[98] A. J. Carpenter, A. Ruttan, and R. S. Varga. Extended numerical computations on the "1/9" conjecture in rational approximation theory. In *Rational Approximation and Interpolation*, P. R. Graves-Morris, E. B. Saff, and R. S. Varga, editors, volume 1105 of *Lecture Notes in Mathematics*, Springer-Verlag, Berlin, 1984, pages 383–411. (Cited on pp. 259, 267.)

[99] Arthur Cayley. A memoir on the theory of matrices. *Philos. Trans. Roy. Soc. London*, 148:17–37, 1858. (Cited on pp. 26, 34, 166, 168, 191.)

[100] Arthur Cayley. On the extraction of the square root of a matrix of the third order. *Proc. Roy. Soc. Edinburgh*, 7:675–682, 1872. (Cited on pp. 26, 166, 168.)

[101] Arthur Cayley. The Newton–Fourier imaginary problem. *Amer. J. Math.*, 2(1):97, 1879. (Cited on pp. 187, 191.)

[102] Elena Celledoni and Arieh Iserles. Approximating the exponential from a Lie algebra to a Lie group. *Math. Comp.*, 69(232):1457–1480, 2000. (Cited on p. 28.)

[103] F. Chaitin-Chatelin and S. Gratton. On the condition numbers associated with the polar factorization of a matrix. *Numer. Linear Algebra Appl.*, 7:337–354, 2000. (Cited on p. 201.)

[104] Raymond H. Chan, Chen Greif, and Dianne P. O'Leary, editors. *Milestones in Matrix Computation: The Selected Works of Gene H. Golub, with Commentaries.* Oxford University Press, 2007. xi+565 pp. ISBN 978-0-19-920681-0. (Cited on p. 391.)

[105] Shivkumar Chandrasekaran and Ilse C. F. Ipsen. Backward errors for eigenvalue and singular value decompositions. *Numer. Math.*, 68:215–223, 1994. (Cited on p. 214.)

[106] Chi-Tsong Chen. *Linear System Theory and Design.* Third edition, Oxford University Press, 1999. xiii+334 pp. ISBN 0-19-511777-8. (Cited on p. 51.)

[107] Sheung Hun Cheng, Nicholas J. Higham, Charles S. Kenney, and Alan J. Laub. Return to the middle ages: A half-angle iteration for the logarithm of a unitary matrix. In *Proceedings of the Fourteenth International Symposium of Mathematical Theory of Networks and Systems, Perpignan, France*, 2000. CD ROM. (Cited on p. 316.)

[108] Sheung Hun Cheng, Nicholas J. Higham, Charles S. Kenney, and Alan J. Laub. Approximating the logarithm of a matrix to specified accuracy. *SIAM J. Matrix Anal. Appl.*, 22(4):1112–1125, 2001. (Cited on pp. 104, 141, 166, 167, 283, 284, 285.)

[109] M. Cipolla. Sulle matrice espressione analitiche di un'altra. *Rendiconti Circolo Matematico de Palermo*, 56:144–154, 1932. (Cited on p. 26.)

[110] W. J. Cody, G. Meinardus, and R. S. Varga. Chebyshev rational approximations to e^{-x} in $[0, +\infty)$ and applications to heat-conduction problems. *J. Approximation Theory*, 2:50–65, 1969. (Cited on p. 259.)

[111] John P. Coleman. Rational approximations for the cosine function; P-acceptability and order. *Numer. Algorithms*, 3:143–158, 1992. (Cited on p. 300.)

[112] A. R. Collar. The first fifty years of aeroelasticity. *Aerospace (Royal Aeronautical Society Journal)*, 5:12–20, 1978. (Cited on pp. 27, 53.)

[113] Samuel D. Conte and Carl de Boor. *Elementary Numerical Analysis: An Algorithmic Approach.* Third edition, McGraw-Hill, Tokyo, 1980. xii+432 pp. ISBN 0-07-066228-2. (Cited on p. 333.)

[114] Robert M. Corless, Hui Ding, Nicholas J. Higham, and David J. Jeffrey. The solution of $S\exp(S) = A$ is not always the Lambert W function of A. In *ISSAC '07: Proceedings of the* 2007 *International Symposium on Symbolic and Algebraic Computation*, New York, 2007, pages 116–121. ACM Press. (Cited on p. 51.)

[115] Robert M. Corless, Gaston H. Gonnet, D. E. G. Hare, David J. Jeffrey, and Donald E. Knuth. On the Lambert W function. *Adv. in Comput. Math.*, 5(4):329–359, 1996. (Cited on pp. 51, 53.)

[116] Robert M. Corless and David J. Jeffrey. The unwinding number. *ACM SIGSAM Bulletin*, 30(2):28–35, 1996. (Cited on p. 283.)

[117] Marius Cornea-Hasegan and Bob Norin. IA-64 floating-point operations and the IEEE standard for binary floating-point arithmetic. *Intel Technology Journal*, 3, 1999. `http://developer.intel.com/technology/itj/`. (Cited on p. 188.)

[118] S. M. Cox and P. C. Matthews. Exponential time differencing for stiff systems. *J. Comput. Phys.*, 176:430–455, 2002. (Cited on p. 37.)

[119] Trevor F. Cox and Michael A. A. Cox. *Multidimensional Scaling.* Chapman and Hall, London, 1994. xi+213 pp. ISBN 0-412-49120-6. (Cited on p. 43.)

[120] Tony Crilly. Cayley's anticipation of a generalised Cayley–Hamilton theorem. *Historia Mathematica*, 5:211–219, 1978. (Cited on p. 30.)

[121] Tony Crilly. *Arthur Cayley: Mathematician Laureate of the Victorian Age.* Johns Hopkins University Press, Baltimore, MD, USA, 2006. xxi+610 pp. ISBN 0-8018-8011-4. (Cited on pp. 26, 30.)

[122] G. W. Cross and P. Lancaster. Square roots of complex matrices. *Linear and Multilinear Algebra*, 1:289–293, 1974. (Cited on p. 16.)

[123] Michel Crouzeix. Bounds for analytical functions of matrices. *Integral Equations and Operator Theory*, 48:461–477, 2004. (Cited on p. 105.)

[124] L. Csanky. Fast parallel matrix inversion algorithms. *SIAM J. Comput.*, 5(4):618–623, 1976. (Cited on p. 90.)

[125] Charles G. Cullen. *Matrices and Linear Transformations.* Second edition, Addison-Wesley, Reading, MA, USA, 1972. xii+318 pp. Reprinted by Dover, New York, 1990. ISBN 0-486-66328-0. (Cited on pp. 28, 29, 52.)

[126] Walter J. Culver. On the existence and uniqueness of the real logarithm of a matrix. *Proc. Amer. Math. Soc.*, 17:1146–1151, 1966. (Cited on p. 17.)

[127] James R. Cuthbert. On uniqueness of the logarithm for Markov semi-groups. *J. London Math. Soc.*, 4:623–630, 1972. (Cited on p. 190.)

[128] Germund Dahlquist. *Stability and Error Bounds in the Numerical Integration of Ordinary Differential Equations.* PhD thesis, Royal Inst. of Technology, Stockholm, Sweden, 1958. Reprinted in Trans. Royal Inst. of Technology, No. 130, Stockholm, Sweden, 1959. (Cited on p. 237.)

[129] Ju. L. Daleckiĭ. Differentiation of non-Hermitian matrix functions depending on a parameter. *Amer. Math. Soc. Transl., Series* 2, 47:73–87, 1965. (Cited on p. 69.)

[130] Ju. L. Daleckiĭ and S. G. Kreĭn. Integration and differentiation of functions of Hermitian operators and applications to the theory of perturbations. *Amer. Math. Soc. Transl., Series* 2, 47:1–30, 1965. (Cited on p. 69.)

[131] E. B. Davies. Approximate diagonalization. *SIAM J. Matrix Anal. Appl.*, 29(4):1051–1064, 2007. (Cited on p. 30.)

[132] E. Brian Davies. *Science in the Looking Glass: What Do Scientists Really Know?* Oxford University Press, 2003. x+295 pp. ISBN 0-19-852543-5. (Cited on p. 34.)

[133] E. Brian Davies. *Linear Operators and their Spectra.* Cambridge University Press, Cambridge, UK, 2007. xii+451 pp. ISBN 978-0-521-86629-3. (Cited on p. 28.)

[134] Philip I. Davies. Structured conditioning of matrix functions. *Electron. J. Linear Algebra*, 11:132–161, 2004. (Cited on p. 315.)

[135] Philip I. Davies and Nicholas J. Higham. A Schur–Parlett algorithm for computing matrix functions. *SIAM J. Matrix Anal. Appl.*, 25(2):464–485, 2003. (Cited on pp. 229, 231.)

[136] Philip I. Davies and Nicholas J. Higham. Computing $f(A)b$ for matrix functions f. In *QCD and Numerical Analysis III*, Artan Boriçi, Andreas Frommer, Báalint Joó, Anthony Kennedy, and Brian Pendleton, editors, volume 47 of *Lecture Notes in Computational Science and Engineering*, Springer-Verlag, Berlin, 2005, pages 15–24. (Cited on pp. 307, 308, 310.)

[137] Philip I. Davies, Nicholas J. Higham, and Françoise Tisseur. Analysis of the Cholesky method with iterative refinement for solving the symmetric definite generalized eigenproblem. *SIAM J. Matrix Anal. Appl.*, 23(2):472–493, 2001. (Cited on p. 35.)

[138] Philip I. Davies and Matthew I. Smith. Updating the singular value decomposition. *J. Comput. Appl. Math.*, 170:145–167, 2004. (Cited on p. 216.)

[139] Chandler Davis. Explicit functional calculus. *Linear Algebra Appl.*, 6:193–199, 1973. (Cited on p. 84.)

[140] George J. Davis. Numerical solution of a quadratic matrix equation. *SIAM J. Sci. Statist. Comput.*, 2(2):164–175, 1981. (Cited on p. 52.)

[141] Philip J. Davis and Aviezri S. Fraenkel. Remembering Philip Rabinowitz. *Notices Amer. Math. Soc.*, 54(11):1502–1506, 2007. (Cited on p. 311.)

[142] Philip J. Davis and Philip Rabinowitz. *Methods of Numerical Integration.* Second edition, Academic Press, London, 1984. xiv+612 pp. ISBN 0-12-206360-0. (Cited on pp. 274, 308.)

[143] Carl de Boor. Divided differences. *Surveys in Approximation Theory*, 1:46–69, 2005. (Cited on pp. 264, 333.)

[144] Lokenath Debnath and Piotr Mikusiński. *Introduction to Hilbert Spaces with Applications.* Second edition, Academic Press, San Diego, CA, USA, 1999. xviii+551 pp. ISBN 0-12-208436-5. (Cited on p. 167.)

[145] James W. Demmel. *Applied Numerical Linear Algebra.* Society for Industrial and Applied Mathematics, Philadelphia, PA, USA, 1997. xi+419 pp. ISBN 0-89871-389-7. (Cited on p. 325.)

[146] Eugene D. Denman and Alex N. Beavers, Jr. The matrix sign function and computations in systems. *Appl. Math. Comput.*, 2:63–94, 1976. (Cited on pp. 52, 141.)

[147] John E. Dennis, Jr., J. F. Traub, and R. P. Weber. The algebraic theory of matrix polynomials. *SIAM J. Numer. Anal.*, 13(6):831–845, 1976. (Cited on p. 52.)

[148] Jean Descloux. Bounds for the spectral norm of functions of matrices. *Numer. Math.*, 15:185–190, 1963. (Cited on pp. 84, 103.)

[149] Robert L. Devaney. *An Introduction to Chaotic Dynamical Systems.* Second edition, Addison-Wesley, Reading, MA, USA, 1989. xvi+336 pp. ISBN 0-201-13046-7. (Cited on p. 156.)

[150] Inderjit S. Dhillon and Joel A. Tropp. Matrix nearness problems with Bregman divergences. *SIAM J. Matrix Anal. Appl.*, 29(4):1120–1146, 2007. (Cited on p. 50.)

[151] Bradley W. Dickinson and Kenneth Steiglitz. Eigenvectors and functions of the discrete Fourier transform. *IEEE Trans. Acoust., Speech, Signal Processing*, ASSP-30(1):25–31, 1982. (Cited on p. 28.)

[152] Luca Dieci. Considerations on computing real logarithms of matrices, Hamiltonian logarithms, and skew-symmetric logarithms. *Linear Algebra Appl.*, 244:35–54, 1996. (Cited on pp. 50, 315, 316.)

[153] Luca Dieci. Real Hamiltonian logarithm of a symplectic matrix. *Linear Algebra Appl.*, 281:227–246, 1998. (Cited on p. 315.)

[154] Luca Dieci, Benedetta Morini, and Alessandra Papini. Computational techniques for real logarithms of matrices. *SIAM J. Matrix Anal. Appl.*, 17(3):570–593, 1996. (Cited on pp. 274, 283, 284.)

[155] Luca Dieci, Benedetta Morini, Alessandra Papini, and Aldo Pasquali. On real logarithms of nearby matrices and structured matrix interpolation. *Appl. Numer. Math.*, 29:145–165, 1999. (Cited on p. 50.)

[156] Luca Dieci and Alessandra Papini. Conditioning and Padé approximation of the logarithm of a matrix. *SIAM J. Matrix Anal. Appl.*, 21(3):913–930, 2000. (Cited on p. 284.)

[157] Luca Dieci and Alessandra Papini. Padé approximation for the exponential of a block triangular matrix. *Linear Algebra Appl.*, 308:183–202, 2000. (Cited on pp. 243, 246, 248, 264, 265.)

[158] Luca Dieci and Alessandra Papini. Conditioning of the exponential of a block triangular matrix. *Numer. Algorithms*, 28:137–150, 2001. (Cited on p. 264.)

[159] J. Dieudonné. *Foundations of Modern Analysis.* Academic Press, New York, 1960. xiv+361 pp. (Cited on p. 58.)

[160] John D. Dixon. Estimating extremal eigenvalues and condition numbers of matrices. *SIAM J. Numer. Anal.*, 20(4):812–814, 1983. (Cited on p. 69.)

[161] Elizabeth D. Dolan and Jorge J. Moré. Benchmarking optimization software with performance profiles. *Math. Programming*, 91:201–213, 2002. (Cited on p. 252.)

[162] Jack J. Dongarra, Iain S. Duff, Danny C. Sorensen, and Henk A. Van der Vorst. *Numerical Linear Algebra for High-Performance Computers.* Society for Industrial and Applied Mathematics, Philadelphia, PA, USA, 1998. xviii+342 pp. ISBN 0-89871-428-1. (Cited on p. 335.)

[163] William F. Donoghue, Jr. *Monotone Matrix Functions and Analytic Continuation.* Springer-Verlag, Berlin, 1974. 182 pp. ISBN 3-540-06543-1. (Cited on p. 315.)

[164] M. P. Drazin, J. W. Dungey, and K. W. Gruenberg. Some theorems on commutative matrices. *J. London Math. Soc.*, 26(2):221–228, 1951. (Cited on p. 29.)

[165] P. G. Drazin. *Nonlinear Systems*. Cambridge University Press, Cambridge, UK, 1992. xiii+317 pp. ISBN 0-521-40668-4. (Cited on p. 359.)

[166] V. Druskin, A. Greenbaum, and L. Knizhnerman. Using nonorthogonal Lanczos vectors in the computation of matrix functions. *SIAM J. Sci. Comput.*, 19(1):38–54, 1998. (Cited on p. 310.)

[167] Vladimir L. Druskin and Leonid A. Knizhnerman. Two polynomial methods of calculating functions of symmetric matrices. *U.S.S.R. Comput. Maths. Math. Phys.*, 29(6): 112–121, 1989. (Cited on p. 309.)

[168] Ian L. Dryden and Kanti V. Mardia. *Statistical Shape Analysis*. Wiley, New York, 1998. xvii+347 pp. ISBN 0-471-95816-6. (Cited on p. 43.)

[169] Augustin A. Dubrulle. An optimum iteration for the matrix polar decomposition. *Electron. Trans. Numer. Anal.*, 8:21–25, 1999. (Cited on pp. 216, 218.)

[170] B. J. Duke. Certification of Algorithm 298 [F1]: Determination of the square root of a positive definite matrix. *Comm. ACM*, 12(6):325–326, 1969. (Cited on p. 167.)

[171] Nelson Dunford and Jacob T. Schwartz. *Linear Operators. Part I: General Theory*. Wiley, New York, 1988. xiv+858 pp. Wiley Classics Library edition. ISBN 0-471-60848-3. (Cited on p. 28.)

[172] Nelson Dunford and Jacob T. Schwartz. *Linear Operators. Part III: Spectral Operators*. Wiley, New York, 1971. xix+1925–2592 pp. (Cited on p. 28.)

[173] Alan Edelman and Gilbert Strang. Pascal matrices. *Amer. Math. Monthly*, 111(3): 189–197, 2004. (Cited on p. 374.)

[174] Michael Eiermann and Oliver G. Ernst. A restarted Krylov subspace method for the evaluation of matrix functions. *SIAM J. Numer. Anal.*, 44(6):2481–2504, 2006. (Cited on p. 305.)

[175] Timo Eirola. A refined polar decomposition: $A = UPD$. *SIAM J. Matrix Anal. Appl.*, 22(3):824–836, 2000. (Cited on p. 214.)

[176] Ludwig Elsner. Iterative Verfahren zur Lösung der Matrizengleichung $X^2 - A = 0$. *Buletinul Institutului Politehnic din Iasi*, xvi(xx):15–24, 1970. (Cited on pp. 106, 159, 167, 169.)

[177] Ivan Erdelyi. On partial isometries in finite-dimensional Euclidean spaces. *SIAM J. Appl. Math.*, 14(3):453–467, 1966. (Cited on p. 326.)

[178] J.-Cl. Evard and F. Jafari. A complex Rolle's theorem. *Amer. Math. Monthly*, 99(9): 858–861, 1992. (Cited on p. 333.)

[179] Jean-Claude Evard and Frank Uhlig. On the matrix equation $f(X) = A$. *Linear Algebra Appl.*, 162–164:447–519, 1992. (Cited on p. 28.)

[180] D. K. Faddeev and V. N. Faddeeva. *Numerische Methoden der Linearen Algebra*. Veb Deutscher Verlag der Wissenschaften, Berlin, 1964. (Cited on p. 370.)

[181] Ky Fan and A. J. Hoffman. Some metric inequalities in the space of matrices. *Proc. Amer. Math. Soc.*, 6:111–116, 1955. (Cited on p. 214.)

[182] Heike Faßbender, D. Steven Mackey, Niloufer Mackey, and Hongguo Xu. Hamiltonian square roots of skew-Hamiltonian matrices. *Linear Algebra Appl.*, 287:125–159, 1999. (Cited on p. 315.)

[183] T. I. Fenner and G. Loizou. Optimally scalable matrices. *Philos. Trans. Roy. Soc. London Ser. A*, 287(1345):307–349, 1977. (Cited on p. 105.)

[184] W. L. Ferrar. *Finite Matrices*. Oxford University Press, 1951. vii+182 pp. (Cited on pp. 27, 34.)

[185] Miroslav Fiedler and Hans Schneider. Analytic functions of M-matrices and generalizations. *Linear and Multilinear Algebra*, 13:185–201, 1983. (Cited on p. 188.)

[186] P. Filipponi. An algorithm for computing functions of triangular matrices. *Computing*, 26:67–71, 1981. (Cited on p. 104.)

[187] Steven R. Finch. *Mathematical Constants*. Cambridge University Press, Cambridge, UK, 2003. xix+602 pp. ISBN 0-521-81805-2. (Cited on p. 259.)

[188] Harley Flanders. Elementary divisors of AB and BA. *Proc. Amer. Math. Soc.*, 2(6): 871–874, 1951. (Cited on p. 28.)

[189] J. Fortiana and C. M. Cuadras. A family of matrices, the discretized Brownian bridge, and distance-based regression. *Linear Algebra Appl.*, 264:173–188, 1997. (Cited on p. 349.)

[190] David Fowler and Eleanor Robson. Square root approximations in old Babylonian mathematics: YBC 7289 in context. *Historia Mathematica*, 25:366–378, 1998. (Cited on p. 166.)

[191] Chris Fraley. Test problems for unconstrained optimization and nonlinear least squares. `http://www.netlib.org/uncon`. (Cited on p. 46.)

[192] Gene F. Franklin, J. David Powell, and Michael L. Workman. *Digital Control of Dynamic Systems*. Third edition, Addison-Wesley, Reading, MA, USA, 1998. xxiii+742 pp. ISBN 0-201-33153-5. (Cited on p. 51.)

[193] R. A. Frazer, W. J. Duncan, and A. R. Collar. *Elementary Matrices and Some Applications to Dynamics and Differential Equations*. Cambridge University Press, 1938. xviii+416 pp. 1963 printing. (Cited on pp. 27, 262.)

[194] Paul Friedland. Algorithm 312: Absolute value and square root of a complex number. *Comm. ACM*, 10(10):665, 1967. (Cited on p. 166.)

[195] G. Frobenius. Über die cogredienten Transformationen der bilinearen Formen. *Sitzungsber K. Preuss. Akad. Wiss. Berlin*, 16:7–16, 1896. (Cited on pp. 26, 28, 52.)

[196] Andreas Frommer, Thomas Lippert, Björn Medeke, and Klaus Schilling, editors. *Numerical Challenges in Lattice Quantum Chromodynamics*, volume 15 of *Lecture Notes in Computational Science and Engineering*. Springer-Verlag, Berlin, 2000. viii+184 pp. ISBN 3-540-67732-1. (Cited on p. 43.)

[197] Andreas Frommer and Valeria Simoncini. Stopping criteria for rational matrix functions of Hermitian and symmetric matrices. Manuscript, 2007. 24 pp. (Cited on p. 306.)

[198] Andreas Frommer and Valeria Simoncini. Matrix functions. In *Model Order Reduction: Theory, Research Aspects and Applications*, W. Schilders and H. A. Van der Vorst, editors, Springer-Verlag, Berlin, 2008. To appear. (Cited on pp. 43, 310.)

[199] Jean Gallier and Dianna Xu. Computing exponentials of skew-symmetric matrices and logarithms of orthogonal matrices. *International Journal of Robotics and Automation*, 17(4):1–11, 2002. (Cited on pp. 50, 266.)

[200] E. Gallopoulos and Y. Saad. Efficient solution of parabolic equations by Krylov approximation methods. *SIAM J. Sci. Statist. Comput.*, 13(5):1236–1264, 1992. (Cited on pp. 260, 263, 309.)

[201] Walter Gander. On Halley's iteration method. *Amer. Math. Monthly*, 92(2):131–134, 1985. (Cited on p. 131.)

[202] Walter Gander. Algorithms for the polar decomposition. *SIAM J. Sci. Statist. Comput.*, 11(6):1102–1115, 1990. (Cited on p. 215.)

[203] F. R. Gantmacher. *The Theory of Matrices*, volume one. Chelsea, New York, 1959. x+374 pp. ISBN 0-8284-0131-4. (Cited on pp. xvii, 3, 28, 51.)

[204] Judith D. Gardiner. A stabilized matrix sign function algorithm for solving algebraic Riccati equations. *SIAM J. Sci. Comput.*, 18(5):1393–1411, 1997. (Cited on p. 51.)

[205] Judith D. Gardiner and Alan J. Laub. A generalization of the matrix-sign-function solution for algebraic Riccati equations. *Internat. J. Control*, 44(3):823–832, 1986. (Cited on p. 130.)

[206] Walter Gautschi. Algorithm 726: ORTHPOL—A package of routines for generating orthogonal polynomials and Gauss-type quadrature rules. *ACM Trans. Math. Software*, 20(1):21–62, 1994. (Cited on p. 274.)

[207] Walter Gautschi. *Numerical Analysis: An Introduction*. Birkhäuser, Boston, MA, USA, 1997. xiii+506 pp. ISBN 0-8176-3895-4. (Cited on p. 263.)

[208] Walter Gautschi. *Orthogonal Polynomials: Computation and Approximation*. Oxford University Press, 2004. viii+301 pp. ISBN 0-19-850672-4. (Cited on p. 274.)

[209] Ivan P. Gavrilyuk, Wolfgang Hackbusch, and Boris N. Khoromskij. $\mathcal{H}$-matrix approximation for the operator exponential with applications. *Numer. Math.*, 92:83–111, 2002. (Cited on p. 316.)

[210] A. O. Gel'fond. *Calculus of Finite Differences*. Hindustan Publishing Corporation, Delhi, 1971. vi+451 pp. Authorized English translation of the third Russian edition. (Cited on p. 333.)

[211] James E. Gentle. *Random Number Generation and Monte Carlo Methods*. Second edition, Springer-Verlag, New York, 2003. xv+381 pp. ISBN 0-387-00178-6. (Cited on p. 118.)

[212] Alan George and Kh. Ikramov. Is the polar decomposition finitely computable? *SIAM J. Matrix Anal. Appl.*, 17(2):348–354, 1996. (Cited on p. 215.)

[213] Alan George and Kh. Ikramov. Addendum: Is the polar decomposition finitely computable? *SIAM J. Matrix Anal. Appl.*, 18(1):264, 1997. (Cited on p. 215.)

[214] Michael Gil. Estimate for the norm of matrix-valued functions. *Linear and Multilinear Algebra*, 35:65–73, 1993. (Cited on p. 105.)

[215] Philip E. Gill, Walter Murray, and Margaret H. Wright. *Practical Optimization*. Academic Press, London, 1981. xvi+401 pp. ISBN 0-12-283952-8. (Cited on p. 199.)

[216] G. Giorgi. Nuove osservazioni sulle funzioni delle matrici. *Atti Accad. Lincei Rend.*, 6 (8):3–8, 1928. (Cited on p. 26.)

[217] Glucat: Generic library of universal Clifford algebra templates. `http://glucat.sourceforge.net/`. (Cited on p. 47.)

[218] GNU MP: Multiple precision arithmetic library. `http://www.swox.com/gmp/`. (Cited on p. 187.)

[219] S. K. Godunov. *Ordinary Differential Equations with Constant Coefficient*, volume 169 of *Translations of Mathematical Monographs*. American Mathematical Society, Providence, RI, USA, 1997. ix+282 pp. ISBN 0-8218-0656-4. (Cited on p. 51.)

[220] I. Gohberg, Peter Lancaster, and Leiba Rodman. *Matrix Polynomials*. Academic Press, New York, 1982. xiv+409 pp. ISBN 0-12-287160-X. (Cited on p. 52.)

[221] Jerome A. Goldstein and Mel Levy. Linear algebra and quantum chemistry. *Amer. Math. Monthly*, 98(8):710–718, 1991. (Cited on p. 42.)

[222] Herman H. Goldstine. *A History of Numerical Analysis from the 16th through the 19th Century*. Springer-Verlag, New York, 1977. xiv+348 pp. ISBN 0-387-90277-5. (Cited on pp. 283, 286.)

[223] Gene H. Golub and Gerard Meurant. Matrices, moments and quadrature. In *Numerical Analysis* 1993, *Proceedings of the* 15*th Dundee Conference*, D. F. Griffiths and G. A. Watson, editors, volume 303 of *Pitman Research Notes in Mathematics*, Addison Wesley Longman, Harlow, Essex, UK, 1994, pages 105–156. Reprinted in [104]. (Cited on p. 318.)

[224] Gene H. Golub and Charles F. Van Loan. *Matrix Computations*. Third edition, Johns Hopkins University Press, Baltimore, MD, USA, 1996. xxvii+694 pp. ISBN 0-8018-5413-X (hardback), 0-8018-5414-8 (paperback). (Cited on pp. xvii, 28, 69, 73, 77, 89, 105, 139, 196, 209, 231, 262, 309, 335, 337.)

[225] Nicholas I. M. Gould, Dominique Orban, and Philippe L. Toint. CUTEr and SifDec: A constrained and unconstrained testing environment, revisited. *ACM Trans. Math. Software*, 29(4):373–394, 2003. (Cited on p. 46.)

[226] John C. Gower and Garmt B. Dijksterhuis. *Procrustes Problems*. Oxford University Press, 2004. xiv+233 pp. ISBN 0-19-851058-6. (Cited on p. 43.)

[227] W. B. Gragg. The Padé table and its relation to certain algorithms of numerical analysis. *SIAM Rev.*, 14(1):1–62, 1972. (Cited on p. 104.)

[228] L. Grasedyck, W. Hackbusch, and B. N. Khoromskij. Solution of large scale algebraic matrix Riccati equations by use of hierarchical matrices. *Computing*, 70:121–165, 2003. (Cited on pp. 51, 316.)

[229] Bert F. Green. The orthogonal approximation of an oblique structure in factor analysis. *Psychometrika*, 17(4):429–440, 1952. (Cited on p. 214.)

[230] Anne Greenbaum. Some theoretical results derived from polynomial numerical hulls of Jordan blocks. *Electron. Trans. Numer. Anal.*, 18:81–90, 2004. (Cited on p. 105.)

[231] R. Grone, C. R. Johnson, E. M. Sá, and H. Wolkowicz. Normal matrices. *Linear Algebra Appl.*, 87:213–225, 1987. (Cited on p. 323.)

[232] Ming Gu. Finding well-conditioned similarities to block-diagonalize nonsymmetric matrices is NP-hard. *Journal of Complexity*, 11(3):377–391, 1995. (Cited on p. 226.)

[233] Chun-Hua Guo and Nicholas J. Higham. A Schur–Newton method for the matrix pth root and its inverse. *SIAM J. Matrix Anal. Appl.*, 28(3):788–804, 2006. (Cited on pp. 51, 186, 188, 190.)

[234] Chun-Hua Guo and Peter Lancaster. Analysis and modification of Newton's method for algebraic Riccati equations. *Math. Comp.*, 67(223):1089–1105, 1998. (Cited on p. 144.)

[235] Hongbin Guo and Rosemary A. Renaut. Estimation of $u^T f(A)v$ for large-scale unsymmetric matrices. *Numer. Linear Algebra Appl.*, 11:75–89, 2004. (Cited on p. 318.)

[236] Markus Haase. *The Functional Calculus for Sectorial Operators*. Number 169 in *Operator Theory: Advances and Applications*. Birkhäuser, Basel, Switzerland, 2006. xiv+392 pp. ISBN 3-7643-7697-X. (Cited on p. 187.)

[237] William W. Hager. Condition estimates. *SIAM J. Sci. Statist. Comput.*, 5(2):311–316, 1984. (Cited on p. 69.)

[238] E. Hairer and G. Wanner. *Analysis by Its History*. Springer-Verlag, New York, 1996. x+374 pp. ISBN 0-387-94551-2. (Cited on p. 283.)

[239] Ernst Hairer, Christian Lubich, and Gerhard Wanner. *Geometric Numerical Integration: Structure-Preserving Algorithms for Ordinary Differential Equations*. Springer-Verlag, Berlin, 2002. xiii+515 pp. ISBN 3-540-43003-2. (Cited on pp. 42, 315.)

[240] Nicholas Hale, Nicholas J. Higham, and Lloyd N. Trefethen. Computing A^{α}, $\log(A)$ and related matrix functions by contour integrals. MIMS EPrint 2007.103, Manchester Institute for Mathematical Sciences, The University of Manchester, UK, August 2007. 19 pp. (Cited on p. 308.)

[241] Brian C. Hall. *Lie Groups, Lie Algebras, and Representations*. Springer-Verlag, New York, 2003. xiv+351 pp. ISBN 0-387-40122-9. (Cited on pp. 262, 263.)

[242] Paul R. Halmos. Positive approximants of operators. *Indiana Univ. Math. J.*, 21(10): 951–960, 1972. (Cited on p. 214.)

[243] Paul R. Halmos. *Finite-Dimensional Vector Spaces.* Springer-Verlag, New York, 1974. viii+200 pp. Reprint of the Second edition published by Van Nostrand, Princeton, NJ, 1958. ISBN 0-387-90093-4. (Cited on p. 214.)

[244] Paul R. Halmos. *A Hilbert Space Problem Book.* Second edition, Springer-Verlag, Berlin, 1982. xvii+369 pp. ISBN 0-387-90685-1. (Cited on pp. 29, 167.)

[245] H. L. Hamburger and M. E. Grimshaw. *Linear Transformations in n-Dimensional Vector Space: An Introduction to the Theory of Hilbert Space.* Cambridge University Press, 1951. x+195 pp. (Cited on p. 27.)

[246] Richard J. Hanson and Michael J. Norris. Analysis of measurements based on the singular value decomposition. *SIAM J. Sci. Statist. Comput.*, 2(3):363–373, 1981. (Cited on p. 368.)

[247] Gareth Hargreaves. *Topics in Matrix Computations: Stability and Efficiency of Algorithms.* PhD thesis, University of Manchester, Manchester, England, 2005. 204 pp. (Cited on p. 74.)

[248] Gareth I. Hargreaves and Nicholas J. Higham. Efficient algorithms for the matrix cosine and sine. *Numer. Algorithms*, 40(4):383–400, 2005. (Cited on pp. 292, 299.)

[249] Lawrence A. Harris. Computation of functions of certain operator matrices. *Linear Algebra Appl.*, 194:31–34, 1993. (Cited on p. 28.)

[250] W. F. Harris. The average eye. *Opthal. Physiol. Opt.*, 24:580–585, 2005. (Cited on p. 50.)

[251] W. F. Harris and J. R. Cardoso. The exponential-mean-log-transference as a possible representation of the optical character of an average eye. *Opthal. Physiol. Opt.*, 26(4): 380–383, 2006. (Cited on p. 50.)

[252] Timothy F. Havel, Igor Najfeld, and Ju-xing Yang. Matrix decompositions of two-dimensional nuclear magnetic resonance spectra. *Proc. Nat. Acad. Sci. USA*, 91:7962–7966, 1994. (Cited on p. 51.)

[253] Thomas Hawkins. The theory of matrices in the 19th century. In *Proceedings of the International Congress of Mathematicians, Vancouver*, volume 2, 1974, pages 561–570. (Cited on p. 26.)

[254] Thomas Hawkins. Another look at Cayley and the theory of matrices. *Arch. Internat. Histoire Sci.*, 27(100):82–112, 1977. (Cited on p. 26.)

[255] Thomas Hawkins. Weierstrass and the theory of matrices. *Archive for History of Exact Sciences*, 12(2):119–163, 1977. (Cited on p. 26.)

[256] Jane M. Heffernan and Robert M. Corless. Solving some delay differential equations with computer algebra. *Mathematical Scientist*, 31(1):21–34, 2006. (Cited on p. 51.)

[257] Peter Henrici. Bounds for iterates, inverses, spectral variation and fields of values of non-normal matrices. *Numer. Math.*, 4:24–40, 1962. (Cited on p. 103.)

[258] Kurt Hensel. Über Potenzreihen von Matrizen. *J. Reine Angew. Math.*, 155(2):107–110, 1926. (Cited on pp. 26, 104.)

[259] Konrad J. Heuvers and Daniel Moak. Matrix solutions of the functional equation of the gamma function. *Aequationes Mathematicae*, 33:1–17, 1987. (Cited on p. 27.)

[260] *The Hewlett-Packard HP-35 Scientific Pocket Calculator.* Hewlett-Packard, Cupertino, CA, USA, 1973. 4 pp. Advertising brochure. 5952-6000 Rev. 7/73. (Cited on p. 286.)

[261] *HP 48 Programmer's Reference Manual.* Hewlett-Packard, Corvallis Division, Corvallis, OR, USA, July 1990. 504 pp. Mfg. No. 00048-90053. (Cited on p. 286.)

[262] Desmond J. Higham. Time-stepping and preserving orthonormality. *BIT*, 37(1):24–36, 1997. (Cited on p. 42.)

[263] Desmond J. Higham and Nicholas J. Higham. *MATLAB Guide*. Second edition, Society for Industrial and Applied Mathematics, Philadelphia, PA, USA, 2005. xxiii+382 pp. ISBN 0-89871-578-4. (Cited on p. 252.)

[264] Nicholas J. Higham. The Matrix Computation Toolbox. `http://www.ma.man.ac.uk/~higham/mctoolbox`. (Cited on pp. 129, 149, 229, 295.)

[265] Nicholas J. Higham. *Nearness Problems in Numerical Linear Algebra*. PhD thesis, University of Manchester, Manchester, England, July 1985. 173 pp. (Cited on p. 216.)

[266] Nicholas J. Higham. Computing the polar decomposition—with applications. *SIAM J. Sci. Statist. Comput.*, 7(4):1160–1174, 1986. (Cited on pp. 167, 215, 216.)

[267] Nicholas J. Higham. Newton's method for the matrix square root. *Math. Comp.*, 46 (174):537–549, 1986. (Cited on pp. 104, 166, 167.)

[268] Nicholas J. Higham. Computing real square roots of a real matrix. *Linear Algebra Appl.*, 88/89:405–430, 1987. (Cited on pp. 28, 166, 167.)

[269] Nicholas J. Higham. Computing a nearest symmetric positive semidefinite matrix. *Linear Algebra Appl.*, 103:103–118, 1988. (Cited on p. 214.)

[270] Nicholas J. Higham. FORTRAN codes for estimating the one-norm of a real or complex matrix, with applications to condition estimation (Algorithm 674). *ACM Trans. Math. Software*, 14(4):381–396, 1988. (Cited on pp. 69, 226.)

[271] Nicholas J. Higham. Experience with a matrix norm estimator. *SIAM J. Sci. Statist. Comput.*, 11(4):804–809, 1990. (Cited on p. 69.)

[272] Nicholas J. Higham. Stability of a method for multiplying complex matrices with three real matrix multiplications. *SIAM J. Matrix Anal. Appl.*, 13(3):681–687, 1992. (Cited on p. 332.)

[273] Nicholas J. Higham. The matrix sign decomposition and its relation to the polar decomposition. *Linear Algebra Appl.*, 212/213:3–20, 1994. (Cited on pp. 129, 214, 215, 216.)

[274] Nicholas J. Higham. Stable iterations for the matrix square root. *Numer. Algorithms*, 15(2):227–242, 1997. (Cited on pp. 108, 167, 363.)

[275] Nicholas J. Higham. Evaluating Padé approximants of the matrix logarithm. *SIAM J. Matrix Anal. Appl.*, 22(4):1126–1135, 2001. (Cited on p. 275.)

[276] Nicholas J. Higham. *Accuracy and Stability of Numerical Algorithms*. Second edition, Society for Industrial and Applied Mathematics, Philadelphia, PA, USA, 2002. xxx+680 pp. ISBN 0-89871-521-0. (Cited on pp. 69, 75, 82, 102, 104, 120, 122, 134, 143, 163, 165, 166, 210, 211, 223, 226, 247, 248, 250, 261, 265, 309, 324, 325, 332, 335, 361, 362.)

[277] Nicholas J. Higham. J-orthogonal matrices: Properties and generation. *SIAM Rev.*, 45(3):504–519, 2003. (Cited on p. 315.)

[278] Nicholas J. Higham. The scaling and squaring method for the matrix exponential revisited. *SIAM J. Matrix Anal. Appl.*, 26(4):1179–1193, 2005. (Cited on pp. 104, 263, 264.)

[279] Nicholas J. Higham and Sheung Hun Cheng. Modifying the inertia of matrices arising in optimization. *Linear Algebra Appl.*, 275–276:261–279, 1998. (Cited on p. 349.)

[280] Nicholas J. Higham and Hyun-Min Kim. Numerical analysis of a quadratic matrix equation. *IMA J. Numer. Anal.*, 20(4):499–519, 2000. (Cited on p. 45.)

[281] Nicholas J. Higham and Hyun-Min Kim. Solving a quadratic matrix equation by Newton's method with exact line searches. *SIAM J. Matrix Anal. Appl.*, 23(2):303–316, 2001. (Cited on p. 45.)

[282] Nicholas J. Higham, D. Steven Mackey, Niloufer Mackey, and Françoise Tisseur. Computing the polar decomposition and the matrix sign decomposition in matrix groups. *SIAM J. Matrix Anal. Appl.*, 25(4):1178–1192, 2004. (Cited on p. 315.)

[283] Nicholas J. Higham, D. Steven Mackey, Niloufer Mackey, and Françoise Tisseur. Functions preserving matrix groups and iterations for the matrix square root. *SIAM J. Matrix Anal. Appl.*, 26(3):849–877, 2005. (Cited on pp. 13, 28, 104, 129, 132, 154, 167, 215, 216, 314, 315, 316, 366.)

[284] Nicholas J. Higham and Pythagoras Papadimitriou. A new parallel algorithm for computing the singular value decomposition. In *Proceedings of the Fifth SIAM Conference on Applied Linear Algebra*, John G. Lewis, editor, Society for Industrial and Applied Mathematics, Philadelphia, PA, USA, 1994, pages 80–84. (Cited on p. 196.)

[285] Nicholas J. Higham and Pythagoras Papadimitriou. A parallel algorithm for computing the polar decomposition. *Parallel Comput.*, 20(8):1161–1173, 1994. (Cited on pp. 207, 216.)

[286] Nicholas J. Higham and Robert S. Schreiber. Fast polar decomposition of an arbitrary matrix. *SIAM J. Sci. Statist. Comput.*, 11(4):648–655, 1990. (Cited on pp. 162, 211, 214.)

[287] Nicholas J. Higham and Matthew I. Smith. Computing the matrix cosine. *Numer. Algorithms*, 34:13–26, 2003. (Cited on pp. 292, 295, 299.)

[288] Nicholas J. Higham and Françoise Tisseur. A block algorithm for matrix 1-norm estimation, with an application to 1-norm pseudospectra. *SIAM J. Matrix Anal. Appl.*, 21(4):1185–1201, 2000. (Cited on p. 67.)

[289] Lester S. Hill. Cryptography in an algebraic alphabet. *Amer. Math. Monthly*, 36: 306–312, 1929. (Cited on p. 165.)

[290] Einar Hille. On roots and logarithms of elements of a complex Banach algebra. *Math. Annalen*, 136(1):46–57, 1958. (Cited on p. 32.)

[291] Marlis Hochbruck and Christian Lubich. On Krylov subspace approximations to the matrix exponential operator. *SIAM J. Numer. Anal.*, 34(5):1911–1925, 1997. (Cited on p. 306.)

[292] Marlis Hochbruck, Christian Lubich, and Hubert Selhofer. Exponential integrators for large systems of differential equations. *SIAM J. Sci. Comput.*, 19(5):1552–1574, 1998. (Cited on pp. 37, 262.)

[293] John H. Hodges. The matrix equation $X^2 - I = 0$ over a finite field. *Amer. Math. Monthly*, 65(7):518–520, 1958. (Cited on p. 168.)

[294] Roger A. Horn. The Hadamard product. In *Matrix Theory and Applications*, Charles R. Johnson, editor, volume 40 of *Proceedings of Symposia in Applied Mathematics*, American Mathematical Society, Providence, RI, USA, 1990, pages 87–169. (Cited on p. 1.)

[295] Roger A. Horn and Charles R. Johnson. *Matrix Analysis*. Cambridge University Press, 1985. xiii+561 pp. ISBN 0-521-30586-1. (Cited on pp. 25, 28, 321, 323, 325, 327, 328, 329, 346.)

[296] Roger A. Horn and Charles R. Johnson. *Topics in Matrix Analysis*. Cambridge University Press, 1991. viii+607 pp. ISBN 0-521-30587-X. (Cited on pp. xvii, 1, 8, 13, 14, 16, 17, 23, 24, 27, 28, 46, 69, 213, 235, 315, 327, 331, 333, 362.)

[297] Roger A. Horn and Dennis I. Merino. Contragredient equivalence: A canonical form and some applications. *Linear Algebra Appl.*, 214:43–92, 1995. (Cited on p. 28.)

[298] Roger A. Horn and Gregory G. Piepmeyer. Two applications of the theory of primary matrix functions. *Linear Algebra Appl.*, 361:99–106, 2003. (Cited on p. 28.)

[299] W. D. Hoskins and D. J. Walton. A faster, more stable method for computing the pth roots of positive definite matrices. *Linear Algebra Appl.*, 26:139–163, 1979. (Cited on p. 188.)

[300] A. S. Householder and John A. Carpenter. The singular values of involutory and of idempotent matrices. *Numer. Math.*, 5:234–237, 1963. (Cited on p. 165.)

[301] Alston S. Householder. *The Numerical Treatment of a Single Nonlinear Equation.* McGraw-Hill, New York, 1970. viii+216 pp. (Cited on p. 90.)

[302] James Lucien Howland. The sign matrix and the separation of matrix eigenvalues. *Linear Algebra Appl.*, 49:221–232, 1983. (Cited on pp. 41, 130.)

[303] John H. Hubbard and Beverly H. West. *Differential Equations: A Dynamical Systems Approach. Higher Dimensional Systems.* Springer-Verlag, New York, 1995. xiv+601 pp. ISBN 0-387-94377-3. (Cited on p. 267.)

[304] Thomas J. R. Hughes, Itzhak Levit, and James Winget. Element-by-element implicit algorithms for heat conduction. *J. Eng. Mech.*, 109(2):576–585, 1983. (Cited on p. 310.)

[305] Bruno Iannazzo. A note on computing the matrix square root. *Calcolo*, 40:273–283, 2003. (Cited on pp. 143, 166, 167, 169.)

[306] Bruno Iannazzo. On the Newton method for the matrix Pth root. *SIAM J. Matrix Anal. Appl.*, 28(2):503–523, 2006. (Cited on pp. 178, 188.)

[307] Bruno Iannazzo. *Numerical Solution of Certain Nonlinear Matrix Equations.* PhD thesis, Università degli studi di Pisa, Pisa, Italy, 2007. 180 pp. (Cited on pp. 95, 104, 130, 167.)

[308] Khakim D. Ikramov. Hamiltonian square roots of skew-Hamiltonian matrices revisited. *Linear Algebra Appl.*, 325(1-3):101–107, 2001. (Cited on p. 315.)

[309] F. Incertis. A skew-symmetric formulation of the algebraic Riccati equation problem. *IEEE Trans. Automat. Control*, AC-29(5):467–470, 1984. (Cited on p. 46.)

[310] Ilse C. F. Ipsen and Dean J. Lee. Determinant approximations. Manuscript, 2003. 13 pp. (Cited on p. 44.)

[311] Eugene Isaacson and Herbert Bishop Keller. *Analysis of Numerical Methods.* Wiley, New York, 1966. xv+541 pp. Reprinted by Dover, New York, 1994. ISBN 0-486-68029-0. (Cited on pp. 90, 333.)

[312] Arieh Iserles. How large is the exponential of a banded matrix? *J. New Zealand Maths Soc.*, 29:177–192, 2000. (Cited on p. 318.)

[313] Arieh Iserles, Hans Z. Munthe-Kaas, Syvert P. Nørsett, and Antonella Zanna. Lie-group methods. *Acta Numerica*, 9:215–365, 2000. (Cited on p. 315.)

[314] Arieh Iserles and Antonella Zanna. Efficient computation of the matrix exponential by generalized polar decompositions. *SIAM J. Numer. Anal.*, 42(5):2218–2256, 2005. (Cited on p. 315.)

[315] Robert B. Israel, Jeffrey S. Rosenthal, and Jason Z. Wei. Finding generators for Markov chains via empirical transition matrices, with applications to credit ratings. *Mathematical Finance*, 11(2):245–265, 2001. (Cited on pp. 38, 190.)

[316] Anzelm Iwanik and Ray Shiflett. The root problem for stochastic and doubly stochastic operators. *J. Math. Anal. and Appl.*, 113(1):93–112, 1986. (Cited on p. 191.)

[317] Drahoslava Janovská and Gerhard Opfer. Computing quaternionic roots by Newton's method. *Electron. Trans. Numer. Anal.*, 26:82–102, 2007. (Cited on p. 189.)

[318] Branislav Jansik, Stinne Høst, Poul Jørgensen, Jeppe Olsen, and Trygve Helgaker. Linear-scaling symmetric square-root decomposition of the overlap matrix. *J. Chem. Phys.*, 126:12404, 2007. (Cited on p. 42.)

[319] A. J. E. M. Janssen and Thomas Strohmer. Characterization and computation of canonical tight windows for Gabor frames. *The Journal of Fourier Analysis and Applications*, 8(1):1–28, 2002. (Cited on p. 168.)

[320] Elias Jarlebring and Tobias Damm. The Lambert W function and the spectrum of some multidimensional time-delay systems. *Automatica*, 43(12):2124–2128, 2007. (Cited on p. 51.)

[321] Charles R. Johnson and Eric Schreiner. The relationship between AB and BA. *Amer. Math. Monthly*, 103(7):578–582, 1996. (Cited on p. 28.)

[322] Bo Kågström. Bounds and perturbation bounds for the matrix exponential. *BIT*, 17: 39–57, 1977. (Cited on p. 263.)

[323] Bo Kågström. Numerical computation of matrix functions. Report UMINF-58.77, Department of Information Processing, University of Umeå, Sweden, July 1977. 109 pp. (Cited on pp. 104, 231.)

[324] Bo Kågström and Peter Poromaa. Distributed and shared memory block algorithms for the triangular Sylvester equation with sep^{-1} estimators. *SIAM J. Matrix Anal. Appl.*, 13(1):90–101, 1992. (Cited on p. 226.)

[325] Bo Kågström and Axel Ruhe. An algorithm for numerical computation of the Jordan normal form of a complex matrix. *ACM Trans. Math. Software*, 6(3):398–419, 1980. (Cited on p. 231.)

[326] W. Kahan. Conserving confluence curbs ill-condition. Technical Report 6, Computer Science Department, University of California, Berkeley, August 1972. 54 pp. (Cited on pp. 88, 106.)

[327] W. Kahan. Branch cuts for complex elementary functions or much ado about nothing's sign bit. In *The State of the Art in Numerical Analysis*, A. Iserles and M. J. D. Powell, editors, Oxford University Press, 1987, pages 165–211. (Cited on pp. 166, 280.)

[328] W. Kahan. Derivatives in the complex z-plane. Course notes available from `http://www.cs.berkeley.edu/~wkahan/Math185/`, September 2006. 16 pp. (Cited on p. 333.)

[329] W. Kahan and Richard J. Fateman. Symbolic computation of divided differences. *ACM SIGSAM Bulletin*, 33(2):7–28, 1999. (Cited on pp. 284, 333.)

[330] Thomas Kailath, Ali H. Sayed, and Babak Hassibi. *Linear Estimation*. Prentice-Hall, Upper Saddle River, NJ, USA, 2000. xxvi+854 pp. ISBN 0-13-022464-2. (Cited on p. 166.)

[331] J. D. Kalbfleisch and J. F. Lawless. The analysis of panel data under a Markov assumption. *J. Amer. Statist. Assoc.*, 80(392):863–871, 1985. (Cited on p. 38.)

[332] G. Kalogeropoulos and Panayiotis Psarrakos. The polar decomposition of block companion matrices. *Computers Math. Applic.*, 50(3):529–537, 2005. (Cited on p. 215.)

[333] Christian Kanzow and Christian Nagel. Semidefinite programs: New search directions, smoothing-type methods, and numerical results. *SIAM J. Optim.*, 13(1):1–23, 2002. (Cited on p. 48.)

[334] Alan H. Karp and Peter Markstein. High-precision division and square root. *ACM Trans. Math. Software*, 23(4):561–589, 1997. (Cited on p. 188.)

[335] Robert Karplus and Julian Schwinger. A note on saturation in microwave spectroscopy. *Phys. Rev.*, 73(9):1020–1026, 1948. (Cited on p. 263.)

[336] Aly-Khan Kassam and Lloyd N. Trefethen. Fourth-order time-stepping for stiff PDEs. *SIAM J. Sci. Comput.*, 26(4):1214–1233, 2005. (Cited on pp. 37, 308.)

[337] Tosio Kato. *Perturbation Theory for Linear Operators*. Second edition, Springer-Verlag, Berlin, 1976. xxi+619 pp. ISBN 3-540-97588-5. (Cited on p. 28.)

[338] A. D. Kennedy. Approximation theory for matrices. *Nuclear Physics B (Proc. Suppl.)*, 128:107–116, 2004. (Cited on p. 130.)

[339] A. D. Kennedy. Fast evaluation of Zolotarev coefficients. In *QCD and Numerical Analysis III*, Artan Boriçi, Andreas Frommer, Báalint Joó, Anthony Kennedy, and Brian Pendleton, editors, volume 47 of *Lecture Notes in Computational Science and Engineering*, Springer-Verlag, Berlin, 2005, pages 169–189. (Cited on p. 130.)

[340] Charles S. Kenney and Alan J. Laub. Condition estimates for matrix functions. *SIAM J. Matrix Anal. Appl.*, 10(2):191–209, 1989. (Cited on pp. 69, 264, 284.)

[341] Charles S. Kenney and Alan J. Laub. Padé error estimates for the logarithm of a matrix. *Internat. J. Control*, 50(3):707–730, 1989. (Cited on pp. 275, 283.)

[342] Charles S. Kenney and Alan J. Laub. Polar decomposition and matrix sign function condition estimates. *SIAM J. Sci. Statist. Comput.*, 12(3):488–504, 1991. (Cited on pp. 129, 131, 200, 215.)

[343] Charles S. Kenney and Alan J. Laub. Rational iterative methods for the matrix sign function. *SIAM J. Matrix Anal. Appl.*, 12(2):273–291, 1991. (Cited on pp. 104, 115, 116, 117, 130.)

[344] Charles S. Kenney and Alan J. Laub. On scaling Newton's method for polar decomposition and the matrix sign function. *SIAM J. Matrix Anal. Appl.*, 13(3):688–706, 1992. (Cited on pp. 130, 216.)

[345] Charles S. Kenney and Alan J. Laub. A hyperbolic tangent identity and the geometry of Padé sign function iterations. *Numer. Algorithms*, 7:111–128, 1994. (Cited on pp. 117, 130.)

[346] Charles S. Kenney and Alan J. Laub. Small-sample statistical condition estimates for general matrix functions. *SIAM J. Sci. Comput.*, 15(1):36–61, 1994. (Cited on p. 68.)

[347] Charles S. Kenney and Alan J. Laub. The matrix sign function. *IEEE Trans. Automat. Control*, 40(8):1330–1348, 1995. (Cited on pp. 110, 129, 130.)

[348] Charles S. Kenney and Alan J. Laub. A Schur–Fréchet algorithm for computing the logarithm and exponential of a matrix. *SIAM J. Matrix Anal. Appl.*, 19(3):640–663, 1998. (Cited on pp. 104, 251, 257, 263, 264, 279, 284.)

[349] Charles S. Kenney, Alan J. Laub, and Edmond A. Jonckheere. Positive and negative solutions of dual Riccati equations by matrix sign function iteration. *Systems and Control Letters*, 13:109–116, 1989. (Cited on p. 40.)

[350] Charles S. Kenney, Alan J. Laub, and P. M. Papadopoulos. A Newton-squaring algorithm for computing the negative invariant subspace of a matrix. *IEEE Trans. Automat. Control*, 38(8):1284–1289, 1993. (Cited on p. 130.)

[351] Andrzej Kiełbasiński, Pawel Zieliński, and Krystyna Ziętak. Numerical experiments with Higham's scaled method for polar decomposition. Technical report, Institute of Mathematics and Computer Science, Wroclaw University of Technology, Wroclaw, Poland, May 2006. 38 pp. (Cited on p. 210.)

[352] Andrzej Kiełbasiński and Krystyna Ziętak. Numerical behaviour of Higham's scaled method for polar decomposition. *Numer. Algorithms*, 32:105–140, 2003. (Cited on pp. 208, 210.)

[353] Fuad Kittaneh. On Lipschitz functions of normal operators. *Proc. Amer. Math. Soc.*, 94(3):416–418, 1985. (Cited on p. 217.)

[354] Fuad Kittaneh. Inequalities for the Schatten p-norm. III. *Commun. Math. Phys.*, 104: 307–310, 1986. (Cited on p. 217.)

[355] Leonard F. Klosinski, Gerald L. Alexanderson, and Loren C. Larson. The fifty-first William Lowell Putnam mathematical competition. *Amer. Math. Monthly*, 98(8):719–727, 1991. (Cited on p. 33.)

[356] Leonid A. Knizhnerman. Calculation of functions of unsymmetric matrices using Arnoldi's method. *U.S.S.R. Comput. Maths. Math. Phys.*, 31(1):1–9, 1991. (Cited on p. 309.)

[357] Donald E. Knuth. *The Art of Computer Programming, Volume 2, Seminumerical Algorithms.* Third edition, Addison-Wesley, Reading, MA, USA, 1998. xiii+762 pp. ISBN 0-201-89684-2. (Cited on p. 118.)

[358] Çetin Kaya Koç and Bertan Bakkaloğlu. Halley's method for the matrix sector function. *IEEE Trans. Automat. Control*, 40(5):944–949, 1995. (Cited on p. 49.)

[359] Plamen Koev and Alan Edelman. The efficient evaluation of the hypergeometric function of a matrix argument. *Math. Comp.*, 75(254):833–846, 2006. (Cited on p. 1.)

[360] S. Koikari. An error analysis of the modified scaling and squaring method. *Computers Math. Applic.*, 53:1293–1305, 2007. (Cited on p. 262.)

[361] Zdislav Kovarik. Some iterative methods for improving orthonormality. *SIAM J. Numer. Anal.*, 7(3):386–389, 1970. (Cited on p. 215.)

[362] Alexander Kreinin and Marina Sidelnikova. Regularization algorithms for transition matrices. *Algo Research Quarterly*, 4(1/2):23–40, 2001. (Cited on p. 38.)

[363] H. Kreis. Auflösung der Gleichung $X^n = A$. *Vierteljschr. Naturforsch. Ges. Zürich*, 53:366–376, 1908. (Cited on p. 28.)

[364] V. N. Kublanovskaya. On certain iteration processes for the symmetrisation of a matrix. *U.S.S.R. Computational Math. and Math. Phys.*, 2(5):859–868, 1963. (Cited on p. 370.)

[365] J. Kuczyński and H. Woźniakowski. Estimating the largest eigenvalue by the power and Lanczos algorithms with a random start. *SIAM J. Matrix Anal. Appl.*, 13(4): 1094–1122, 1992. (Cited on p. 69.)

[366] Pentti Laasonen. On the iterative solution of the matrix equation $AX^2 - I = 0$. *M.T.A.C.*, 12:109–116, 1958. (Cited on pp. 104, 166, 167.)

[367] Edmond Nicolas Laguerre. Le calcul des systèmes linéaires, extrait d'une lettre adressé à M. Hermite. In *Oeuvres de Laguerre*, Ch. Hermite, H. Poincaré, and E. Rouché, editors, volume 1, Gauthier–Villars, Paris, 1898, pages 221–267. The article is dated 1867 and is "Extrait du Journal de l'École Polytechnique, LXIIe Cahier". (Cited on p. 26.)

[368] S. Lakić. On the computation of the matrix k-th root. *Z. Angew. Math. Mech.*, 78(3): 167–172, 1998. (Cited on p. 188.)

[369] Peter Lancaster. *Lambda-Matrices and Vibrating Systems.* Pergamon Press, Oxford, 1966. xiii+196 pp. Reprinted by Dover, New York, 2002. ISBN 0-486-42546-0. (Cited on pp. 45, 52.)

[370] Peter Lancaster and Leiba Rodman. *Algebraic Riccati Equations.* Oxford University Press, 1995. xvii+480 pp. ISBN 0-19-853795-6. (Cited on pp. 40, 51, 169.)

[371] Peter Lancaster and Miron Tismenetsky. *The Theory of Matrices.* Second edition, Academic Press, London, 1985. xv+570 pp. ISBN 0-12-435560-9. (Cited on pp. xvii, 3, 18, 23, 24, 27, 28, 52, 321, 329, 331.)

[372] B. Laszkiewicz and Krystyna Ziętak. Approximation of matrices and a family of Gander methods for polar decomposition. *BIT*, 46(2):345–366, 2006. (Cited on pp. 214, 215, 217, 409.)

[373] Guy Latouche and V. Ramaswami. *Introduction to Matrix Analytic Methods in Stochastic Modeling.* Society for Industrial and Applied Mathematics, Philadelphia, PA, USA, 1999. xiv+334 pp. ISBN 0-89871-425-7. (Cited on p. 45.)

[374] Alan J. Laub. Invariant subspace methods for the numerical solution of Riccati equations. In *The Riccati Equation*, Sergio Bittanti, Alan J. Laub, and Jan C. Willems, editors, Springer-Verlag, Berlin, 1991, pages 163–196. (Cited on pp. 51, 53.)

[375] P.-F. Lavallée, A. Malyshev, and M. Sadkane. Spectral portrait of matrices by block diagonalization. In *Numerical Analysis and Its Applications*, Lubin Vulkov, Jerzy Waśniewski, and Plamen Yalamov, editors, volume 1196 of *Lecture Notes in Computer Science*, Springer-Verlag, Berlin, 1997, pages 266–273. (Cited on p. 89.)

[376] J. Douglas Lawson. Generalized Runge-Kutta processes for stable systems with large Lipschitz constants. *SIAM J. Numer. Anal.*, 4(3):372–380, 1967. (Cited on p. 263.)

[377] Jimmie D. Lawson and Yongdo Lim. The geometric mean, matrices, metrics, and more. *Amer. Math. Monthly*, 108(9):797–812, 2001. (Cited on p. 52.)

[378] Dean J. Lee and Ilse C. F. Ipsen. Zone determinant expansions for nuclear lattice simulations. *Physical Review C*, 68:064003, 2003. (Cited on p. 44.)

[379] R. B. Leipnik. Rapidly convergent recursive solution of quadratic operator equations. *Numer. Math.*, 17:1–16, 1971. (Cited on p. 215.)

[380] Randall J. LeVeque. *Finite Difference Methods for Ordinary and Partial Differential Equations: Steady-State and Time-Dependent Problems.* Society for Industrial and Applied Mathematics, Philadelphia, PA, USA, 2007. xv+341 pp. ISBN 978-0-898716-29-0. (Cited on p. 37.)

[381] Jack Levine and H. M. Nahikian. On the construction of involutory matrices. *Amer. Math. Monthly*, 69(4):267–272, 1962. (Cited on p. 165.)

[382] Bernard W. Levinger. The square root of a 2×2 matrix. *Math. Mag.*, 53(4):222–224, 1980. (Cited on p. 168.)

[383] Malcolm H. Levitt. *Spin Dynamics: Basics of Nuclear Magnetic Resonance.* Wiley, Chichester, UK, 2001. xxiv+686 pp. ISBN 0-471-48921-2. (Cited on p. 51.)

[384] Jing Li. *An Algorithm for Computing the Matrix Exponential.* PhD thesis, Mathematics Department, University of California, Berkeley, CA, USA, 1988. 77 pp. (Cited on p. 251.)

[385] Ren-Cang Li. A perturbation bound for the generalized polar decomposition. *BIT*, 33:304–308, 1993. (Cited on p. 215.)

[386] Ren-Cang Li. New perturbation bounds for the unitary polar factor. *SIAM J. Matrix Anal. Appl.*, 16(1):327–332, 1995. (Cited on pp. 201, 218.)

[387] Ren-Cang Li. Relative perturbation bounds for the unitary polar factor. *BIT*, 37(1): 67–75, 1997. (Cited on p. 215.)

[388] Ren-Cang Li. Relative perturbation bounds for positive polar factors of graded matrices. *SIAM J. Matrix Anal. Appl.*, 27(2):424–433, 2005. (Cited on p. 215.)

[389] Wen Li and Weiwei Sun. Perturbation bounds of unitary and subunitary polar factors. *SIAM J. Matrix Anal. Appl.*, 23(4):1183–1193, 2002. (Cited on pp. 201, 215.)

[390] Pietr Liebl. Einige Bemerkungen zur numerischen Stabilität von Matrizeniterationen. *Aplikace Matematiky*, 10(3):249–254, 1965. (Cited on p. 167.)

[391] Yongdo Lim. The matrix golden mean and its applications to Riccati matrix equations. *SIAM J. Matrix Anal. Appl.*, 29(1):54–66, 2007. (Cited on p. 47.)

[392] Chih-Chang Lin and Earl Zmijewski. A parallel algorithm for computing the eigenvalues of an unsymmetric matrix on an SIMD mesh of processors. Report TRCS 91-15, Department of Computer Science, University of California, Santa Barbara, July 1991. (Cited on p. 51.)

[393] Lu Lin and Zhong-Yun Liu. On the square root of an H-matrix with positive diagonal elements. *Annals of Operations Research*, 103:339–350, 2001. (Cited on p. 167.)

[394] Ya Yan Lu. Computing the logarithm of a symmetric positive definite matrix. *Appl. Numer. Math.*, 26:483–496, 1998. (Cited on p. 284.)

[395] Ya Yan Lu. Exponentials of symmetric matrices through tridiagonal reductions. *Linear Algebra Appl.*, 279:317–324, 1998. (Cited on p. 260.)

[396] Ya Yan Lu. A Padé approximation method for square roots of symmetric positive definite matrices. *SIAM J. Matrix Anal. Appl.*, 19(3):833–845, 1998. (Cited on p. 164.)

[397] Ya Yan Lu. Computing a matrix function for exponential integrators. *J. Comput. Appl. Math.*, 161:203–216, 2003. (Cited on p. 262.)

[398] Cyrus Colton MacDuffee. *Vectors and Matrices.* Number 7 in *The Carus Mathematical Monographs*. Mathematical Association of America, 1943. xi+203 pp. (Cited on p. 34.)

[399] Cyrus Colton MacDuffee. *The Theory of Matrices.* Chelsea, New York, 1946. v+110 pp. Corrected reprint of first edition (J. Springer, Berlin, 1933). Also available as Dover edition, 2004. (Cited on p. 27.)

[400] D. Steven Mackey, Niloufer Mackey, and Françoise Tisseur. Structured factorizations in scalar product spaces. *SIAM J. Matrix Anal. Appl.*, 27:821–850, 2006. (Cited on p. 315.)

[401] Arne Magnus and Jan Wynn. On the Padé table of $\cos z$. *Proc. Amer. Math. Soc.*, 47 (2):361–367, 1975. (Cited on pp. 262, 290, 296.)

[402] Jan R. Magnus and Heinz Neudecker. *Matrix Differential Calculus with Applications in Statistics and Econometrics.* Revised edition, Wiley, Chichester, UK, 1999. xviii+395 pp. ISBN 0-471-98633-X. (Cited on p. 69.)

[403] Wilhelm Magnus. On the exponential solution of differential equations for a linear operator. *Comm. Pure Appl. Math.*, 7:649–673, 1954. (Cited on p. 263.)

[404] P. J. Maher. Partially isometric approximation of positive operators. *Illinois Journal of Mathematics*, 33(2):227–243, 1989. (Cited on p. 214.)

[405] Jianqin Mao. Optimal orthonormalization of the strapdown matrix by using singular value decomposition. *Computers Math. Applic.*, 12A(3):353–362, 1986. (Cited on p. 42.)

[406] Marvin Marcus and Henryk Minc. Some results on doubly stochastic matrices. *Proc. Amer. Math. Soc.*, 13(4):571–579, 1962. (Cited on p. 191.)

[407] Jerrold E. Marsden and Tudor S. Ratiu. *Introduction to Mechanics and Symmetry.* Second edition, Springer-Verlag, New York, 1999. xviii+582 pp. ISBN 0-387-98643-X. (Cited on p. 266.)

[408] Roy Mathias. Evaluating the Frechet derivative of the matrix exponential. *Numer. Math.*, 63:213–226, 1992. (Cited on pp. 66, 256.)

[409] Roy Mathias. Approximation of matrix-valued functions. *SIAM J. Matrix Anal. Appl.*, 14(4):1061–1063, 1993. (Cited on p. 77.)

[410] Roy Mathias. Perturbation bounds for the polar decomposition. *SIAM J. Matrix Anal. Appl.*, 14(2):588–597, 1993. (Cited on p. 214.)

[411] Roy Mathias. Condition estimation for matrix functions via the Schur decomposition. *SIAM J. Matrix Anal. Appl.*, 16(2):565–578, 1995. (Cited on pp. 68, 69.)

[412] Roy Mathias. A chain rule for matrix functions and applications. *SIAM J. Matrix Anal. Appl.*, 17(3):610–620, 1996. (Cited on pp. 13, 69, 355.)

[413] *Control Systems Toolbox Documentation.* The MathWorks, Inc., Natick, MA, USA. Online version. (Cited on p. 39.)

[414] *MATLAB.* The MathWorks, Inc., Natick, MA, USA. `http://www.mathworks.com`. (Cited on p. 100.)

[415] Neal H. McCoy. On the characteristic roots of matric polynomials. *Bull. Amer. Math. Soc.*, 42:592–600, 1936. (Cited on p. 29.)

[416] A. McCurdy, K. C. Ng, and B. N. Parlett. Accurate computation of divided differences of the exponential function. *Math. Comp.*, 43(168):501–528, 1984. (Cited on pp. 250, 264.)

[417] Robert I. McLachlan and G. Reinout W. Quispel. Splitting methods. *Acta Numerica*, 11:341–434, 2002. (Cited on p. 263.)

[418] Volker Mehrmann. *The Autonomous Linear Quadratic Control Problem: Theory and Numerical Solution*, volume 163 of *Lecture Notes in Control and Information Sciences*. Springer-Verlag, Berlin, 1991. 177 pp. ISBN 0-340-54170-5. (Cited on pp. 132, 169.)

[419] Volker Mehrmann and Werner Rath. Numerical methods for the computation of analytic singular value decompositions. *Electron. Trans. Numer. Anal.*, 1:72–88, 1993. (Cited on pp. 43, 214.)

[420] Madan Lal Mehta. *Matrix Theory: Selected Topics and Useful Results*. Second edition, Hindustan Publishing Company, Delhi, 1989. xvii+376 pp. ISBN 81-7075-012-1. (Cited on p. 187.)

[421] Beatrice Meini. The matrix square root from a new functional perspective: Theoretical results and computational issues. *SIAM J. Matrix Anal. Appl.*, 26(2):362–376, 2004. (Cited on pp. 142, 160, 166.)

[422] Brian J. Melloy and G. Kemble Bennett. Computing the exponential of an intensity matrix. *J. Comput. Appl. Math.*, 46:405–413, 1993. (Cited on pp. 105, 263.)

[423] Michael Metcalf and John K. Reid. *Fortran 90/95 Explained*. Second edition, Oxford University Press, 1999. xv+341 pp. ISBN 0-19-850558-2. (Cited on p. 1.)

[424] W. H. Metzler. On the roots of matrices. *Amer. J. Math.*, 14(4):326–377, 1892. (Cited on pp. 26, 28.)

[425] Gérard Meurant. A review on the inverse of symmetric tridiagonal and block tridiagonal matrices. *SIAM J. Matrix Anal. Appl.*, 13(3):707–728, 1992. (Cited on p. 316.)

[426] Carl D. Meyer. *Matrix Analysis and Applied Linear Algebra*. Society for Industrial and Applied Mathematics, Philadelphia, PA, USA, 2000. xii+718 pp. ISBN 0-89871-454-0. (Cited on pp. 28, 170, 326.)

[427] Charles A. Micchelli and R. A. Willoughby. On functions which preserve the class of Stieltjes matrices. *Linear Algebra Appl.*, 23:141–156, 1979. (Cited on p. 315.)

[428] L. M. Milne-Thompson. *The Calculus of Finite Differences*. Macmillan, London, 1933. xxiii+558 pp. (Cited on p. 333.)

[429] Henryk Minc. *Nonnegative Matrices*. Wiley, New York, 1988. xiii+206 pp. ISBN 0-471-83966-3. (Cited on p. 191.)

[430] Borislav V. Minchev. Computing analytic matrix functions for a class of exponential integrators. Reports in Informatics 278, Department of Informatics, University of Bergen, Norway, June 2004. 11 pp. (Cited on p. 262.)

[431] Borislav V. Minchev and Will M. Wright. A review of exponential integrators for first order semi-linear problems. Preprint 2/2005, Norwegian University of Science and Technology, Trondheim, Norway, 2005. 44 pp. (Cited on pp. 37, 53.)

[432] L. Mirsky. Symmetric gauge functions and unitarily invariant norms. *Quart. J. Math.*, 11:50–59, 1960. (Cited on p. 328.)

[433] L. Mirsky. *An Introduction to Linear Algebra*. Oxford University Press, 1961. viii+440 pp. Reprinted by Dover, New York, 1990. ISBN 0-486-66434-1. (Cited on p. 328.)

[434] Maher Moakher. Means and averaging in the group of rotations. *SIAM J. Matrix Anal. Appl.*, 24(1):1–16, 2002. (Cited on pp. 42, 216.)

[435] Maher Moakher. A differential geometric approach to the geometric mean of symmetric positive-definite matrices. *SIAM J. Matrix Anal. Appl.*, 26(3):735–747, 2005. (Cited on p. 52.)

[436] Cleve B. Moler. *Numerical Computing with MATLAB*. Society for Industrial and Applied Mathematics, Philadelphia, PA, USA, 2004. xi+336 pp. Also available electronically from www.mathworks.com. ISBN 0-89871-560-1. (Cited on p. 165.)

[437] Cleve B. Moler and Charles F. Van Loan. Nineteen dubious ways to compute the exponential of a matrix. *SIAM Rev.*, 20(4):801–836, 1978. (Cited on pp. 27, 233, 262, 267.)

[438] Cleve B. Moler and Charles F. Van Loan. Nineteen dubious ways to compute the exponential of a matrix, twenty-five years later. *SIAM Rev.*, 45(1):3–49, 2003. (Cited on pp. 27, 231, 233, 237, 241, 243, 246, 262, 263, 265, 267.)

[439] Pierre Montagnier, Christopher C. Paige, and Raymond J. Spiteri. Real Floquet factors of linear time-periodic systems. *Systems and Control Letters*, 50:251–262, 2003. (Cited on p. 286.)

[440] I. Moret and P. Novati. The computation of functions of matrices by truncated Faber series. *Numer. Funct. Anal. Optim.*, 22(5&6):697–719, 2001. (Cited on p. 309.)

[441] Jean-Michel Muller. *Elementary Functions: Algorithms and Implementation*. Birkhäuser, Boston, MA, USA, 1997. xv+204 pp. ISBN 0-8176-3990-X. (Cited on p. 100.)

[442] David Mumford, Caroline Series, and David Wright. *Indra's Pearls: The Vision of Felix Klein*. Cambridge University Press, 2002. xix+396 pp. ISBN 0-521-35253-3. (Cited on p. 188.)

[443] H. Z. Munthe-Kaas, G. R. W. Quispel, and A. Zanna. Generalized polar decompositions on Lie groups with involutive automorphisms. *Found. Comput. Math.*, 1(3): 297–324, 2001. (Cited on p. 315.)

[444] Noël M. Nachtigal, Satish C. Reddy, and Lloyd N. Trefethen. How fast are nonsymmetric matrix iterations? *SIAM J. Matrix Anal. Appl.*, 13(3):778–795, 1992. (Cited on p. 75.)

[445] Igor Najfeld and Timothy F. Havel. Derivatives of the matrix exponential and their computation. *Advances in Applied Mathematics*, 16:321–375, 1995. (Cited on pp. 51, 69, 263, 264.)

[446] Herbert Neuberger. Exactly massless quarks on the lattice. *Phys. Lett. B*, 417(1-2): 141–144, 1998. (Cited on p. 43.)

[447] Arnold Neumaier. *Introduction to Numerical Analysis*. Cambridge University Press, 2001. viii+356 pp. ISBN 0-521-33610-4. (Cited on p. 333.)

[448] Kwok Choi Ng. Contributions to the computation of the matrix exponential. Technical Report PAM-212, Center for Pure and Applied Mathematics, University of California, Berkeley, February 1984. 72 pp. PhD thesis. (Cited on pp. 104, 227, 251.)

[449] Jorge Nocedal and Stephen J. Wright. *Numerical Optimization*. Springer-Verlag, New York, 1999. xx+636 pp. ISBN 0-387-98793-2. (Cited on p. 199.)

[450] Paolo Novati. A polynomial method based on Fejér points for the computation of functions of unsymmetric matrices. *Appl. Numer. Math.*, 44:201–224, 2003. (Cited on p. 309.)

[451] Jeffrey Nunemacher. Which matrices have real logarithms? *Math. Mag.*, 62(2):132–135, 1989. (Cited on p. 17.)

[452] G. Opitz. Steigungsmatrizen. *Z. Angew. Math. Mech.*, 44:T52–T54, 1964. (Cited on p. 264.)

[453] James M. Ortega and Werner C. Rheinboldt. *Iterative Solution of Nonlinear Equations in Several Variables.* Society for Industrial and Applied Mathematics, Philadelphia, PA, USA, 2000. xxvi+572 pp. Republication of work first published by Academic Press in 1970. ISBN 0-89871-461-3. (Cited on pp. 69, 167.)

[454] E. E. Osborne. On pre-conditioning of matrices. *J. Assoc. Comput. Mach.*, 7:338–345, 1960. (Cited on p. 100.)

[455] A. M. Ostrowski. *Solution of Equations in Euclidean and Banach Spaces.* Academic Press, New York, 1973. xx+412 pp. Third edition of Solution of Equations and Systems of Equations. ISBN 0-12530260-6. (Cited on pp. 333, 357.)

[456] C. C. Paige. Krylov subspace processes, Krylov subspace methods, and iteration polynomials. In *Proceedings of the Cornelius Lanczos International Centenary Conference*, J. David Brown, Moody T. Chu, Donald C. Ellison, and Robert J. Plemmons, editors, Society for Industrial and Applied Mathematics, Philadelphia, PA, USA, 1994, pages 83–92. (Cited on p. 311.)

[457] Pradeep Pandey, Charles S. Kenney, and Alan J. Laub. A parallel algorithm for the matrix sign function. *Int. J. High Speed Computing*, 2(2):181–191, 1990. (Cited on p. 130.)

[458] Michael James Parks. *A Study of Algorithms to Compute the Matrix Exponential.* PhD thesis, Mathematics Department, University of California, Berkeley, CA, USA, 1994. 53 pp. (Cited on p. 263.)

[459] Beresford N. Parlett. Computation of functions of triangular matrices. Memorandum ERL-M481, Electronics Research Laboratory, College of Engineering, University of California, Berkeley, November 1974. 18 pp. (Cited on pp. 86, 106.)

[460] Beresford N. Parlett. A recurrence among the elements of functions of triangular matrices. *Linear Algebra Appl.*, 14:117–121, 1976. (Cited on p. 85.)

[461] Beresford N. Parlett. *The Symmetric Eigenvalue Problem.* Society for Industrial and Applied Mathematics, Philadelphia, PA, USA, 1998. xxiv+398 pp. Unabridged, amended version of book first published by Prentice-Hall in 1980. ISBN 0-89871-402-8. (Cited on p. 35.)

[462] Beresford N. Parlett and Kwok Choi Ng. Development of an accurate algorithm for $\exp(Bt)$. Technical Report PAM-294, Center for Pure and Applied Mathematics, University of California, Berkeley, August 1985. 23 pp. Fortran program listings are given in an appendix with the same report number printed separately. (Cited on pp. 89, 228, 250.)

[463] Karen Hunger Parshall. Joseph H. M. Wedderburn and the structure theory of algebras. *Archive for History of Exact Sciences*, 32(3-4):223–349, 1985. (Cited on p. 26.)

[464] Karen Hunger Parshall. *James Joseph Sylvester. Life and Work in Letters.* Oxford University Press, 1998. xv+321 pp. ISBN 0-19-850391-1. (Cited on p. 30.)

[465] Karen Hunger Parshall. *James Joseph Sylvester. Jewish Mathematician in a Victorian World.* Johns Hopkins University Press, Baltimore, MD, USA, 2006. xiii+461 pp. ISBN 0-8018-8291-5. (Cited on p. 26.)

[466] Michael S. Paterson and Larry J. Stockmeyer. On the number of nonscalar multiplications necessary to evaluate polynomials. *SIAM J. Comput.*, 2(1):60–66, 1973. (Cited on p. 73.)

[467] G. Peano. Intégration par séries des équations différentielles linéaires. *Math. Annalen*, 32:450–456, 1888. (Cited on p. 26.)

[468] Heinz-Otto Peitgen, Hartmut Jürgens, and Dietmar Saupe. *Fractals for the Classroom. Part Two: Complex Systems and Mandelbrot Set.* Springer-Verlag, New York, 1992. xii+500 pp. ISBN 0-387-97722-8. (Cited on p. 178.)

[469] R. Penrose. A generalized inverse for matrices. *Proc. Cambridge Philos. Soc.*, 51(3): 406–413, 1955. (Cited on p. 214.)

[470] P. P. Petrushev and V. A. Popov. *Rational Approximation of Real Functions.* Cambridge University Press, Cambridge, UK, 1987. ISBN 0-521-33107-2. (Cited on p. 130.)

[471] Bernard Philippe. An algorithm to improve nearly orthonormal sets of vectors on a vector processor. *SIAM J. Alg. Discrete Methods*, 8(3):396–403, 1987. (Cited on p. 188.)

[472] George M. Phillips. *Two Millennia of Mathematics: From Archimedes to Gauss.* Springer-Verlag, New York, 2000. xii+223 pp. ISBN 0-387-95022-2. (Cited on pp. 283, 286.)

[473] H. Poincaré. Sur les groupes continus. *Trans. Cambridge Phil. Soc.*, 18:220–255, 1899. (Cited on p. 26.)

[474] George Pólya and Gabor Szegö. *Problems and Theorems in Analysis I. Series. Integral Calculus. Theory of Functions.* Springer-Verlag, New York, 1998. xix+389 pp. Reprint of the 1978 edition. ISBN 3-540-63640-4. (Cited on p. 343.)

[475] George Pólya and Gabor Szegö. *Problems and Theorems in Analysis II. Theory of Functions. Zeros. Polynomials. Determinants. Number Theory. Geometry.* Springer-Verlag, New York, 1998. xi+392 pp. Reprint of the 1976 edition. ISBN 3-540-63686-2. (Cited on p. 300.)

[476] Renfrey B. Potts. Symmetric square roots of the finite identity matrix. *Utilitas Mathematica*, 9:73–86, 1976. (Cited on p. 165.)

[477] M. J. D. Powell. *Approximation Theory and Methods.* Cambridge University Press, Cambridge, UK, 1981. x+339 pp. ISBN 0-521-29514-9. (Cited on p. 79.)

[478] William H. Press, Saul A. Teukolsky, William T. Vetterling, and Brian P. Flannery. *Numerical Recipes: The Art of Scientific Computing.* Third edition, Cambridge University Press, 2007. xxi+1235 pp. ISBN 978-0-521-88068-8. (Cited on p. 274.)

[479] Panayiotis J. Psarrakos. On the mth roots of a complex matrix. *Electron. J. Linear Algebra*, 9:32–41, 2002. (Cited on p. 174.)

[480] Péter Pulay. An iterative method for the determination of the square root of a positive definite matrix. *Z. Angew. Math. Mech.*, 46:151, 1966. (Cited on pp. 167, 171.)

[481] Norman J. Pullman. *Matrix Theory and its Applications: Selected Topics.* Marcel Dekker, New York, 1976. vi+240 pp. ISBN 0-8247-6420-X. (Cited on pp. 28, 29.)

[482] W. Pusz and S. L. Woronowicz. Functional calculus for sesquilinear forms and the purification map. *Reports on Mathematical Physics*, 8(2):159–170, 1975. (Cited on p. 52.)

[483] Heydar Radjavi and Peter Rosenthal. *Simultaneous Triangularization.* Springer-Verlag, New York, 2000. xii+318 pp. ISBN 0-387-98466-6. (Cited on p. 29.)

[484] C. R. Rao. Matrix approximations and reduction of dimensionality in multivariate statistical analysis. In *Multivariate Analysis—V*, P. R. Krishnaiah, editor, North Holland, Amsterdam, 1980, pages 3–22. (Cited on p. 214.)

[485] Lothar Reichel. The application of Leja points to Richardson iteration and polynomial preconditioning. *Linear Algebra Appl.*, 154/156:389–414, 1991. (Cited on p. 75.)

[486] Matthias W. Reinsch. A simple expression for the terms in the Baker–Campbell–Hausdorff series. *J. Math. Phys.*, 41(4):2434–2442, 2000. (Cited on p. 263.)

[487] John R. Rice. A theory of condition. *SIAM J. Numer. Anal.*, 3(2):287–310, 1966. (Cited on p. 69.)

[488] Norman M. Rice. On nth roots of positive operators. *Amer. Math. Monthly*, 89(5): 313–314, 1982. (Cited on p. 189.)

[489] A. N. Richmond. Expansions for the exponential of a sum of matrices. In *Applications of Matrix Theory*, M. J. C. Gover and S. Barnett, editors, Oxford University Press, 1989, pages 283–289. (Cited on p. 236.)

[490] Hans Richter. Zum Logarithmus einer Matrix. *Archiv der Mathematik*, 2(5):360–363, 1949. (Cited on p. 283.)

[491] Hans Richter. Über Matrixfunktionen. *Math. Ann.*, 22(1):16–34, 1950. (Cited on p. 28.)

[492] Frigyes Riesz and Béla Sz.-Nagy. *Functional Analysis*. Second edition, Blackie & Son, London and Glasgow, 1956. xii+468 pp. (Cited on p. 167.)

[493] R. F. Rinehart. The equivalence of definitions of a matric function. *Amer. Math. Monthly*, 62:395–414, 1955. (Cited on pp. 27, 34.)

[494] R. F. Rinehart. The derivative of a matric function. *Proc. Amer. Math. Soc.*, 7:2–5, 1956. (Cited on p. 70.)

[495] R. F. Rinehart. Elements of a theory of intrinsic functions on algebras. *Duke Math. J.*, 27:1–19, 1960. (Cited on p. 28.)

[496] J. D. Roberts. Linear model reduction and solution of the algebraic Riccati equation by use of the sign function. *Internat. J. Control*, 32(4):677–687, 1980. First issued as report CUED/B-Control/TR13, Department of Engineering, University of Cambridge, 1971. (Cited on pp. 39, 49, 129, 132.)

[497] Gian-Carlo Rota. *Indiscrete Thoughts*. Birkhäuser, Boston, 1997. xxii+280 pp. Edited by Fabrizio Palombi. ISBN 0-8176-3866-0. (Cited on p. 53.)

[498] Y. Saad. Analysis of some Krylov subspace approximations to the matrix exponential operator. *SIAM J. Numer. Anal.*, 29(1):209–228, 1992. (Cited on pp. 309, 311.)

[499] Youcef Saad. *Numerical Methods for Large Eigenvalue Problems*. Manchester University Press, Manchester, and Halsted Press, New York, 1992. 346 pp. ISBN 0-7190-3386-1. (Cited on p. 306.)

[500] Yousef Saad. *Iterative Methods for Sparse Linear Systems*. Second edition, Society for Industrial and Applied Mathematics, Philadelphia, PA, USA, 2003. xviii+528 pp. ISBN 0-89871-534-2. (Cited on p. 309.)

[501] E. B. Saff. On the degree of best rational approximation to the exponential function. *J. Approximation Theory*, 9:97–101, 1973. (Cited on p. 241.)

[502] Martin Schechter. *Principles of Functional Analysis*. Academic Press, New York, 1971. xix+383 pp. ISBN 0-12-622750-0. (Cited on p. 167.)

[503] Thomas Schmelzer and Lloyd N. Trefethen. Evaluating matrix functions for exponential integrators via Carathéodory-Fejér approximation and contour integrals. Report Number 06/20, Numerical Analysis Group, Oxford University Computing Laboratory, Oxford, UK, 2006. 19 pp. (Cited on pp. 37, 308.)

[504] Bernhard A. Schmitt. An algebraic approximation for the matrix exponential in singularly perturbed boundary value problems. *SIAM J. Numer. Anal.*, 27(1):51–66, 1990. (Cited on p. 48.)

[505] Christoph Schmoeger. On the operator equation $e^A = e^B$. *Linear Algebra Appl.*, 359:169–179, 2003. (Cited on p. 350.)

[506] Daniel Scholz and Michael Weyrauch. A note on the Zassenhaus product formula. *J. Math. Phys.*, 47:033505.1–033505.7, 2006. (Cited on p. 263.)

[507] Peter H. Schönemann. A generalized solution of the orthogonal Procrustes problem. *Psychometrika*, 31(1):1–10, 1966. (Cited on p. 214.)

[508] Robert S. Schreiber and Beresford N. Parlett. Block reflectors: Theory and computation. *SIAM J. Numer. Anal.*, 25(1):189–205, 1988. (Cited on p. 213.)

[509] Ernst Schröder. Ueber unendliche viele Algorithmen zur Auflösung der Gleichungen. *Math. Annalen*, 2:317–365, 1870. (Cited on pp. 130, 407.)

[510] Ernst Schröder. On infinitely many algorithms for solving equations. Technical Report TR-92-121, Department of Computer Science, University of Maryland, College Park, MD, USA, November 1992. 57 pp. Translation of [509] by G. W. Stewart. (Cited on p. 130.)

[511] Manfred Schroeder. *Fractals, Chaos, Power Laws: Minutes from an Infinite Paradise.* W. H. Freeman, New York, 1991. xviii+429 pp. ISBN 0-7167-2136-8. (Cited on pp. 131, 178.)

[512] Günther Schulz. Iterative Berechnung der reziproken Matrix. *Z. Angew. Math. Mech.*, 13:57–59, 1933. (Cited on pp. 114, 181.)

[513] Hans Schwerdtfeger. *Les Fonctions de Matrices. I. Les Fonctions Univalentes.* Number 649 in *Actualités Scientifiques et Industrielles*. Hermann, Paris, France, 1938. 58 pp. (Cited on pp. 26, 30, 51.)

[514] Steven M. Serbin. Rational approximations of trigonometric matrices with application to second-order systems of differential equations. *Appl. Math. Comput.*, 5(1):75–92, 1979. (Cited on p. 287.)

[515] Steven M. Serbin and Sybil A. Blalock. An algorithm for computing the matrix cosine. *SIAM J. Sci. Statist. Comput.*, 1(2):198–204, 1980. (Cited on p. 299.)

[516] Lawrence F. Shampine and Mark W. Reichelt. The MATLAB ODE suite. *SIAM J. Sci. Comput.*, 18(1):1–22, 1997. (Cited on p. 165.)

[517] Wyatt D. Sharpa and Edward J. Allen. Stochastic neutron transport equations for rod and plane geometries. *Annals of Nuclear Energy*, 27(2):99–116, 2000. (Cited on p. 167.)

[518] N. Sherif. On the computation of a matrix inverse square root. *Computing*, 46:295–305, 1991. (Cited on p. 169.)

[519] L. S. Shieh, Y. T. Tsay, and C. T. Wang. Matrix sector functions and their applications to system theory. *IEE Proc.*, 131(5):171–181, 1984. (Cited on p. 48.)

[520] Ken Shoemake and Tom Duff. Matrix animation and polar decomposition. In *Proceedings of the Conference on Graphics Interface '92*, Morgan Kaufmann Publishers Inc., San Francisco, CA, USA, 1992, pages 258–264. (Cited on p. 42.)

[521] Avram Sidi. *Practical Extrapolation Methods; Theory and Applications.* Cambridge University Press, Cambridge, UK, 2003. xxii+519 pp. ISBN 0-521-66159-5. (Cited on p. 104.)

[522] Roger B. Sidje. Expokit: A software package for computing matrix exponentials. *ACM Trans. Math. Software*, 24(1):130–156, 1998. (Cited on pp. 262, 263, 264.)

[523] Roger B. Sidje, Kevin Burrage, and Shev MacNamara. Inexact uniformization method for computing transient distributions of Markov chains. *SIAM J. Sci. Comput.*, 29(6): 2562–2580, 2007. (Cited on p. 260.)

[524] Roger B. Sidje and William J. Stewart. A numerical study of large sparse matrix exponentials arising in Markov chains. *Computational Statistics & Data Analysis*, 29: 345–368, 1999. (Cited on p. 260.)

[525] Burton Singer and Seymour Spilerman. The representation of social processes by Markov models. *Amer. J. Sociology*, 82(1):1–54, 1976. (Cited on p. 38.)

[526] Abraham Sinkov. *Elementary Cryptanalysis: A Mathematical Approach.* Mathematical Association of America, Washington, D.C., 1966. ix+222 pp. ISBN 0-88385-622-0. (Cited on p. 171.)

[527] Bård Skaflestad and Will M. Wright. The scaling and modified squaring method for matrix functions related to the exponential. Preprint, Norwegian University of Science and Technology, Trondheim, Norway, 2006. 17 pp. (Cited on p. 262.)

[528] David M. Smith. Algorithm 693: A FORTRAN package for floating-point multiple-precision arithmetic. *ACM Trans. Math. Software*, 17(2):273–283, 1991. (Cited on p. 286.)

[529] Matthew I. Smith. *Numerical Computation of Matrix Functions.* PhD thesis, University of Manchester, Manchester, England, September 2002. 157 pp. (Cited on p. 370.)

[530] Matthew I. Smith. A Schur algorithm for computing matrix pth roots. *SIAM J. Matrix Anal. Appl.*, 24(4):971–989, 2003. (Cited on pp. 174, 176, 187, 188, 190.)

[531] R. A. Smith. Infinite product expansions for matrix n-th roots. *J. Austral. Math. Soc.*, 8:242–249, 1968. (Cited on p. 188.)

[532] Alicja Smoktunowicz, Jesse L. Barlow, and Julien Langou. A note on the error analysis of classical Gram–Schmidt. *Numer. Math.*, 105:299–313, 2006. (Cited on p. 309.)

[533] Inge Söderkvist. Perturbation analysis of the orthogonal Procrustes problem. *BIT*, 33: 687–694, 1993. (Cited on p. 201.)

[534] Mark Sofroniou and Giulia Spaletta. Solving orthogonal matrix differential systems in Mathematica. In *Computational Science—ICCS* 2002 *Proceedings, Part III*, Peter M. A. Sloot, C. J. Kenneth Tan, Jack J. Dongarra, and Alfons G. Hoekstra, editors, volume 2002 of *Lecture Notes in Computer Science*, Springer-Verlag, Berlin, 2002, pages 496–505. (Cited on pp. 42, 215.)

[535] Irene A. Stegun and Milton Abramowitz. Pitfalls in computation. *J. Soc. Indust. Appl. Math.*, 4(4):207–219, 1956. (Cited on p. 74.)

[536] G. W. Stewart. Error and perturbation bounds for subspaces associated with certain eigenvalue problems. *SIAM Rev.*, 15(4):727–764, 1973. (Cited on p. 231.)

[537] G. W. Stewart. *Matrix Algorithms. Volume I: Basic Decompositions.* Society for Industrial and Applied Mathematics, Philadelphia, PA, USA, 1998. xx+458 pp. ISBN 0-89871-414-1. (Cited on p. 357.)

[538] G. W. Stewart. *Matrix Algorithms. Volume II: Eigensystems.* Society for Industrial and Applied Mathematics, Philadelphia, PA, USA, 2001. xix+469 pp. ISBN 0-89871-503-2. (Cited on pp. 35, 69, 309, 323, 325.)

[539] G. W. Stewart and Ji-guang Sun. *Matrix Perturbation Theory.* Academic Press, London, 1990. xv+365 pp. ISBN 0-12-670230-6. (Cited on p. 327.)

[540] William J. Stewart. *Introduction to the Numerical Solution of Markov Chains.* Princeton University Press, Princeton, NJ, USA, 1994. xix+539 pp. ISBN 0-691-03699-3. (Cited on p. 260.)

[541] Eberhard Stickel. On the Fréchet derivative of matrix functions. *Linear Algebra Appl.*, 91:83–88, 1987. (Cited on p. 70.)

[542] Josef Stoer and R. Bulirsch. *Introduction to Numerical Analysis.* Third edition, Springer-Verlag, New York, 2002. xv+744 pp. ISBN 0-387-95452-X. (Cited on pp. 28, 154, 333.)

[543] David R. Stoutemyer. Crimes and misdemeanors in the computer algebra trade. *Notices Amer. Math. Soc.*, 38(7):778–785, 1991. (Cited on p. 286.)

[544] Gilbert Strang. On the construction and comparison of difference schemes. *SIAM J. Numer. Anal.*, 5(3):506–517, 1968. (Cited on p. 263.)

[545] Torsten Ström. Minimization of norms and logarithmic norms by diagonal similarities. *Computing*, 10:1–7, 1972. (Cited on p. 105.)

[546] Torsten Ström. On logarithmic norms. *SIAM J. Numer. Anal.*, 12(5):741–753, 1975. (Cited on p. 237.)

[547] Ji-guang Sun. A note on backward perturbations for the Hermitian eigenvalue problem. *BIT*, 35:385–393, 1995. (Cited on p. 214.)

[548] Ji-guang Sun. Perturbation analysis of the matrix sign function. *Linear Algebra Appl.*, 250:177–206, 1997. (Cited on p. 129.)

[549] Ji-guang Sun and C.-H. Chen. Generalized polar decomposition. *Math. Numer. Sinica*, 11:262–273, 1989. In Chinese. Cited in [372]. (Cited on p. 214.)

[550] Xiaobai Sun and Enrique S. Quintana-Ortí. The generalized Newton iteration for the matrix sign function. *SIAM J. Sci. Comput.*, 24(2):669–683, 2002. (Cited on p. 130.)

[551] Xiaobai Sun and Enrique S. Quintana-Ortí. Spectral division methods for block generalized Schur decompositions. *Math. Comp.*, 73(248):1827–1847, 2004. (Cited on p. 49.)

[552] Masuo Suzuki. Generalized Trotter's formula and systematic approximants of exponential operators and inner derivations with applications to many-body problems. *Commun. Math. Phys.*, 51(2):183–190, 1976. (Cited on pp. 262, 263.)

[553] J. J. Sylvester. Additions to the articles, "On a New Class of Theorems," and "On Pascal's Theorem". *Philosophical Magazine*, 37:363–370, 1850. Reprinted in [558, pp. 1451–151]. (Cited on p. 26.)

[554] J. J. Sylvester. Note on the theory of simultaneous linear differential or difference equations with constant coefficients. *Amer. J. Math.*, 4(1):321–326, 1881. Reprinted in [559, pp. 551-556]. (Cited on pp. 188, 191.)

[555] J. J. Sylvester. Sur les puissances et les racines de substitutions linéaires. *Comptes Rendus de l'Académie des Sciences*, 94:55–59, 1882. Reprinted in [559, pp. 562–564]. (Cited on pp. 187, 188.)

[556] J. J. Sylvester. Sur les racines des matrices unitaires. *Comptes Rendus de l'Académie des Sciences*, 94:396–399, 1882. Reprinted in [559, pp. 565–567]. (Cited on p. 168.)

[557] J. J. Sylvester. On the equation to the secular inequalities in the planetary theory. *Philosophical Magazine*, 16:267–269, 1883. Reprinted in [560, pp. 110–111]. (Cited on pp. 26, 34, 187.)

[558] *The Collected Mathematical Papers of James Joseph Sylvester*, volume 1 (1837–1853). Cambridge University Press, 1904. xii+650 pp. (Cited on p. 409.)

[559] *The Collected Mathematical Papers of James Joseph Sylvester*, volume III (1870–1883). Chelsea, New York, 1973. xv+697 pp. ISBN 0-8284-0253-1. (Cited on p. 409.)

[560] *The Collected Mathematical Papers of James Joseph Sylvester*, volume IV (1882–1897). Chelsea, New York, 1973. xxxvii+756 pp. ISBN 0-8284-0253-1. (Cited on p. 409.)

[561] Henry Taber. On the theory of matrices. *Amer. J. Math.*, 12(4):337–396, 1890. (Cited on p. 28.)

[562] Ping Tak Peter Tang. Table-driven implementation of the expm1 function in IEEE floating-point arithmetic. *ACM Trans. Math. Software*, 18(2):211–222, 1992. (Cited on p. 261.)

[563] Pham Dinh Tao. Convergence of a subgradient method for computing the bound norm of matrices. *Linear Algebra Appl.*, 62:163–182, 1984. In French. (Cited on p. 69.)

[564] Olga Taussky. Commutativity in finite matrices. *Amer. Math. Monthly*, 64(4):229–235, 1957. (Cited on p. 29.)

[565] Olga Taussky. How I became a torchbearer for matrix theory. *Amer. Math. Monthly*, 95(9):801–812, 1988. (Cited on p. 29.)

[566] R. C. Thompson. On the matrices AB and BA. *Linear Algebra Appl.*, 1:43–58, 1968. (Cited on p. 28.)

[567] C. Thron, S. J. Dong, K. F. Liu, and H. P. Ying. Padé-Z_2 estimator of determinants. *Physical Review D*, 57(3):1642–1653, 1997. (Cited on p. 44.)

[568] Françoise Tisseur. Parallel implementation of the Yau and Lu method for eigenvalue computation. *International Journal of Supercomputer Applications and High Performance Computing*, 11(3):197–204, 1997. (Cited on p. 300.)

[569] Françoise Tisseur. Newton's method in floating point arithmetic and iterative refinement of generalized eigenvalue problems. *SIAM J. Matrix Anal. Appl.*, 22(4): 1038–1057, 2001. (Cited on p. 166.)

[570] Françoise Tisseur and Karl Meerbergen. The quadratic eigenvalue problem. *SIAM Rev.*, 43(2):235–286, 2001. (Cited on p. 45.)

[571] L. N. Trefethen and J. A. C. Weideman. The fast trapezoid rule in scientific computing. Paper in preparation, 2007. (Cited on p. 308.)

[572] Lloyd N. Trefethen and David Bau III. *Numerical Linear Algebra.* Society for Industrial and Applied Mathematics, Philadelphia, PA, USA, 1997. xii+361 pp. ISBN 0-89871-361-7. (Cited on pp. 325, 335.)

[573] Lloyd N. Trefethen and Mark Embree. *Spectra and Pseudospectra: The Behavior of Nonnormal Matrices and Operators.* Princeton University Press, Princeton, NJ, USA, 2005. xvii+606 pp. ISBN 0-691-11946-5. (Cited on pp. 47, 105, 263.)

[574] Lloyd N. Trefethen, J. A. C. Weideman, and Thomas Schmelzer. Talbot quadratures and rational approximations. *BIT*, 46(3):653–670, 2006. (Cited on pp. 259, 308.)

[575] H. F. Trotter. On the product of semi-groups of operators. *Proc. Amer. Math. Soc.*, 10(4):545–551, 1959. (Cited on p. 263.)

[576] J. S. H. Tsai, L. S. Shieh, and R. E. Yates. Fast and stable algorithms for computing the principal nth root of a complex matrix and the matrix sector function. *Comput. Math. Applic.*, 15(11):903–913, 1988. (Cited on p. 129.)

[577] H. W. Turnbull. The matrix square and cube roots of unity. *J. London Math. Soc.*, 2 (8):242–244, 1927. (Cited on p. 188.)

[578] H. W. Turnbull. *The Theory of Determinants, Matrices, and Invariants.* Blackie, London and Glasgow, 1929. xvi+338 pp. (Cited on p. 188.)

[579] H. W. Turnbull and A. C. Aitken. *An Introduction to the Theory of Canonical Matrices.* Blackie, London and Glasgow, 1932. xiii+200 pp. Reprinted with appendix, 1952. (Cited on pp. 27, 29, 52.)

[580] G. M. Tuynman. The derivation of the exponential map of matrices. *Amer. Math. Monthly*, 102(9):818–820, 1995. (Cited on p. 263.)

[581] Frank Uhlig. Explicit polar decomposition and a near-characteristic polynomial: The 2×2 case. *Linear Algebra Appl.*, 38:239–249, 1981. (Cited on p. 216.)

[582] R. Vaidyanathaswamy. Integer-roots of the unit matrix. *J. London Math. Soc.*, 3(12): 121–124, 1928. (Cited on p. 189.)

[583] R. Vaidyanathaswamy. On the possible periods of integer matrices. *J. London Math. Soc.*, 3(12):268–272, 1928. (Cited on p. 189.)

[584] P. Van Den Driessche and H. K. Wimmer. Explicit polar decomposition of companion matrices. *Electron. J. Linear Algebra*, 1:64–69, 1996. (Cited on p. 215.)

[585] J. van den Eshof, A. Frommer, Th. Lippert, K. Schilling, and H. A. Van der Vorst. Numerical methods for the QCD overlap operator. I. Sign-function and error bounds. *Computer Physics Communications*, 146:203–224, 2002. (Cited on pp. 43, 130.)

[586] Jasper van den Eshof. *Nested Iteration methods for Nonlinear Matrix Problems.* PhD thesis, Utrecht University, Utrecht, Netherlands, September 2003. vii+183 pp. (Cited on p. 130.)

[587] Henk A. Van der Vorst. An iterative solution method for solving $f(A)x = b$, using Krylov subspace information obtained for the symmetric positive definite matrix A. *J. Comput. Appl. Math.*, 18:249–263, 1987. (Cited on p. 309.)

[588] Henk A. Van der Vorst. Solution of $f(A)x = b$ with projection methods for the matrix A. In *Numerical Challenges in Lattice Quantum Chromodynamics*, Andreas Frommer, Thomas Lippert, Björn Medeke, and Klaus Schilling, editors, volume 15 of *Lecture Notes in Computational Science and Engineering*, Springer-Verlag, Berlin, 2000, pages 18–28. (Cited on p. 309.)

[589] Henk A. Van der Vorst. *Iterative Krylov Methods for Large Linear Systems.* Cambridge University Press, 2003. xiii+221 pp. ISBN 0-521-81828-1. (Cited on p. 309.)

[590] Jos L. M. van Dorsselaer, Michiel E. Hochstenbach, and Henk A. Van der Vorst. Computing probabilistic bounds for extreme eigenvalues of symmetric matrices with the Lanczos method. *SIAM J. Matrix Anal. Appl.*, 22(3):837–852, 2000. (Cited on p. 69.)

[591] J. L. van Hemmen and T. Ando. An inequality for trace ideals. *Commun. Math. Phys.*, 76:143–148, 1980. (Cited on p. 135.)

[592] Charles F. Van Loan. A study of the matrix exponential. Numerical Analysis Report No. 10, University of Manchester, Manchester, UK, August 1975. Reissued as MIMS EPrint 2006.397, Manchester Institute for Mathematical Sciences, The University of Manchester, UK, November 2006. (Cited on pp. 84, 105, 262.)

[593] Charles F. Van Loan. On the limitation and application of Padé approximation to the matrix exponential. In *Padé and Rational Approximation: Theory and Applications*, E. B. Saff and R. S. Varga, editors, Academic Press, New York, 1977, pages 439–448. (Cited on p. 263.)

[594] Charles F. Van Loan. The sensitivity of the matrix exponential. *SIAM J. Numer. Anal.*, 14(6):971–981, 1977. (Cited on pp. 238, 263.)

[595] Charles F. Van Loan. Computing integrals involving the matrix exponential. *IEEE Trans. Automat. Control*, AC-23(3):395–404, 1978. (Cited on p. 264.)

[596] Charles F. Van Loan. A note on the evaluation of matrix polynomials. *IEEE Trans. Automat. Control*, AC-24(2):320–321, 1979. (Cited on p. 74.)

[597] R. Vandebril, M. Van Barel, G. H. Golub, and N. Mastronardi. A bibliography on semiseparable matrices. *Calcolo*, 42:249–70, 2005. (Cited on p. 316.)

[598] V. S. Varadarajan. *Lie Groups, Lie Algebras, and Their Representations.* Prentice-Hall, Englewood Cliffs, NJ, USA, 1974. xiii+430 pp. ISBN 0-13-535732-2. (Cited on p. 262.)

[599] J. M. Varah. On the separation of two matrices. *SIAM J. Numer. Anal.*, 16(2):216–222, 1979. (Cited on p. 231.)

[600] Richard S. Varga. *Matrix Iterative Analysis.* Second edition, Springer-Verlag, Berlin, 2000. x+358 pp. ISBN 3-540-66321-5. (Cited on pp. 242, 260, 264.)

[601] R. Vertechy and V. Parenti-Castelli. Real-time direct position analysis of parallel spherical wrists by using extra sensors. *Journal of Mechanical Design*, 128:288–294, 2006. (Cited on pp. 42, 208.)

[602] Cornelis Visser. Note on linear operators. *Proc. Kon. Akad. Wet. Amsterdam*, 40(3): 270–272, 1937. (Cited on p. 167.)

[603] John von Neumann. Über Adjungierte Funktionaloperatoren. *Ann. of Math.* (2), 33 (2):294–310, 1923. (Cited on p. 214.)

[604] John von Neumann. Über die analytischen Eigenschaften von Gruppen linearer Transformationen und ihrer Darstellungen. *Math. Zeit.*, 30:3–42, 1929. (Cited on p. 283.)

[605] Grace Wahba. Problem 65-1, a least squares estimate of satellite attitude. *SIAM Rev.*, 7(3):409, 1965. Solutions in 8(3):384–386, 1966. (Cited on p. 214.)

[606] Robert C. Ward. Numerical computation of the matrix exponential with accuracy estimate. *SIAM J. Numer. Anal.*, 14(4):600–610, 1977. (Cited on pp. 104, 246, 263, 264.)

[607] David S. Watkins. *Fundamentals of Matrix Computations.* Second edition, Wiley, New York, 2002. xiii+618 pp. ISBN 0-471-21394-2. (Cited on pp. 69, 323, 325, 335.)

[608] David S. Watkins. *The Matrix Eigenvalue Problem: GR and Krylov Subspace Methods.* Society for Industrial and Applied Mathematics, Philadelphia, PA, USA, 2007. x+442 pp. ISBN 978-0-898716-41-2. (Cited on pp. 309, 323.)

[609] G. A. Watson. *Approximation Theory and Numerical Methods.* Wiley, Chichester, UK, 1980. x+229 pp. ISBN 0-471-27706-1. (Cited on p. 79.)

[610] Frederick V. Waugh and Martin E. Abel. On fractional powers of a matrix. *J. Amer. Statist. Assoc.*, 62:1018–1021, 1967. (Cited on pp. 38, 167.)

[611] J. H. M. Wedderburn. *Lectures on Matrices*, volume 17 of *American Mathematical Society Colloquium Publications.* American Mathematical Society, Providence, RI, USA, 1934. vii+205 pp. (Cited on pp. 27, 28.)

[612] G. H. Weiss and A. A. Maradudin. The Baker-Hausdorff formula and a problem in crystal physics. *J. Math. Phys.*, 3(4):771–777, 1962. (Cited on p. 263.)

[613] Edgar M. E. Wermuth. Two remarks on matrix exponentials. *Linear Algebra Appl.*, 117:127–132, 1989. (Cited on p. 262.)

[614] Edouard Weyr. Note sur la théorie de quantités complexes formées avec n unités principales. *Bull. Sci. Math. II*, 11:205–215, 1887. (Cited on pp. 26, 104.)

[615] R. M. Wilcox. Exponential operators and parameter differentiation in quantum physics. *J. Math. Phys.*, 8(4):962–982, 1967. (Cited on p. 263.)

[616] J. H. Wilkinson. *The Algebraic Eigenvalue Problem.* Oxford University Press, 1965. xviii+662 pp. ISBN 0-19-853403-5 (hardback), 0-19-853418-3 (paperback). (Cited on p. 69.)

[617] Arthur Wouk. Integral representation of the logarithm of matrices and operators. *J. Math. Anal. and Appl.*, 11:131–138, 1965. (Cited on p. 283.)

[618] Pei Yuan Wu. Approximation by partial isometries. *Proc. Edinburgh Math. Soc.*, 29: 255–261, 1986. (Cited on p. 214.)

[619] Shing-Tung Yau and Ya Yan Lu. Reducing the symmetric matrix eigenvalue problem to matrix multiplications. *SIAM J. Sci. Comput.*, 14(1):121–136, 1993. (Cited on p. 300.)

[620] N. J. Young. A bound for norms of functions of matrices. *Linear Algebra Appl.*, 37: 181–186, 1981. (Cited on p. 105.)

[621] V. Zakian. Properties of I_{MN} and J_{MN} approximants and applications to numerical inversion of Laplace transforms and initial value problems. *J. Inst. Maths. Applics.*, 50:191–222, 1975. (Cited on p. 263.)

[622] Hongyuan Zha and Zhenyue Zhang. Fast parallelizable methods for the Hermitian eigenvalue problem. Technical Report CSE-96-041, Department of Computer Science and Engineering, Pennsylvania State University, University Park, PA, May 1996. 19 pp. (Cited on p. 216.)

[623] Xingzhi Zhan. *Matrix Inequalities*, volume 1790 of *Lecture Notes in Mathematics.* Springer-Verlag, Berlin, 2002. ISBN 3-540-43798-3. (Cited on p. 330.)

[624] Fuzhen Zhang, editor. *The Schur Complement and Its Applications.* Springer-Verlag, New York, 2005. xvi+295 pp. ISBN 0-387-24271-6. (Cited on p. 329.)

[625] Pawel Zieliński and Krystyna Ziętak. The polar decomposition—properties, applications and algorithms. *Applied Mathematics, Ann. Pol. Math. Soc.*, 38:23–49, 1995. (Cited on p. 213.)

Index

Sherlock Holmes sat moodily at one side of the fireplace cross-indexing his records of crime.
— ARTHUR CONAN DOYLE, *The Five Orange Pips* (1891)

A suffix "t" after a page number denotes a table, "n" a footnote, and "q" a quotation. Mathematical symbols and Greek letters are indexed as if they were spelled out. The solution to a problem is not usually indexed if the problem itself is indexed under the same term.